Subcellular Biochemistry

Volume 84

The book series SUBCELLULAR BIOCHEMISTRY is a renowned and well recognized forum for disseminating advances of emerging topics in Cell Biology and related subjects. All volumes are edited by established scientists and the individual chapters are written by experts on the relevant topic. The individual chapters of each volume are fully citable and indexed in Medline/Pubmed to ensure maximum visibility of the work.

More information about this series at http://www.springer.com/series/6515

Jan Löwe • Linda A. Amos
Editors

Prokaryotic Cytoskeletons

Filamentous Protein Polymers Active in the Cytoplasm of Bacterial and Archaeal Cells

Springer

Editors
Jan Löwe
Structural Studies Division
MRC Laboratory of Molecular Biology
Cambridge, UK

Linda A. Amos
Structural Studies Division
MRC Laboratory of Molecular Biology
Cambridge, UK

ISSN 0306-0225
Subcellular Biochemistry
ISBN 978-3-319-53045-1 ISBN 978-3-319-53047-5 (eBook)
DOI 10.1007/978-3-319-53047-5

Library of Congress Control Number: 2017939597

Printed on acid-free paper

This Springer imprint is published by Springer Nature
The registered company is Springer International Publishing AG
The registered company address is: Gewerbestrasse 11, 6330 Cham, Switzerland

Preface

Until the 1990s, it was assumed that the absence of a cytoskeleton, comprised of intracellular dynamic protein filaments that generate long-range order, is a defining characteristic of all archaea and bacteria, distinguishing them from eukaryotic cells.

Bacterial DNA was thought to be attached to the cell wall, so that, after replication, a chromosome pair would be segregated by wall extension. Division of the cell into two would then occur by construction of a new central partition. Plasmids, small pieces of DNA and other molecular entities such as protein complexes were thought to travel through prokaryotic cells without guidance, through diffusion.

These views began to break down when FtsZ protein was found to locate, as a ringlike filamentous structure, at the cell division site, and this widely distributed protein was shown to be a homologue of eukaryotic tubulin. The absence of a cytoskeleton in archaea and bacteria was finally abandoned in 2003, when other bacterial proteins, already known to have amino acid sequence homology to eukaryotic actin, were shown to form intracellular filaments with similarity to F-actin. Soon after that, bacterial plasmid systems were discovered that used actin-like or tubulin-like protein filaments to effect segregation. Since then, many other types of filament have been discovered; some are unique to prokaryotes, in a surprising variety of biological contexts, but many are related to eukaryotic cytoskeletal filaments.

With important discoveries being made in archaea as well as bacteria, the field has expanded enormously to warrant the dedication of a whole book to known prokaryotic filament structures, together with current theories and controversies regarding the mechanisms by which the filaments carry out their diverse functions, that go well beyond what similarities with eukaryotic homologues would suggest.

What has emerged, in particular, is that prokaryotes lack the motor proteins associated with eukaryotic cytoskeletal filaments and that intracellular motility often depends on the activity and dynamics of the filaments themselves. As described in the various chapters of the book, these cytomotive filaments have been found to exhibit treadmilling, dynamic instability and relative sliding. In other words, our view of a bacterium has undergone a truly remarkable revolution.

Cambridge, UK

Jan Löwe
Linda A. Amos

Contents

About the Editors

Jan Löwe gained his PhD in 1996 at the Max Planck Institute of Biochemistry, Martinsried, with Robert Huber, having helped to solve the crystal structures of large protein complexes known as proteasomes and thermosomes. He then moved to Cambridge to work in LAA's group at MRC-LMB, where he solved the structure of FtsZ, a bacterial tubulin homologue. In 1998, he set up his own group at MRC-LMB and began working on the structures and functions of a wide range of bacterial proteins involved in cell division and organisation. In particular, he has investigated the prokaryotic actin-related proteins, as well as a range of tubulin-related filaments. In addition to X-ray crystallography, his group uses electron microscopy and electron cryo-tomography to determine the structures of bacterial cytoskeletal filaments, together with light microscopy to follow their activity. He is a fellow of the Royal Society (London), a member of EMBO and a fellow of Leopoldina (German National Academy of Sciences). In 2010, he was promoted to be joint head of the Structural Studies Division of MRC-LMB in Cambridge, and in 2016, he became deputy director of the entire institute.

http://scholar.google.co.uk/citations?user=0XTmVmoAAAAJ&hl=en
http://orcid.org/0000-0002-5218-6615

Linda A. Amos began working at MRC-LMB in Aaron Klug's group in 1969, helping to develop techniques for 3D structure analysis of macromolecular complexes with icosahedral or helical symmetry from EM images. While working on the structure of microtubules, she became very interested in their activity and continued to research in this field. Her publications on this topic include the 3D structures of tubulin sheets and of microtubules decorated with a variety of accessory proteins such as motors of the kinesin family. Her interest in bacterial proteins that form filaments began when JL came to work in her group and solved the crystal structure of FtsZ, proving its homology to tubulin. She continued to collaborate with him on a number of related projects, including bacterial actin-related proteins, until retiring from active research. LAA is a member of EMBO and a faculty member of F1000.

http://orcid.org/0000-0001-8305-5346

Selected Publications

Alonso MC, Drummond DR, Kain S, Hoeng J, Amos L, Cross RA (2007) An ATP-gate controls tubulin binding by the tethered head of kinesin-1. Science 316:120–123

Amos LA, Löwe J (2014) The subtle allostery of microtubule dynamics. Nat Struct Mol Biol 21:505–506

Amos LA, van den Ent F, Löwe J (2004) Structural/functional homology between the bacterial and eukaryotic cytoskeletons. Curr Opin Cell Biol 16:24–31

Aylett CH, Löwe J (2012) Superstructure of the centromeric complex of TubZRC plasmid partitioning systems. Proc Natl Acad Sci U S A 109:16522–16527

Aylett CH, Wang Q, Michie KA, Amos LA, Löwe J (2010) Filament structure of bacterial tubulin hom0ologue TubZ. Proc Natl Acad Sci U S A 107:19766–19771

Aylett CH, Löwe J, Amos LA (2011) New insights into the mechanisms of cytomotive actin and tubulin filaments. Int Rev Cell Mol Biol 292:1–71

Aylett CH, Izoré T, Amos LA, Löwe J (2013) Structure of the tubulin/FtsZ-like protein TubZ from Pseudomonas bacteriophage ΦKZ. J Mol Biol 425:2164–2173

Bharat TA, Murshudov GN, Sachse C, Löwe J (2015) Structures of actin-like ParM filaments show architecture of plasmid-segregating spindles. Nature 523:106–110

Bharat TA, Russo CJ, Löwe J, Passmore LA, Scheres SH (2015) Advances in single-particle electron cryomicroscopy structure determination applied to sub-tomogram averaging. Structure 23:1743–1753

Cordell SC, Löwe J (2001) Crystal structure of the bacterial cell division regulator MinD. FEBS Lett 492:160–165

Cordell SC, Anderson RE, Löwe J (2001) Crystal structure of the bacterial cell division inhibitor MinC. EMBO J 20:2454–2461

des Georges A, Katsuki M, Drummond DR, Osei M, Cross RA, Amos LA (2008) Mal3, the S. pombe homologue of EB1, changes the microtubule lattice. Nat Struct Mol Biol 15:1102–1108

Ditzel L, Löwe J, Stock D, Stetter KO, Huber H, Huber R, Steinbacher S (1998) Crystal structure of the thermosome, the archaeal chaperonin and homolog of CCT. Cell 93:125–138

Duggin IG, Aylett CH, Walsh JC, Michie KA, Wang Q, Turnbull L, Dawson EM, Harry EJ, Whitchurch CB, Amos LA, Löwe J (2015) CetZ tubulin-like proteins control archaeal cell shape. Nature 519:362–365

Duman RE, Löwe J (2010) Crystal structures of Bacillus subtilis Lon protease. J Mol Biol 401:653–670

Duman R, Ishikawa S, Celik I, Strahl H, Ogasawara N, Troc P, Löwe J, Hamoen LW (2013) Structural and genetic analyses reveal the protein SepF as a new membrane anchor for the Z ring. Proc Natl Acad Sci U S A 110:E4601-10

Fink G, Löwe J (2015) Reconstitution of a prokaryotic minus end-tracking system using TubRC centromeric complexes and tubulin-like protein TubZ filaments. Proc Natl Acad Sci U S A 112:E1845–50

Gayathri P, Fujii T, Møller-Jensen J, van den Ent F, Namba K, Löwe J (2012) A bipolar spindle of antiparallel ParM filaments drives bacterial plasmid segregation. Science 338:1334–1337

Gayathri P, Fujii T, Namba, K, Löwe J (2013) Structure of the ParM filament at 8.5Å resolution. J Struct Biol 184:33–42

Ghosal D, Trambaiolo D, Amos LA, Löwe J (2014) MinCD cell division proteins form alternating copolymeric cytomotive filaments. Nat Commun 5:5341

Ghosal D, Löwe J, Amos LA (2015) Bacterial and archaeal cytoskeletons. In Encyclopedia of cell biology, Vol 2. Elsevier, Amsterdam

Groll M, Ditzel L, Löwe J, Stock D, Bochtler M, Bartunik HD, Huber R (1997) Structure of 20S proteasome from yeast at 2.4 A resolution. Nature 386:463–471

Haering CH, Schoffnegger D, Nishino T, Helmhart W, Nasmyth K, Löwe J (2004) Structure and stability of cohesin's Smc1-kleisin interaction. Mol Cell 15:951–964

Hirose K, Lockhart A, Cross RA, Amos LA (1995) Nucleotide-dependent angular change in kinesin motor domain bound to tubulin. Nature (London) 376:277–279

Hirose K, Akimaru E, Akiba T, Endow SA, Amos LA (2006) Large conformational changes in a Kinesin motor catalysed by interaction with microtubules. Mol Cell 23:913–923

Izoré T, Duman R, Kureisaite-Ciziene D, Löwe J (2014) Crenactin from Pyrobaculum calidifontis is closely related to actin in structure and forms steep helical filaments. FEBS Lett 588:776–782

Kar S, Fan J, Smith MJ, Goedert M, Amos LA (2003) Repeat motifs of tau bind to the insides of microtubules in the absence of taxol. EMBO J 22:70–77

Kiekebusch D, Michie KA, Essen LO, Löwe J, Thanbichler M (2012) Localized dimerization and nucleoid binding drive gradient formation by the bacterial cell division inhibitor MipZ. Mol Cell 46:245–259

Lee BG, Roig MB, Jansma M, Petela N, Metson J, Nasmyth K, Löwe J (2016) Crystal structure of the cohesin Gatekeeper Pds5 and in complex with Kleisin Scc1. Cell Rep 14:2108–2115

Leonard TA, Butler PJ, Löwe J (2004) Structural analysis of the chromosome segregation protein Spo0J from Thermus thermophilus. Mol Microbiol 53:419–432

Leonard TA, Butler PJ, Löwe J (2005) Bacterial chromosome segregation: structure and DNA binding of the Soj dimer--a conserved biological switch. EMBO J 24:270–282

Leonard TA, Møller-Jensen J, Löwe J (2005) Towards understanding the molecular basis of bacterial DNA segregation. Philos Trans R Soc Lond B Biol Sci 360:523–535

Low HH, Sachse C, Amos LA, Löwe J (2009) Structure of a bacterial dynamin-like protein lipid tube provides a mechanism for assembly and membrane curving. Cell 139:1342–1352

Löwe J, Amos LA (1998) Crystal structure of the bacterial cell-division protein FtsZ. Nature 391:203–206

Löwe J, Amos LA (1999) Tubulin-like protofilaments in Ca2+-induced FtsZ sheets. EMBO J 18:2364–2371.

Löwe J, Amos LA (2009) Evolution of cytomotive filaments: the cytoskeleton from prokaryotes to eukaryotes. Int J Biochem Cell Biol 41:323–329

Löwe J, Stock D, Jap B, Zwickl P, Baumeister W, Huber R (1995) Crystal structure of the 20S proteasome from the archaeon T. acidophilum at 3.4 A resolution. Science 268:533–539

Löwe J, Li H, Downing KH, Nogales E (2001) Refined structure of alpha beta-tubulin at 3.5 A resolution. J Mol Biol 313:1045–1057

Löwe J, van den Ent F, Amos LA (2004) Molecules of the bacterial cytoskeleton. Annu Rev Biophys Biomol Struct 33:177–198

Löwe J, Ellonen A, Allen MD, Atkinson C, Sherratt DJ, Grainge I (2008) Molecular mechanism of sequence-directed DNA loading and translocation by FtsK. Mol Cell 31:498–509

Magnani E, Fan J, Gasparini L, Golding M, Williams M, Schiavo G, Goedert M, Amos LA, Spillantini MG (2007) Interaction of tau protein with the dynactin complex. EMBO J 26:4546–4554

Michie KA, Löwe J (2006) Dynamic filaments of the bacterial cytoskeleton. Annu Rev Biochem 75:467–492

Michie KA, Boysen A, Low HH, Møller-Jensen J, Löwe J (2014) LeoA, B and C from enterotoxigenic Escherichia coli (ETEC) are bacterial dynamins. PLoS One 9:e107211

Møller-Jensen J, Ringgaard S, Mercogliano CP, Gerdes K, Löwe J (2007) Structural analysis of the ParR/parC plasmid partition complex. EMBO J 26:4413–4422

Nogales E, Downing KH, Amos LA, Löwe J (1998) Tubulin and FtsZ form a distinct family of GTPases. Nat Struct Biol 5:451–4580

Oliva MA, Cordell SC, Löwe J (2004) Structural insights into FtsZ protofilament formation. Nat Struct Mol Biol 11:1243–1250

Oliva MA, Trambaiolo D, Löwe J (2007) Structural insights into the conformational variability of FtsZ. J Mol Biol 373:1229–1242

Oliva MA, Halbedel S, Freund SM, Dutow P, Leonard TA, Veprintsev DB, Hamoen LW, Löwe J (2010) Features critical for membrane binding revealed by DivIVA crystal structure. EMBO J 29:1988–2001

Ortiz C, Kureisaite-Ciziene D, Schmitz F, McLaughlin SH, Vicente M, Löwe J (2015) Crystal structure of the Z-ring associated cell division protein ZapC from Escherichia coli. FEBS Lett 589:3822–3828

Palmer CM, Löwe J (2014) A cylindrical specimen holder for electron cryo-tomography. Ultramicroscopy 137:20–29

Ringgaard S, Löwe J, Gerdes K (2007) Centromere pairing by a plasmid-encoded type I ParB protein. J Biol Chem 282:28216–28225

Salje J, Löwe J (2008) Bacterial actin: architecture of the ParMRC plasmid DNA partitioning complex. EMBO J 27:2230–2238

Salje J, van den Ent F, de Boer P, Löwe J (2011) Direct membrane binding by bacterial actin MreB. Mol Cell 43:478–487

Schlieper D, Oliva MA, Andreu JM, Löwe J (2005) Structure of bacterial tubulin BtubA/B: evidence for horizontal gene transfer. Proc Natl Acad Sci U S A 102:9170–9175

Seemüller E, Lupas A, Stock D, Löwe J, Huber R, Baumeister W (1995) Proteasome from Thermoplasma acidophilum: a threonine protease. Science 268:579–582

Szwedziak P, Löwe J (2013) Do the divisome and elongasome share a common evolutionary past. Curr Opin Microbiol 16:745–751

Szwedziak P, Wang Q, Freund SM, Löwe J (2012) FtsA forms actin-like protofilaments. EMBO J 31:2249–2260

Szwedziak P, Wang Q, Bharat TA, Tsim M, Löwe J (2014) Architecture of the ring formed by the tubulin homologue FtsZ in bacterial cell division. Elife 3:e04601

van den Ent F, Löwe J (2000) Crystal structure of the cell division protein FtsA from Thermotoga maritima. EMBO J 19:5300–5307

van den Ent F, Amos LA, Löwe J (2001) Prokaryotic origin of the actin cytoskeleton. Nature 413:39–44

van den Ent F, Møller-Jensen J, Amos LA, Gerdes K, Löwe J (2002) F-actin-like filaments formed by plasmid segregation protein ParM. EMBO J 21:6935–6943

van den Ent F, Leaver M, Bendezu F, Errington J, de Boer P, Löwe J (2006) Dimeric structure of the cell shape protein MreC and its functional implications. Mol Microbiol 62:1631–1642

van den Ent F, Vinkenvleugel TM, Ind A, West P, Veprintsev D, Nanninga N, den Blaauwen T, Löwe J (2008) Structural and mutational analysis of the cell division protein FtsQ. Mol Microbiol 68:110–123

van den Ent F, Johnson CM, Persons L, de Boer P, Löwe J (2010) Bacterial actin MreB assembles in complex with cell shape protein RodZ. EMBO J 29:1081–1090

van den Ent F, Izoré T, Bharat TA, Johnson CM, Löwe J (2014) Bacterial actin MreB forms antiparallel double filaments. Elife 3:e02634

Chapter 1
Overview of the Diverse Roles of Bacterial and Archaeal Cytoskeletons

Linda A. Amos and Jan Löwe

Abstract As discovered over the past 25 years, the cytoskeletons of bacteria and archaea are complex systems of proteins whose central components are dynamic cytomotive filaments. They perform roles in cell division, DNA partitioning, cell shape determination and the organisation of intracellular components. The protofilament structures and polymerisation activities of various actin-like, tubulin-like and ESCRT-like proteins of prokaryotes closely resemble their eukaryotic counterparts but show greater diversity. Their activities are modulated by a wide range of accessory proteins but these do not include homologues of the motor proteins that supplement filament dynamics to aid eukaryotic cell motility. Numerous other filamentous proteins, some related to eukaryotic IF-proteins/lamins and dynamins etc, seem to perform structural roles similar to those in eukaryotes.

Keywords FtsZ • Tubulin • MreB • ParM • Actin • TubZ • ESCRT • Bacterial cell division • Archaea • Cytomotive filaments • Plasmid segregation • Cell constriction • Sliding filament motility • Dynamic instability • Treadmilling

Introduction

Life, as we know it, is based on self-replicating cells. Although membrane-bounded cells were presumably predated by self-sustaining systems of complex chemical reactions that may have taken place within confined spaces (Martin and Sousa 2015), all of the living cells known to us are enclosed in a lipid bilayer. Free-living eukaryotic cells have systems for expelling excess water which allows them to live surrounded by water without swelling up and bursting. However, except under particularly favourable conditions, independent bacterial or archaeal cells all need a protective outer wall to survive (Errington 2013; Chap. 3). The cells of bacteria and

L.A. Amos (✉) • J. Löwe (✉)
MRC Laboratory of Molecular Biology, Francis Crick Avenue, Cambridge CB2 0QH, UK
e-mail: laa@mrc-lmb.cam.ac.uk; jyl@mrc-lmb.cam.ac.uk

J. Löwe, L.A. Amos (eds.), *Prokaryotic Cytoskeletons*, Subcellular Biochemistry 84, DOI 10.1007/978-3-319-53047-5_1

archaea also differ from eukaryotes in lacking a dedicated nuclear compartment enclosed in a conserved double membrane, punctuated by pores, that separates DNA from the cytoplasm in eukaryotic cells. Instead, the boundary between the prokaryotic nucleoid and cytoplasm is usually indistinct. Thus, it was long assumed that the contents of these cells were randomly-placed particles, moved around by diffusion. However, visualisation of the bacterial cytoskeleton and investigation of the relevant proteins in the 1990s totally changed this view. Even bacteria depend on dedicated systems of cytoplasmic filaments to carry out many of the activities involved in their replication. Studies over the past two decades have revealed that the key filaments are remarkably well conserved in all cells, including bacteria and archaea, and it is clear that systems of filamentous proteins belonging to a few superfamilies are present in some form throughout all kingdoms of life. Moreover, subtle variations in the protein subunits and the filaments they form apparently tailor the filaments to their respective tasks. Studying the filaments provides some unique insights into the evolutionary relationships between different organisms currently in existence. The cytoskeleton appears to provide a better evolutionary history than, for example, metabolism (Koonin 2015), because of its multitude of interactions that makes exchange between different organisms more difficult than the exchange of an enzyme that works on common small molecule substrates.

A variety of highly dynamic filament-forming proteins exist in the many types of Gram-positive bacteria [such as rod-shaped *Bacillus subtilis* (see Chap. 3), globular *Staphylococcus aureus* (see Chap. 9, section "Noc function in other bacteria: Noc is essential for NO in *S. aureus*"; Turner et al. 2010) or multicellular streptomycetes (see Chap. 6, section "Bacterial intermediate filament-like proteins"; Chap. 9, section "The control of septal placement in sporulating *Streptomyces*"; Bush et al. 2015), in Gram-negative bacteria [such as *Escherichia coli* (Chap. 2) or C*aulobacter crescentus* (Chap. 4)], in pathogenic chlamydiae (see Jacquier et al. 2015; Ouellette et al. 2012), and also in the three major archaeal phyla, Euryarchaeota, Crenarchaeota and Thaumarchaeota (Brochier-Armanet et al. 2008; see Chaps. 12, 13, and 14). Filaments present in all these cells, are used there for membrane-shaping, cell polarisation and cell division (Shaevitz and Gitae 2010; Aylett et al. 2011); the great advantage of having polymers perform such roles is in achieving long range order from small subunits. In most species, there are filaments closely related to eukaryotic cytoskeletal filaments, in particular, to the strands of F-actin and microtubules (Fig. 1.1). Coiled-coil protein filaments are also needed by species such as *C. crescentus* to shape the wall during growth and duplication (Mercier et al. 2014; Chap. 4). From these findings, it is evident that the cytoskeleton of the last common ancestor of all extant cells, including the 'simplest' independent bacteria now known, was already highly sophisticated. Comparisons among different bacteria, archaea and eukarya, of their cytoskeletal proteins as well as a variety of other genetically-encoded components, suggest it is most likely that archaea emerged from early Gram-positive bacteria (Valas and Bourne 2011; Cavalier-Smith 2010; Forterre 2015), while eukaryotes seem to be derived from archaea-like host-cells (see e.g. Baum and Baum 2014; Spang et al. 2015; Koonin 2015) that encompassed endosymbiotic organelles (notably mitochondria and chloroplasts), derived from

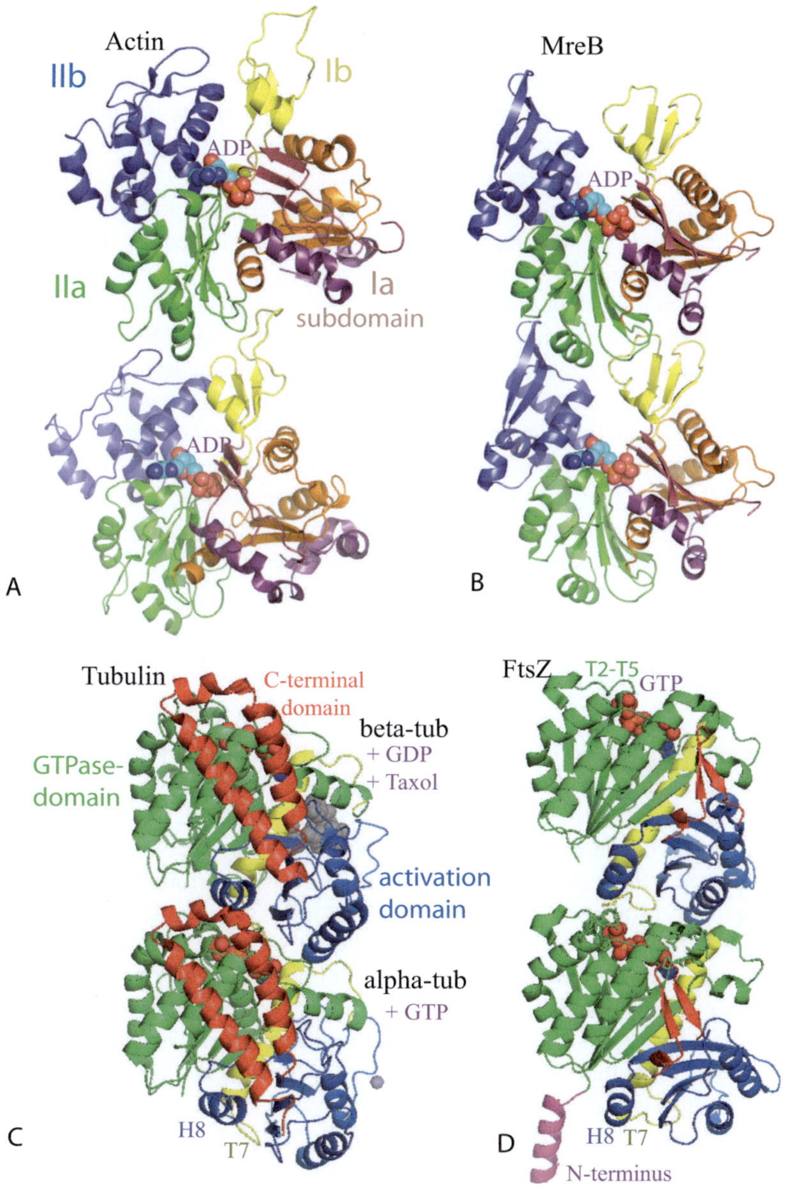

Fig. 1.1 The cytomotive filament proteins found in all kingdoms of life: ribbon-model structures of subunit dimers of the youngest (eukaryotic) and oldest (bacterial) protein species of (**a**, **b**) the actin family (PDB codes 2ZWH, 1JCG; Oda et al. 2009; van den Ent et al. 2001) and (**c**, **d**) the tubulin family (PDB codes; 1JFF, 1WSB; Löwe et al. 2001; Oliva et al. 2004). In **a**, **b** the four subdomains are distinguished by colour; in **c**, **d** the GTPase domains are coloured *green*, GTPase-activation domains (including loop T7 and helix H8) *blue*, central helices (H7) *yellow*, C-terminal hairpin structures *red*; nucleotides and Taxol are shown as ball models. Related proteins are shown in Chaps. 7, 8, 10, 11 and 14

Gram-negative bacteria and mostly retain their original cytomotive filaments (Yooshida et al. 2012; Leger et al. 2015).

A wide range of techniques has been used, in vivo and in vitro, to reach our present view of bacterial and archaeal cytoplasmic filaments, with various kinds of structural investigation playing particularly important parts (e.g. Chen et al. 2010; Erickson et al. 2010; Aylett et al. 2011; Bharat et al. 2015b). The distinguishing characteristics of the filament-forming proteins in different prokaryotes will become clearer in the varied chapters of this book. It is important to mention that EM images of filamentous material are not sufficient evidence that any particular protein is capable of assembly into polymers (see supplementary data of Ghosal et al. 2014), unless a reproducible longitudinal periodicity from the subunits is clearly detectable; this is because dust always includes microscopic filamentous material, so filaments can usually be found on any EM grid that is searched sufficiently; also, if negative stain is used, this can provide extra artefactual objects (Griffith and Bonner 1973). However, most of the images presented in these reviews have been supported in published studies by light-scattering, ultracentrifugation, gel filtration, near-atomic resolution structures obtained by X-ray crystallography or by electron microscopy of frozen specimens (cryo-EM) and other data.

Proteins that Form the Filaments

Many dynamic members of nucleotide-dependent filamentous proteins work as linear motors, pushing or pulling objects through the cell and are thus known as *cytomotive filaments.* The basic properties of these filaments are so well adapted for intracellular movement that they have been conserved throughout evolution, though they have been subtly modified in different species to perform a wide variety of tasks (Fig. 1.2). Movement may be driven by adding subunits to a filament "plus" end (pushing the "cargo", Chap. 10) or removing them from the "minus" end (pulling the cargo, Chap. 11; Gerdes et al. 2010), or both activities together may lead to treadmilling (Chaps. 11 and 16). Another particularly common theme is to have one repeating row of subunits simply sliding relative to another row, which may be identical to the first or completely different (Chaps 7, 10 and 11); this applies equally to these cytomotive filaments and to the various types of rotary motors (Islam and Mignot 2015; Minamino and Imada 2015; Wilkens 2015; Grüber et al. 2014), which are not covered in this book.

Although the structure of the so-called "protofilaments" (strands) and the "longitudinal" interactions (along the strands) between subunits have remained clearly recognisable (Fig. 1.1), the "lateral" interactions between different protofilaments (strands), needed to provide an adequate level of structural stability, are quite varied. All of the various possibilities have been found (see e.g. Popp and Robinson 2012; Ghosal and Löwe 2015), from pairs, polar or apolar, to large bundles or sheets or even cylinders (Jiang et al. 2016; Pilhofer et al. 2011). Protofilaments may run straight or be helical; in the latter case, pairs are usually twisted around one-another. A greater variation within each family of the conserved filament-forming proteins

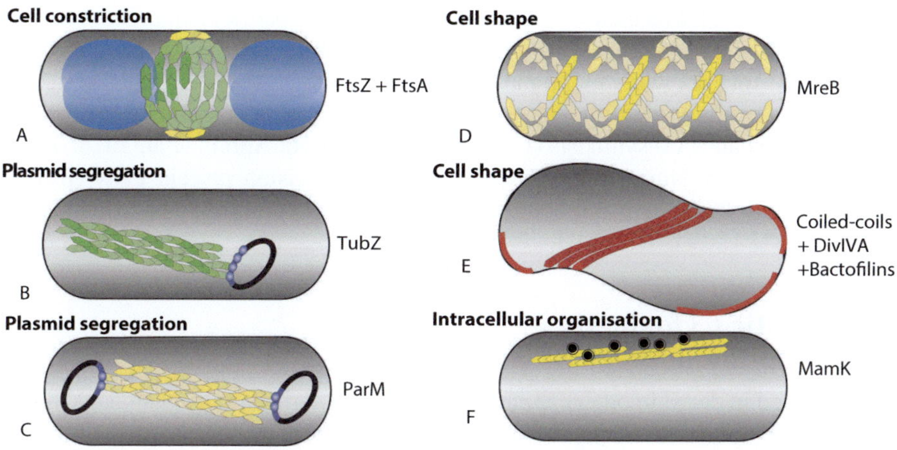

Fig. 1.2 Diverse functions of cytoskeletal elements in bacteria. Bacterial cytoskeletal filaments perform roles in cell division, DNA segregation, cell shape maintenance and cellular compartment organization. Bacterial *tubulin homologues* shown in *green*: FtsZ has a role in cytokinesis (**a**), TubZ executes plasmid segregation (**b**). Bacterial *actin homologues* shown in *yellow*: FtsA is needed for cell division (**a**), ParM executes plasmid segregation (**c**), MamK organises magnetosome chains (**f**), MreB maintains cell shape (**d**). Various other types of proteins self-assemble into filaments on membrane to regulate cell shape (including Intermediate Filament-like coiled-coil proteins, like crescentin) (**e**). Nucleoids shown in *blue* (**a**)

in bacteria and archaea than in eukarya reflects the long evolutionary history of prokaryotes compared with the relatively brief time that eukaryotes have been in existence and the stronger evolutionary pressure on genome size in prokaryotes. The confusing variety of core proteins in bacteria is compounded because some bacteria have acquired additional genes, only distantly related to their own typical sequences, apparently through horizontal gene transfer (HGT). Thus, many bacterial species contain plasmids or large phages that code for cytomotive filaments (e.g. actin-family ParM – Chap. 10 – or tubulin-family TubZ – Chap. 11) that resemble archaeal proteins. *Verrucomicrobia* may have acquired its tubulin-like BtubA/B heterodimers that assemble into narrow microtubules with only a few protofilaments (Pilhofer et al. 2011) from an early form of eukaryote. In contrast, the finding that the archaeal *Lokiarchaeum* genome encodes hypothetical short proteins containing gelsolin-like domains, not found in any bacterial or any other archaeal genomes, is thought to mean that these organisms are themselves related to the vertical line linking archaea and eukarya (Spang et al. 2015). They are however distinct from all known eukarya in lacking motor proteins such as kinesin, myosin or dynein.

Accessory proteins that link prokaryotic cytomotive filaments to their cargoes or to the cell membrane and/or act as catalysts of the cytomotive activity may also strongly modulate the behaviour of the core filaments in different cells; some appear to work in similar ways to known eukaryotic proteins but, unlike the conserved filaments, mostly seem to result from convergent evolution rather than being directly related. So far, no motor proteins related to kinesin, myosin or dynein, that tend to use the filaments as static cytoskeletal tracks in eukaryotic cytoskeletal networks, have been found. Nevertheless, it is becoming clear (see below) that prokaryotes do make use of sliding filaments for pushing, pulling and constricting. How the filaments may interact with membrane-associated motors, such as flagellar rotary motors and their homologues (Nan et al., 2014), or with protein complexes involved in peptidoglycan (PG) synthesis (Laddomada et al. 2016; Agrebi et al. 2015; Typas et al. 2012; Perez-Nuñez et al. 2011) or with cell division protein FtsK (Reyes-Lamothe et al. 2012; Trip and Scheffers 2015) is only beginning to be investigated.

Prokaryotic Cytokinesis

As described in detail in other chapters, almost all bacteria and many archaea divide at mid-cell by processes dependent on a ring that contains protofilaments of tubulin-related FtsZ (Fig. 1.1a); the names of the Fts proteins derive from the finding that temperature-sensitive mutants of *B. subtilis or E. coli* fail to divide at non-permissive temperatures, leading to filamenting cells (Lutkenhaus and Donachie 1979; Beall and Lutkenhaus 1989). However, some groups of archaea divide using ESCRT filaments (Chap. 12) instead of FtsZ and the Chlamidiae group of bacteria seem to use actin-like filaments (Erickson and Osawa 2010; Ouellette et al. 2012). Cytokinesis can take place without the aid of filaments in mutants lacking a cell wall (L-forms), depending only on the production of sufficient membrane to enclose two cells (Mercier et al. 2014), but in normal circumstances some active mechanism is required to ensure that division occurs in an orderly way. In the case of cells protected by an outer wall, its growth by insertion of new material is controlled by the filaments on the cytoplasmic side of the membrane.

Cell Constriction by a Ring of FtsZ Filaments

A ring of FtsZ filaments, the so-called Z-ring (Ma et al. 1996), controls the assembly of a new septum between a pair of daughter cells (see Egan et al. 2015) and also appears to exercise a constrictive force on the underlying cell membrane. Until very recently, the most popular theory for membrane constriction was an "iterative pinching" mechanism (Erickson et al. 2010; Chap. 5) involving nucleotide-dependent bending of dynamic membrane-tethered FtsZ protofilaments; this fits in with the fact that FtsZ protofilaments observed *in vitro* can be straight filaments, single or bundled, or can be curved to form arcs and mini-rings, especially in the presence of ADP. A more recent alternative idea (see Chap. 7) is that a ring of overlapping, though perhaps individually dynamic, protofilaments shrinks in diameter simply by FtsZ potofilaments sliding relative to one another. Of course, iterative-bending of dynamic filaments and a sliding-filament mechanism are not mutually exclusive and their relative importance could even vary among different species. The assembled Z ring organises the machinery that synthesises the PG-based septum (see Egan et al.

2015; Chap. 2). There is evidence that septum assembly is the rate-limiting process of cytokinesis (Coltharp et al. 2016) but is unlikely to be the primary driving force. The Z-centric hypothesis is supported by work showing that FtsZ can assemble Z rings in vitro and generate a force that constricts liposomes (Chaps. 5 and 7).

Control of FtsZ Filaments by Associated Proteins

Associated proteins are needed to dynamically connect FtsZ to the membrane. FtsA and ZipA are the currently best-characterised Z-ring membrane-anchoring proteins but are absent from some bacteria that have FtsZ. Other proteins, such as SepF (Duman et al. 2013), link FtsZ to membrane in these cells. Adjacent filaments in a ring may also be cross-bridged and stabilised by SepF or other proteins such as ZapA (Dajkovic et al. 2010; Bailey et al. 2014), ZapC (Ortiz et al. 2015; Schumacher et al. 2016) or ZipA (Skoog and Daley 2012). Although ZapA, ZapB, ZapC and ZapD are all dispensable for division in *E. coli*, it is likely that, when present, they contribute to its overall robustness. Other proteins control ring-assembly in a variety of ways, depending on species. The stress-response protein SulA that binds to the minus end of FtsZ (Cordell et al. 2003) may inhibit ring formation either by preventing treadmilling or by sequestering subunits (Dajkovic et al. 2008; Chen et al. 2012).

The Z-ring that primarily defines the site of division also recruits other conserved proteins of the "divisome complex" needed for exporting materials to the periplasm (Glas et al. 2015; LaPointe et al. 2013; Villanelo et al. 2011; Goley et al. 2011), where new cell wall is assembled to partition the daughter cells. As the diameter of the ring decreases, divisome components on either side of the membrane constriction insert new peptidoglycan into the ingrowing cell septum. This process is quite closely homologous to that carried out by the "elongasome complex" during cell wall growth (see Szwedziak and Löwe 2013). An early model of constriction involved just the pair of ingrowing partition walls pushing on the cell membrane but this now seems unlikely to be the primary force in species that have FtsZ, because of observations of wall-less liposomes being constricted by filaments (Osawa and Erickson 2013; Szwedziak et al. 2014; Chaps. 5 and 7). It is possibly the sole constrictive force for division of *Chlamydia* cells, which have no FtsZ although, despite being round, they do have actin-like MreB (Ouellette et al. 2012), which is usually required only for shape-maintenance in rod-shaped cells. In any case, constriction needs to take place at exactly the same rate as PG assembly so the latter process is likely to restrict the speed of filament-ring constriction (Coltharp et al. 2016). In plant cells, the division of chloroplasts, endosymbiotic bacteria that have mostly lost their PG cell wall, appears to be driven by a contracting ring of FtsZ (Osteryoung and Pyke 2014; Miyagishima et al. 2014).

Also to be considered in relation to the constriction mechanism by a Z-ring is the *in vitro* observation that rings of filaments assembled from both FtsZ and FtsA on to a lipid layer undergo continuous chiral treadmilling (Loose and Mitchison 2014; Chap. 15), rather than the stochastic cycles of assembly and disassembly that would be expected for iterative pinching due to filament bending (Osama and Erickson 2011; Chap. 5). Thus, once a dynamic ring has been set up *in vivo,* it is possible that constriction is driven by relative sliding movements that continually maximise filament-overlap (Szwedziak et al. 2014, Chap. 7); this would probably involve every plus end sliding (instead of growing, as during setting-up of a ring) to keep up with the shrinking minus end of a neighbouring filament. FtsA appears to play a crucial role in regulating activity at filament ends (Beuria et al. 2009); if the membrane anchor used *in vitro* is ZipA instead of FtsA, the FtsZ filaments do not treadmill (Loose and Mitchison 2014; Chap. 15).

FtsA, which is related to actin family member MreB (Fig. 1.1 and Chap. 8), apart from the replacement of one subdomain by another in a different position (Szwedziak et al. 2012; Chap. 7), is essential to assemble the mid-cell Z-ring in many species, including *E. coli*; the C-terminus of FtsZ binds to FtsA and the C-terminus of FtsA binds to lipid. As mentioned already, these interactions are sufficiently dynamic to allow the FtsZ filaments to treadmill, at least before there is any constriction of the Z-ring. FtsA itself is capable of assembly into actin-like protofilaments. During initial establishment of the ring at mid-cell, the mismatch between subunit length (~5 nm vs ~4 nm) in the two polymers will inhibit FtsA self-assembly at low curvature but interacting *pairs* of FtsA monomers may still exert a tension on FtsZ filaments, to favour a higher curvature and help to orient the ring. However, some components of the divisome can only bind to FtsA subunits that are not blocked by self-association (Chaps. 2, 7 and 8). Thus, as the ring curvature increases and the difference in radii compensates more for the subunit mismatch, longer FtsA polymers become possible and an increasing number of divisome components will be gradually displaced; this may be important for finishing off the septum neatly.

The activity of the DNA-translocase FtsK, an integral membrane protein component of the divisome, is essential during bacterial cell division, particularly for the final stage in the segregation of the chromosomes into the daughter cells (Bailey et al. 2014; Reyes-Lamothe et al. 2012; Chap. 9, section "NO: nucleoid equidistribution by random cell division in *Mycobacterium*"). The structurally-related HerA protein is thought to similarly mediate DNA pumping into daughter cells during archaeal cell division (Iyer et al. 2004).

Restriction of Z Ring Assembly to Mid-Cell

A surprising variety of mechanisms prevent assembly of FtsZ filaments in the wrong places, to ensure positioning of the Z ring on the membrane at mid-cell and also to help co-ordinate cell division with chromosome segregation. In *E coli*, for example, nucleoid occlusion factor SlmA, on the surface of each nucleoid, specifically inhibits Z ring assembly nearby and thus protects the nucleoid from being bisected (Tonthat et al. 2013; Chap. 9). Similarly, in *C. crescentus,* an ATPase called MipZ binds to the chromosomal replication origin (*oriC*) on a nucleoid and inhibits Z-ring assembly there (Kiekebusch et al. 2012; Kiekebusch and Thanbichler 2014). In *B.*

subtilis, however, the nucleoid occlusion protein Noc binds simultaneously to specific DNA sequences and to the cell membrane via an N-terminal amphipathic helix (Adams et al. 2015; Chaps. 3, 7 and 9); the close association of the nucleoprotein complexes with membrane is thought to physically occlude assembly of division machinery.

Other factors may positively favour location of a Z ring at midcell, especially in round cells where FtsZ assembles poorly alone; for example, some cocci have a transmembrane protein called LocZ or MapZ that, in some unknown way, localizes to midcell to help guide Z ring positioning and cell shape (Fleurie et al. 2014; Holečková et al. 2015). In the round-celled *Staphylococcus aureus*, there is evidence for direct spatial cues left in the cell wall by previous septations (Turner et al. 2010). Even in rod-shaped *E coli*, the divisome is given direct guidance to midcell (Bailey et al. 2014), including that by MatP, a small protein associated with the origin of DNA replication (Dupaigne et al. 2012; Durand et al. 2012; Espéli et al. 2012). In rod-shaped *M. xanthus* cells, a protein called PomZ, a homologue of ParA (see below, section on WACAs – Walker A Cytoskeletal ATPases), must localize first to midcell, to prevent defects such as the formation of minicells or long filamentous cells (Treuner-Lange et al. 2013). Similarly, in the rod-shaped Actinobacterium *Corynebacterium glutamicum*, the protein PldP moves to midcell early in the cell cycle and seems to play a role in localizing the Z ring (Donovan and Bramkamp 2014).

Prevention of Ring Assembly Near the Poles

MinC and MinD associate on the membrane in many rod-shaped species to prevent the formation of empty mini-cells at the poles (see e.g. Bisicchia et al. 2013). MinD, like other proteins in the WACA family, undergoes ATP-dependent dimerisation. In the case of MinD, this change also enables the protein to bind to membrane. MinC uses its C-terminal domain to bind to MinD dimers and its N-terminal domain to inhibit the assembly of FtsZ into a ring. There is some doubt about the precise mechanism of inhibition. Some researchers advocate severing of FtsZ filaments by both MinC and SlmA (Chap. 2); others propose that each binds to filaments and orients them so that they cannot function (Chaps. 7 and 9). What actually happens may depend on the conditions; filaments prevented from associating with the membrane or with each other may be more prone to depolymerisation at low FtsZ concentrations. Related to this is the question of how well-organised MinCD molecules are when bound to membrane. Lutkenhaus and colleagues argue that individual MinCD complexes, or possibly dimers, bind to the membrane (Park et al. 2015; Chap. 2) and, by actively severing filaments, are enough to inhibit Z-ring formation. However, Ghosal et al. (2014) assembled copolymers of alternating MinD dimers and MinC dimers, forming well-organised patches on membrane in liposomes and propose that the proteins also do this in vivo (Chapter 7). The proteins seem perfectly designed to form such copolymers but, puzzlingly, the cellular concentration of MinC is much lower than that of MinD. Further studies are needed of membrane-bound MinD alone, which is already known to induce membrane tubulation and bind with the same longitudinal periodicity as MinCD copolymers (Ghosal et al. 2014). Possibly MinC inserts into a minor proportion of sites along a membrane-associated MinD filament and the two tethered N-domains of each MinC dimer can move around to inhibit Z-ring-assembly over an extended area. The membrane-bound MinD filaments, with or without associated MinC-N-domain dimers, associate laterally to form fairly regular 2D arrays (Ghosal et al. 2014).

In Gram negative bacteria, the effector protein MinE (not present in most Gram positives) displaces MinC and also releases MinD from the membrane as ADP-bound monomers. Free MinD and MinC can then diffuse to the other pole. A long-lasting ring of MinE at mid-cell seems to prevent binding of MinCD there (Zheng et al. 2014). Thus the Min proteins in species such as *E. coli* are observed to oscillate back and forth between the two available binding regions near the two poles. In vitro, MinC/D/E mixtures, supplied with ATP, have been observed to bind in dynamic fashion to a lipid-coated surface, producing a variety of waves characteristic of reversible reaction-diffusion systems, and the back-and-forth oscillation has been reconstituted in membrane-coated compartments (Loose et al. 2008, 2011; Chap. 15; Vecchiarelli et al. 2016).

Constriction by ESCRT-III-Like Filaments

Proteins related to eukaryotic ESCRT-III are found in some archaea (Chap. 12) but not in any bacteria. ESCRT-III filaments play an important role in eukaryotic cytokinesis, controlling the final cut to the mid-body (Alonso et al. 2016; Agromayor and Martin-Serrano 2013), at the stage when acto-myosin cytokinetic force has reduced the cell's diameter to the size of a prokaryote; ESCRT filaments also play a critical role in membrane budding and abscission in the eukaryotic endosome system for transporting membrane proteins between the plasma membrane, the *trans*-Golgi network and the lysosome/vacuole (Schuh and Audhya 2014). How the final cut is implemented between these eukaryotic membrane compartments is still unclear but it occurs in the absence of anything equivalent to an archaeal cell wall.

The filaments can self-assemble alone but disassemble with the help of an AAA (ATPase Associated with various cellular Activities) protein, such as eukaryotic Vps4 (Vacuolar protein sorting 4) that catalyses dissociation of the ESCRT machinery from the endosomal membranes (Yang et al. 2015). Vps4-like proteins (called CdvC) are also present in the archaea (Moriscot et al. 2011; Samson et al. 2011). In Crenarchaea and Thaumarchaea, the ESCRT system seems to be entirely responsible for cell division (Chap. 12). A membrane-binding protein called CdvA provides a link between the cell membrane and the ESCRT-III-like CdvB filaments. The latter have been visualised in vivo as cone-shaped spiral assemblies that gradually constrict the membrane between 2 daughter cells (Dobro et al. 2013). It is not yet known whether these filaments undergo relative sliding or simply keep growing spirally inside a gradually-constricting circle.

Because FtsZ is ubiquitous in bacteria and is active in cell division in some archaea but in no eukarya, whereas cell division in some archaea and in many eukarya involves ESCRT-III (Lindås et al. 2008; Samson et al. 2008), it seems likely that FtsZ and FtsZ-based cell division are ancient (Davis 2002; Faguy and Doolittle 1998) whereas ESCRT-III is more recent; also that eukaryotic cells are derived from archaea rather than directly from bacteria.

DNA Segregation

After, and to some extent during, DNA duplication, copies of chromosomal or plasmid DNA are separated to opposite halves of the mother cell by a variety of co-operating mechanisms, many of which employ cytomotive filaments (see eg. Szardenings et al. 2011; Eun et al. 2015). Some bacteriophages also move their DNA around inside cells using filaments (eg. Guo et al. 2014).

Segregation of Plasmids by Actin-Like Filaments

Actin-family proteins that effect plasmid partitioning by assembling into filaments in the bacterial cytoplasm ('type II' partitioning systems), include ParM, AlfA (Shaevitz and Gitai 2010; Polka et al. 2014) and possibly some or all of Alps1-6 (Rivera et al. 2011; Donovan et al. 2015). A similar protein has recently been found encoded by a lytic bacteriophage (Yuan et al. 2015). ParM, whose sequence and crystal structure are closer to some archaeal actin-like proteins, such as Ta0583 found in *Thermoplasma acidophilum* (Roeben et al. 2006), than to bacterial MreB, is currently the best understood cytomotive filament in prokaryotes (Gayathri et al, 2012; Bharat et al. 2015a; Chap. 10).

ParMRC (ParM/ParR/*parC*) is responsible for the equal distribution of R1 plasmids in the two daughter cells during a cell division. Two ParM protofilaments form a left-handed double helical filament, similar, at low resolution, to a mirror image of F-actin. Dynamic instability of ParM filaments assembling alone involves rapid filament formation and breakup, approximately 200 times faster than actin, causing them to disappear unless they are usefully occupied. Plus ends are stabilised when bound to ParR, a protein complex that, in turn, binds to a centromere-like region of the plasmid DNA (*parC*) (Møller-Jensen et al. 2007). Gayathri et al. (2012) found that the C-terminus of ParR binds to the barbed (plus) end of ParM. A ParM filament attached by its plus end to ParR, either after capture or initiating assembly there, grows steadily and more slowly than a free end. It became clear from this work that ParM double-stranded filaments associate side-by-side in vitro to form an anti-parallel mitotic-spindle-like bundle, with ParRC bound at each end; thus, ParM filament elongation pushes plasmids apart (cf. eukaryotic anaphase B). A curved polymer of ParR connecting each plasmid to its end of the growing spindle is functionally similar to a eukaryotic formin complex (e.g. Skau and Watermann 2015).

More recently, Bharat et al. (2015a), found that just pairs of double helical filaments come together to form spindles in *E. coli* cells, where they imaged them directly by cryo-EM tomography. Since only doublets and no bundles were seen in cells, a ParMRC-dependent mechanism of segregating individual pairs of plasmids asynchronously was proposed, obviating the need for cohesion of plasmid pairs or for spindle checkpoints. It is most likely that the spindles are seeded right after replication when the *parC* regions of the plasmids have been doubled into two; the two *parC* regions will each seed one unipolar ParM filament and these will readily associate into an antiparallel arrangement. Dynamic instability may remove filaments if they are not paired to form a spindle, ensuring (checkpoint-like) that pairs of plasmids are pushed to opposite ends of the cell.

Segregation of Plasmids and Phage by Tubulin-Like TubZ

TubZRC (a 'type III' partitioning system, Larson et al. 2007; see Chap. 11) has many parallels with the ParMRC system, a clear example of convergent evolution, suggesting that small bundles of dynamic filaments are very good for distributing objects within a cell. The plasmid-encoded tubulin-family protein, TubZ, found in *Bacilli* appears to be more closely-related to archaeal CetZs than to FtsZ (Aylett et al. 2011); CetZs are also more similar to tubulins than FtsZs are (Duggin et al. 2015; Chap. 14). However, although the tubulin-like longitudinal contacts are preserved in TubZ, a slight kink within each monomer produces a twist in an individual protofilament (unlike the straight protofilaments of FtsZ and tubulin); thus filaments consisting of paired protofilaments are also twisted and look superficially like the structure of F-actin (Aylett et al. 2010). A pair of double filaments or twisted bundle of four single protofilaments seem to be needed for TubZ/TubR/*tubC* plasmid partitioning (Montabana and Agard 2014), whereby shrinking TubZ filament ends may pull replicated plasmids into each daughter cell (cf. eukaryotic anaphase A). TubR is a protein that, like ParR, forms a short curved polymer and binds to a centromere-like region of the DNA (*tubC*) (Ni et al. 2010; Aylett and Löwe 2012). The C-termini of TubZ subunits interact with the TubR complex; this usually happens only at one end of a filament (Fink and Löwe 2015), since the C-termini of TubZ subunits in a filament attach to the surface of the next subunit (Aylett et al. 2010; Montabana and Agard 2014).

A variety of bacteriophages have also been found to encode tubulin/TubZ-like proteins, sometimes known as PhuZ (Kraemer et al. 2012; Oliva et al. 2012; Aylett et al. 2013). Rather than separating duplicated DNA, assembling filaments are believed to be essential for centring all of the replicated bacteriophage virions together within the bacterial host (Erb et al. 2014). The lateral interactions between different protofilaments are also unlike any seen in other members of the tubulin family (Zehr et al. 2014)

Segregation of Chromosomal DNA or Plasmids by WACAs

A large class of proteins known as WACAs (Walker A Cytoskeletal ATPases, having sequences that are placed in the 'P loop' superfamily but have a deviant Walker A motif (Koonin 1993), which promotes the formation of "sandwich" dimers (e.g. Soj, Leonard et al. 2005). They may therefore assemble to form bipolar filaments like eukaryotic septins, to which Soj monomers show some structural similarity (Löwe and Amos 2009). However, there is currently no structural information regarding the possible form of assembly of Soj and similar proteins. Like actin and tubulin, WACAs use nucleotide-dependent polymerisation to carry out complex tasks. Their roles, often performed on a specific substrate such as DNA or membrane, include chromosome segregation (Soj), plasmid segregation 'type I' (ParA, Motallebi-Veshareh et al. 1990; Szardenings et al. 2011) or controlling the position of the Z-ring (MinD). Many have a partner protein that is needed in activating the ATPase cycle. Plasmid ParAs bind nonspecifically to nucleoid DNA in a nucleotide-dependent reversible way. The activating partner protein, ParB, with cargo attached (via the centromere site *parS* on the plasmid) stimulates the ATPase cycle. ParA2 has been shown to form filaments in vitro (Hui et al. 2010) but other ParAs may form filaments only when interacting with the nucleoid. Replicated plasmids might then be segregated by being transported over the nucleoid surface by filament depolymerisation (Hu et al. 2015). Alternatively, ParA dimers may bind all over the surface and dynamic relocation of each plasmid might be explained by a 'burnt bridge' molecular ratchet mechanism (Ietswaart et al. 2014; Ptacin et al. 2010; Vecchiarelli et al. 2014): interaction of the ParB/*parS* complex with a ParA dimer bound to the nucleoid would release the ParA monomers and the ParB/*parS* complex would then be free to bind to a neighbouring ParA dimer. Even a random walk would tend to separate plasmid copies.

ParB binds tightly only to *parS* on the plasmid but other ParB dimers may associate with the complex and bind weakly to adjacent stretches of DNA (Baek et al. 2014; Chen et al. 2015). Spreading of the complex in 2 or 3 dimensions is thought to help the DNA to condense into a compact form, with SMC (structural maintenance of chromosomes) proteins being recruited once folding has begun (Graham et al. 2014; Wang and Rudner 2014).

Cell Shape Determination

Bacteria mostly have simple shapes, spheres, rods, plates, helices and crescents; non-spherical cells depend on filaments associated with the membrane to maintain their shapes. Motile cells have elongated shapes, usually with one or more flagella attached to one pole. Curved or helical shapes seem to be particularly good, both for swimming in liquid and for gliding over surfaces. Some bacteria can even change shape, moving by contractility (Trachtenberg et al. 2014; Ku et al. 2014).

Actin-Like MreB in Rod-Shaped Cells

The longitudinal protofilament contacts between MreB subunits (see Fig. 1.1b and Chap. 8) are almost identical to those of polymerised actin but pairs of protofilaments are antiparallel, rather than parallel, and straight, rather than twisting around each other. MreB is found in all walled bacteria with elongated shapes and its filaments lie close to the cell membrane, in a similar manner to that of actin in the eukaryotic submembrane "cortex" Unlike F-actin, MreB filaments seem to assemble naturally into bends or helices (see images in van den Ent et al. 2001) and may assemble preferentially on to curved membrane. Association of MreB with cell membrane can be via accessory proteins such as RodZ (van den Ent et al. 2010; Morgenstein et al. 2015) but can also be direct (van den Ent et al. 2014; Chap. 8).

Consistent with behaviour found in vitro, MreB was first visualised in cells as long continuous helical filaments stretching along the cell (Jones et al. 2001) but this view was challenged when new imaging experiments suggested MreB formed only small patches or short filaments (reviewed in Reimold et al. 2013; Chap. 8) and it was claimed that long helical filaments are artefacts (Swulius and Jensen 2012). However, the extended filament model has been quite robustly defended (Errington 2015 and Chap. 3; Nan et al. 2014). Certainly, it is unclear why a protein with such versatile cytomotive activity in vitro would have evolved simply to serve as a scaffold for cell wall extension driven by the elongasome in rod-shaped bacteria. FtsZ filaments provide all the direction needed for wall growth in round bacteria. It seems possible that, even if individual MreB filaments are normally short and unstable in vivo, they may overlap intermittently to form short-lived continuous bands. Continuous MreB structures seem to be more stable in vivo in the gliding bacteria (*Myxococcus*: Mauriello et al. 2010). MreB homologues are also found, unexpectedly, in Spiroplasma and Haloplasma, which are wall-less and contractile (Ku et al. 2014).

Whether or not MreB filaments actively shape cell membranes of rod-shaped cells, they appear to organise their elongasome complexes that are responsible for cell wall synthesis and cell growth, with insertion of new peptidoglycan into the cell wall in a helical pattern (Kawai et al. 2009). Normally, this process seems to cause patches of MreB to rotate dynamically around the cell circumference (van Teeffelen et al. 2011). However, Morgenstein et al. (2015) found some *mreB* mutants that suppressed the shape defect resulting from deletion of RodZ, but without restoring rotation. The uncoupling of rotation and growth showed that MreB rotation is not essential for building a rod-shaped wall; but much more research is needed to understand the details of this complex process.

Chlamydia cocci have MreB, which is absent from other round cells, but they lack FtsZ; Ouellette et al. (2012) found that these species require MreB for cell division, as well as FtsI/Pbp3 (a component of all divisomes, see Weiss et al. 1999) and Pbp2 (part of both the elongasome, the PG synthesis machinery required for elongation in rod-shaped bacteria, and of the divisome machinery that builds the cell walls between daughter-cells, see Szwedziak and Löwe 2013). Whether MreB filaments are actively involved in constricting the membrane between *Chlamydia* daughter cells, as well as positioning the PG-synthesis machinery, is currently unclear.

Filamentous Proteins That Promote Curvature

A variety of filamentous proteins are directly or indirectly involved in curvature control (Bohuszewicz et al. 2016). A prominent type with long coiled-coil segments, similar to those in metazoan intermediate (IF) filaments (Chernyatina et al. 2015), tektins present in eukaryotic cilia and flagellar microtubules (Amos 2008; Linck et al. 2014) and the lamins found in all eukaryotic nuclei, has been identified in many different bacteria (Bagchi et al. 2008). As suggested in Chap. 6, the definition of the IF-like family needs to be relaxed a little to allow the inclusion of bacterial proteins with *bona fide* IF functions, while still excluding most of the large number of other proteins that can form coiled-coil assemblies. The series of predicted *coiled coil* domains in crescentin (Cabeen et al. 2011) does resemble that of eukaryotic nuclear lamins and IF proteins such as vimentin (Chernyatina et al. 2012) but no crescentin atomic structure is currently available to confirm the similarity.

Most of the bacterial IF-like filaments, like lamin filaments, seem to associate closely with membrane, carrying out supporting and shaping roles. Some, like PopZ (Chaps. 6, 7 and 9; Ptacin et al. 2014) are found only at cell poles. However, crescentin filaments in *Caulobacter crescentus* (Chap. 4) and AglZ in *Myxococcus xanthus* (Yang et al. 2004; Mauriello et al. 2010), extend along the whole cell length. Filaments of crescentin produce the typical crescent-shape of *Caulobacter crescentus* by forming a membrane-stiffening band of polymers on one side of the cell, where it interacts with MreB on the membrane (Dye et al. 2011).

Bactofilins are bacteria-specific proteins that are quite widespread in both Gram-positive and Gram-negative bacteria (Kühn et al. 2010) but have no exact eukaryotic counterpart apart from, possibly, the amyloid-like proteins that support certain organelles such as gas vesicles. The bactofilins of *C. crescentus* form filamentous structures involved in the regulation of murrein synthesis and are normally restricted to the more curved regions of the cell membrane, at the poles. When overexpressed, however, or expressed in a heterologous cell system, bactofilins can form long filamentous structures. They also form multimers as purified proteins but cannot be studied by x-ray crystallography or solution nuclear magnetic resonance (NMR) because of their propensity to spontaneously polymerise. However, the atomic structure of *C. crescentus* BacA has recently been calculated to 1.0 Å precision from solid-state NMR data, leading to a model in which the core domain forms a *right-handed* β helix made up of 3 β-sheets, with a triangular hydrophobic core (Shi et al. 2015). Head-to-tail assembly of these domains yields 3 nm diameter filaments and, at physiological salt concentrations, the filaments associate as ribbons or two-dimensional sheets.

Myxococcus xanthus, a predatory soil bacterium (Nan and Zusman, 2011), possesses four bactofilins of which one, BacM, is known to play an important role in cell shape maintenance (Zuckerman et al. 2015). Electron and fluorescence light microscopy, as well as studies of purified BacM, indicated that this protein polymerises in vivo and in vitro into ~3 nm wide filaments that further associate into fibers of about 10 nm. Zuckerman et al reported that BacM forms an extended *left-handed* β-helix and further hypothesised that β-sheet domains polymerise head-to-tail into 3 nm filaments via hydrophobic patches. Surprisingly, protein folding and polymerisation occurred even in the presence of chaotropic agents such as one molar urea.

Bactofilins BacE and BacF are essential for motility in *B. subtilis* cells, where they have been seen as complexes of defined size that show a dynamic localisation pattern (El Andari et al. 2015). They play a role in flagellar assembly, being required for the formation of flagellar hook and filament structures (Shimogonya et al. 2015) but not of basal bodies. Functional YFP fusions to BacE and to BacF were seen to localise on the cell membrane as discrete assemblies with a diameter of 60–70 nm. BacF assemblies were relatively static, partially colocalising with basal bodies, while BacE assemblies were fewer per cell and highly mobile. Tracking of BacE foci showed the assemblies arresting at single points for a few hundred milliseconds, so their interaction with flagellar structures must be transient.

The DivIVA peripheral membrane protein, found in Gram positive bacteria, is similar to the bar-domain proteins of eukaryotes (Mim and Unger 2012) and also locates to curved regions of membrane (Ramamurthy and Losick 2009; Oliva et al. 2010). DivIVA plays a role in locating MinCD to polar regions as well as promoting membrane curvature (Bach et al. 2014). Flotillins are widespread membrane proteins that have been shown to form rafts in the membrane of *B. subtilis* and influence its curvature (Bramkamp and Lopez 2015).

Many other filaments have been noticed in the cytoplasms of individual species (Barry and Gitai 2011; Lin and Thanbichler 2013) and their functions are usually unknown. However, the presence of the enzyme CTP-synthase (CtpS), which forms filaments in eukaryotes as well as bacteria (Ingerson-Mahar and Gitai 2012), affects the curvature of *C. crescentus* (Chap. 4). There are, of course, several types of extracellular filaments, flagella and pili used in bacterial and archaeal motility (Islam and Mignot 2015; Wilde and Mullineaux 2015; Shrivastava et al. 2015).

Organisation of Intracellular Membrane Compartments

Although there are few membranous organelles in bacteria, large intracellular compartments may exist as invaginations of the cell membrane, especially in more specialised bacteria, such as the Magnetotactic bacteria or Cyanobacteria ('blue-green algae') (see Bohuszewicz et al. 2016). Magnetotactic bacteria, such as *Magnetospirillum,* have bag-like cell-membrane invaginations called magnetosomes that contain iron oxide or iron sulphide crystals. MamK, a specialised actin-like protein, forms straight double filaments that organise a linear arrangement of magnetosomes to act as a magnetic field sensor (Draper et al. 2011; Ozyamak et al. 2013; Chap. 8). Cyanobacteria have their photosynthetic machinery embedded in extensive inward folds of the cell membrane (Nierzwicki-Bauer et al. 1983) and some contain gas vacuoles to provide buoyancy; there is evidence that filaments are involved in organising such membranes (Walsby 1994). In some phyla, the invaginations may be pinched off completely by some unknown mechanism. In fast-growing species, invaginations may be small and transient and thus currently undetected. The roles of bacterial dynamin-like proteins (BDLPs, see Bramkamp 2012; Low and Löwe 2010) are still unclear but it seems likely that they include the creation and maintenance of some such membrane compartments; *in vitro*, BDLPs form polymers on lipid tubes like their eukaryotic counterparts (Low et al. 2009; Michie et al. 2014).

Evolutionary Relationships

Studies of prokaryotic cytomotive filaments are providing a rich source of information about cell evolution (Wickstead and Gull 2011; Yutin and Koonin 2012; Baum and Baum 2014; Spang et al. 2015). Actin-family proteins in some archaea are very closely-related to eukaryotic actin (see Chap. 13). As mentioned above, many types of archaea divide their cells using filaments of FtsZ proteins that are virtually indistinguishable from those in bacteria, although archaea commonly have two genes coding for FtsZs, as compared with a single gene in a bacterial genome; it is not yet known why. Some archaeal groups also have, in addition, one or more genes coding for CetZs, proteins that appear to be intermediate in sequence and structure between FtsZ and tubulin (Aylett et al. 2011; Duggin et al. 2015, see Chap. 14). They assemble into protofilaments in vitro but appear to have in vivo roles not required for cytokinesis. Bacteria contain structurally-similar proteins, but only in association with plasmids (see Chap. 11) and these were presumably acquired through HGT. Thus, evidence from the tubulin-family proteins is consistent with archaea having evolved from a fully-functional bacterium-like ancestor and the original eukaryote host cell having been an archaeon. The few *Verrucomicrobia* species that have tubulin-like BtubA/B microtubules are thought to have gained the sequences via HGT (Schlieper et al. 2005; Pilhofer et al. 2007; Koonin 2015), possibly from an extinct primitive eukaryote (Martin-Galiano et al. 2011).

Archaea that use ESCRT proteins for cytokinesis (Chap. 12) are likely to be descended from an ancestor that lost its unnecessary FtsZ after developing ESCRTs. Eukaryotic cells may share an ancestor with these archaea or may have acquired ESCRT proteins horizontally at some early stage in their evolution. Further evidence for evolution having progressed from a bacterium-like ancestor to an archaeon-like stage, then to a eukaryotic cell, is provided by the atomic structures and filament assemblies of actin-family proteins.,As detailed in Chap. 13, archaeal species express actin-family proteins called crenactin that closely resemble eukaryotic actin (Lindås et al. 2014; Izoré et al. 2014; Braun et al. 2015). In bacteria, plasmid proteins, such as ParM (Chap. 10), which resembles an actin-family protein of an archaeon (*Thermoplasma*, Roeben et al. 2006) much more than genomic bacterial MreB, would have arrived via HGT. MreB itself appears to have been derived originally from a family of non-filamentous enzymatic proteins (Kabsch and Holmes 1995) and resembles the heat-shock protein HSP-70 even more than eukaryotic actin does. MreB is much less widespread in bacteria than FtsZ. It may either have appeared later than FtsZ or have been lost by cells that did not need MreB after gaining FtsZ; its use by *Chlamydia* in cytokinetic constriction, in place of FtsZ, may be a secondary development. Nevertheless, both families of very active cytomotive filamentous proteins have been highly preserved in the vertical heritage from early bacteria to complex eukaryotes. In other words, descendents of the cells in which these versatile proteins evolved have completely taken over the Earth.

References

Adams DW, Wu LJ, Errington J (2015) Nucleoid occlusion protein Noc recruits DNA to the bacterial cell membrane. EMBO J 34:491–501

Agrebi R, Wartel M, Brochier-Armanet C, Mignot T (2015) An evolutionary link between capsular biogenesis and surface motility in bacteria. Nat Rev Microbiol 13:318–326

Agromayor M, Martin-Serrano J (2013) Knowing when to cut and run: mechanisms that control cytokinetic abscission. Trends Cell Biol 23:433–441

Alonso Y, Adell M, Migliano SM, Teis D (2016) ESCRT-III and Vps4: a dynamic multipurpose tool for membrane budding and scission. FEBS J 283(18):3288–3302

Amos LA (2008) The tektin family of microtubule-stabilizing proteins. Genome Biol 9:229

Aylett CH, Löwe J (2012) Superstructure of the centromeric complex of TubZRC plasmid partitioning systems. Proc Natl Acad Sci U S A 109:16522–16527

Aylett CH, Wang Q, Michie KA, Amos LA, Löwe J (2010) Filament structure of bacterial tubulin homologue TubZ. Proc Natl Acad Sci U S A 107:19766–19771

Aylett CH, Löwe J, Amos LA (2011) New insights into the mechanisms of cytomotive actin and tubulin filaments. Int Rev Cell Mol Biol 292:1–71

Aylett CH, Izoré T, Amos LA, Löwe J (2013) Structure of the tubulin/FtsZ-like protein TubZ from Pseudomonas bacteriophage ΦKZ. J Mol Biol 425:2164–2173

Bach JN, Albrecht N, Bramkamp M (2014) Imaging DivIVA dynamics using photo-convertible and activatable fluorophores in Bacillus subtilis. Front Microbiol 5:59

Baek JH, Rajagopala SV, Chattoraj DK (2014) Chromosome segregation proteins of Vibrio cholerae as transcription regulators. MBio 5:e01061–e01014
Bagchi S, Tomenius H, Belova LM, Ausmees N (2008) Intermediate filament-like proteins in bacteria and a cytoskeletal function in Streptomyces. Mol Microbiol 70:1037–1050
Bailey MW, Bisicchia P, Warren BT, Sherratt DJ, Männik J (2014) Evidence for divisome localization mechanisms independent of the Min system and SlmA in Escherichia coli. PLoS Genet 10:e1004504
Barry RM, Gitai Z (2011) Self-assembling enzymes and the origins of the cytoskeleton. Curr Opin Microbiol 14:704–711
Baum DA, Baum B (2014) An inside-out origin for the eukaryotic cell. BMC Biol 12:76
Beall B, Lutkenhaus J (1989) Nucleotide sequence and insertional inactivation of a Bacillus subtilis gene that affects cell division, sporulation, and temperature sensitivity. J Bacteriol 171:6821–6834
Beuria TK et al (2009) Adenine nucleotide-dependent regulation of assembly of bacterial tubulin-like FtsZ by a hypermorph of bacterial actin-like FtsA. J Biol Chem 284:14079–14086
Bharat TA, Murshudov GN, Sachse C, Löwe J (2015a) Structures of actin-like ParM filaments show architecture of plasmid-segregating spindles. Nature 523:106–110
Bharat TA, Russo CJ, Löwe J, Passmore LA, Scheres SH (2015b) Advances in single-particle electron cryomicroscopy structure determination applied to sub-tomogram averaging. Structure 23:1743–1753
Bisicchia P, Arumugam S, Schwille P, Sherratt D (2013) MinC, MinD, and MinE drive counter-oscillation of early-cell-division proteins prior to Escherichia coli septum formation. MBio 4:e00856–e00813
Bohuszewicz O, Liu J, Low HH (2016) Membrane remodelling in bacteria. J Struct Biol 196(1):3–14
Bramkamp M (2012) Structure and function of bacterial dynamin-like proteins. Biol Chem 393:1203–1214
Bramkamp M, Lopez D (2015) Exploring the existence of lipid rafts in bacteria. Microbiol Mol Biol Rev 79:81–100
Braun T et al (2015) Archaeal actin from a hyperthermophile forms a single-stranded filament. Proc Natl Acad Sci U S A 112:9340–9345
Brochier-Armanet C, Boussau B, Gribaldo S, Forterre P (2008) Mesophilic Crenarchaeota: proposal for a third archaeal phylum, the Thaumarchaeota. Nat Rev Microbiol 6:245–252
Bush MJ, Tschowri N, Schlimpert S, Flärdh K, Buttner MJ (2015) c-di-GMP signalling and the regulation of developmental transitions in streptomycetes. Nat Rev Microbiol 13:749–760
Cabeen MT, Herrmann H, Jacobs-Wagner C (2011) The domain organization of the bacterial intermediate filament-like protein crescentin is important for assembly and function. Cytoskeleton (Hoboken) 68:205–219
Cavalier-Smith T (2010) Kingdoms Protozoa and Chromista and the eozoan root of the eukaryotic tree. Biol Lett 6:342–345
Chen S et al (2010) Electron cryotomography of bacterial cells. J Vis Exp 39:1943
Chen Y, Milam SL, Erickson HP (2012) SulA inhibits assembly of FtsZ by a simple sequestration mechanism. Biochemistry 51:3100–3109
Chen BW, Lin MH, Chu CH, Hsu CE, Sun YJ (2015) Insights into ParB spreading from the complex structure of Spo0J and parS. Proc Natl Acad Sci U S A 112:6613–6618
Chernyatina AA, Nicolet S, Aebi U, Herrmann H, Strelkov SV (2012) Atomic structure of the vimentin central α-helical domain and its implications for intermediate filament assembly. Proc Natl Acad Sci U S A 109:13620–13625
Chernyatina AA, Guzenko D, Strelkov SV (2015) Intermediate filament structure: the bottom-up approach. Curr Opin Cell Biol 32:65–72
Coltharp C, Buss J, Plumer TM, Xiao J (2016) Defining the rate-limiting processes of bacterial cytokinesis. Proc Natl Acad Sci U S A 113(8):E1044–E1053

Cordell SC, Robinson EJ, Lowe J (2003) Crystal structure of the SOS cell division inhibitor SulA and in complex with FtsZ. Proc Natl Acad Sci U S A 100:7889–7894
Dajkovic A, Mukherjee A, Lutkenhaus J (2008) Investigation of regulation of FtsZ assembly by SulA and development of a model for FtsZ polymerization. J Bacteriol 190:2513–2526
Dajkovic A, Pichoff S, Lutkenhaus J, Wirtz D (2010) Cross-linking FtsZ polymers into coherent Z rings. Mol Microbiol 78:651–668
Davis BK (2002) Molecular evolution before the origin of species. Prog Biophys Mol Biol 79:77–133
Dobro MJ et al (2013) Electron cryotomography of ESCRT assemblies and dividing Sulfolobus cells suggests that spiraling filaments are involved in membrane scission. Mol Biol Cell 24:2319–2327
Donovan C, Bramkamp M (2014) Cell division in Corynebacterineae. Front Microbiol 5:132
Donovan C et al (2015) A prophage-encoded actin-like protein required for efficient viral DNA replication in bacteria. Nucleic Acids Res 43:5002–5016
Draper O et al (2011) MamK, a bacterial actin, forms dynamic filaments in vivo that are regulated by the acidic proteins MamJ and LimJ. Mol Microbiol 82:342–354
Duggin IG et al (2015) CetZ tubulin-like proteins control archaeal cell shape. Nature 519:362–365
Duman R et al (2013) Structural and genetic analyses reveal the protein SepF as a new membrane anchor for the Z ring. Proc Natl Acad Sci U S A 110:E4601–E4610
Dupaigne P et al (2012) Molecular basis for a protein-mediated DNA-bridging mechanism that functions in condensation of the E. coli chromosome. Mol Cell 48:560–571
Durand D et al (2012) Expression, purification and preliminary structural analysis of Escherichia coli MatP in complex with the matS DNA site. Acta Crystallogr Sect F Struct Biol Cryst Commun 68:638–643
Dye NA, Pincus Z, Fisher IC, Shapiro L, Theriot JA (2011) Mutations in the nucleotide binding pocket of MreB can alter cell curvature and polar morphology in Caulobacter. Mol Microbiol 81(2):368–394
Egan AJ, Biboy J, van't Veer I, Breukink E, Vollmer W (2015) Activities and regulation of peptidoglycan synthases. Philos Trans R Soc Lond Ser B Biol Sci 370
El Andari J, Altegoer F, Bange G, Graumann PL (2015) Bacillus subtilis bactofilins are essential for flagellar Hook- and filament assembly and dynamically localize into structures of less than 100 nm diameter underneath the cell membrane. PLoS One 10:e0141546
Erb ML et al (2014) A bacteriophage tubulin harnesses dynamic instability to center DNA in infected cells. Elife 3
Erickson HP, Osawa M (2010) Cell division without FtsZ–a variety of redundant mechanisms. Mol Microbiol 78:267–270
Erickson HP, Anderson DE, Osawa M (2010) FtsZ in bacterial cytokinesis: cytoskeleton and force generator all in one. Microbiol Mol Biol Rev 74:504–528
Errington J (2013) L-form bacteria, cell walls and the origins of life. Open Biol 3:120143
Errington JAJ-W (2015) Bacterial morphogenesis and the enigmatic MreB helix. Nat Rev Microbiol 13:241–248
Espéli O et al (2012) A MatP-divisome interaction coordinates chromosome segregation with cell division in E. coli. EMBO J 31:3198–3211
Eun YJ, Kapoor M, Hussain S, Garner EC (2015) Bacterial filament systems: toward understanding their emergent behavior and cellular functions. J Biol Chem 290:17181–17189
Faguy DM, Doolittle WF (1998) Cytoskeletal proteins: the evolution of cell division. Curr Biol 8:R338–R341
Fink G, Löwe J (2015) Reconstitution of a prokaryotic minus end-tracking system using TubRC centromeric complexes and tubulin-like protein TubZ filaments. Proc Natl Acad Sci U S A 112:E1845–E1850
Fleurie A et al (2014) MapZ marks the division sites and positions FtsZ rings in Streptococcus pneumoniae. Nature 516:259–262

Forterre P (2015) The universal tree of life: an update. Front Microbiol 6:717
Gayathri P et al (2012) A bipolar spindle of antiparallel ParM filaments drives bacterial plasmid segregation. Science 338:1334–1337
Gerdes K, Howard M, Szardenings F (2010) Pushing and pulling in prokaryotic DNA segregation. Cell 141:927–942
Ghosal D, Löwe J (2015) Collaborative protein filaments. EMBO J 34:2312–2320
Ghosal D, Trambaiolo D, Amos LA, Lowe J (2014) MinCD cell division proteins form alternating copolymeric cytomotive filaments. Nat Commun 5:5341
Glas M et al (2015) The Soluble periplasmic domains of Escherichia coli cell division proteins FtsQ/FtsB/FtsL form a trimeric complex with submicromolar affinity. J Biol Chem 290:21498–21509
Goley ED et al (2011) Assembly of the caulobacter cell division machine. Mol Microbiol 80:1680–1698
Graham TG et al (2014) ParB spreading requires DNA bridging. Genes Dev 28:1228–1238
Griffith JD, Bonner JF (1973) Chromatin-like aggregates of uranyl acetate. Nat New Biol 244:80–81
Grüber G, Manimekalai MS, Mayer F, Müller V (2014) ATP synthases from archaea: the beauty of a molecular motor. Biochim Biophys Acta 1837:940–952
Guo P et al (2014) Common mechanisms of DNA translocation motors in bacteria and viruses using one-way revolution mechanism without rotation. Biotechnol Adv 32:853–872
Holečková N et al (2015) LocZ is a new cell division protein involved in proper septum placement in Streptococcus pneumoniae. MBio 6:e01700–e01714
Hu L, Vecchiarelli AG, Mizuuchi K, Neuman KC, Liu J (2015) Directed and persistent movement arises from mechanochemistry of the ParA/ParB system. Proc Natl Acad Sci U S A 112:E7055–E7064
Hui MP et al (2010) ParA2, a Vibrio cholerae chromosome partitioning protein, forms left-handed helical filaments on DNA. Proc Natl Acad Sci U S A 107:4590–4595
Ietswaart R, Szardenings F, Gerdes K, Howard M (2014) Competing ParA structures space bacterial plasmids equally over the nucleoid. PLoS Comput Biol 10:e1004009
Ingerson-Mahar M, Gitai Z (2012) A growing family: the expanding universe of the bacterial cytoskeleton. FEMS Microbiol Rev 36:256–266
Islam ST, Mignot T (2015) The mysterious nature of bacterial surface (gliding) motility: a focal adhesion-based mechanism in Myxococcus xanthus. Semin Cell Dev Biol 46:143–154
Iyer LM, Makarova KS, Koonin EV, Aravind L (2004) Comparative genomics of the FtsK-HerA superfamily of pumping ATPases: implications for the origins of chromosome segregation, cell division and viral capsid packaging. Nucleic Acids Res 32:5260–5279
Izoré T, Duman R, Kureisaite-Ciziene D, Löwe J (2014) Crenactin from Pyrobaculum calidifontis is closely related to actin in structure and forms steep helical filaments. FEBS Lett 588:776–782
Jacquier N, Viollier PH, Greub G (2015) The role of peptidoglycan in chlamydial cell division: towards resolving the chlamydial anomaly. FEMS Microbiol Rev 39:262–275
Jiang S et al (2016) Novel actin filaments from Bacillus thuringiensis form nanotubules for plasmid DNA segregation. Proc Natl Acad Sci U S A 113(9):E1200–E1205
Jones LJ, Carballido-López R, Errington J (2001) Control of cell shape in bacteria: helical, actin-like filaments in Bacillus subtilis. Cell 104:913–922
Kabsch W, Holmes KC (1995) The actin fold. FASEB J 9:167–174
Kawai Y, Daniel RA, Errington J (2009) Regulation of cell wall morphogenesis in Bacillus subtilis by recruitment of PBP1 to the MreB helix. Mol Microbiol 71:1131–1144
Kiekebusch D, Thanbichler M (2014) Spatiotemporal organization of microbial cells by protein concentration gradients. Trends Microbiol 22:65–73
Kiekebusch D, Michie KA, Essen LO, Löwe J, Thanbichler M (2012) Localized dimerization and nucleoid binding drive gradient formation by the bacterial cell division inhibitor MipZ. Mol Cell 46:245–259

Koonin EV (1993) A superfamily of ATPases with diverse functions containing either classical or deviant ATP-binding motif. J Mol Biol 229:1165–1174

Koonin EV (2015) Origin of eukaryotes from within archaea, archaeal eukaryome and bursts of gene gain: eukaryogenesis just made easier. Philos Trans R Soc Lond Ser B Biol Sci 370:20140333

Kraemer JA et al (2012) A phage tubulin assembles dynamic filaments by an atypical mechanism to center viral DNA within the host cell. Cell 149:1488–1499

Ku C, Lo WS, Kuo CH (2014) Molecular evolution of the actin-like MreB protein gene family in wall-less bacteria. Biochem Biophys Res Commun 446:927–932

Kühn J et al (2010) Bactofilins, a ubiquitous class of cytoskeletal proteins mediating polar localization of a cell wall synthase in Caulobacter crescentus. EMBO J 29:327–339

Laddomada F, Miyachiro MM, Dessen A (2016) Structural insights into protein-protein interactions involved in bacterial cell wall biogenesis. Antibiotics (Basel) 5:14

LaPointe LM et al (2013) Structural organization of FtsB, a transmembrane protein of the bacterial divisome. Biochemistry 52:2574–2585

Larsen RA et al (2007) Treadmilling of a prokaryotic tubulin-like protein, TubZ, required for plasmid stability in Bacillus thuringiensis. Genes Dev 21:1340–1352

Leger MM et al (2015) An ancestral bacterial division system is widespread in eukaryotic mitochondria. Proc Natl Acad Sci U S A 112:10239–10246

Leonard TA, Butler PJ, Löwe J (2005) Bacterial chromosome segregation: structure and DNA binding of the Soj dimer–a conserved biological switch. EMBO J 24:270–282

Lin L, Thanbichler M (2013) Nucleotide-independent cytoskeletal scaffolds in bacteria. Cytoskeleton (Hoboken) 70:409–423

Linck R et al (2014) Insights into the structure and function of ciliary and flagellar doublet microtubules: tektins, Ca2+-binding proteins, and stable protofilaments. J Biol Chem 289:17427–17444

Lindås AC, Karlsson EA, Lindgren MT, Ettema TJ, Bernander R (2008) A unique cell division machinery in the Archaea. Proc Natl Acad Sci U S A 105(48):18942–18946

Lindås AC, Chruszcz M, Bernander R, Valegård K (2014) Structure of crenactin, an archaeal actin homologue active at 90°C. Acta Crystallogr D Biol Crystallogr 70:492–500

Loose M, Mitchison TJ (2014) The bacterial cell division proteins FtsA and FtsZ self-organize into dynamic cytoskeletal patterns. Nat Cell Biol 16:38–46

Loose M, Fischer-Friedrich E, Ries J, Kruse K, Schwille P (2008) Spatial regulators for bacterial cell division self-organize into surface waves in vitro. Science 320:789–792

Loose M, Fischer-Friedrich E, Herold C, Kruse K, Schwille P (2011) Min protein patterns emerge from rapid rebinding and membrane interaction of MinE. Nat Struct Mol Biol 18:577–583

Low HH, Löwe J (2010) Dynamin architecture–from monomer to polymer. Curr Opin Struct Biol 20:791–798

Low HH, Sachse C, Amos LA, Löwe J (2009) Structure of a bacterial dynamin-like protein lipid tube provides a mechanism for assembly and membrane curving. Cell 139:1342–1352

Löwe J, Amos LA (2009) Evolution of cytomotive filaments: the cytoskeleton from prokaryotes to eukaryotes. Int J Biochem Cell Biol 41:323–329

Löwe J, Li H, Downing KH, Nogales E (2001) Refined structure of alpha beta-tubulin at 3.5 A resolution. J Mol Biol 313:1045–1057

Lutkenhaus JF, Donachie WD (1979) Identification of the ftsA gene product. J Bacteriol 137:1088–1094

Ma X, Ehrhardt DW, Margolin W (1996) Colocalization of cell division proteins FtsZ and FtsA to cytoskeletal structures in living Escherichia coli cells by using green fluorescent protein. Proc Natl Acad Sci U S A 93:12998–13003

Martin WF, Sousa FL (2015) Early microbial evolution: the age of anaerobes. Cold Spring Harb Perspect Biol 8(2)

Martin-Galiano AJ et al (2011) Bacterial tubulin distinct loop sequences and primitive assembly properties support its origin from a eukaryotic tubulin ancestor. J Biol Chem 286:19789–19803
Mauriello EM et al (2010) Bacterial motility complexes require the actin-like protein, MreB and the Ras homologue, MglA. EMBO J 29:315–326
Mercier R, Kawai Y, Errington J (2014) General principles for the formation and proliferation of a wall-free (L-form) state in bacteria. Elife 3
Michie KA, Boysen A, Low HH, Møller-Jensen J, Löwe J (2014) LeoA, B and C from enterotoxigenic Escherichia coli (ETEC) are bacterial dynamins. PLoS One 9:e107211
Mim C, Unger VM (2012) Membrane curvature and its generation by BAR proteins. Trends Biochem Sci 37:526–533
Minamino T, Imada K (2015) The bacterial flagellar motor and its structural diversity. Trends Microbiol 23:267–274
Miyagishima SY, Nakamura M, Uzuka A, Era A (2014) FtsZ-less prokaryotic cell division as well as FtsZ- and dynamin-less chloroplast and non-photosynthetic plastid division. Front Plant Sci 5:459
Møller-Jensen J, Ringgaard S, Mercogliano CP, Gerdes K, Löwe J (2007) Structural analysis of the ParR/parC plasmid partition complex. EMBO J 26:4413–4422
Montabana EA, Agard DA (2014) Bacterial tubulin TubZ-Bt transitions between a two-stranded intermediate and a four-stranded filament upon GTP hydrolysis. Proc Natl Acad Sci U S A 111:3407–3412
Morgenstein RM et al (2015) RodZ links MreB to cell wall synthesis to mediate MreB rotation and robust morphogenesis. Proc Natl Acad Sci U S A 112:12510–12515
Moriscot C et al (2011) Crenarchaeal CdvA forms double-helical filaments containing DNA and interacts with ESCRT-III-like CdvB. PLoS One 6:e21921
Motallebi-Veshareh M, Rouch DA, Thomas CM (1990) A family of ATPases involv2ed in active partitioning of diverse bacterial plasmids. Mol Microbiol 4:1455–1463
Nan B, Zusman DR (2011) Uncovering the mystery of gliding motility in the myxobacteria. Annu Rev Genet 45:21–39
Nan B, McBride MJ, Chen J, Zusman DR, Oster G (2014) Bacteria that glide with helical tracks. Curr Biol 24:R169–R173
Ni L, Xu W, Kumaraswami M, Schumacher MA (2010) Plasmid protein TubR uses a distinct mode of HTH-DNA binding and recruits the prokaryotic tubulin homolog TubZ to effect DNA partition. Proc Natl Acad Sci U S A 107:11763–11768
Nierzwicki-Bauer SA, Balkwill DL, Stevens SE (1983) Three-dimensional ultrastructure of a unicellular cyanobacterium. J Cell Biol 97:713–722
Oda T, Iwasa M, Aihara T, Maéda Y, Narita A (2009) The nature of the globular- to fibrous-actin transition. Nature 457:441–445
Oliva MA, Cordell SC, Lowe J (2004) Structural insights into FtsZ protofilament formation. Nat Struct Mol Biol 11:1243–1250
Oliva MA et al (2010) Features critical for membrane binding revealed by DivIVA crystal structure. EMBO J 29:1988–2001
Oliva MA, Martin-Galiano AJ, Sakaguchi Y, Andreu JM (2012) Tubulin homolog TubZ in a phage-encoded partition system. Proc Natl Acad Sci U S A 109:7711–7716
Ortiz C et al (2015) Crystal structure of the Z-ring associated cell division protein ZapC from Escherichia coli. FEBS Lett 589:3822–3828
Osawa M, Erickson HP (2011) Inside-out Z rings–constriction with and without GTP hydrolysis. Mol Microbiol 81:571–579
Osawa M, Erickson HP (2013) Liposome division by a simple bacterial division machinery. Proc Natl Acad Sci U S A 110:11000–11004
Osteryoung KW, Pyke KA (2014) Division and dynamic morphology of plastids. Annu Rev Plant Biol 65:443–472

Ouellette SP, Karimova G, Subtil A, Ladant D (2012) Chlamydia co-opts the rod shape-determining proteins MreB and Pbp2 for cell division. Mol Microbiol 85:164–178

Ozyamak E, Kollman J, Agard DA, Komeili A (2013) The bacterial actin MamK: in vitro assembly behavior and filament architecture. J Biol Chem 288:4265–4277

Park KT, Du S, Lutkenhaus J (2015) MinC/MinD copolymers are not required for Min function. Mol Microbiol 98:895–909

Pérez-Núñez D et al (2011) A new morphogenesis pathway in bacteria: unbalanced activity of cell wall synthesis machineries leads to coccus-to-rod transition and filamentation in ovococci. Mol Microbiol 79:759–771

Pilhofer M, Rosati G, Ludwig W, Schleifer KH, Petroni G (2007) Coexistence of tubulins and ftsZ in different Prosthecobacter species. Mol Biol Evol 24:1439–1442

Pilhofer M, Ladinsky MS, McDowall AW, Petroni G, Jensen GJ (2011) Microtubules in bacteria: ancient tubulins build a five-protofilament homolog of the eukaryotic cytoskeleton. PLoS Biol 9:e1001213

Polka JK, Kollman JM, Mullins RD (2014) Accessory factors promote AlfA-dependent plasmid segregation by regulating filament nucleation, disassembly, and bundling. Proc Natl Acad Sci U S A 111:2176–2181

Popp D, Robinson RC (2012) Supramolecular cellular filament systems: how and why do they form. Cytoskeleton (Hoboken) 69:71–87

Ptacin JL et al (2010) A spindle-like apparatus guides bacterial chromosome segregation. Nat Cell Biol 12:791–798

Ptacin JL et al (2014) Bacterial scaffold directs pole-specific centromere segregation. Proc Natl Acad Sci U S A 111:E2046–E2055

Ramamurthi KS, Losick R (2009) Negative membrane curvature as a cue for subcellular localization of a bacterial protein. Proc Natl Acad Sci U S A 106:13541–13545

Reimold C, Defeu Soufo HJ, Dempwolff F, Graumann PL (2013) Motion of variable-length MreB filaments at the bacterial cell membrane influences cell morphology. Mol Biol Cell 24:2340–2349

Reyes-Lamothe R, Nicolas E, Sherratt DJ (2012) Chromosome replication and segregation in bacteria. Annu Rev Genet 46:121–143

Rivera CR, Kollman JM, Polka JK, Agard DA, Mullins RD (2011) Architecture and assembly of a divergent member of the ParM family of bacterial actin-like proteins. J Biol Chem 286:14282–14290

Roeben A et al (2006) Crystal structure of an archaeal actin homolog. J Mol Biol 358:145–156

Samson RY, Obita T, Freund SM, Williams RL, Bell SD (2008) A role for the ESCRT system in cell division in archaea. Science 322(5908):1710–1713

Samson RY et al (2011) Molecular and structural basis of ESCRT-III recruitment to membranes during archaeal cell division. Mol Cell 41:186–196

Schlieper D, Oliva MA, Andreu JM, Löwe J (2005) Structure of bacterial tubulin BtubA/B: evidence for horizontal gene transfer. Proc Natl Acad Sci U S A 102:9170–9175

Schuh AL, Audhya A (2014) The ESCRT machinery: from the plasma membrane to endosomes and back again. Crit Rev Biochem Mol Biol 49:242–261

Schumacher MA, Zeng W, Huang KH, Tchorzewski L, Janakiraman A (2016) Structural and functional analyses reveal insights into the molecular properties of the Escherichia coli Z ring stabilizing protein, ZapC. J Biol Chem 291:2485–2498

Shaevitz JW, Gitai Z (2010) The structure and function of bacterial actin homologs. Cold Spring Harb Perspect Biol 2:a000364

Shi C et al (2015) Atomic-resolution structure of cytoskeletal bactofilin by solid-state NMR. Sci Adv 1:e1501087

Shimogonya Y et al (2015) Torque-induced precession of bacterial flagella. Sci Rep 5:18488

Shrivastava A, Lele PP, Berg HC (2015) A rotary motor drives Flavobacterium gliding. Curr Biol 25:338–341

Skau CT, Waterman CM (2015) Specification of architecture and function of actin structures by actin nucleation factors. Annu Rev Biophys 44:285–310
Skoog K, Daley DO (2012) The Escherichia coli cell division protein ZipA forms homodimers prior to association with FtsZ. Biochemistry 51:1407–1415
Spang A et al (2015) Complex archaea that bridge the gap between prokaryotes and eukaryotes. Nature 521:173–179
Swulius MT, Jensen GJ (2012) The helical MreB cytoskeleton in Escherichia coli MC1000/pLE7 is an artifact of the N-Terminal yellow fluorescent protein tag. J Bacteriol 194:6382–6386
Szardenings F, Guymer D, Gerdes K (2011) ParA ATPases can move and position DNA and subcellular structures. Curr Opin Microbiol 14:712–718
Szwedziak P, Löwe J (2013) Do the divisome and elongasome share a common evolutionary past. Curr Opin Microbiol 16:745–751
Szwedziak P, Wang Q, Freund SM, Löwe J (2012) FtsA forms actin-like protofilaments. EMBO J 31:2249–2260
Szwedziak P, Wang Q, Bharat TA, Tsim M, Löwe J (2014) Architecture of the ring formed by the tubulin homologue FtsZ in bacterial cell division. Elife 3:e04601
Tonthat NK et al (2013) SlmA forms a higher-order structure on DNA that inhibits cytokinetic Z-ring formation over the nucleoid. Proc Natl Acad Sci U S A 110:10586–10591
Trachtenberg S, Schuck P, Phillips TM, Andrews SB, Leapman RD (2014) A structural framework for a near-minimal form of life: mass and compositional analysis of the helical mollicute Spiroplasma melliferum BC3. PLoS One 9:e87921
Treuner-Lange A et al (2013) PomZ, a ParA-like protein, regulates Z-ring formation and cell division in Myxococcus xanthus. Mol Microbiol 87:235–253
Trip EN, Scheffers DJ (2015) A 1 MDa protein complex containing critical components of the Escherichia coli divisome. Sci Rep 5:18190
Turner RD et al (2010) Peptidoglycan architecture can specify division planes in Staphylococcus aureus. Nat Commun 1:26
Typas A, Banzhaf M, Gross CA, Vollmer W (2012) From the regulation of peptidoglycan synthesis to bacterial growth and morphology. Nat Rev Microbiol 10:123–136
Valas RE, Bourne PE (2011) The origin of a derived superkingdom: how a gram-positive bacterium crossed the desert to become an archaeon. Biol Direct 6:16
van den Ent F, Amos LA, Löwe J (2001) Prokaryotic origin of the actin cytoskeleton. Nature 413:39–44
van den Ent F, Johnson CM, Persons L, de Boer P, Löwe J (2010) Bacterial actin MreB assembles in complex with cell shape protein RodZ. EMBO J 29:1081–1090
van den Ent F, Izoré T, Bharat TA, Johnson CM, Löwe J (2014) Bacterial actin MreB forms antiparallel double filaments. Elife 3:e02634
van Teeffelen S et al (2011) The bacterial actin MreB rotates, and rotation depends on cell-wall assembly. Proc Natl Acad Sci U S A 108:15822–15827
Vecchiarelli AG, Neuman KC, Mizuuchi K (2014) A propagating ATPase gradient drives transport of surface-confined cellular cargo. Proc Natl Acad Sci U S A 111:4880–4885
Vecchiarelli AG, Lia M, Mizuuchia M, Hwanga LC, Seol Y, Neuman KC, Mizuuchia K (2016) Membrane-bound MinDE complex acts as a toggle switch that drives Min oscillation coupled to cytoplasmic depletion of MinD. Proc Natl Acad Sci U S A 113:E1479–E1488
Villanelo F, Ordenes A, Brunet J, Lagos R, Monasterio O (2011) A model for the Escherichia coli FtsB/FtsL/FtsQ cell division complex. BMC Struct Biol 11:28
Walsby AE (1994) Gas vesicles. Microbiol Rev 58:94–144
Wang X, Rudner DZ (2014) Spatial organization of bacterial chromosomes. Curr Opin Microbiol 22:66–72
Weiss DS, Chen JC, Ghigo JM, Boyd D, Beckwith J (1999) Localization of FtsI (PBP3) to the septal ring requires its membrane anchor, the Z ring, FtsA, FtsQ, and FtsL. J Bacteriol 181:508–520
Wickstead B, Gull K (2011) The evolution of the cytoskeleton. J Cell Biol 194:513–525

Wilde A, Mullineaux CW (2015) Motility in cyanobacteria: polysaccharide tracks and Type IV pilus motors. Mol Microbiol 98:998–1001
Wilkens S (2015) Structure and mechanism of ABC transporters. F1000Prime Rep 7:14
Yang R et al (2004) AglZ is a filament-forming coiled-coil protein required for adventurous gliding motility of Myxococcus xanthus. J Bacteriol 186:6168–6178
Yang B, Stjepanovic G, Shen Q, Martin A, Hurley JH (2015) Vps4 disassembles an ESCRT-III filament by global unfolding and processive translocation. Nat Struct Mol Biol 22:492–498
Yooshida Y, Miyagishima SY, Kuroiwa H, Kuroiwa T (2012) The plastid-dividing machinery: formation, constriction and fission. Curr Opin Plant Biol 15:714–721
Yuan Y et al (2015) Effects of actin-like proteins encoded by two Bacillus pumilus phages on unstable lysogeny, revealed by genomic analysis. Appl Environ Microbiol 81:339–350
Yutin N, Koonin EV (2012) Archaeal origin of tubulin. Biol Direct 7:10
Zehr EA et al (2014) The structure and assembly mechanism of a novel three-stranded tubulin filament that centers phage DNA. Structure 22:539–548
Zheng M et al (2014) Self-assembly of MinE on the membrane underlies formation of the MinE ring to sustain function of the Escherichia coli Min system. J Biol Chem 289:21252–21266
Zuckerman DM et al (2015) The bactofilin cytoskeleton protein BacM of Myxococcus xanthus forms an extended β-sheet structure likely mediated by hydrophobic interactions. PLoS One 10:e0121074

Website

Slides (including some movies) can be downloaded from the Löwe group website: http://www2.mrc-lmb.cam.ac.uk/groups/jyl/slideshows/encyclopedia2014/index.html

Chapter 2
E. coli Cell Cycle Machinery

Joe Lutkenhaus and Shishen Du

Abstract Cytokinesis in *E. coli* is organized by a cytoskeletal element designated the Z ring. The Z ring is formed at midcell by the coalescence of FtsZ filaments tethered to the membrane by interaction of FtsZ's conserved C-terminal peptide (CCTP) with two membrane-associated proteins, FtsA and ZipA. Although interaction between an FtsZ monomer and either of these proteins is of low affinity, high affinity is achieved through avidity – polymerization linked CCTPs interacting with the membrane tethers. The placement of the Z ring at midcell is ensured by antagonists of FtsZ polymerization that are positioned within the cell and target FtsZ filaments through the CCTP. The placement of the ring is reinforced by a protein network that extends from the terminus (Ter) region of the chromosome to the Z ring. Once the Z ring is established, additional proteins are recruited through interaction with FtsA, to form the divisome. The assembled divisome is then activated by FtsN to carry out septal peptidoglycan synthesis, with a dynamic Z ring serving as a guide for septum formation. As the septum forms, the cell wall is split by spatially regulated hydrolases and the outer membrane invaginates in step with the aid of a transenvelope complex to yield progeny cells.

Keywords *E. coli* • FtsZ • Z ring • FtsA • ZipA • Zap proteins • Cytokinetic machinery • Divisome • Polymerization driven avidity • Min system • Oscillatíon • Nucleoid occlusion • Ter linkage • Septal PG synthesis • FtsEX • FtsN

Overview of Cell Cycle Regulation – Two Key Proteins

The bacterial cell cycle is primarily regulated at the initiation of two major events, DNA replication and cytokinesis (septation). Studies in *E. coli* indicate that the regulatory inputs that control these two events converge on just two proteins, DnaA for DNA replication and FtsZ for cytokinesis (Fig. 2.1). DnaA , which assembles into an oligomer on *oriC*, is required to unwind the DNA so that DnaB, the replicative helicase, can be loaded and the replication forks established (Bramhill and

J. Lutkenhaus (✉) • S. Du
University of Kansas Medical Center, Kansas City, KS, USA
e-mail: Jlutkenh@kumc.edu

J. Löwe, L.A. Amos (eds.), *Prokaryotic Cytoskeletons*, Subcellular Biochemistry 84, DOI 10.1007/978-3-319-53047-5_2

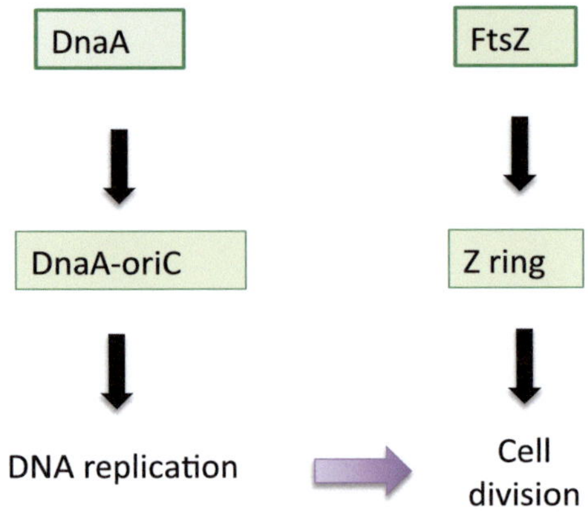

Fig. 2.1 Regulatory inputs for cell cycle control converge on two key proteins, DnaA and FtsZ. DnaA-ATP assembles on *oriC* to initiate DNA replication. FtsZ assembles into a Z ring that determines the division plane by organizing the machinery to synthesize the septum. Whereas DnaA-ATP assembles on the *oriC* template, the Z ring does not have a landmark and is a self organizing organelle that assembles where conditions are favorable

Kornberg 1988; Erzberger et al. 2006). FtsZ assembles into the Z ring, a cytoskeletal element that determines the site of cytokinesis and functions as a scaffold to recruit additional division proteins to synthesize septal peptidoglycan (PG) (Bi and Lutkenhaus 1991). These two major events are not obligatorily coupled, since DNA replication and segregation can continue in the absence of cytokinesis. This chapter is focused on *E. coli,* with occasional references to results from other organisms as indicated.

DnaA and Initiation of Replication

Much of the control of the cell cycle operates at the level of initiation of DNA replication. Donachie found that the ratio of cell mass to DNA origins was a constant, independent of growth rate, and proposed that replication was initiated when cells double their critical size, with cytokinesis occurring a fixed time later (Donachie 1968). The consequences can be readily observed upon a shift up in growth rate. When a culture is shifted from a slow to a fast growth rate, the rate of mass increase immediately shifts to the new growth rate and new rounds of DNA replication initiate as the mass doubles and before the previous round is finished (Cooper and Helmstetter 1968). The first cytokinetic event, however, is delayed because the time required to replicate the chromosome and to divide are constant and longer than the fast doubling time. As a consequence, the cells are larger at the first division and then reach a new steady state. Under these conditions, cytokinesis follows a fixed time after DNA replication is initiated.

Initiation occurs when DnaA-ATP binds to sites located within *oriC*, a unique 245 base pair region on the chromosome. *oriC* has three high affinity sites, as well as many low affinity sites, for DnaA-ATP (Leonard and Grimwade 2015). Adjacent to these sites is a region designated DUE, the DNA unwinding element, which is converted to a single-stranded region when all DnaA binding sites are occupied and where DnaB is loaded (Bramhill and Kornberg 1988). Although the high affinity sites are usually occupied, the loading of the low affinity sites is highly regulated, as their occupation results in the triggering of initiation (Mott and Berger 2007). Occupancy of the low affinity sites each cell cycle requires the synthesis of DnaA to generate DnaA-ATP since DnaA-ADP, which is generated following an initiation event, does not readily exchange with ATP (Kato and Katayama 2001). Thus, DnaA must be synthesized each cell cycle to obtain the ATP-bound form, consistent with the old observation that each new round of initiation of replication requires protein synthesis (Helmstetter 1974). It has been suggested that it is the ratio of DnaA-ATP to DnaA-ADP that is critical for the firing of *oriC* but this is controversial (Donachie and Blakely 2003; Vadia and Levin 2015). The reader is referred to several excellent reviews on the regulation of replication, as this article will focus on cytokinesis (Mott and Berger 2007; Leonard and Grimwade 2015; Katayama et al. 2010).

FtsZ and the Z Ring

Assembly of the Z ring at the division site is the first step in bacterial cytokinesis (Bi and Lutkenhaus 1991). The Z ring was the first cytoskeletal element to be described in bacteria and is assembled from FtsZ filaments formed by the polymerization of FtsZ (Ma and Margolin 1999), the ancestral homologue of eukaryotic tubulin (Lowe and Amos 1998). It is a very dynamic structure ($T1/2 < 10$ s) (Chen and Erickson 2005), formed by the coalescence of FtsZ filaments attached to the membrane. This process is under spatial regulation to ensure that the Z ring is assembled at midcell, between segregated chromosomes (Lutkenhaus 2007).

The mechanisms of spatial regulation appear quite different among diverse bacteria and can include negative as well as positive systems (Lutkenhaus 2007; Monahan et al. 2014; Mannik and Bailey 2015). In addition, multiple systems can exist within the same organism and each contribute to spatial regulation. These systems are usually not essential but their absence often leads to altered morphology due to misplacement of the septum. In *E. coli* the negative regulatory systems include Min (minicell) and NO (nucleoid occlusion). The Min system is highly conserved and widely distributed across diverse bacterial species (Rothfield et al. 2005). Loss of spatial regulation in Min mutants leads to assembly of the Z ring near the cell poles, resulting in the formation of anucleate minicells, thus emphasizing that the position of the Z ring dictates the site of cytokinesis (Bi and Lutkenhaus 1993). NO, which prevents Z ring assembly over the nucleoid, is present in many bacteria and loss of this system leads to Z rings forming over chromosomes delayed for segregation (Bernhardt and De Boer 2005). In addition, a positive regulatory system in *E. coli* involves linkage between the Ter macrodomain of the chromosome and the Z ring (Bailey et al. 2014). This latter system is revealed when the other two are removed.

In contrast to DnaA, which binds to defined sequences in *oriC* to produce a helical filament on the DNA that can initiate replication (Mott and Berger 2007), the Z ring does not have a template. It is a dynamic self-organizing structure that is positioned by spatial regulation and, once established, determines the division site (Lutkenhaus 2007). Importantly, FtsZ is expressed at a constant rate and the major control of Z ring assembly is due to the spatial regulation that positions the Z ring to midcell (Weart and Levin 2003).

Components of the Cytokinetic Machinery – Cell Division Genes

The identification of cell division genes in *E. coli* started in the 1960s with the isolation of mutants with a filamenting temperature sensitive phenotype (*fts*) (Van De Putte et al. 1964; Hirota et al. 1968). Such mutants continue to replicate and segregate their DNA at the nonpermissive temperature but grow as long nonseptate filaments that are unable to form colonies. This conditional lethality allowed the cloning of the respective genes, which in turn led to the characterization of their gene products. Although mutations in many non cell division genes can lead to a filamentous phenotype, these were eventually ruled out as their effect on cell division was indirect.

Key steps in the characterization of a cell division gene are the demonstration that it is essential and that the gene product localizes to the division site. This has led to the realization that the complex process of cytokinesis is carried out by only 12 essential genes (*ftsZ, ftsA, zipA, ftsE, ftsX, ftsK, ftsQ, ftsL, ftsB, ftsW, ftsI* and *ftsN*) (De Boer 2010; Lutkenhaus et al. 2012). FtsE/FtsX are essential even though loss of these genes can be suppressed by increased osmolarity (Reddy 2007). Many other proteins localize to the division site but their genes are not essential because their function is redundant (for example, genes for Zap proteins and amidases) or they only contribute to the efficiency of division (Tol-Pal complex). Other genes are essential, but are not specific for cell division, because they are also involved in cell elongation (genes for PG synthesis).

Additional insight into the process of cell division came from investigating the phenotypic effect of penicillin derivatives (Spratt 1975). Among the derivatives are some (cephalexin and piperacillin) that selectively block cytokinesis without affecting cell elongation. These antibiotics specifically target penicillin-binding protein 3 (PBP3), the product of *ftsI*, supporting its role in cytokinesis. Among the 12 or so penicillin binding proteins in *E. coli*, only PBP3 is used exclusively in cytokinesis (Young 2001). Less specific penicillins, which target multiple PBPs, lead to lysis at the division site, as this is a major site of cell wall synthesis (Spratt 1975).

Assembly of the Z Rings

The assembly of the divisome occurs in two temporally distinct steps (Aarsman et al. 2005). In the first step, the Z ring (also referred to as the proto-ring (Rico et al. 2013)) is formed at midcell (Bi and Lutkenhaus 1991). Proteins that interact with FtsZ coassemble with FtsZ to form the ring. Key among these are ZipA and FtsA, which function as membrane tethers for FtsZ filaments (Hale and De Boer 1997; Pichoff and Lutkenhaus 2002). In addition, various Zap proteins (ZapA, ZapC and ZapD) interact with FtsZ and can crosslink FtsZ filaments (Gueiros-Filho and Losick 2002; Hale et al. 2011; Durand-Heredia et al. 2011; Huang et al. 2013). FtsEX also assembles early and FtsE is reported to interact with FtsZ (Corbin et al. 2007). The formation of the Z ring is under spatial regulation, ensuring that it is formed at midcell (Lutkenhaus 2007). After a delay that can be up to 1/3 of a cell cycle, the remaining proteins are recruited to form a complete divisome (Aarsman et al. 2005).

FtsZ and Tubulin Form Dynamic Structures

The globular domains of FtsZ and tubulin are remarkably similar (Lowe and Amos 1998; Nogales et al. 1998). They belong to a distinct family of GTPases that undergo dynamic polymerization dependent upon GTP hydrolysis (Nogales et al. 1998; Mukherjee and Lutkenhaus 1998). Whereas tubulin assembles into a 13-stranded microtubule, FtsZ assembles into a linear filament, equivalent to one strand of a microtubule, and undergoes treadmilling driven by GTP hydrolysis (Mukherjee and Lutkenhaus 1994, 1998; Chen and Erickson 2005; Loose and Mitchison 2014). Despite FtsZ forming a linear filament, assembly, like that of microtubules, is cooperative, with a critical concentration near 1 μM (Mukherjee and Lutkenhaus 1998; Chen et al. 2005). Importantly, this family of proteins can use the dynamic capability provided by GTP hydrolysis to explore intracellular space. Microtubules use the assembly dynamics to search for kinetochores, in the process of forming a spindle (Kirschner and Mitchison 1986; Heald and Khodjakov 2015), whereas FtsZ uses dynamics to search for conditions that are favorable for forming the Z ring (Lutkenhaus 2007). These conditions are where the concentration of factors that antagonize FtsZ polymerization is the lowest.

CCTP – High Affinity Through Polymerization Driven Avidity

Both FtsZ and tubulin have long C-terminal segments which, although quite different, mediate interaction with a number of interacting proteins (Erickson et al. 2010; Roll-Mecak 2015). The FtsZ C-terminal segment is composed of a long

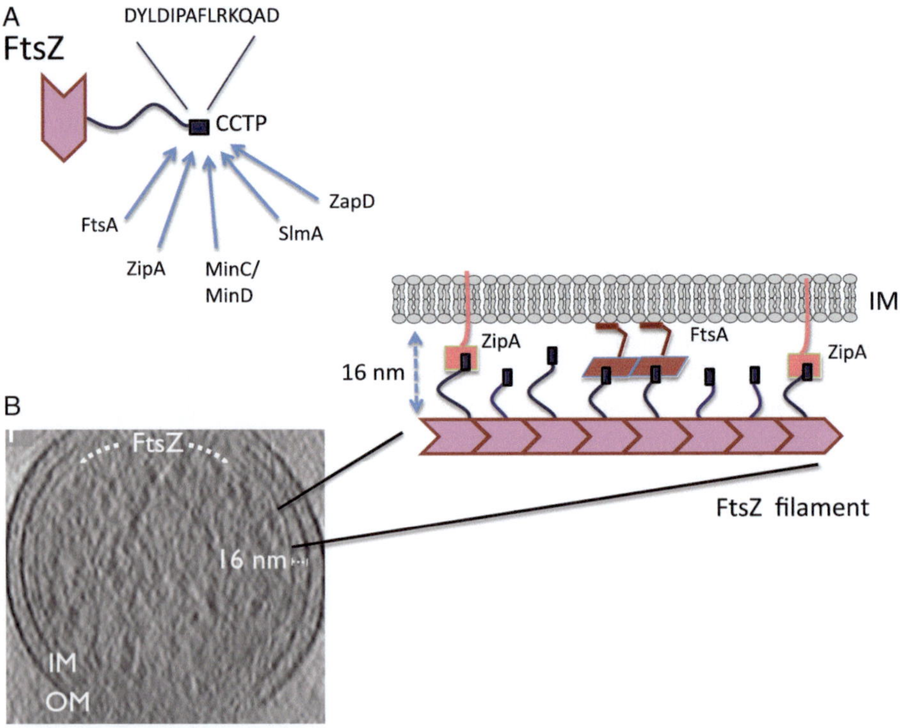

Fig. 2.2 FtsZ interacts with many partners through the CCTP. (**a**) FtsZ contains a globular tubulin domain attached to the CCTP by a long linker (50 amino acids in *E. coli*). It is a rare example in prokaryotes of a short linear peptide embedded within a region of intrinsic disorder (the linker) that drives the interaction of a protein with variety of unrelated proteins. (**b**) Image and cartoon of an FtsZ filament *in vivo*. Electron cryotomogram of an *E. coli* cell in which FtsZ (D212N) was expressed shows an FtsZ filament 16 nm from the cytoplasmic membrane (Szwedziak et al. 2014). Similar filaments are observed in WT cells. The accompanying cartoon depicts the FtsZ filament tethered to the cytoplasmic membrane by ZipA and FtsA. High affinity of the filaments for the membrane is due to polymerization driven avidity as multiple CCTPs contact multiple membrane partners

non-conserved linker (~50 amino acids in *E. coli*) that connects the globular (GTP-polymerizing) domain to a very conserved short region of about 14 residues (Fig. 2.2a). This short region has been termed the C-terminal tail (CTT) or CCTP (conserved C-terminal peptide) (Du et al. 2015). It mediates interaction between FtsZ and many of its partners, including the membrane anchors, ZipA and FtsA, and the spatial regulators, MinC/MinD and SlmA (Ma and Margolin 1999; Haney et al. 2001). The CCTP is an example of a short linear peptide embedded within a region of intrinsic disorder (the linker) that drives the interaction of a protein with a variety of unrelated proteins; something more common among eukaryotic proteins (Uversky 2013).

The binding sites for the CCTP on ZipA and FtsA have little in common and there are quite different side chain contacts between residues in the CCTP and these partners (Szwedziak et al. 2012; Mosyak et al. 2000; Schumacher and Zeng 2016). The CCTP has relatively weak affinity for these partners (with a K_D in the 30–50 μM range), indicating that an FtsZ monomer is unlikely to localize to the membrane. However, an FtsZ filament has high affinity for the membrane anchors due to avidity (Du et al. 2015). Thus, FtsZ filaments are at the membrane (Fig. 2.2b) and GTP-hydrolysis releases FtsZ monomers into the cytoplasm.

In addition to FtsA and ZipA, FtsZ filaments interact with several regulatory proteins (MinC/MinD and SlmA) and at least one Zap protein, ZapD, through the CCTP (Lutkenhaus et al. 2012; Durand-Heredia et al. 2012; Schumacher and Zeng 2016). MinC/MinD and SlmA are antagonists of FtsZ assembly that become positioned in the cell by interacting with the membrane and DNA respectively. Once in position, they use a two-pronged mechanism to disrupt FtsZ filaments. Both antagonists bind to an FtsZ filament through interaction with the CCTP and, in a second step, sever the filament (Cho et al. 2011; Shen and Lutkenhaus 2010; Du and Lutkenhaus 2014). The interaction between the CCTP and SlmA follows the same principle as for ZipA; monomers bind weakly and filaments bind strongly due to avidity (Du et al. 2015). These differential affinities for monomers and filaments, along with a putative severing mechanism, enable MinC/MinD and SlmA to effectively disrupt FtsZ filaments, even though they are present in the cell at much lower concentrations than FtsZ.

Membrane-Tethered FtsZ Filaments Coalesce at Midcell with the Aid of Zap Proteins

The two membrane associated proteins that tether FtsZ filaments to the membrane are quite different, although both bind to the CCTP of FtsZ (Pichoff and Lutkenhaus 2002; Ma and Margolin 1999; Haney et al. 2001). ZipA is a Type 1b transmembrane protein with a long linker connecting the transmembrane domain to a globular domain that binds the CCTP of FtsZ (Hale and De Boer 1997). In contrast, FtsA is a peripheral membrane protein that binds to the membrane through a C-terminal amphipathic helix (Pichoff and Lutkenhaus 2005). It is an actin related protein, with an unusual domain structure but assembles into actin-like filaments on a lipid bilayer *in vitro* (Szwedziak et al. 2012). So far, no ATPase activity has been associated with assembly and the polymers are not dynamic (Lara et al. 2005; Szwedziak et al. 2012). FtsA polymers have also been observed *in vivo* at the membrane when overexpressed (Szwedziak et al. 2012). In one model, the C-terminal amphipathic helix obstructs polymerization and binding to the membrane relieves this inhibition (Krupka et al. 2014). In the absence of the C-terminal amphipathic helix, polymerization occurs in the cytoplasm but is less efficient, suggesting that the membrane enhances assembly, which is consistent with the model (Pichoff and Lutkenhaus 2005).

FtsZ filaments are formed at the membrane throughout the cell and even oscillate between the ends of the cell under the influence of the Min system (Thanedar and Margolin 2004; Bisicchia et al. 2013a). Eventually these membrane bound filaments coalesce into a Z ring at midcell with the aid of the Zap proteins. Three of the Zap proteins (ZapA, C and D) crosslink FtsZ filaments (Hale et al. 2011; Durand-Heredia et al. 2011, 2012; Gueiros-Filho and Losick 2002; Mohammadi et al. 2009; Dajkovic et al. 2010), whereas a fourth, ZapB, does not interact directly with FtsZ but interacts with ZapA (Galli and Gerdes 2010). The loss of all three Zaps that interact directly with FtsZ results in increased cell length and poor viability, whereas loss of any one has less effect, indicating some functional overlap (Durand-Heredia et al. 2012).

However, the three Zap proteins crosslink FtsZ filaments through different mechanisms. ZapA forms dimers and tetramers and is believed to crosslink filaments by interacting with the lateral sides of filaments (Pacheco-Gomez et al. 2013), whereas ZapD is a dimer and binds to CCTPs on adjacent filaments (Durand-Heredia et al. 2012). In contrast, ZapC is a monomer that binds to the FtsZ globular domain, but can crosslink FtsZ filaments since each monomer has two unique FtsZ-binding sites (Schumacher et al. 2015).

Although crosslinking filaments by Zap proteins facilitates Z ring formation, it is not clear if direct lateral interaction between FtsZ filaments also has a role. FtsZ filaments are readily bundled *in vitro,* depending upon cations and pH, but the bundles lack repetitive protein-protein interactions (Erickson et al. 2010). One FtsZ mutant displays many interesting phenotypes, similar to FtsA* (see later). This mutant displays increased bundling *in vitro* and it is likely that the bundling contributes to the phenotypes but how is not clear (Haeusser et al. 2015).

Structure of the Z Ring

One of the most intriguing questions is the structure of the Z ring, or more specifically the arrangement of filaments in the ring. Two approaches have been used to try and resolve this, super resolution fluorescence microscopy and electron cryotomography. By super high-resolution fluorescence microscopy the Z ring appears patchy, which has been interpreted as a discontinuous ring (Fu et al. 2010; Strauss et al. 2012). In nonconstricting cells, the ring appears to be 115 nm in width and ~700 nm in diameter. Interestingly, increasing the FtsZ concentration does not affect the dimensions of the midcell ring but leads to formation of additional misplaced rings, suggesting a defined structure (Fu et al. 2010). About 500 molecules of FtsZ are required to form a filament (with a 4.3 nm intersubunit distance) sufficient in length to encircle the division site in a nonconstricting cell. Since *E. coli* contains ~6500 molecules of FtsZ per cell (fast growth rate) (Li et al. 2014) and ~30% of FtsZ is in the ring (Stricker et al. 2002), there is sufficient FtsZ assembled to encircle the septum about four times.

FtsZ filaments are difficult to detect by electron microscopy due to the density of the cytoplasm, their low numbers (lack of a regular lateral array) and their proximity to the membrane. However, FtsZ filaments have been clearly observed by electron cryotomography, approximately 16 nm from the membrane (Li et al. 2007; Szwedziak et al. 2014) (Fig. 2.2b). Furthermore, varying the length of the linker changes the distance of the filaments from the membrane, confirming that the observed filaments are FtsZ (Szwedziak et al. 2014). In addition, when FtsA was overproduced along with a hydrolysis-deficient mutant of FtsZ (D212N), an FtsA filament was observed half way between the membrane and the FtsZ filament. Although this lends support to the idea that FtsA polymers can form *in vivo*, FtsA filaments are not observed in wild type cells (Fig. 2.2b). Presumably, this is because they are very short and less abundant than FtsZ filaments, consistent with a ratio of FtsZ to FtsA that is ~7 to 1 (Li et al. 2014). When overproduced, FtsZ (D212N) filaments were observed as doublets with an interfilament spacing (center-to-center distance) of 6.8 nm, too large to arise from lateral interactions (Szwedziak et al. 2014). Whether these doublets can also occur under wild type conditions is not clear but, even under these overproduction conditions, bundles of FtsZ filaments are not observed, indicating bundling due to lateral interactions is unlikely.

The Ter region of the chromosome is connected to the Z ring by a protein network (ZapA, ZapB and MatP), designated the Ter linkage. Conventional fluorescence microscopy revealed that ZapB is located interior to the Z ring, whereas ZapA colocalized with FtsZ (Galli and Gerdes 2010). Since MatP condenses the Ter region into a macrodomain and ZapB links MatP to the Z ring through ZapA, a model emerged in which there was a continuous link between the Ter region and the Z ring (Mercier et al. 2008; Espeli et al. 2012). Higher resolution imaging revealed the dimensions of ZapA rings were similar to Z rings whereas ZapB rings are thicker and have a smaller diameter, with MatP interior to ZapB,which supports the linkage model (Buss et al. 2015). This linkage is not essential for division, as ZapA or ZapB mutants are viable, although Z rings appear to have difficulty in forming and are sometimes askew. Once formed, however, Z rings lacking ZapA or ZapB appear similar to rings in wild type cells (Buss et al. 2013).

In Vitro *Reconstruction*

One of the goals in biology is *in vitro* reconstruction of a system to demonstrate that the identified components reconstitute the behavior observed *in vivo*. Although this is a big challenge in the field of cytokinesis, especially considering the important role of the cell wall, attempts have been made to study the interaction of division components with a lipid bilayer. As an initial approach, the CCTP of FtsZ was replaced with an amphipathic helix to bypass the need for membrane anchors. Such an FtsZ construct was able to bind to a lipid bilayer and, when placed inside vesicles, caused constriction (Osawa et al. 2008). Although GTP was necessary, its hydrolysis was not, indicating that just formation of the filaments on the membrane was responsible for the observed deformation of the vesicles.

Second generation reconstitution experiments used FtsZ and one of the natural membrane anchors, ZipA or FtsA. Reconstitution on a flat lipid bilayer showed recruitment of FtsZ filaments to the bilayer, with the proteins self-organizing into complex patterns (Loose and Mitchison 2014). Whereas ZipA led to FtsZ forming seemingly static but dynamic bundles, FtsA caused FtsZ to form dynamic vortices that underwent rapid reorganization; the dynamics of the system were due to treadmilling of FtsZ filaments. Although GTP was required (for FtsZ), surprisingly ATP was not (ADP being sufficient for FtsA). Reconstitution experiments have also been done with the proteins inside vesicles. By using an FtsA hyperactive mutant (FtsA*), which appears to behave better *in vitro* than the wild type protein, Z rings were occasionally observed within unilamellar vesicles and observed to constrict in an ATP and GTP dependent manner (Osawa and Erickson 2013). ZipA-containing vesicles have also been observed to shrink when FtsZ and GTP were included (Cabre et al. 2013). Together, these experiments demonstrated the remarkable ability of FtsA, ZipA and FtsZ to self-organize and raised the possibility that, together, they could provide a constrictive force *in vivo*.

Additional Roles of the Z Ring

Since the *E. coli* cell is under ~3 atmospheres of pressure, there has been speculation as to the force that drives septation. It is clear that peptidoglycan synthesis is required, as any block to PG synthesis halts septation (Spratt 1975). It was suggested that, in addition to a well-characterized scaffolding function, the Z ring provides a constrictive force at the leading edge of the septum (Erickson et al. 2010). This suggestion arose from the presence of FtsZ in mycoplasma, which lack a cell wall, and also from the ability of FtsZ to assemble both curved and straight filaments. This possible activity is supported by in *vitro* reconstitution studies, where FtsZ containing a membrane tether was observed to invaginate lipid vesicles (Osawa et al. 2008). However, the lack of an effect of varying FtsZ's GTPase activity on the rate of constriction *in vivo* has been used to argue against this proposal (Coltharp et al. 2016) but this is still controversial. Even if the Z ring provides force *in vivo,* the constriction would be limited by the rate of PG synthesis. In any event, the Z ring is not the only contributor, as FtsZ leaves the invaginating septum before cytokinesis is complete (Soderstrom et al. 2014). Another role for the Z ring is as a guide for the invaginating septum. FtsZ mutants deficient in GTPase activity result in a twisted septum (Bi and Lutkenhaus 1992). Such mutants assemble what appears to be a normal ring but the stable FtsZ filaments become twisted into a spiral induced by cell wall growth. This observation has led to the idea that rapidly reorganizing FtsZ filaments are necessary to guide formation of a symmetrical septum (Dajkovic and Lutkenhaus 2006).

Spatial Regulation of the Z Ring

The Z ring is placed at midcell with great precision without the use of any known landmarks. So far two negative systems (Min and NO [nucleoid occlusion, mediated by SlmA]), and one positive system (Ter linkage) have been identified that contribute to the placement of the Z ring in *E. coli* (Lutkenhaus 2007; Mannik and Bailey 2015). The Min system prevents formation of the Z ring away from midcell while SlmA, responsible at least in part for NO, prevents formation of the Z ring over the nucleoid (Lutkenhaus 2007). In addition to this negative regulation, the Ter linkage promotes Z ring assembly in the vicinity of the Ter region on the nucleoids, near midcell (Bailey et al. 2014). The contribution of each system is revealed following its inactivation alone or in combination (Fig. 2.3). In some cases, the effect is growth rate dependent (Min⁻ SlmA⁻) and at slow growth rates all three systems can be deleted and the Z ring still has a preference for midcell, indicating at least one additional mechanism must exist (Bailey et al. 2014).

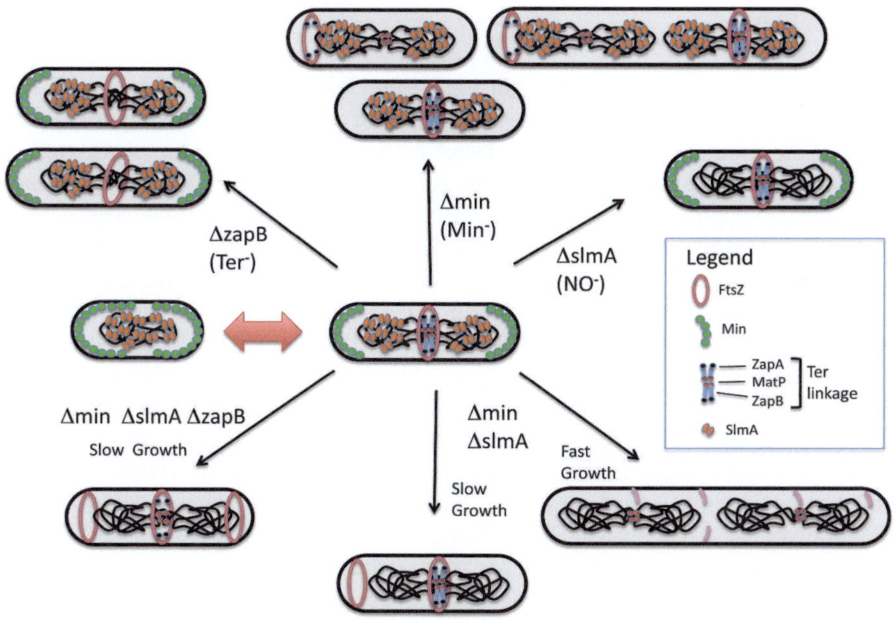

Fig. 2.3 Spatial regulation of the Z ring. Three systems contribute to the spatial regulation of the Z ring – Min, NO and Ter linkage and their influence is revealed following inactivation of a system alone or in combination. Under normal growth conditions inactivation of Min results in a dramatic phenotype with minicells and nucleoid containing cells of heterogeneous cell length, whereas a more subtle phenotype is observed with loss of the Ter linkage (Δ*zapB*) and none with loss of NO (Δ*slmA*). The phenotype due to deleting two or more of the systems depends upon growth rate. At fast growth rates loss of Min and SlmA leads to lethal filamentation whereas at slow growth rates cells survive but with central and polar rings. Loss of all three systems leads to additional polar rings

Min System

The Min system was recognized early-on as having an important role in spatial regulation of cytokinesis, since its absence (Δ*min* mutants) results in cytokinesis also occurring at the poles of the cell and, consequently, minicell formation (Adler et al. 1967) (Fig. 2.3). This regulation of cytokinesis operates at the level of Z ring formation (Bi and Lutkenhaus 1993; Pichoff and Lutkenhaus 2001). Consistent with this, the Min system consists of an antagonist of FtsZ assembly, MinC, which is recruited to the membrane by MinD and directed away from midcell by MinE (De Boer et al. 1989; Hu et al. 1999; Raskin and De Boer 1999a). This system in not static and the Min proteins rapidly oscillate between the ends of the cell, with the Z ring forming at midcell where their influence is at a minimum (Meinhardt and De Boer 2001). Two elements of the Min system have been under intense scrutiny. One is the mechanism of the oscillation and the other is the mechanism by which the Min system antagonizes FtsZ assembly.

Min Oscillation

The Min oscillator is a geometry sensing system with a preferred wavelength (on the order of 3–5 microns) (Varma et al. 2008; Wu et al. 2015). The important features of the oscillation were largely determined by manipulating the expression of the Min proteins *in vivo* and the mechanism was derived from studying their biochemistry, including *in vitro* reconstitution and computer simulations. The *in vivo* expression demonstrated that MinD and MinE were sufficient to establish the oscillation, that the ratio between them was critical, and that MinC was the division inhibitor and a passenger in the oscillation (De Boer et al. 1989; Raskin and De Boer 1999b; Hu and Lutkenhaus 1999). During the oscillation, MinD and MinC accumulate at one pole and are flanked by a MinE ring. As this ring moves closer to the pole, MinC and MinD are released and they re-assemble at the other pole, flanked again by a MinE ring (Fu et al. 2001; Hale et al. 2001) (Fig. 2.4a).

Biochemical and genetic studies revealed MinC acted directly on FtsZ filaments and that the dynamic interplay of the Min proteins with the membrane that underlies the oscillation is driven by MinD's ATPase cycle (Hu et al. 1999; Hu and Lutkenhaus 2001). MinD dimerizes in the presence of ATP and binds to the membrane through a C-terminal amphipathic helix (Fig. 2.4b)(Szeto et al. 2002; Hu and Lutkenhaus 2003). Dimerization of MinD is necessary to provide sufficient membrane affinity and to generate binding sites for MinE and MinC, which overlap at the dimer interface (Wu et al. 2011). MinE binding to MinD stimulates the ATPase activity causing the release of MinD from the membrane (Hu et al. 2003; Lackner et al. 2003). The interaction of MinE with MinD is quite complex as MinE is in a latent form (6-beta strands) that must convert to the active form (4 –beta strands) to bind MinD (Park et al. 2011). The conversion of MinE to the active form also releases N-terminal amphipathic helices that allows MinE to bind the membrane as it binds MinD (Fig.

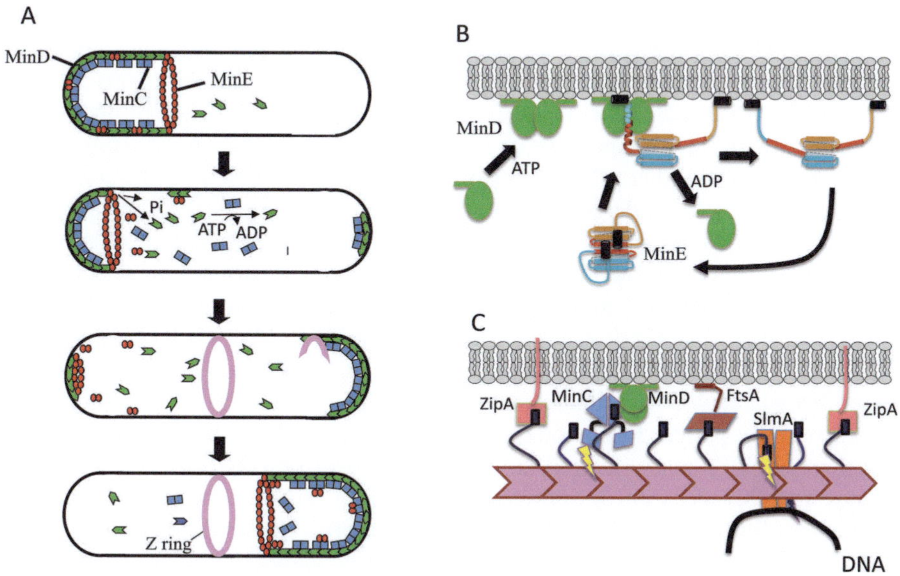

Fig. 2.4 Min system. (**a**) The Min proteins oscillate between the poles of the cell as described in the text. (**b**) The dynamic interaction of MinD and MinE with the membrane is fueled by MinE stimulated ATP hydrolysis by MinD. (**c**) An FtsZ filament tethered to the membrane is attacked by MinCD and SlmA using a two-pronged mechanism (lightning bolts). These antagonists are recruited to FtsZ filaments through interaction with the CCTP resulting in a second interaction to break the filament

2.4b). Although MinE is a dimer and has two MinD binding sites the geometry of the complex only allows MinE to bind one side of a MinD dimer. After stimulating MinD's ATPase and causing the release of MinD from the membrane, MinE can linger transiently on the membrane before returning to the latent form in the cytoplasm. This step gives MinE a chance to search for another membrane bound MinD to cooperatively remove MinD from the membrane before dissociating. Although genetic and biochemical studies indicate that MinE binding to one side of a MinD dimer is sufficient to stimulate the ATPase activity (Park et al. 2012), a recent model invokes a MinE dimer binding on each side of MinD to stimulate ATP hydrolysis (Vecchiarelli et al. 2016). This 2:1 ratio may explain the existence of the MinE ring.

The recruitment of MinC to the membrane by MinD generates a complex that can antagonize FtsZ assembly (De Boer et al. 1989; Pichoff and Lutkenhaus 2001). Since MinD forms a symmetric dimer, it has a binding site for MinC on each side (Fig. 2.4c). Also, since MinC is a dimer it has the potential to bridge MinD dimers to form an alternating copolymer attached to the membrane. In fact, such unusual copolymers form *in vitro* and *in vivo* when overexpressed (Ghosal et al. 2014; Conti et al. 2015). However, mutations that separate copolymer formation from MinC-MinD interaction do not affect Min function indicating that such copolymers are not necessary for Min function (Park et al. 2015). Also, the excess of MinD over MinC *in vivo* (7 to 1) does not favor copolymers (Li et al. 2014).

Due to the relative simplicity the Min oscillation, it was modeled to determine how the oscillatory behavior could be achieved. Many models were developed that generate the oscillatory behavior *in silico* (Meinhardt and De Boer 2001; Huang et al. 2003; Kruse et al. 2007; Fange and Elf 2006). A general feature of the models is that MinD displays cooperative attachment to the membrane and recruits MinE, which stimulates MinD's ATPase causing the release of MinD. The MinD released from a polar region undergoes nucleotide exchange before it can rebind, which gives it a chance to escape the pole. If it does not diffuse away, it can readily rebind due to the presence of membrane bound MinD, but if it diffuses away it takes a higher concentration to bind to the other pole, as it has to overcome the absence of membrane-bound MinD (Fig. 2.4a). This property of MinD, along with the sequestration of MinE in a MinE ring at the edge of the MinD polar zone, drives the oscillation (Huang et al. 2003; Halatek and Frey 2012).

Importantly, pattern formation by the Min system has been reconstituted *in vitro* on planar lipid bilayers as well as in 3-dimensional objects (Loose et al. 2008; Ivanov and Mizuuchi 2010; Schweizer et al. 2012). The reconstitution confirmed that pattern formation by the Min system only required MinD, MinE, a lipid bilayer and ATP. On a planar bilayer MinD and MinE form waves that migrate across the surface with the concentration of MinE highest at the rear of the wave. For reasons that are not clear the wavelength is much larger than observed *in vivo*. MinC also oscillates when it was included and these reconstituted Min patterns alternating with FtsZ assemblies (using FtsZ with a membrane tether) on the lipid bilayers (Arumugam et al. 2014). These patterns form even though the densities of the Min proteins on the membrane are quite high and Min proteins are not depleted from the overlying solution. Modeling, however, indicates that depletion of the Min proteins from the cytoplasm plays an important role (Huang et al. 2003). In a more recent reconstitution system, conditions were found in which pattern formation was accompanied by Min protein depletion from the overlying solution. Under these conditions radially enlarging disks of MinD were formed that were surrounded by a MinE ring, which then rapidly decreased in diameter (Vecchiarelli et al. 2016). These disks oscillated with neighboring disks and are likely to better reflect the *in vivo* situation.

MinC/MinD Antagonism of Z Ring Formation

MinC has two functional domains, an N-terminal domain ($MinC^N$) that interacts directly with the globular domain of FtsZ and antagonizes polymerization *in vitro* and a C- terminal domain ($MinC^C$) that binds MinD and the CCTP of FtsZ (Hu and Lutkenhaus 2000; Cordell et al. 2001; Dajkovic et al. 2008). Based upon the ability of $MinC^N$ to antagonize assembly *in vitro* and the isolation of FtsZ mutants resistant to the two domains of MinC, a model emerged for MinC antagonism (Shen and Lutkenhaus 2009, 2010). In this model the interaction of $MinC^C$/MinD with the CCTP and the membrane positions $MinC^N$ near the FtsZ filament, leading to breakage of the filament (Fig. 2.4c). Based upon the location of residues altered by resistance mutations, the target of $MinC^N$ is helix 10 of FtsZ, which is near the junction of two subunits in an FtsZ filament.

$MinC^C$/MinD binds to the CCTP of FtsZ and can compete with FtsA and ZipA (Shen and Lutkenhaus 2009). Thus, it is possible that the oscillation of the Min proteins delays formation of the Z ring through competition for the CCTP. In smaller cells there is likely to be less of an area at midcell free of the oscillating MinC/MinD than in longer cells. Interestingly, FtsA* has higher affinity for the CCTP, which may explain why it has some resistance to MinC/MinD (Geissler et al. 2007; Pichoff et al. 2012). This model may also explain why FtsA* cells are able to assemble a Z ring at a smaller cell size than wild type cells (Geissler et al. 2007).

Min and DNA Segregation

There are reports that the inactivation of the Min system affects DNA segregation (Akerlund et al. 2002; Di Ventura et al. 2013). One effect is due to the lack of resolution of chromosome dimers in a fraction of cells in a Min mutant.

The main phenotype of Min mutants is the formation of minicells and the heterogeneous length of nucleoid containing cells. This is due to minicell divisions occurring at the expense of medial divisions resulting in an increased average cell length of the nucleated cells. Chromosome dimers form in approximately 10% of cells per cell-cycle, due to recombination between newly formed daughter chromosomes (Steiner et al. 1999). Separation of chromosome dimers is effected by the divisome protein FtsK (see section "FtsK") and therefore occurs during division at midcell. However, a chromosome dimer present in a nucleated cell that produces a minicell and skips division at midcell cannot be rescued by the dimer resolution system; this results in a dramatic segregation defect in that cell. Other reports indicate that loss of Min causes a slight delay in bulk chromosome segregation in all cells in the population (Akerlund et al. 2002; Di Ventura et al. 2013). The basis for such an effect is unclear, although it was recently reported that MinD binds DNA and that the Min oscillation may serve to assist DNA segregation. However, the *minD* mutation studied affected membrane binding as well as DNA binding, questioning such a possibility (Di Ventura et al. 2013).

Nucleoid Occlusion and SlmA

A long-standing observation in bacterial cell biology is that cytokinesis over the nucleoid, which would result in guillotining of the chromosome, is rarely observed. This led to the concept of nucleoid occlusion – that, somehow, the nucleoid prevented cytokinesis in its vicinity (Woldringh et al. 1990). Later it was shown that the Z ring does not form over the nucleoid in cells where division was blocked, resulting in the hypothesis that the nucleoid exerted a negative effect on Z ring assembly (Yu and Margolin 1999). Factors responsible were identified in both *E. coli* and *B.*

subtilis. In both cases a protein was identified that bound to sites (~50) that were present in the *oriC* proximal two third of the chromosome but absent in the Ter region (Wu et al. 2009; Bernhardt and De Boer 2005; Cho et al. 2011).

In *E. coli* the SlmA protein, a member of the TetR repressor family, was found to mediate NO and interact directly with FtsZ (Cho et al. 2011). An unrelated protein in *B. subtilis* (Noc) is functionally analogous but does not interact directly with FtsZ (Adams et al. 2015).

In the absence of DNA, SlmA weakly antagonizes FtsZ assembly, but this activity is dramatically enhanced upon binding to DNA containing a SlmA binding site (SBS) (Cho et al. 2011; Cho and Bernhardt 2013). The structure of free SlmA and SlmA bound to an SBS revealed that SlmA undergoes a conformational change upon DNA binding that is typical of TetR family members, leading to increased affinity for FtsZ (Tonthat et al. 2011; Tonthat et al. 2013). Residues affected by the conformational change are required for SlmA function and the recent structure of DNA/SlmA/CCTP peptide complex confirms these residues are involved in binding to the CCTP of FtsZ (Schumacher and Zeng 2016). SlmA is a dimer and binds cooperatively to an SBS site as a dimer of dimers, with the possibility that additional dimers are also recruited. A cell contains about 5 SlmA dimers per SBS binding site on the chromosome (Li et al. 2014).

The ability of SlmA bound to an SBS to depolymerize FtsZ requires the CCTP (Du and Lutkenhaus 2014). Furthermore, the interaction of SlmA with a monomer of FtsZ is of low affinity, similar to the interaction of the CCTP with ZipA and FtsA (Du et al. 2015). However, as observed with ZipA, the affinity is dramatically enhanced by avidity, that is, by a multimer of FtsZ binding to SlmA dimers. Thus, SlmA bound to an SBS preferentially binds to polymers of FtsZ. Once SlmA is bound to a polymer of FtsZ it is thought to break the polymer, although the mechanism is not clear (Cho and Bernhardt 2013; Cho et al. 2011). A mutation altering a residue in the long helix 7, which connects the two globular domains of FtsZ, is resistant to the action of SlmA (Du and Lutkenhaus 2014). This suggests that SlmA bound to an FtsZ filament through interaction with the CCTP makes an additional contact to stimulate GTP hydrolysis (Fig. 2.4c). Thus, SlmA, like MinC/MinD, uses a two-pronged mechanism to attack FtsZ filaments (Du and Lutkenhaus 2014).

During the cell cycle SlmA bound to the chromosome is transported away from midcell by segregation of the chromosome (Fig. 2.3, center). This movement creates a void at midcell making it permissive for FtsZ filaments to persist at the membrane, whereas delays in segregation, for example, caused by disruption of DNA replication or segregation, prevent this movement and formation of the Z ring at midcell (Bernhardt and De Boer 2005). Although SlmA can clearly prevent assembly of the Z ring in the vicinity of an unsegregated nucleoid, its role in the normal placement of the Z ring is not clear. Deletion of SlmA does not affect cell morphology, due to the dominance of the Min system, however, when both Min and SlmA are deleted cell division is prevented at fast growth rates (Bernhardt and De Boer 2005). Under these conditions FtsZ filaments are distributed throughout the cell and sufficient filaments are unable to accumulate at any one position in the cell to form a complete Z ring (Fig. 2.3). Consistent with this, raising the level of FtsZ rescues such cells, and results in Z rings at midcell and also at the poles. Cells lacking Min and SlmA are also rescued by growth at high temperature or in minimal medium (Bernhardt and De Boer 2005; Shen and Lutkenhaus 2010). Under this latter condition, cytokinesis occurs mostly at midcell and few minicells are produced, indicating additional mechanisms of directing Z rings to midcell (Fig. 2.3). This led to the discovery of the Ter linkage and its role in Z ring formation.

Ter Linkage

The impetus for looking for positive Z ring regulation in addition to the two negative systems came from work in both *E. coli* and *B. subtilis*. Work in *B. subtilis* revealed that in the absence of the two negative regulatory systems (Min and NO) the Z ring is placed at midcell in outgrowing spores with good precision but with reduced efficiency (Rodrigues and Harry 2012). Examination of *E. coli* forced to grow at reduced diameter revealed misshapen cells, but with Z rings placed between the nucleoids, even in the absence of Min and SlmA, suggesting the chromosome itself may provide positional information (Mannik et al. 2012). This result, combined with the observation that Z rings still form at midcell (as well as the poles) in the absence of Min and SlmA at slow growth rates, led to a search for a possible mechanism (Fig. 2.3).

Time-lapse microscopy of a mutant lacking both Min and SlmA growing in minimal medium (which suppresses the otherwise lethal effect of deleting both of these systems) revealed that the position of Z rings correlated more precisely with the center of segregating chromosomes than with the middle of the cell (Bailey et al. 2014). In these cells, Z ring formation over the nucleoid was preceded by the arrival of the Ter region. In newborn cells, the Ter region is located at the poles but rapidly snaps to the cell center as the Ter region is replicated (Espeli et al. 2012). Since the Ter region is organized into a distinct macrodomain by MatP (Mercier et al. 2008), this led to a model in which the MatP-Ter region is linked to the Z ring through ZapA and ZapB, a model supported by super resolution microscopy (Buss et al. 2015) (Fig. 2.3).

At slow growth, a *Δmin ΔslmA* double mutant is rescued and rings form at midcell and the poles but the polar rings infrequently lead to minicells (Bailey et al. 2014) (Fig. 2.3). However, disruption of the Ter linkage by deleting one of the components disrupts the precision of the placement of the Z ring at midcell and dramatically increases the frequency by which polar rings form and constrict. This observation indicates that the Ter linkage aids the positioning of the Z ring and promotes its usage at midcell.

Overlap in Spatial Regulation

As described above, at least three mechanisms have been identified that contribute to the positioning of the Z ring in *E. coli* (Fig. 2.3). Together, these systems ensure that the Z ring is formed precisely at midcell. Several observations indicate the Min system is the major determinant of Z ring positioning under normal growth conditions. First, deletion of the Min system has a dramatic morphological effect as polar rings form, minicells are produced and the average cell length is increased due to minicell divisions occurring at the expense of medial divisions (Teather et al. 1974; Bi and Lutkenhaus 1990). Second, in wild type cells the Z ring can occasionally form before Ter relocates from the pole to midcell (Bailey et al. 2014). Third, in anucleate cells (therefore lacking SlmA and Ter) the Z ring is positioned at midcell (presumably by Min) with only slightly less precision than in WT cells (Sun et al. 1998). In contrast, deletion of just SlmA has no obvious effect on Z ring assembly or cell morphology during normal growth (Bernhardt and De Boer 2005) and disrupting the Ter linkage (deletion of ZapA or ZapB) only results in a 15% increase in the average cell length and some skewed Z rings (Fig. 2.3) (Hale et al. 2011).

From Z Ring to Divisome, Recruitment and Activation

Once the Z ring is established, additional essential proteins are recruited to form the divisome (Fig. 2.5a). These proteins must carry out septal PG synthesis. How these proteins are recruited to the Z ring and activated is an active area of investigation.

Downstream Proteins and Septal PG Synthesis

The Z ring functions as a scaffold to organize the septal PG synthesis, analogous to MreB functioning as an organizer of a PG machine for cell elongation (elongasome) (Egan et al. 2015, Szwedziak and Lowe 2013). Whereas the recruitment pathway for the elongasome is difficult to assess (MreB and components are dispersed throughout the length of the cell), the recruitment pathway for the divisome is accessible to study due to its unique occurrence in the cell. A summary of these divisome proteins and their function precedes a discussion of their recruitment.

FtsEX

FtsEX is a member of the ABC transporter family, although FtsEX is not thought to transport small molecules (Schmidt et al. 2004). The closest homologue to FtsEX is the LolCDE system, which extracts lipidated proteins from the cytoplasmic

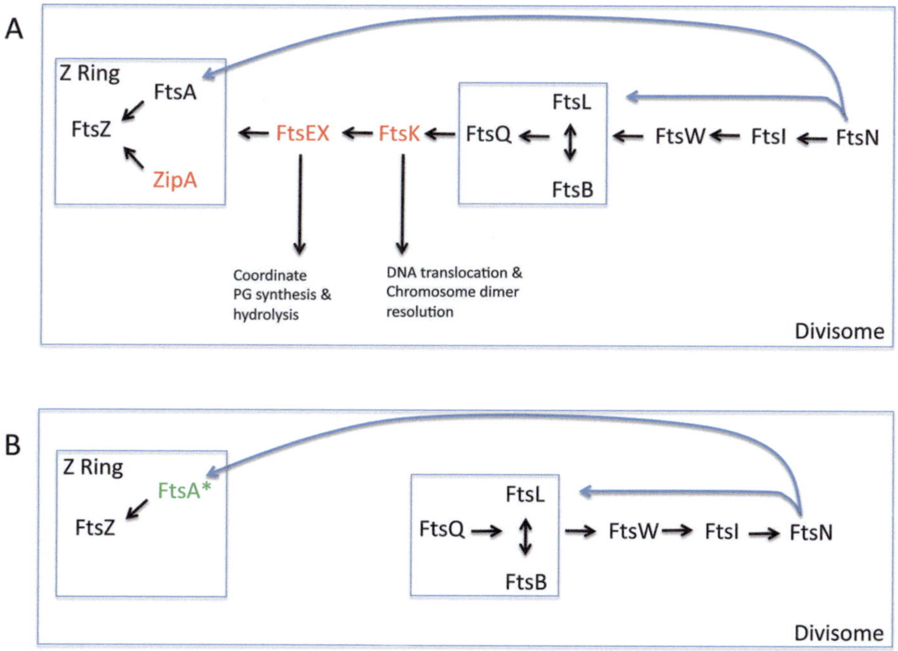

Fig. 2.5 The Z ring organizes the divisome. (**a**) FtsZ, FtsA and ZipA form the Z ring and FtsEX is added. After a delay the remaining proteins are added to form the divisome. Although all are essential under standard growth conditions they have been divided into core (*black*) and non-core (*red*). (**b**) The non-core proteins can be bypassed by FtsA* mutations (*green*). Under these conditions (for example, loss of ZipA) the interaction of FtsN with FtsA may be responsible for back recruiting the other division proteins

membrane for transfer to the outer membrane (Narita and Tokuda 2006). FtsEX localizes relatively early to the Z ring and appears to play three roles in cell division (Corbin et al. 2007). Two roles are essential: (1) FtsEX recruits downstream proteins and (2) FtsEX's ATPase activity is required for the initiation of constriction (Arends et al. 2009). The requirement for FtsEX in these two roles can be suppressed by high osmolarity, overexpression of FtsN or by mutations in *ftsA* (*ftsA**) (Reddy 2007). FtsE mutants unable to bind or hydrolyze ATP are able to carry out the recruitment function, but are unable to promote the start of constriction (Arends et al. 2009). Recent evidence indicates FtsEX carries out these two functions by modulating the activity of FtsA, possibly by altering its polymerization state (Du et al. 2016).

In addition to these two essential roles, FtsEX also recruits the periplasmic protein EnvC and together they activate two periplasmic amidases, in a step requiring ATP hydrolysis by FtsE (Yang et al. 2011). Thus, FtsEX couples ATP hydrolysis in the cytoplasm to the start of constriction and to the splitting of the bonds cleaved by the amidases. High osmolarity does not suppress the requirement for FtsEX for activation of the two amidases and cells have a mild chaining defect. A third amidase can carry out the cleavage step, but does so less efficiently. In the absence of all three amidases, cell chaining is quite dramatic but is not lethal, indicating other hydrolytic enzymes can do this inefficiently (Heidrich et al. 2001).

FtsK

FtsK is a DNA translocase that has 4 N-terminal transmembrane segments connected to a translocase domain through a long linker (Begg et al. 1995). The only essential role of FtsK is its role in the recruitment pathway, which only requires the N-terminal membrane region (Wang and Lutkenhaus 1998). Two separate domains within the long linker interact with the Z ring, possibly by interacting directly with FtsZ (Dubarry et al. 2010). FtsK is able to translocate DNA and this C-terminal domain is required for the resolution of chromosome dimers (formed in ~10% of cells by homologous recombination between daughter chromosomes) by stimulating the XerCD recombinase. The localization of FtsK to the septum spatially restricts dimer resolution to the septal region (Steiner et al. 1999).

FtsQ, FtsL and FtsB

The FtsQLB proteins form a complex that has been extensively characterized by genetic and biochemical approaches (Buddelmeijer and Beckwith 2004). All three proteins are single-pass membrane proteins. Although the proteins show poor conservation at the amino acid level, a careful phylogenetic analysis revealed they are highly conserved throughout bacteria with a cell wall (Gonzalez et al. 2010). FtsL and FtsB form coiled-coil proteins with the transmembrane region contributing to the formation of the alpha helical region in the periplasm (Gonzalez and Beckwith 2009; Gonzalez et al. 2010; Glas et al. 2015). The three proteins interact in the periplasm through their extreme C-terminal regions. Mutations in *ftsL* and *ftsB* that reduce or eliminate the requirement for FtsN map to a region in the middle of the coiled-coil domain of each protein (possibly a kink in the coil) and these regions were designated CCD, for control of cell division (Tsang and Bernhardt 2015; Liu et al. 2015).

FtsW

FtsW is a multipass transmembrane protein (10 transmembrane segments) that is highly conserved and a member of the SEDS (shape, elongation, division and sporulation) family (Boyle et al. 1997; Pastoret et al. 2004). Other family members include RodA, involved in cell elongation, and SpoVE, involved in spore formation in *B. subtilis* (not present in *E. coli*). Some biochemical evidence suggests that FtsW may serve as a flippase for lipid II (required for peptidoglycan synthesis) (Mohammadi et al. 2011); however, genetic evidence indicates that the flippase is

MurJ (Meeske et al. 2015). Members of the SEDS family are each associated with a specific penicillin-binding protein (FtsW with FtsI [PBP3] and RodA with PBP2) (Typas et al. 2012). Recent biochemical and genetic evidence indicates RodA is a transglycosylace suggesting that FBW is also a transglycosylace (Meeske et al. 2016; Cho et al. 2016). It so then syptal PG synthesis is carried out by FBW and FBI(PBP3) acting out together to carry out the two necessary enzymatic activities(transglycosylace and transpeptidation. The role of the higher molecular weight PBPs (PBP la and PBP lb), long thought to be the primary synthesis is unclear, but might be needed to remade and strengthen PG made by FBW and FBI.

FtsI

FtsI encodes penicillin-binding protein 3 (PBP3), which is dedicated to cytokinesis (Spratt 1975). It is a single pass integral membrane protein with the periplasmic domain composed of two domains, a C-terminal transpeptidase domain preceded by a domain of unknown function. FtsI depends upon FtsW for localization to the division site (Mercer and Weiss 2002). The transmembrane domain is responsible for recruitment to the septum and mutations in the domain of unknown function prevent FtsN recruitment (Wissel and Weiss 2004). Biochemical studies indicate FtsI interacts with a number of proteins involved in cell wall synthesis, including PBP1b and FtsN (Muller et al. 2007).

FtsN

FtsN is a single pass integral membrane protein and is the last essential protein to arrive at the division site (Dai et al. 1993, 1996; Addinall et al. 1997). The cytoplasmic N-terminal domain interacts with FtsA (Busiek et al. 2012; Pichoff et al. 2015; Liu et al. 2015) and the C-terminal domain constitutes a SPOR domain that binds to denuded glycan chains (Gerding et al. 2009; Yahashiri et al. 2015). These chains are formed transiently at the septum by amidases that cleave the peptide side chains during processing of septal PG. Connecting these two domains is a region that is predicted to contain 3 small helical regions (Yang et al. 2004). One of these constitutes the essential region of FtsN (E domain) that activates septal PG synthesis (Gerding et al. 2009). This region is the only absolutely essential region of FtsN, since it can complement an FtsN depletion strain when it is overexpressed and exported to the periplasm. However, at the physiological level the N-terminal cytoplasmic and C-terminal SPOR domains are essential in order to efficiently localize the E domain to the septum (Busiek and Margolin 2014; Pichoff et al. 2015).

Septal PG Machine

Estimates of protein copy numbers indicate there are ~150 molecules per cell of FtsQ, FtsL, FtsB, FtsW, FtsI, FtsN at slow growth rates (Li et al. 2014). As mentioned, FtsQLB form a well-documented complex that persists even outside of the Z ring. The similar copy numbers of these divisome proteins and the web of interactions among them suggest they are in a stoichiometric complex, which would constitute a septal PG biosynthetic machine (Li et al. 2014). In fact, a large complex containing many of these proteins was detected by native gel electrophoresis (Trip and Scheffers 2015). *E. coli* has 3 class A PBPs – 1a, 1b and 1c (have both transglycosylase and transpeptidase activities). Either PBP1a or PBP1b is sufficient for survival, suggesting overlap in the function of these two PBPs (Yousif et al. 1985). However, cells growing with only PBP1a are less robust, indicating PBP1b is the more important synthetic enzyme. PBP1b, but not PBP1a, is known to interact with FtsN and FtsI and is also present in a similar copy number (Muller et al. 2007). Since PBP1b appears to be a dimer it was suggested that the septal PG machine exists as a dimeric complex, with PBP1b at the core and two copies of each of the Fts proteins (FtsQ-N) (Egan and Vollmer 2015). This means that there are only ~75 of these dimeric septal PG machines present in slow growing cells. The level of FtsK is about the same as the others, but it is known to form a hexamer and, thus, may not be a stoichiometric part of the machine (Bisicchia et al. 2013b). Consistent with this, FtsK can be bypassed with little cost to the cell by mutations in *ftsA* (Geissler and Margolin 2005).

Recruitment of Divisome Components to the Z Ring

Once the Z ring is established, the remaining proteins are recruited. The initial approaches used to examine the recruitment pathway resulted in a mostly linear dependency pathway; however, subsequent studies have found a web of interactions that can contribute to localization.

Depletion Studies Leading to a Linear Dependency Pathway

A mostly linear dependency pathway for the recruitment of downstream division proteins to the Z ring arose from depleting individual components and seeing which proteins are still recruited (Fig. 2.5a). For example, following depletion of FtsL the Z ring forms and FtsK and FtsQ localize but the other essential proteins do not (Ghigo et al. 1999). This sort of analysis resulted in the elaboration of the linear sequential pathway. With the realization that FtsQLB form a complex outside of the Z ring, it is clear that depletion of FtsL disrupts this complex (Buddelmeijer and Beckwith 2004). Nonetheless, FtsQ localizes in the absence of FtsL demonstrating that FtsQ can localize to the Z ring in the absence of its partners, and thus contains the information for recruitment (Gonzalez et al. 2010). Targeting FtsL directly to the Z ring (by fusion to ZapA, see below) revealed that it was able to recruit FtsB and downstream proteins even in the absence of FtsQ (Goehring et al. 2006). Thus, the FtsLB complex can recruit the downstream proteins in the absence of FtsQ, indicating that the information for recruiting downstream proteins lies within FtsL and FtsB, whereas FtsQ is required to bring FtsLB to the Z ring.

Lessons from Fusions Leading to Forced Localization

Another approach used to study protein recruitment to the division site is to fuse a protein in the middle of the pathway to a protein that binds directly to FtsZ, such as ZapA (Goehring et al. 2005). Such a fusion can recruit proteins downstream, but can also back-recruit upstream proteins. This recruitment occurred even if FtsA, which interacts with several downstream proteins, is depleted. Thus, for example, ZapA-FtsQ can recruit both the upstream protein FtsK and downstream proteins like FtsI to the Z ring in the absence of FtsA, suggesting FtsA's role is indirect. One notable exception among the downstream proteins was FtsN, which is not recruited by ZapA fusions if FtsA was removed. Also, FtsN is not recruited by ZapA fusions to other proteins (FtsL, FtsW) if FtsQ was removed (Goehring et al. 2006). Thus, the recruitment of FtsN is complex and dependent upon FtsI, as shown in the linear recruitment assay, but is also dependent upon FtsQ and FtsA, as revealed by the ZapA fusions.

In another approach, the long linker of FtsK (that connects the membrane region to the DNA translocase domain) was fused to a number of integral membrane division proteins (Dubarry et al. 2010). Fusion of just the linker region to FtsW resulted in normal cellular morphology (in the absence of FtsK), indicating that the other regions of FtsK are dispensable and that the only essential function of FtsK is participation in the recruitment pathway. This analysis also revealed that the linker region had two separable regions that interacted with the Z ring and was capable of bringing FtsW, and all other proteins, to the Z ring.

Bypass Mutations Suggest Some Components Are Core Whereas Others Are Non-core

In yet another approach to explore divisome assembly, conditions were explored that bypass one or more of the normally essential genes. The initial approach showed that ZipA could be bypassed by a mutation in *ftsA*, designated *ftsA** (Geissler et al. 2003) (Fig. 2.5b). This mutation could also bypass FtsEX and FtsK without much of an effect on cell morphology (Geissler and Margolin 2005; Reddy 2007). Subsequent work showed that many *ftsA* mutations could bypass ZipA and that almost all of these mutations reduced the ability of FtsA to self-interact (Pichoff et al. 2012). A reasonable possibility is that these mutations in *ftsA* decrease its oligomerization status and increase the affinity between FtsA and some downstream component, making ZipA, FtsK or FtsEX (normally part of the linear recruitment pathway) dispensable.

The relative ease at which ZipA, FtsK and FtsEX can be bypassed suggests that their addition to the core divisome (FtsZ, FtsA, FtsQLB, FtsI, FtsW, FtsN) improves coordination of chromosome segregation with cytokinesis (FtsK) or enhances coordination of septal PG synthesis and hydrolysis (FtsEX). ZipA may have been added to link additional (nonessential) proteins to the Z ring. FtsN can also be bypassed by some mutations in other components but there is more cost associated with the bypass, as cells are elongated (Tsang and Bernhardt 2015; Liu et al. 2015). Such mutations map to *ftsL*, *ftsB* and *ftsA* and lead to a more activated divisome that is less dependent upon FtsN. Most of the genes for the core components map to a large cluster of genes at the 2 min region (all except for *ftsB* and *ftsN*) and encode the machinery for the minimal Z ring (FtsZ and FtsA), septal PG synthesis (FtsW and FtsI), the connector (FtsQLB) and the activator (FtsN).

Role of FtsA and ZipA in Recruitment of Downstream Proteins

Although a Z ring can form with either FtsA or ZipA providing the membrane connection, both are required for downstream proteins to be recruited (Pichoff and Lutkenhaus 2002). Of the two, FtsA is the more important player in divisome assembly, since mutations in *ftsA* can bypass ZipA (Geissler et al. 2003) and FtsA, but not ZipA, is known to interact with a number of downstream proteins (Karimova et al. 2005). Since the *ftsA* mutations that bypass ZipA were found to reduce the ability of FtsA to self-interact, it led to a model in which FtsA monomers are the form of FtsA active in the recruitment pathway (Pichoff et al. 2012). In this model ZipA's essential role is to reduce FtsA's oligomerization state, possibly by competing for the CCTP of FtsZ (Fig. 2.6). Also, in this model FtsA's self-interaction competes with the interaction of downstream proteins for FtsA; domain 1C of FtsA is involved in both self- interaction and interaction with FtsN, but the two interactions are proposed to be mutually exclusive. Overproduction of FtsN also bypasses ZipA and the FtsA-FtsN interaction is required (Pichoff et al. 2015). It is likely that the overproduced FtsN promotes the FtsA interaction. Overproduction of FtsN also bypasses FtsEX and this bypass also requires the FtsA-FtsN interaction. Since loss of FtsEX disrupts the normal recruitment pathway, it raises the possibility that the FtsA-FtsN interaction is responsible for back recruiting all divisome components to the ring under these conditions (Fig. 2.5b). Back recruitment by FtsN may also explain why FtsA*, impaired for self-interaction, can bypass the need for FtsEX (Reddy 2007), as it may favor the FtsA-FtsN interaction.

The Role of FtsN in Activation of the Divisome

The arrival of FtsN to the divisome signals that the divisome is complete and septal PG biosynthesis can be initiated (Gerding et al. 2009; Lutkenhaus 2009; Weiss 2015). Recruitment of FtsN to the septum is quite complex and requires FtsA, FtsQ and FtsI (section "Lessons from fusions leading to forced localization") (Addinall

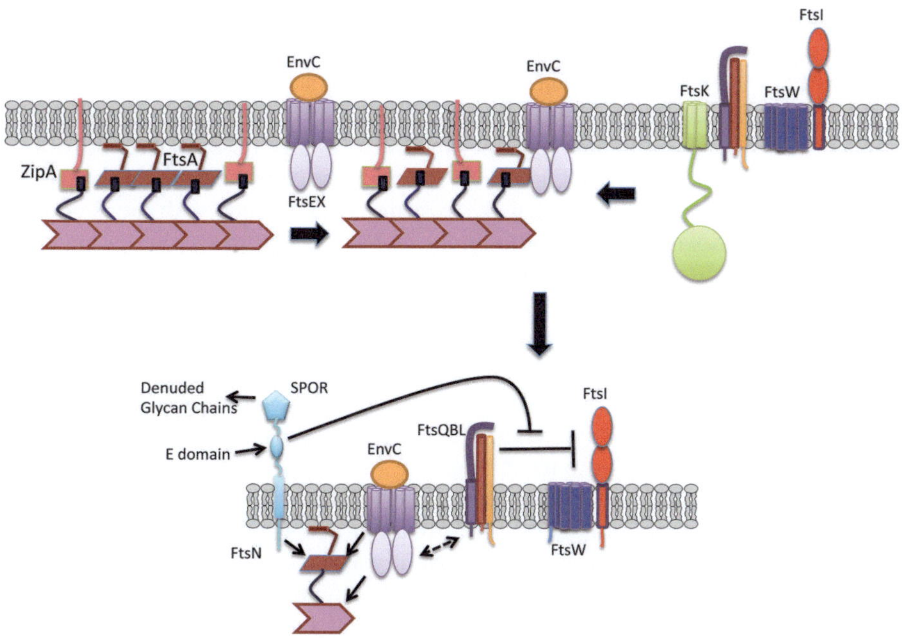

Fig. 2.6 Assembly of the divisome and activation of septal PG synthesis. (**a**) Divisome assembly. FtsZ filaments are tethered to the membrane by FtsA and ZipA. The arrival of FtsEX promotes monomeric FtsA at the Z ring, which results in recruitment of downstream proteins. These proteins are held in an inactive state by FtsQLB, FtsA and FtsW. (**b**) Divisome activation. The arrival of FtsN at the septum activates the divisome. The N-terminus of FtsN interacts with nonpolymerized FtsA in the cytoplasm and the E domain acts in the periplasm to overcome the inhibition of PG synthesis by FtsQLB and W. FtsN at the septum is reinforced by the SPOR domain interacting with denuded glycan chains formed by the action of amidases controlled by FtsEX. FtsEX acts on FtsA and continuous ATP hydrolysis by FtsEX is required for both PG synthesis and hydrolysis

et al. 1997; Goehring et al. 2006). If FtsN interacts with FtsA, why does it depend upon FtsI and FtsQ and why is it not recruited to the Z ring as soon as it is formed? Overexpression of the N-terminal domain of FtsN (cyto and transmembrane domains) leads to relatively early recruitment to the ring (Busiek and Margolin 2014). This recruitment is dependent upon FtsA but, in contrast to full length FtsN, is independent of FtsQ and FtsI. Possibly, FtsN is in a complex with other proteins (it is known to interact with PBP1b and FtsI) that impedes its recruitment to the septum, constraints which don't apply to the N-terminal domain.

In the current self-enhancing model (Gerding et al. 2009), the recruitment of FtsN to the septum is initiated by the cytoplasmic N-terminus interacting with an FtsA monomer in order to bring the E domain to the divisome to activate septal PG synthesis (Liu et al. 2015; Gerding et al. 2009). The activation of PG synthesis ultimately leads to amidase activation through FtsEX and EnvC (Yang et al. 2011). This is turn reinforces the recruitment of FtsN to the septal region, by the SPOR domain binding to PG strands processed by the amidases (Yahashiri et al. 2015). Many additional proteins are then recruited to the septum in an FtsN-dependent fashion, although it is likely to be indirect, since it is the activation of septal PG synthesis that is required (Gerding et al. 2009).

How PG synthesis is activated by the E domain is not clear; however, the isolation of mutations that bypass the E domain provides some possible clues and implicates FtsQLB and FtsA in the regulatory pathway (Tsang and Bernhardt 2015; Liu et al. 2015). In this model, FtsQLB and FtsA are recruited to the septum in the off state and the arrival of FtsN pushes the system to the on state with the E domain acting in the periplasm and the N-terminus of FtsN acting in the cytoplasm (Fig. 2.6). In this model FtsN acting in the cytoplasm interacts with FtsA that has a reduced oligomerization state and the E domain in the periplasm activates septal PG synthesis. Originally it was thought that the E domain activated PG synthesis directly (possibly by acting on PBP3); however, the isolation of the E bypass mutations in *ftsL* and *ftsB* suggests that the E domain relieves inhibition of PG synthesis imposed by FtsQLB. FtsN tethered to FtsA in the cytoplasm may also act as a link to tether septal PG synthesis to the Z ring.

A recent investigation into the essential roles of FtsEX indicate that FtsEX acts on FtsA to promote recruitment of downstream proteins and appears to do this by promoting FtsA monomers (Du et al. 2016). It was also shown that continual ATP hydrolysis by FtsEX was required for septal PG synthesis to occur (Yang et al. 2011). Since the role of FtsEX in regulating septal PG hydrolysis is well established, these results indicate that FtsEX coordinates septal PG synthesis with septal PG hydrolysis.

One of the striking phenotypes of a strong mutation in *ftsB* or *ftsL* (that bypasses the E domain of FtsN) is the smaller cell size associated with these mutations. Although they bypass FtsN, FtsN is still required for the small cell phenotype (Tsang and Bernhardt 2015; Liu et al. 2015). Cell cycle analysis suggests that the smaller cell size is due to FtsQLB (with one of the strong mutations) overcoming the delay normally observed between formation of the Z ring and arrival of the late divisome components (such as FtsQLB). It is not clear why there is a delay between the establishment of the Z ring and the formation of the mature divisome in wild type cells, but understanding how these mutations work should illuminate the mechanism behind this delay.

In *Caulobacter crescentus,* cytokinesis is inhibited following DNA damage, by the production of small integral membrane peptides that bind to and inhibit FtsN or FtsW (Modell et al. 2011, 2014). A smaller cell size is observed with mutations in *ftsW* that provide resistance to these inhibitors. In *E. coli*, FtsW is also implicated in the regulation of the onset of cytokinesis (Du et al. 2016). This suggests that the whole divisome (FtsQLBWI and PBP1b) is held in an inactive state and needs to be activated by FtsN.

Enigmatic Role of MreB and the Cell Elongation System in Cytokinesis

MreB is more closely related to actin than FtsA and is necessary for the maintenance of rod shape in many bacteria (Van Den Ent et al. 2001). MreB assembles into filaments at the membrane that organize the machinery for synthesizing peptidoglycan around the cylinder of the cell, necessary for elongation (elongasome) (Typas et al. 2012). This system has some components that are functionally analogous to some components of the cell division machinery. For example, RodA is related to FtsW and PBP2 is related to PBP3 and presumably they carry out analogous biochemical activities (Typas et al. 2012). A role for MreB in cytokinesis is not obvious, as genes of the elongation machinery can be deleted and cells grow and divide with a spherical morphology. Deletion of RodA, PBP2 or MreB is well tolerated at slow growth rates; however, at faster growth rates intracellular vesicles accumulate due to excess membrane synthesis and growth is impaired (Bendezu and De Boer 2008).

The fact that the elongation genes can be deleted indicates they are not essential for cytokinesis; however, a role is suggested by several results. MreB and PBP2 are recruited relatively early to the Z ring, suggesting some role in division (Van Der Ploeg et al. 2013). In at least some bacteria the Z ring can drive cell elongation before switching to a constriction mode. However, in *Caulobacter crescentus*, where this step is quite extensive, MreB is not required (Aaron et al. 2007). Notably, MreB interacts directly with FtsZ and mutants that fail to interact fail to divide (Fenton and Gerdes 2013). This raises the possibility that MreB is required for the division apparatus to make the transition from a cylindrical wall to initiate constriction.

Splitting the Septum

In *E. coli* the cell wall is split by amidases as the septum forms, resulting in a constriction. Deletion of the three amidase genes (*amiA*, *amiB* and *amiC*), but not just two, results in a severe chaining phenotype, but is not lethal (Heidrich et al. 2001). This result indicates that the amidases are primarily responsible for splitting the septum but other enzymes can do so inefficiently. The amidases remove the peptide stems from peptidoglycan, resulting in denuded glycan chains that exist transiently at the septum. The existence of denuded glycan chains was deduced from the biochemical binding properties of SPOR domains (Ursinus et al. 2004) and the amidase dependent localization of SPOR domain-containing proteins, of which, *E. coli* has four (FtsN, RplA, DamX and DedD) (Gerding et al. 2009; Arends et al. 2010). This was further supported by demonstrating the loss of SPOR-binding after treating PG ghosts with an enzyme that is specific to denuded glycan chains (Yahashiri et al. 2015).

Importantly, the amidases require activators; AmiA and AmiB require EnvC and AmiC requires NlpD (Uehara et al. 2010). EnvC is recruited to the septum rather early by FtsEX, whereas the amidases depend on FtsN to activate the divisome before they localize (Peters et al. 2011). The recruitment of EnvC by FtsEX occurs in the absence of ATP, whereas the activation of the amidase requires ATP hydrolysis. These results suggest that hydrolysis of ATP by FtsEX modulates EnvC so that the autoinhibition of AmiB is relieved (Yang et al. 2012). Mutants defective in the Tol-Pal complex display defects in outer membrane integrity and in cell division. Subsequent studies revealed that these proteins (TolQARB and Pal) localize to the septum and have an important role in ensuring that the outer membrane follows the growth of the invaginating cell wall (Gerding et al. 2007).

Cell Size Regulation

Cell size in *E. coli* increases with growth rate, increasing in both width and length (Pierucci 1978). This increase in cell size accommodates the increased DNA content of fast growing cells due to multifork replication (Cooper and Helmstetter 1968). The doubling time of fast growing cells is less than the time to replicate the chromosome (~40 min); however, this problem is solved by initiation of another round of replication before the previous round is finished.

Metabolic Regulation

How is the increase in cell size as a function to growth rate regulated? In *B. subtilis* it was found that division is transiently delayed in fast growing cells to achieve an increased cell size and that the level of UDP-glucose provided a metabolic signal (Weart et al. 2007). The UDP-glucose effect is mediated by an enzyme (UgtP) in a pathway for synthesis of a cell wall sugar. At higher growth rates the increased level of UDP-glucose stimulates the interaction between UgtP and FtsZ, delaying assembly of the Z ring. Consistent with this, inactivation of the UgtP pathway does not affect the growth rate but results in cells that are ~25% smaller. A similar nutrient-sensing pathway also exists in *E. coli* and also involves UDP-glucose but in a pathway for a periplasmic polysaccharide (Hill et al. 2013). The enzyme (OpgH) in this case is unrelated to UgtP but also acts on FtsZ in a UDP-glucose dependent fashion. Knocking out the relevant genes in the pathway also results in a size reduction. Thus, these two evolutionarily distant organisms both use UDP-glucose utilizing enzymes to couple growth rate to cell size.

Mutations Reducing Cell Size

In addition to knocking out the UDP-glucose sensing pathway, several mutations in essential cell division genes have been isolated that unexpectedly cause *E. coli* to grow at reduced cell size without affecting the growth rate. The first of these is the *ftsA** mutation, which reduces cell size at fast growth rates by 25% (Geissler et al. 2007). More recently, mutations in *ftsB* and *ftsL* have also been shown to reduce cell size by a similar amount (Tsang and Bernhardt 2015; Liu et al. 2015). The mechanisms responsible for the size reduction by the various *fts* mutations could be different. For *fts*B and *ftsL* the mutations appear to eliminate the delay between the establishment of the Z ring and the formation and activation of the complete divisome. Thus, newborn cells increase in cell size before assembling a Z ring but, as soon as it appears, it immediately matures into an active divisome. These mutations appear to activate the divisome complex but why this eliminates the delay in recruitment of the downstream division proteins is not clear. In contrast, the *ftsA** mutation appears to lead to the assembly of Z rings at a smaller cell size (Geissler et al. 2007). The reason the Z ring assembles at a smaller cell size is not known but may be due to the higher affinity that FtsA* displays for the CCTP of FtsZ (Pichoff et al. 2012). This higher affinity may promote Z ring assembly by competing better with negative regulators (MinC/MinD and SlmA) that also bind the CCTP.

Size Control and the Cell Cycle

Under steady state growth, *E. coli* cells fall within a relatively narrow size range. Recent work from several labs revealed that this homeostasis is achieved by a relatively simple mechanism. By monitoring the growth of individual cells it was revealed that cells grow a constant amount (incremental or adder rule) between divisions, regardless of cell size at birth (Campos et al. 2014; Taheri-Araghi et al. 2015). How this is achieved is not clear but constant extension between divisions, along with division at midcell, is sufficient for a population of cells to maintain a narrow cell length distribution. Interestingly, it appears that it is division itself that resets the mechanism, as a polar division that generates a minicell in a MinC mutant is sufficient to do so (Campos et al. 2014).

Summary

The discovery of the Z ring initiated the field of the bacterial cytoskeleton and led to a simple model in which the Z ring orchestrated the synthesis of septal PG, resulting in cytokinesis (Bi and Lutkenhaus 1991). Subsequent studies revealed that FtsZ is the ancestral homologue of tubulin and FtsA is an actin related protein that tethers

FtsZ filaments to the membrane. FtsZ and FtsA, along with ZipA and various Zap proteins, coassemble into the Z ring. This step is spatially regulated by at least 3 systems to ensure the Z ring is assembled at midcell. After a delay additional proteins are recruited to the Z ring to form a functional divisome. These additional proteins constitute the septal PG machine, which is activated by FtsN through integration of signals in the cytoplasm and the periplasm. The divisome is a complex and sophisticated machine that is highly regulated and we are only beginning to understand how it is regulated.

References

Aaron M, Charbon G, Lam H, Schwarz H, Vollmer W, Jacobs-Wagner C (2007) The tubulin homologue FtsZ contributes to cell elongation by guiding cell wall precursor synthesis in Caulobacter crescentus. Mol Microbiol 64:938–952

Aarsman ME, Piette A, Fraipont C, Vinkenvleugel TM, Nguyen-Disteche M, den Blaauwen T (2005) Maturation of the Escherichia coli divisome occurs in two steps. Mol Microbiol 55:1631–1645

Adams DW, Wu LJ, Errington J (2015) Nucleoid occlusion protein Noc recruits DNA to the bacterial cell membrane. EMBO J 34:491–501

Addinall SG, Cao C, Lutkenhaus J (1997) FtsN, a late recruit to the septum in Escherichia coli. Mol Microbiol 25:303–309

Adler HI, Fisher WD, Cohen A, Hardigree AA (1967) Miniature Escherichia coli cells deficient in DNA. Proc Natl Acad Sci U S A 57:321–326

Akerlund T, Gullbrand B, Nordstrom K (2002) Effects of the Min system on nucleoid segregation in Escherichia coli. Microbiology 148:3213–3222

Arends SJ, Kustusch RJ, Weiss DS (2009) ATP-binding site lesions in FtsE impair cell division. J Bacteriol 191:3772–3784

Arends SJ, Williams K, Scott RJ, Rolong S, Popham DL, Weiss DS (2010) Discovery and characterization of three new Escherichia coli septal ring proteins that contain a SPOR domain: DamX, DedD, and RlpA. J Bacteriol 192:242–255

Arumugam S, Petrasek Z, Schwille P (2014) MinCDE exploits the dynamic nature of FtsZ filaments for its spatial regulation. Proc Natl Acad Sci U S A 111:E1192–E1200

Bailey MW, Bisicchia P, Warren BT, Sherratt DJ, Mannik J (2014) Evidence for divisome localization mechanisms independent of the Min system and SlmA in Escherichia coli. PLoS Genet 10:e1004504

Begg KJ, Dewar SJ, Donachie WD (1995) A new Escherichia coli cell division gene, ftsK. J Bacteriol 177:6211–6222

Bendezu FO, De Boer PA (2008) Conditional lethality, division defects, membrane involution, and endocytosis in mre and mrd shape mutants of Escherichia coli. J Bacteriol 190:1792–1811

Bernhardt TG, De Boer PA (2005) SlmA, a nucleoid-associated, FtsZ binding protein required for blocking septal ring assembly over Chromosomes in E. coli. Mol Cell 18:555–564

Bi E, Lutkenhaus J (1990) FtsZ regulates frequency of cell division in Escherichia coli. J Bacteriol 172:2765–2768

Bi EF, Lutkenhaus J (1991) FtsZ ring structure associated with division in Escherichia coli. Nature 354:161–164

Bi E, Lutkenhaus J (1992) Isolation and characterization of ftsZ alleles that affect septal morphology. J Bacteriol 174:5414–5423

Bi E, Lutkenhaus J (1993) Cell division inhibitors SulA and MinCD prevent formation of the FtsZ ring. J Bacteriol 175:1118–1125

Bisicchia P, Arumugam S, Schwille P, Sherratt D (2013a) MinC, MinD, and MinE drive counter-oscillation of early-cell-division proteins prior to Escherichia coli septum formation. MBio 4:e00856–e00813

Bisicchia P, Steel B, Mariam Debela MH, Lowe J, Sherratt D (2013b) The N-terminal membrane-spanning domain of the Escherichia coli DNA translocase FtsK hexamerizes at midcell. MBio 4:e00800–e00813

Boyle DS, Khattar MM, Addinall SG, Lutkenhaus J, Donachie WD (1997) ftsW is an essential cell-division gene in Escherichia coli. Mol Microbiol 24:1263–1273

Bramhill D, Kornberg A (1988) Duplex opening by dnaA protein at novel sequences in initiation of replication at the origin of the E. coli chromosome. Cell 52:743–755

Buddelmeijer N, Beckwith J (2004) A complex of the Escherichia coli cell division proteins FtsL, FtsB and FtsQ forms independently of its localization to the septal region. Mol Microbiol 52:1315–1327

Busiek KK, Margolin W (2014) A role for FtsA in SPOR-independent localization of the essential Escherichia coli cell division protein FtsN. Mol Microbiol 92:1212–1226

Busiek KK, Eraso JM, Wang Y, Margolin W (2012) The early divisome protein FtsA interacts directly through its 1c subdomain with the cytoplasmic domain of the late divisome protein FtsN. J Bacteriol 194:1989–2000

Buss J, Coltharp C, Huang T, Pohlmeyer C, Wang SC, Hatem C, Xiao J (2013) In vivo organization of the FtsZ-ring by ZapA and ZapB revealed by quantitative super-resolution microscopy. Mol Microbiol 89:1099–1120

Buss J, Coltharp C, Shtengel G, Yang X, Hess H, Xiao J (2015) A multi-layered protein network stabilizes the Escherichia coli FtsZ-ring and modulates constriction dynamics. PLoS Genet 11:e1005128

Cabre EJ, Sanchez-Gorostiaga A, Carrara P, Ropero N, Casanova M, Palacios P, Stano P, Jimenez M, Rivas G, Vicente M (2013) Bacterial division proteins FtsZ and ZipA induce vesicle shrinkage and cell membrane invagination. J Biol Chem 288:26625–26634

Campos M, Surovtsev IV, Kato S, Paintdakhi A, Beltran B, Ebmeier SE, Jacobs-Wagner C (2014) A constant size extension drives bacterial cell size homeostasis. Cell 159:1433–1446

Chen Y, Erickson HP (2005) Rapid in vitro assembly dynamics and subunit turnover of FtsZ demonstrated by fluorescence resonance energy transfer. J Biol Chem 280:22549–22554

Chen Y, Bjornson K, Redick SD, Erickson HP (2005) A rapid fluorescence assay for FtsZ assembly indicates cooperative assembly with a dimer nucleus. Biophys J 88:505–514

Cho H, Bernhardt TG (2013) Identification of the SlmA active site responsible for blocking bacterial cytokinetic ring assembly over the chromosome. PLoS Genet 9:e1003304

Cho H, Mcmanus HR, Dove SL, Bernhardt TG (2011) Nucleoid occlusion factor SlmA is a DNA-activated FtsZ polymerization antagonist. Proc Natl Acad Sci U S A 108:3773–3778

Cho H, Wivagg CW, Kapoor M, Barry Z, Rohs PD, Shu M, Marto JA, Garner TC, Bernhardt TG (2016) Bacterial cell wall biogenesis is mediated by SEDS and PBP polymerase families functioning semi – autonomously. Nat Microbiol 1:16172

Coltharp C, Buss J, Plumer TM, Xiao J (2016) Defining the rate-limiting processes of bacterial cytokinesis. Proc Natl Acad Sci U S A 113:E1044–E1053

Conti J, Viola MG, Camberg JL (2015) The bacterial cell division regulators MinD and MinC form polymers in the presence of nucleotide. FEBS Lett 589:201–206

Cooper S, Helmstetter CE (1968) Chromosome replication and the division cycle of Escherichia coli B/r. J Mol Biol 31:519–540

Corbin BD, Wang Y, Beuria TK, Margolin W (2007) Interaction between cell division proteins FtsE and FtsZ. J Bacteriol 189:3026–3035

Cordell SC, Anderson RE, Lowe J (2001) Crystal structure of the bacterial cell division inhibitor MinC. EMBO J 20:2454–2461

Dai K, Xu Y, Lutkenhaus J (1993) Cloning and characterization of ftsN, an essential cell division gene in Escherichia coli isolated as a multicopy suppressor of ftsA12(Ts). J Bacteriol 175:3790–3797

Dai K, Xu Y, Lutkenhaus J (1996) Topological characterization of the essential Escherichia coli cell division protein FtsN. J Bacteriol 178:1328–1334
Dajkovic A, Lutkenhaus J (2006) Z ring as executor of bacterial cell division. J Mol Microbial Biotechnol 11:140–151
Dajkovic A, Lan G, Sun SX, Wirtz D, Lutkenhaus J (2008) MinC spatially controls bacterial cytokinesis by antagonizing the scaffolding function of FtsZ. Curr Biol 18:235–244
Dajkovic A, Pichoff S, Lutkenhaus J, Wirtz D (2010) Cross-linking FtsZ polymers into coherent Z rings. Mol Microbiol 78:651–668
De Boer PA (2010) Advances in understanding E. coli cell fission. Curr Opin Microbiol 13:730–737
De Boer PA, Crossley RE, Rothfield LI (1989) A division inhibitor and a topological specificity factor coded for by the minicell locus determine proper placement of the division septum in E. coli. Cell 56:641–649
Di Ventura B, Knecht B, Andreas H, Godinez WJ, Fritsche M, Rohr K, Nickel W, Heermann DW, Sourjik V (2013) Chromosome segregation by the Escherichia coli Min system. Mol Syst Biol 9:686
Donachie WD (1968) Relationship between cell size and time of initiation of DNA replication. Nature 219:1077–1079
Donachie WD, Blakely GW (2003) Coupling the initiation of chromosome replication to cell size in Escherichia coli. Curr Opin Microbiol 6:146–150
Du S, Lutkenhaus J (2014) SlmA antagonism of FtsZ assembly employs a two-pronged mechanism like MinCD. PLoS Genet 10:e1004460
Du S, Park KT, Lutkenhaus J (2015) Oligomerization of FtsZ converts the FtsZ tail motif (conserved carboxy-terminal peptide) into a multivalent ligand with high avidity for partners ZipA and SlmA. Mol Microbiol 95:173–188
Du S, Pichoff S, Lutkenhaus J (2016) FtsEX acts on FtsA to regulate divisome assembly and activity. Proc Natl Acad Sci U S A 113. doi:10.1073/pnas.1606656113
Dubarry N, Possoz C, Barre FX (2010) Multiple regions along the Escherichia coli FtsK protein are implicated in cell division. Mol Microbiol 78:1088–1100
Durand-Heredia JM, Yu HH, de Carlo S, Lesser CF, Janakiraman A (2011) Identification and characterization of ZapC, a stabilizer of the FtsZ ring in Escherichia coli. J Bacteriol 193:1405–1413
Durand-Heredia J, Rivkin E, Fan G, Morales J, Janakiraman A (2012) Identification of ZapD as a cell division factor that promotes the assembly of FtsZ in Escherichia coli. J Bacteriol 194:3189–3198
Egan AJ, Vollmer W (2015) The stoichiometric divisome: a hypothesis. Front Microbiol 6:455
Egan AJ, Biboy J, Van't Veer I, Breukink E, Vollmer W (2015) Activities and regulation of peptidoglycan synthases. Philos Trans R Soc Lond Ser B Biol Sci 370
Erickson HP, Anderson DE, Osawa M (2010) FtsZ in bacterial cytokinesis: cytoskeleton and force generator all in one. Microbiol Mol Biol Rev 74:504–528
Erzberger JP, Mott ML, Berger JM (2006) Structural basis for ATP-dependent DnaA assembly and replication-origin remodeling. Nat Struct Mol Biol 13:676–683
Espeli O, Borne R, Dupaigne P, Thiel A, Gigant E, Mercier R, Boccard F (2012) A MatP-divisome interaction coordinates chromosome segregation with cell division in E. coli. EMBO J 31:3198–3211
Fange D, Elf J (2006) Noise-induced Min phenotypes in E coli. PLoS Comput Biol 2:e80
Fenton AK, Gerdes K (2013) Direct interaction of FtsZ and MreB is required for septum synthesis and cell division in Escherichia coli. EMBO J 32:1953–1965
Fu X, Shih YL, Zhang Y, Rothfield LI (2001) The MinE ring required for proper placement of the division site is a mobile structure that changes its cellular location during the Escherichia coli division cycle. Proc Natl Acad Sci U S A 98:980–985
Fu G, Huang T, Buss J, Coltharp C, Hensel Z, Xiao J (2010) In vivo structure of the E. coli FtsZ-ring revealed by photoactivated localization microscopy (PALM). PLoS One 5:e12682

Galli E, Gerdes K (2010) Spatial resolution of two bacterial cell division proteins: ZapA recruits ZapB to the inner face of the Z-ring. Mol Microbiol 76:1514–1526

Geissler B, Margolin W (2005) Evidence for functional overlap among multiple bacterial cell division proteins: compensating for the loss of FtsK. Mol Microbiol 58:596–612

Geissler B, Elraheb D, Margolin W (2003) A gain-of-function mutation in ftsA bypasses the requirement for the essential cell division gene zipA in Escherichia coli. Proc Natl Acad Sci U S A 100:4197–4202

Geissler B, Shiomi D, Margolin W (2007) The ftsA* gain-of-function allele of Escherichia coli and its effects on the stability and dynamics of the Z ring. Microbiology 153:814–825

Gerding MA, Ogata Y, Pecora ND, Niki H, De Boer PA (2007) The trans-envelope Tol-Pal complex is part of the cell division machinery and required for proper outer-membrane invagination during cell constriction in E. coli. Mol Microbiol 63:1008–1025

Gerding MA, Liu B, Bendezu FO, Hale CA, Bernhardt TG, De Boer PA (2009) Self-enhanced accumulation of FtsN at division sites and roles for other proteins with a SPOR domain (DamX, DedD, and RlpA) in Escherichia coli cell constriction. J Bacteriol 191:7383–7401

Ghigo JM, Weiss DS, Chen JC, Yarrow JC, Beckwith J (1999) Localization of FtsL to the Escherichia coli septal ring. Mol Microbiol 31:725–737

Ghosal D, Trambaiolo D, Amos LA, Lowe J (2014) MinCD cell division proteins form alternating copolymeric cytomotive filaments. Nat Commun 5:5341

Glas M, van den Berg Van Saparoea HB, Mclaughlin SH, Roseboom W, Liu F, Koningstein GM, Fish A, den Blaauwen T, Heck AJ, de Jong L, Bitter W, de Esch IJ, Luirink J (2015) The soluble periplasmic domains of Escherichia coli cell division proteins FtsQ/FtsB/FtsL form a trimeric complex with submicromolar affinity. J Biol Chem 290:21498–21509

Goehring NW, Gueiros-Filho F, Beckwith J (2005) Premature targeting of a cell division protein to midcell allows dissection of divisome assembly in Escherichia coli. Genes Dev 19:127–137

Goehring NW, Gonzalez MD, Beckwith J (2006) Premature targeting of cell division proteins to midcell reveals hierarchies of protein interactions involved in divisome assembly. Mol Microbiol 61:33–45

Gonzalez MD, Beckwith J (2009) Divisome under construction: distinct domains of the small membrane protein FtsB are necessary for interaction with multiple cell division proteins. J Bacteriol 191:2815–2825

Gonzalez MD, Akbay EA, Boyd D, Beckwith J (2010) Multiple interaction domains in FtsL, a protein component of the widely conserved bacterial FtsLBQ cell division complex. J Bacteriol 192:2757–2768

Gueiros-Filho FJ, Losick R (2002) A widely conserved bacterial cell division protein that promotes assembly of the tubulin-like protein FtsZ. Genes Dev 16:2544–2556

Haeusser DP, Rowlett VW, Margolin W (2015) A mutation in Escherichia coli ftsZ bypasses the requirement for the essential division gene zipA and confers resistance to FtsZ assembly inhibitors by stabilizing protofilament bundling. Mol Microbiol 97:988–1005

Halatek J, Frey E (2012) Highly canalized MinD transfer and MinE sequestration explain the origin of robust MinCDE-protein dynamics. Cell Rep 1:741–752

Hale CA, De Boer PA (1997) Direct binding of FtsZ to ZipA, an essential component of the septal ring structure that mediates cell division in E. coli. Cell 88:175–185

Hale CA, Meinhardt H, De Boer PA (2001) Dynamic localization cycle of the cell division regulator MinE in Escherichia coli. EMBO J 20:1563–1572

Hale CA, Shiomi D, Liu B, Bernhardt TG, Margolin W, Niki H, De Boer PA (2011) Identification of Escherichia coli ZapC (YcbW) as a component of the division apparatus that binds and bundles FtsZ polymers. J Bacteriol 193:1393–1404

Haney SA, Glasfeld E, Hale C, Keeney D, He Z, de Boer P (2001) Genetic analysis of the Escherichia coli FtsZ.ZipA interaction in the yeast two-hybrid system. Characterization of FtsZ residues essential for the interactions with ZipA and with FtsA. J Biol Chem 276:11980–11987

Heald R, Khodjakov A (2015) Thirty years of search and capture: the complex simplicity of mitotic spindle assembly. J Cell Biol 211:1103–1111

Heidrich C, Templin MF, Ursinus A, Merdanovic M, Berger J, Schwarz H, De Pedro MA, Holtje JV (2001) Involvement of N-acetylmuramyl-L-alanine amidases in cell separation and antibiotic-induced autolysis of Escherichia coli. Mol Microbiol 41:167–178

Helmstetter CE (1974) Initiation of chromosome replication in Escherichia coli. I. Requirements for RNA and protein synthesis at different growth rates. J Mol Biol 84:1–19

Hill NS, Buske PJ, Shi Y, Levin PA (2013) A moonlighting enzyme links Escherichia coli cell size with central metabolism. PLoS Genet 9:e1003663

Hirota Y, Jacob F, Ryter A, Buttin G, Nakai T (1968) On the process of cellular division in Escherichia coli. I. Asymmetrical cell division and production of deoxyribonucleic acid-less bacteria. J Mol Biol 35:175–192

Hu Z, Lutkenhaus J (1999) Topological regulation of cell division in Escherichia coli involves rapid pole to pole oscillation of the division inhibitor MinC under the control of MinD and MinE. Mol Microbiol 34:82–90

Hu Z, Lutkenhaus J (2000) Analysis of MinC reveals two independent domains involved in interaction with MinD and FtsZ. J Bacteriol 182:3965–3971

Hu Z, Lutkenhaus J (2001) Topological regulation of cell division in E. coli. spatiotemporal oscillation of MinD requires stimulation of its ATPase by MinE and phospholipid. Mol Cell 7:1337–1343

Hu Z, Lutkenhaus J (2003) A conserved sequence at the C-terminus of MinD is required for binding to the membrane and targeting MinC to the septum. Mol Microbiol 47:345–355

Hu Z, Mukherjee A, Pichoff S, Lutkenhaus J (1999) The MinC component of the division site selection system in Escherichia coli interacts with FtsZ to prevent polymerization. Proc Natl Acad Sci U S A 96:14819–14824

Hu Z, Saez C, Lutkenhaus J (2003) Recruitment of MinC, an inhibitor of Z-ring formation, to the membrane in Escherichia coli: role of MinD and MinE. J Bacteriol 185:196–203

Huang KC, Meir Y, Wingreen NS (2003) Dynamic structures in Escherichia coli: spontaneous formation of MinE rings and MinD polar zones. Proc Natl Acad Sci U S A 100:12724–12728

Huang KH, Durand-Heredia J, Janakiraman A (2013) FtsZ ring stability: of bundles, tubules, crosslinks, and curves. J Bacteriol 195:1859–1868

Ivanov V, Mizuuchi K (2010) Multiple modes of interconverting dynamic pattern formation by bacterial cell division proteins. Proc Natl Acad Sci U S A 107:8071–8078

Karimova G, Dautin N, Ladant D (2005) Interaction network among Escherichia coli membrane proteins involved in cell division as revealed by bacterial two-hybrid analysis. J Bacteriol 187:2233–2243

Katayama T, Ozaki S, Keyamura K, Fujimitsu K (2010) Regulation of the replication cycle: conserved and diverse regulatory systems for DnaA and oriC. Nat Rev Microbiol 8:163–170

Kato J, Katayama T (2001) Hda, a novel DnaA- related protein, regulates the replication cycle in Escherichia coli. EMBO J 20:4253–4262

Kirschner M, Mitchison T (1986) Beyond self- assembly: from microtubules to morphogenesis. Cell 45:329–342

Krupka M, Cabre EJ, Jimenez M, Rivas G, Rico AI, Vicente M (2014) Role of the FtsA C terminus as a switch for polymerization and membrane association. MBio 5:e02221

Kruse K, Howard M, Margolin W (2007) An experimentalist's guide to computational modelling of the Min system. Mol Microbiol 63:1279–1284

Lackner LL, Raskin DM, De Boer PA (2003) ATP-dependent interactions between Escherichia coli Min proteins and the phospholipid membrane in vitro. J Bacteriol 185:735–749

Lara B, Rico AI, Petruzzelli S, Santona A, Dumas J, Biton J, Vicente M, Mingorance J, Massidda O (2005) Cell division in cocci: localization and properties of the Streptococcus pneumoniae FtsA protein. Mol Microbiol 55:699–711

Leonard AC, Grimwade JE (2015) The orisome: structure and function. Front Microbiol 6:545

Li Z, Trimble MJ, Brun YV, Jensen GJ (2007) The structure of FtsZ filaments in vivo suggests a force-generating role in cell division. EMBO J 26:4694–4708
Li GW, Burkhardt D, Gross C, Weissman JS (2014) Quantifying absolute protein synthesis rates reveals principles underlying allocation of cellular resources. Cell 157:624–635
Liu B, Persons L, Lee L, De Boer PA (2015) Roles for both FtsA and the FtsBLQ subcomplex in FtsN-stimulated cell constriction in Escherichia coli. Mol Microbiol 95:945–970
Loose M, Mitchison TJ (2014) The bacterial cell division proteins FtsA and FtsZ self-organize into dynamic cytoskeletal patterns. Nat Cell Biol 16:38–46
Loose M, Fischer-Friedrich E, Ries J, Kruse K, Schwille P (2008) Spatial regulators for bacterial cell division self-organize into surface waves in vitro. Science 320:789–792
Lowe J, Amos LA (1998) Crystal structure of the bacterial cell-division protein FtsZ. Nature 391:203–206
Lutkenhaus J (2007) Assembly dynamics of the bacterial MinCDE system and spatial regulation of the Z ring. Annu Rev Biochem 76:539–562
Lutkenhaus J (2009) FtsN–trigger for septation. J Bacteriol 191:7381–7382
Lutkenhaus J, Pichoff S, Du S (2012) Bacterial cytokinesis: from Z ring to divisome. Cytoskeleton (Hoboken) 69:778–790
Ma X, Margolin W (1999) Genetic and functional analyses of the conserved C-terminal core domain of Escherichia coli FtsZ. J Bacteriol 181:7531–7544
Mannik J, Bailey MW (2015) Spatial coordination between chromosomes and cell division proteins in Escherichia coli. Front Microbiol 6:306
Mannik J, Wu F, Hol FJ, Bisicchia P, Sherratt DJ, Keymer JE, Dekker C (2012) Robustness and accuracy of cell division in Escherichia coli in diverse cell shapes. Proc Natl Acad Sci U S A 109:6957–6962
Meeske AJ, Sham LT, Kimsey H, Koo BM, Gross CA, Bernhardt TG, Rudner DZ (2015) MurJ and a novel lipid II flippase are required for cell wall biogenesis in Bacillus subtilis. Proc Natl Acad Sci U S A 112:6437–6442
Meeske AJ, Riley EP, Robins WP, Uehara T, Mekalanos JJ, Kahne D, Walker S, Kruse AC, Bernhardt TG, Rudner D (2016) SEDS proteins are a widespread family at bacterial cell wall polymerases. Nature 357:634–6384
Meinhardt H, De Boer PA (2001) Pattern formation in Escherichia coli: a model for the pole-to-pole oscillations of Min proteins and the localization pole oscillations of Min proteins and the localization of the division site. Proc Natl Acad Sci U S A 98:14202–14207
Mercer KL, Weiss DS (2002) The Escherichia coli cell division protein FtsW is required to recruit its cognate transpeptidase, FtsI (PBP3), to the division site. J Bacteriol 184:904–912
Mercier R, Petit MA, Schbath S, Robin S, El Karoui M, Boccard F, Espeli O (2008) The MatP/matS site-specific system organizes the terminus region of the E. coli chromosome into a macrodomain. Cell 135:475–485
Modell JW, Hopkins AC, Laub MT (2011) A DNA damage Checkpoint in Caulobacter crescentus inhibits cell division through a direct interaction with FtsW. Genes Dev 25:1328–1343
Modell JW, Kambara TK, Perchuk BS, Laub MT (2014) A DNA damage-induced, SOS-independent checkpoint regulates cell division in Caulobacter crescentus. PLoS Biol 12:e1001977
Mohammadi T, Ploeger GE, Verheul J, Comvalius AD, Martos A, Alfonso C, Van Marle J, Rivas G, Den Blaauwen T (2009) The GTPase activity of Escherichia coli FtsZ determines the magnitude of the FtsZ polymer bundling by ZapA in vitro. Biochemistry 48:11056–11066
Mohammadi T, Van Dam V, Sijbrandi R, Vernet T, Zapun A, Bouhss A, Diepeveen-De Bruin M, Nguyen-Disteche M, de Kruijff B, Breukink E (2011) Identification of FtsW as a transporter of lipid-linked cell wall precursors across the membrane. EMBO J 30:1425–1432
Monahan LG, Liew AT, Bottomley AL, Harry EJ (2014) Division site positioning in bacteria: one size does not fit all. Front Microbiol 5:19

Mosyak L, Zhang Y, Glasfeld E, Haney S, Stahl M, Seehra J, Somers WS (2000) The bacterial cell-division protein ZipA and its interaction with an FtsZ fragment revealed by X-ray crystallography. EMBO J 19:3179–3191
Mott ML, Berger JM (2007) DNA replication initiation: mechanisms and regulation in bacteria. Nat Rev Microbiol 5:343–354
Mukherjee A, Lutkenhaus J (1994) Guanine nucleotide-dependent assembly of FtsZ into filaments. J Bacteriol 176:2754–2758
Mukherjee A, Lutkenhaus J (1998) Dynamic assembly of FtsZ regulated by GTP hydrolysis. EMBO J 17:462–469
Muller P, Ewers C, Bertsche U, Anstett M, Kallis T, Breukink E, Fraipont C, Terrak M, Nguyen-Disteche M, Vollmer W (2007) The essential cell division protein FtsN interacts with the murein (peptidoglycan) synthase PBP1B in Escherichia coli. J Biol Chem 282:36394–36402
Narita S, Tokuda H (2006) An ABC transporter mediating the membrane detachment of bacterial lipoproteins depending on their sorting signals. FEBS Lett 580:1164–1170
Nogales E, Downing KH, Amos LA, Lowe J (1998) Tubulin and FtsZ form a distinct family of GTPases. Nat Struct Biol 5:451–458
Osawa M, Erickson HP (2013) Liposome division by a simple bacterial division machinery. Proc Natl Acad Sci U S A 110:11000–11004
Osawa M, Anderson DE, Erickson HP (2008) Reconstitution of contractile FtsZ rings in liposomes. Science 320:792–794
Pacheco-Gomez R, Cheng X, Hicks MR, Smith CJ, Roper DI, Addinall S, Rodger A, Dafforn TR (2013) Tetramerization of ZapA is required for FtsZ bundling. Biochem J 449:795–802
Park KT, Wu W, Battaile KP, Lovell S, Holyoak T, Lutkenhaus J (2011) The Min oscillator uses MinD-dependent conformational changes in MinE to spatially regulate cytokinesis. Cell 146:396–407
Park KT, Wu W, Lovell S, Lutkenhaus J (2012) Mechanism of the asymmetric activation of the MinD ATPase by MinE. Mol Microbiol 85:271–281
Park KT, Du S, Lutkenhaus J (2015) MinC/MinD copolymers are not required for Min function. Mol Microbiol 98(5):895–909
Pastoret S, Fraipont C, den Blaauwen T, Wolf B, Aarsman ME, Piette A, Thomas A, Brasseur R, Nguyen-Disteche M (2004) Functional analysis of the cell division protein FtsW of Escherichia coli. J Bacteriol 186:8370–8379
Peters NT, Dinh T, Bernhardt TG (2011) A fail-safe mechanism in the septal ring assembly pathway generated by the sequential recruitment of cell separation amidases and their activators. J Bacteriol 193:4973–4983
Pichoff S, Lutkenhaus J (2001) Escherichia coli division inhibitor MinCD blocks septation by preventing Z-ring formation. J Bacteriol 183:6630–6635
Pichoff S, Lutkenhaus J (2002) Unique and overlapping roles for ZipA and FtsA in septal ring assembly in Escherichia coli. EMBO J 21:685–693
Pichoff S, Lutkenhaus J (2005) Tethering the Z ring to the membrane through a conserved membrane targeting sequence in FtsA. Mol Microbiol 55:1722–1734
Pichoff S, Shen B, Sullivan B, Lutkenhaus J (2012) FtsA mutants impaired for self-interaction bypass ZipA suggesting a model in which FtsA's self-interaction competes with its ability to recruit downstream division proteins. Mol Microbiol 83:151–167
Pichoff S, Du S, Lutkenhaus J (2015) The bypass of ZipA by overexpression of FtsN requires a previously unknown conserved FtsN motif essential for FtsA-FtsN interaction supporting a model in which FtsA monomers recruit late cell division proteins to the Z ring. Mol Microbiol 95:971–987
Pierucci O (1978) Dimensions of Escherichia coli at various growth rates: model for envelope growth. J Bacteriol 135:559–574
Raskin DM, De Boer PA (1999a) MinDE-dependent pole-to-pole oscillation of division inhibitor MinC in Escherichia coli. J Bacteriol 181:6419–6424

Raskin DM, De Boer PA (1999b) Rapid pole-to- pole oscillation of a protein required for directing division to the middle of Escherichia coli. Proc Natl Acad Sci U S A 96:4971–4976
Reddy M (2007) Role of FtsEX in cell division of Escherichia coli: viability of ftsEX mutants is dependent on functional SufI or high osmotic strength. J Bacteriol 189:98–108
Rico AI, Krupka M, Vicente M (2013) In the beginning, Escherichia coli assembled the proto-ring: an initial phase of division. J Biol Chem 288:20830–20836
Rodrigues CD, Harry EJ (2012) The Min system and nucleoid occlusion are not required for identifying the division site in Bacillus subtilis but ensure its efficient utilization. PLoS Genet 8:e1002561
Roll-Mecak A (2015) Intrinsically disordered tubulin tails: complex tuners of microtubule functions? Semin Cell Dev Biol 37:11–19
Rothfield L, Taghbalout A, Shih YL (2005) Spatial control of bacterial division-site placement. Nat Rev Microbiol 3:959–968
Schmidt KL, Peterson ND, Kustusch RJ, Wissel MC, Graham B, Phillips GJ, Weiss DS (2004) A predicted ABC transporter, FtsEX, is needed for cell division in Escherichia coli. J Bacteriol 186:785–793
Schumacher MA, Zeng W (2016) Structures of the nucleoid occlusion protein SlmA bound to DNA and the C-terminal domain of the cytoskeletal protein FtsZ. Proc Natl Acad Sci U S A 113:4988–4993
Schumacher MA, Zeng W, Huang KH, Tchorzewski L, Janakiraman A (2015) Structural and functional analyses reveal insights into the molecular properties of the E. coli Z ring stabilizing protein, ZapC. J Biol Chem 291:2485
Schweizer J, Loose M, Bonny M, Kruse K, Monch I, Schwille P (2012) Geometry sensing by self-organized protein patterns. Proc Natl Acad Sci U S A 109:15283–15288
Shen B, Lutkenhaus J (2009) The conserved C- terminal tail of FtsZ is required for the septal localization and division inhibitory activity of MinC(C)/MinD. Mol Microbiol 72:410–424
Shen B, Lutkenhaus J (2010) Examination of the interaction between FtsZ and MinCN in E. coli suggests how MinC disrupts Z rings. Mol Microbiol 75:1285–1298
Soderstrom B, Skoog K, Blom H, Weiss DS, Von Heijne G, Daley DO (2014) Disassembly of the divisome in Escherichia coli: evidence that FtsZ dissociates before compartmentalization. Mol Microbiol 92:1–9
Spratt BG (1975) Distinct penicillin binding proteins involved in the division, elongation, and shape of Escherichia coli K12. Proc Natl Acad Sci U S A 72:2999–3003
Steiner W, Liu G, Donachie WD, Kuempel P (1999) The cytoplasmic domain of FtsK protein is required for resolution of chromosome dimers. Mol Microbiol 31:579–583
Strauss MP, Liew AT, Turnbull L, Whitchurch CB, Monahan LG, Harry EJ (2012) 3D-SIM super resolution microscopy reveals a bead-like arrangement for FtsZ and the division machinery: implications for triggering cytokinesis. PLoS Biol 10:e1001389
Stricker J, Maddox P, Salmon ED, Erickson HP (2002) Rapid assembly dynamics of the Escherichia coli FtsZ-ring demonstrated by fluorescence recovery after photobleaching. Proc Natl Acad Sci U S A 99:3171–3175
Sun Q, Yu XC, Margolin W (1998) Assembly of the FtsZ ring at the central division site in the absence of the chromosome. Mol Microbiol 29:491–503
Szeto TH, Rowland SL, Rothfield LI, King GF (2002) Membrane localization of MinD is mediated by a C-terminal motif that is conserved across eubacteria, archaea, and chloroplasts. Proc Natl Acad Sci U S A 99:15693–15698
Szwedziak P, Lowe J (2013) Do the divisome and elongasome share a common evolutionary past? Curr Opin Microbiol 16:745–751
Szwedziak P, Wang Q, Freund SM, Lowe J (2012) FtsA forms actin-like protofilaments. EMBO J 31:2249–2260
Szwedziak P, Wang Q, Bharat TA, Tsim M, Lowe J (2014) Architecture of the ring formed by the tubulin homologue FtsZ in bacterial cell division. Elife 3:e04601

Taheri-Araghi S, Bradde S, Sauls JT, Hill NS, Levin PA, Paulsson J, Vergassola M, Jun S (2015) Cell-size control and homeostasis in bacteria. Curr Biol 25:385–391

Teather RM, Collins JF, Donachie WD (1974) Quantal behavior of a diffusible factor which initiates septum formation at potential division sites in Escherichia coli. J Bacteriol 118:407–413

Thanedar S, Margolin W (2004) FtsZ exhibits rapid movement and oscillation waves in helix-like patterns in Escherichia coli. Curr Biol 14:1167–1173

Tonthat NK, Arold ST, Pickering BF, Van Dyke MW, Liang S, Lu Y, Beuria TK, Margolin W, Schumacher MA (2011) Molecular mechanism by which the nucleoid occlusion factor, SlmA, keeps cytokinesis in check. EMBO J 30:154–164

Tonthat NK, Milam SL, Chinnam N, Whitfill T, Margolin W, Schumacher MA (2013) SlmA forms a higher-order structure on DNA that inhibits cytokinetic Z-ring formation over the nucleoid. Proc Natl Acad Sci U S A 110:10586–10591

Trip EN, Scheffers DJ (2015) A 1 MDa protein complex containing critical components of the Escherichia coli divisome. Sci Rep 5:18190

Tsang MJ, Bernhardt TG (2015) A role for the FtsQLB complex in cytokinetic ring activation revealed by an ftsL allele that accelerates division. Mol Microbiol 95:925–944

Typas A, Banzhaf M, Gross CA, Vollmer W (2012) From the regulation of peptidoglycan synthesis to bacterial growth and morphology. Nat Rev Microbiol 10:123–136

Uehara T, Parzych KR, Dinh T, Bernhardt TG (2010) Daughter cell separation is controlled by cytokinetic ring-activated cell wall hydrolysis. EMBO J 29:1412–1422

Ursinus A, Van Den Ent F, Brechtel S, De Pedro M, Holtje JV, Lowe J, Vollmer W (2004) Murein (peptidoglycan) binding property of the essential cell division protein FtsN from Escherichia coli. J Bacteriol 186:6728–6737

Uversky VN (2013) The most important thing is the tail: multitudinous functionalities of intrinsically disordered protein termini. FEBS Lett 587:1891–1901

Vadia S, Levin PA (2015) Growth rate and cell size: a re-examination of the growth law. Curr Opin Microbiol 24:96–103

Van De Putte P, Van D, Roersch A (1964) The selection of mutants of Escherichia coli with impaired cell division at elevated temperature. Mutat Res 106:121–128

Van Den Ent F, Amos LA, Lowe J (2001) Prokaryotic origin of the actin cytoskeleton. Nature 413:39–44

Van Der Ploeg R, Verheul J, Vischer NO, Alexeeva S, Hoogendoorn E, Postma M, Banzhaf M, Vollmer W, Den Blaauwen T (2013) Colocalization and interaction between elongasome and divisome during a preparative cell division phase in Escherichia coli. Mol Microbiol 87:1074–1087

Varma A, Huang KC, Young KD (2008) The Min system as a general cell geometry detection mechanism: branch lengths in Y-shaped Escherichia coli cells affect Min oscillation patterns and division dynamics. J Bacteriol 190:2106–2117

Vecchiarelli AG, Li M, Mizuuchi M, Hwang LC, Seol Y, Neuman KC, Mizuuchi K (2016) Membrane-bound MinDE complex acts as a toggle switch that drives Min oscillation coupled to cytoplasmic depletion of MinD. Proc Natl Acad Sci U S A 113:E1479–E1488

Wang L, Lutkenhaus J (1998) FtsK is an essential cell division protein that is localized to the septum and induced as part of the SOS response. Mol Microbiol 29:731–740

Weart RB, Levin PA (2003) Growth rate- dependent regulation of medial FtsZ ring formation. J Bacteriol 185:2826–2834

Weart RB, Lee AH, Chien AC, Haeusser DP, Hill NS, Levin PA (2007) A metabolic sensor governing cell size in bacteria. Cell 130:335–347

Weiss DS (2015) Last but not least: new insights into how FtsN triggers constriction during Escherichia coli cell division. Mol Microbiol 95:903–909

Wissel MC, Weiss DS (2004) Genetic analysis of the cell division protein FtsI (PBP3): amino acid substitutions that impair septal localization of FtsI and recruitment of FtsN. J Bacteriol 186:490–502

Woldringh CL, Mulder E, Valkenburg JA, Wientjes FB, Zaritsky A, Nanninga N (1990) Role of the nucleoid in the toporegulation of division. Res Microbiol 141:39–49
Wu LJ, Ishikawa S, Kawai Y, Oshima T, Ogasawara N, Errington J (2009) Noc protein binds to specific DNA sequences to coordinate cell division with chromosome segregation. EMBO J 28:1940–1952
Wu W, Park KT, Holyoak T, Lutkenhaus J (2011) Determination of the structure of the MinD- ATP complex reveals the orientation of MinD on the membrane and the relative location of the binding sites for MinE and MinC. Mol Microbiol 79:1515–1528
Wu F, van Schie BG, Keymer JE, Dekker C (2015) Symmetry and scale orient Min protein patterns in shaped bacterial sculptures. Nat Nanotechnol 10:719–726
Yahashiri A, Jorgenson MA, Weiss DS (2015) Bacterial SPOR domains are recruited to septal peptidoglycan by binding to glycan strands that lack stem peptides. Proc Natl Acad Sci U S A 112:11347–11352
Yang JC, Van Den Ent F, Neuhaus D, Brevier J, Lowe J (2004) Solution structure and domain architecture of the divisome protein FtsN. Mol Microbiol 52:651–660
Yang DC, Peters NT, Parzych KR, Uehara T, Markovski M, Bernhardt TG (2011) An ATP-binding cassette transporter-like complex governs cell-wall hydrolysis at the bacterial cytokinetic ring. Proc Natl Acad Sci U S A 108:E1052–E1060
Yang DC, Tan K, Joachimiak A, Bernhardt TG (2012) A conformational switch controls cell wall-remodelling enzymes required for bacterial cell division. Mol Microbiol 85:768–781
Young KD (2001) Approaching the physiological functions of penicillin-binding proteins in Escherichia coli. Biochimie 83:99–102
Yousif SY, Broome-Smith JK, Spratt BG (1985) Lysis of Escherichia coli by beta-lactam antibiotics: deletion analysis of the role of penicillin- binding proteins 1A and 1B. J Gen Microbiol 131:2839–2845
Yu XC, Margolin W (1999) FtsZ ring clusters in min and partition mutants: role of both the Min system and the nucleoid in regulating FtsZ ring localization. Mol Microbiol 32:315–326

Chapter 3
Cell Cycle Machinery in *Bacillus subtilis*

Jeff Errington and Ling Juan Wu

Abstract *Bacillus subtilis* is the best described member of the Gram positive bacteria. It is a typical rod shaped bacterium and grows by elongation in its long axis, before dividing at mid cell to generate two similar daughter cells. *B. subtilis* is a particularly interesting model for cell cycle studies because it also carries out a modified, asymmetrical division during endospore formation, which can be simply induced by starvation. Cell growth occurs strictly by elongation of the rod, which maintains a constant diameter at all growth rates. This process involves expansion of the cell wall, requiring intercalation of new peptidoglycan and teichoic acid material, as well as controlled hydrolysis of existing wall material. Actin-like MreB proteins are the key spatial regulators that orchestrate the plethora of enzymes needed for cell elongation, many of which are thought to assemble into functional complexes called elongasomes. Cell division requires a switch in the orientation of cell wall synthesis and is organised by a tubulin-like protein FtsZ. FtsZ forms a ring-like structure at the site of impending division, which is specified by a range of mainly negative regulators. There it recruits a set of dedicated division proteins to form a structure called the divisome, which brings about the process of division. During sporulation, both the positioning and fine structure of the division septum are altered, and again, several dedicated proteins that contribute specifically to this process have been identified. This chapter summarises our current understanding of elongation and division in *B. subtilis*, with particular emphasis on the cytoskeletal proteins MreB and FtsZ, and highlights where the major gaps in our understanding remain.

Keywords Bacillus • *B. subtilis* • MreB • MreB homologues • Bacterial cell shape • Helical filaments • Circumferential motion • Cell elongation machinery • Peptidoglycan synthesis • PG • Divisome • Min system • MinJ • FtsZ • Z ring • Sporulation • SpoIIE • L-form bacteria

J. Errington (✉) • L.J. Wu
Centre for Bacterial Cell Biology, Institute for Cell and Molecular Biosciences, Faculty of Medical Sciences, Newcastle University, Newcastle upon Tyne NE2 4AX, UK
e-mail: jeff.errington@ncl.ac.uk

J. Löwe, L.A. Amos (eds.), *Prokaryotic Cytoskeletons*, Subcellular Biochemistry 84, DOI 10.1007/978-3-319-53047-5_3

Introduction to *B. subtilis*

Bacillus subtilis is an aerobic, Gram positive, endospore forming bacterium of the phylum Firmicutes. It is by far the best characterised Gram positive organism and basic knowledge about *B. subtilis* is frequently used to guide and inform thinking about other Gram positive organisms. Historically, interest in *B. subtilis* was based largely on three features of its biology: early success in achieving natural transformation with linear DNA, which greatly facilitated genetic analysis of the organism (Anagnostopoulos and Spizizen 1961); its ability to form endospores, which was used as a simple model for cellular development and differentiation (Errington 1993, 2003; Tan and ramamurthi 2014); and industrial interest in its prodigious ability to secrete certain valuable hydrolytic enzymes (e.g. proteases and amylases) directly into the growth medium (Pohl and Harwood 2010).

The biggest driver for study of *B. subtilis*, at least in the 1960s to 1990s, was probably interest in endospore (spore) formation (Fig. 3.1). Sporulation of *B. subtilis* is triggered essentially by nutrient stress. The process begins with a modified, highly asymmetric cell division. This generates a small prespore (sometimes called forespore) cell, destined to become the mature endospore, and a much larger mother cell. The mother cell engulfs the prespore, forming a cell within a cell. The two cells then cooperate in a complex developmental process in which the prespore becomes highly differentiated and covered in protective layers. Eventually, the mother cell lyses to release the now dormant endospore. Endospores are incredibly resistant and can remain dormant for extremely long periods of time, before germinating in response to specific chemical signal molecules (germinants). The process of sporulation in *B. subtilis* is now understood in great detail (Errington 1993, 2003; Tan and Ramamurthi 2014).

Research on spore formation contributed considerably to the development of methods for studying the sub-cellular distribution of proteins and other important macromolecules in bacteria, laying the foundations for modern bacterial cell biology (Shapiro and Losick 2000; Errington 2003). These imaging methods, together with the exceptionally powerful molecular genetics of *B. subtilis*, stimulated a new era of studies on the cell cycle and cell morphogenesis. FtsZ, a tubulin homologue that is the key player in bacterial cell division, and MreB, an actin homologue that governs cell shape in many rod shaped bacteria, will be the main topics for discussion in this chapter. As the main focus of the chapter lies on *B. subtilis*, reference to work on other bacteria will be limited to situations where the contrast or additional information is helpful. For more detail on the *E. coli* system and on FtsZ and MreB proteins generally, the reader is directed to Chaps. 2, 5, 7 and 8.

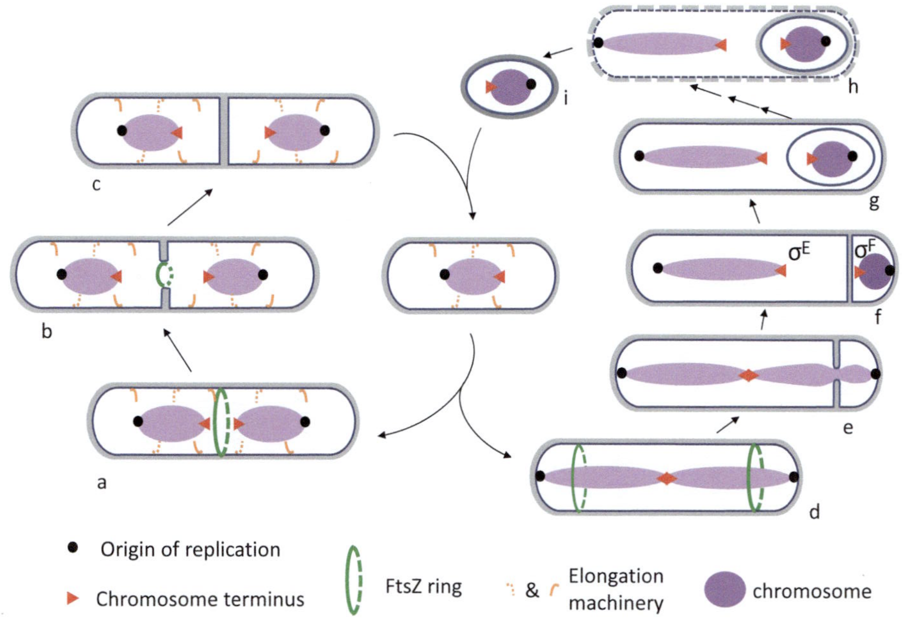

Fig. 3.1 *B. subtilis* cell cycle. The *left half* represents the vegetative cycle, where a new born cell (*centre*) grows in length, controlled by the elongation machinery (*orange curved lines*), in the meantime the chromosome is replicated and FtsZ (*green ring*) assembles between segregated chromosomes at mid cell (*a*). As cell division progresses and septum grows inwards, the Z ring contracts (*b*). Upon completion of septation, which generates two identical daughter cells, the Z ring and the divisome disassemble (*c*) and the dividing wall splits to allow separation of the new born cells (*centre*). Under starvation conditions *B. subtilis* cells initiate spore development (*right half*). Instead of segregating the replicated chromosomes to quarter positions, the sister chromosomes undergo a conformational change to form an elongated structure called the axial filament, which extends from pole to pole. FtsZ assembles at two sub-polar positions, one at each cell pole (*d*). Only one of the two Z rings develops into a septum, which forms over one of the chromosomes (*e*). Following the completion of asymmetric septation, which generates two unequal sized daughter cells, the small prespore or forespore and the large mother cell, transcription factor Sigma F (σ^F) is activated in the prespore and the remaining part of the prespore chromosome is segregated. Activation of σ^F in the prespore leads to the activation of Sigma E (σ^E) in the mother cell (*f*). The different programmes of gene expression in the prespore and the mother cell direct the engulfment of the prespore by the mother cell (*g*). Finally, the spore undergoes maturation, and the mother cell lyses (*h*) to release a highly resistant, dormant spore (*i*), which can germinate and start growing (*centre*) when nutrients become available

MreB and the Cell Elongation Machinery

Organization of the B. subtilis *Cell Wall*

Peptidoglycan (PG) is the major component of virtually all bacteria (Typas et al. 2012). It comprises a single huge macromolecule that covers the entire surface of the cell. Lying just outside the cytoplasmic membrane it acts as a protective layer but it also constrains the membrane against the outward turgor pressure imposed by the high osmolarity of the cytoplasm. PG is of considerable practical significance as its synthesis is the target for many useful antibiotics, and fragments of the wall are recognised by innate immune receptors during infection. The PG contributes to the shape of the cell but has no intrinsic 3D shape, so it must be sculpted by synthetic enzymes into the correct form.

PG is composed of long glycan strands with alternating N-acetylglucosamine and N-acetylmuramic acid sugars, cross linked by peptide bridges made up of a mixture of L- and D-amino acids (De Pedro and Cava 2015). The precursor for PG synthesis, called lipid II, is a disaccharide pentapeptide coupled to a C_{55} isoprenoid lipid (bactoprenol) and is synthesised in the cytosol by a well characterised series of enzymes. Lipid II is flipped to the exterior and assembled into the existing cell wall sacculus by a multiplicity of synthetic enzymes called penicillin-binding proteins (PBPs), which possess the glycosyltransferase and transpeptidase activities needed to extend the glycan strands and create peptide cross bridges (Lovering et al. 2012; Scheffers and Tol 2015). Recently the RodA protein was identified as a possible monofunctional glycosyltransferase (Meeske et al. 2016 and Emami et al. 2017). Extracellular autolytic enzymes are required to allow expansion of the wall by breaking bonds in pre-existing material. Their activities need to be tightly regulated to enable controlled expansion of the wall during growth, while avoiding potentially catastrophic turgor-driven lysis (Vollmer et al. 2008).

Gram positive bacteria lack the outer membrane characteristic of Gram negatives. However, Gram positive walls typically contain a second major class of polymers called teichoic acids (TAs) (Sewell and Brown 2014; Percy and Grundling 2014). In many Gram positives there are two major forms: wall teichoic acids (WTA), which are covalently linked to the PG; and lipoteichoic acids (LTA), which are coupled to a lipid carrier. In *B. subtilis* WTA and LTA have the same general composition, of poly-[glycerol-phosphate]. TAs have been implicated in many functions. Metal homeostasis is probably a central role – scavenging divalent cations and maintaining surface charge.

B. subtilis cells have a characteristic morphology; in essence, an elongated regular cylindrical tube with hemispherical poles. Growth occurs by elongation along the long axis of the cell, and division occurs approximately each time a doubling in length occurs. The cells have a typical diameter of about 850 nm, irrespective of the growth rate. Changes in growth rate are accommodated by alterations in cell length, with faster growing cells (average up to about 5 μm) being longer than slower growing cells (average minimally about 2 μm) (Sharpe et al. 1998).

Cell shape determination is a central problem in biology. In *B. subtilis* and many other rod-shaped bacteria, shape is thought to be determined and maintained by the action of "cytoskeletal" proteins of the MreB family (see below). These proteins are structurally and biochemically related to eukaryotic actins. Like actin proteins, they undergo reversible polymerization, which is regulated by binding and hydrolysis of ATP. However, recent work has highlighted certain intrinsic differences, most notably the tight direct association of MreB polymers with the cytoplasmic membrane (Salje et al. 2011). Mutations in the *mreB* genes of many bacteria abolish proper cell shape control. The possibility that they have a direct role in cell shape determination and or maintenance comes in part from the analogy with shape control by actin filaments in eukaryotes but also from the view that the organization of a large scale (μm) structure requires long range topographical instructions, such as might be provided by elongated MreB filaments. Early experiments examining the localization of MreB proteins in *B. subtilis* provided strong support for this view through the observation of elongated helical filaments that appeared to wrap around the long axis of the cell (Jones et al. 2001). However, the significance of these elongated filaments and even their existence have been the subject of much recent debate (see below).

B. subtilis *Has Three Actin Like MreB Homologues*

The *mreB* gene was first defined by mutations altering cell shape in *E. coli* (Wachi et al. 1987). Early genome sequence analyses revealed that *B. subtilis* has three *mreB* paralogous genes (Levin et al. 1992; Varley and Stewart 1992; Abhayawardhane and Stewart 1995). The gene designated *mreB* has an equivalent chromosomal location to that of *mreB* genes of most other bacteria, in lying immediately upstream of homologues of *mreC* and *mreD* genes that are also involved in cell elongation. The other homologues – *mbl* (MreB like) and *mreBH* (MreB homologue) are located in distant parts of the chromosome. Early mutational studies of *mreB* and *mbl* were hampered by the presence of the important *mreC* and *mreD* genes downstream from *mreB* and the lethal nature of the mutations. The latter problem was simplified by the finding that for both paralogues, the viability of null mutants can be rescued by addition of high concentrations (e.g. 20 mM) of Mg^{2+} to the culture medium, for reasons that are not clear (Formstone and Errington 2005). To summarise the results of several papers, the three paralogues appear to have partially redundant functions, and overexpression of any one of the genes can enable growth and reasonably normal morphology of an otherwise triple null mutant. Single mutations tend to have subtly different effects on morphology: *mreB* mutants have an increased diameter but remain able to grow in a straight line; *mbl* mutants are highly twisted with some bulging and lysis; *mreBH* mutants have a narrow cell phenotype, especially under low Mg^{2+} conditions (Jones et al. 2001; Carballido-López et al. 2006; Kawai et al. 2009a; Defeu Soufo and Graumann 2006).

Interestingly, Schirner and Errington (2009) found that upregulation of an ECF sigma factor (σ^I) involved in cell envelope stress was sufficient to suppress the lethality that normally occurs in attempting to construct an *mreB*, *mbl*, *mreBH* triple mutant. The suppressed mutant grew well (though requiring high Mg^{2+}) but with a spherical morphology. Thus, it seems that, as in many other organisms, MreB proteins play a crucial role in rod-shape morphogenesis and cylindrical cell wall extension.

Filaments, Foci and Movement

Early imaging experiments with all three *B. subtilis* MreB family proteins appeared to reveal extended helical filaments localized close to the cell periphery and thus presumably close to the inner surface of the cytoplasmic membrane (Jones et al. 2001). All three MreB family proteins exhibited roughly similar patterns of localization and seemed to co-localize, at least under some conditions (Carballido-López et al. 2006; Defeu Soufo and Graumann 2006). The localization of various cell elongation proteins (Leaver and Errington 2005; Formstone et al. 2008) and the use of labelling methods designed to identify nascent cell wall synthesis (Daniel and Errington 2003; Tiyanont et al. 2006) also seemed compatible with a helical mode of wall synthesis. Some experiments suggested a degree of remodelling or active movement of the filaments during cell elongation (Carballido-López and Errington 2003; Defeu Soufo and Graumann 2004). The filamentous helical view of MreB localization was exciting because, in principle, the filaments could act as a geometric guide for the synthetic machinery directly defining the morphology of rod-shaped cells.

In 2011, however, three groups described experiments that revealed the circumferential movement of relatively short filaments or foci (rather than long filaments) of MreB proteins in *B. subtilis* (Garner et al. 2011; Domínguez-Escobar et al. 2011) and *E. coli* (Van Teeffelen and Gitai 2011). Importantly, the circumferential movement was dependent on active cell wall synthesis, suggesting that MreB follows rather than leads the synthetic process. The problem with the short filament or patchy view of MreB organization lies in the question of what its role is in cell elongation. Domínguez-Escobar et al. (2011) proposed that MreB acts as a scaffold to help assemble the complex of proteins needed to coordinate the synthesis of PG, WTA and other wall associated elements. The complex was proposed to use existing glycan strands as a template for insertion of new material. However, this presumably only works while the template strands have the correct geometry and could not account for the long term fidelity of shape maintenance and especially the restoration of shape in cells with any shape abnormality.

Meanwhile, electron cryotomography failed to detect elongated filaments of native MreB in flash frozen intact *E. coli* cells (Swulius et al. 2011), and demonstrated that the very prominent helical cytoplasmic filaments made by one particular *E. coli* MreB/GFP fusion protein were an artefact specific to that fusion protein

(Swulius and Jensen 2012). However, Salje et al. (2011) potentially solved the conundrum of the missing filaments of native protein in the cryo-EM experiments through the discovery that MreB protein polymers have a high affinity for membranes and in vivo this tight membrane association would likely hide the filaments from detection in the low contrast cryo-EM images. This also explained the problem with the *E. coli* GFP-MreB fusion protein mentioned above because the membrane targeting sequence of *E. coli* MreB is close to the N-terminus and thus probably occluded by the GFP fusion. Meanwhile, Dempwolff et al. (2011) obtained supporting evidence for membrane association of MreB and Mbl, though interestingly not MreBH, by expressing GFP fusions of the 3 proteins in eukaryotic cells.

The most recent in vivo imaging experiments on *B. subtilis*, using various super-resolution methods may have resolved some of the confusion around MreB localization, by demonstrating that the proteins (at least MreB and Mbl) are able to form relatively extended helical filaments, at least under some conditions, but that the whole filament systems undergo overall near circumferential movement, which is dependent on (and presumably driven by) PG synthesis (Reimold et al. 2013; Olshausen et al. 2013). Olshausen et al. (2013) suggested that the elongated filaments could serve to coordinate the synthetic activities of multiple wall synthetic complexes, providing a plausible mechanistic explanation for the function of long MreB filaments.

Two recent lines of work on the *E. coli* MreB system suggest that MreB may have more than one mode of action in cell shape regulation, which may clarify some of the conflicting data described previously. First, Ursell et al. (2014) and Billings et al. (2014) found that MreB filaments or patches have affinity for regions of a particular (aberrant) curvature. Recruitment of the cell wall machinery to those sites could help to correct the curvature leading to the restoration of shape. Morgenstein et al. (2015) then discovered that in certain mutational backgrounds, MreB motion is not required for maintenance of a rod shape. The authors proposed that MreB can help specify cell shape by two distinct mechanisms: first, by a motion-independent mechanism that relies on recruitment of the synthetic machinery to sites of inappropriate curvature, effectively a "repair" mechanism; and second, by a motion-dependent mechanism that helps distribute synthesis over a greater proportion of the surface and is used when cells are growing rapidly and presumably are relatively unperturbed.

The significance of movement and the possible roles of filaments in shape determination have been reviewed in detail recently (Errington 2015).

A Complex Web of Interactions Between MreB Proteins and Cell Wall Effectors

The models discussed above mainly assume that MreB proteins work by controlling the spatial activity of the enzymes responsible for cell wall expansion. This view is supported by numerous reports describing interactions between MreB and various components of the cell wall machinery. The list of possible MreB interacting proteins identified by various methods as candidate members of the *B. subtilis* elongation machine or "elongasome" are summarised in Table 3.1. Similar collections of interacting proteins have been identified for several other rod-shaped organisms, particularly *E. coli* and *Caulobacter crescentus* (reviewed by (Errington 2015). The methods used to identify these proteins include various indirect approaches, such as co-localization, localization dependence, and genetic epistasis, as well as more direct methods of bacterial or yeast 2-hybrid approaches and biochemical pull downs. Although some of the data are not entirely convincing, for example, the "helix-like" localization patterns, taken together, these methods point to the existence of large elongasome complexes containing multiple proteins involved in synthesis of PG and WTA, as well as potentially extracellular factors, such as autolytic enzymes or their regulators. At the moment, little is known about the stoichiometry of these complexes or about whether they are stable or transient.

The Future

Much remains to be learned about the detailed functions of the MreB family proteins of *B. subtilis*. It is by no means clear why *B. subtilis* possesses three paralogous genes. At one level, it reflects the general complexity of the cell wall synthetic machinery of the organism. Thus, it also carries multiple copies of many other synthetic genes, including, for example: 4 class A (TPase and GTase) PBPs (Popham and Setlow 1996), 3 LTA synthases (Grundling and Schneewind 2007a, b; Schirner et al. 2009), 3 WTA transferases (Kawai et al. 2011), and at least 2 families of lipid II flippases (Meeske et al. 2015). The overlapping semi-redundant functions of the 3 MreB proteins may reflect that they interact differentially with subsets of cell envelope proteins in order to adapt cell envelope properties to changing environmental conditions. Perhaps "chemical warfare" between organisms in complex and highly competitive environments such as soil, drives adaptability in cell envelope synthesis and organization. Although some aspects of the differential activities of the 3 MreB proteins are beginning to be worked out (Carballido-López et al. 2006; Domínguez-Cuevas et al. 2013), much more probably remains to be elucidated.

An important related question concerns how the many interactions between MreB proteins and the various other components of the cell envelop synthetic machinery (PG synthases, PBPs, autolysins, WTA synthases, *etc*) are mediated,

Table 3.1 Possible MreB interacting poteins identified as candidate components of the elongasome

Protein[a]	MW (kDa)	Localization[b]	Comment	References
MreC	32	I/E	Cell shape function. Encoded by gene immediately downstream from *mreB*	Leaver and Errington (2005), Kawai et al. (2009b), Garner et al. (2011) and Domínguez-Escobar et al. (2011)
MreD	19	I	Cell shape function. Encoded by gene immediately downstream from *mreBC*	Leaver and Errington (2005), Garner et al. (2011), Domínguez-Escobar et al. (2011) and Muchova et al. (2013)
RodZ	23	I/C	Required for normal cell shape	Domínguez-Escobar et al. (2011) and Muchova et al. (2013)
CwlO	50	E	Autolytic enzyme, regulated by FtsEX	Domínguez-Cuevas et al. (2013)
LytE	37	E	Autolytic enzyme. Export regulated by MreBH?	Carballido-López et al. (2006)
FtsE	25	C	ABC-transporter (ATP-binding protein). With FtsX regulates CwlO. Controlled specifically by Mbl?	Domínguez-Cuevas et al. (2013)
FtsX	32	I	ABC-transporter (membrane protein). With FtsE regulates CwlO. Controlled specifically by Mbl?	Domínguez-Cuevas et al. (2013)
PBP 1	99	E	Major bifunctional PBP. Important for both cell elongation and division	Van Den Ent et al. (2006) and Kawai et al. (2009a, b)
PBP 2A	79	E	Major TPase with specific role in elongation. Partially redundant to PBP H	Van Den Ent et al. (2006), Kawai et al. (2009a, b), Garner et al. (2011) and Domínguez-Escobar et al. (2011)
PBP 2B	79	E	Major TPase with specific role in division	Van Den Ent et al. (2006) and Kawai et al. (2009b)
PBP 2C	79	E	Bifunctional PBP with unknown function	Van Den Ent et al. (2006) and Kawai et al. (2009b)
PBP 2D	71	E	Transpeptidase with unknown function	Van Den Ent et al. (2006) and Kawai et al. (2009b)

(continued)

Table 3.1 (continued)

Protein[a]	MW (kDa)	Localization[b]	Comment	References
PBP 3	74	E	Accessory TPase that can rescue cell division in the absence of PBP 2B activity	Kawai et al. (2009b)
PBP 4	70	E	Bifunctional PBP with unknown function	Kawai et al. (2009a, b)
PBP H	76	E	Major TPase with specific role in elongation. Partially redundant to PBP 2A	Van Den Ent et al. (2006), Kawai et al. (2009b), Domínguez-Escobar et al. (2011)
PBP I	65	E	TPase of unknown function.	Van Den Ent et al. (2006) and Kawai et al. (2009b)
RodA	43	I	PG synthesis. Possible monofunctional GTase	Domínguez-Escobar et al. (2011), Meeske et al. (2016), Emami et al. (2017)
DapI	41	C	N-acetyl-diaminopimelate deacetylase. PG synthesis	Rueff et al. (2014)
TagA	29	C	Teichoic acid synthesis. UDP-N-acetyl-D-mannosamine transferase	Formstone et al. (2008)
TagB	44	C	Teichoic acid synthesis. Putative CDP-glycerol:glycerol phosphate glycerophosphotransferase	Formstone et al. (2008)
TagF	87	C	Teichoic acid synthesis. CDP-glycerol:polyglycerol phosphate glycero-phosphotransferase	Formstone et al. (2008)
TagG	32	I	ABC transporter for teichoic acid translocation (permease)	Formstone et al. (2008)
TagH	59	C	ABC transporter for teichoic acid translocation (ATP-binding protein)	Formstone et al. (2008)
TagO	39	C	Teichoic acid synthesis. Undecaprenyl-phosphate-GlcNAc-1-phosphate transferase	Formstone et al. (2008)
TagT	35	E	Transfer of anionic cell wall polymers from lipid-linked precursors to peptidoglycan	Kawai et al. (2011)
TagU	34	E	Transfer of anionic cell wall polymers from lipid-linked precursors to peptidoglycan	Kawai et al. (2011)
YvcK	34	C	Required for normal localization of PBP 1	Foulquier et al. (2011)
GpsB	11	C	Regulation of PBP 1 localization, especially its switch between elongation and division sites.	Claessen et al. (2008)
EF-Tu	43	C	Translation elongation factor	Defeu Soufo et al. (2015)

[a]In addition to the above, Kawai et al. (2011) identified many additional MreB-associated proteins by pull-down mass spectrometry

[b]*I* integral membrane, *E* extracellular, *C* cytoplasmic

particularly whether they are static or dynamic and the extent to which they are hierarchical and mutually permissive or exclusive.

A final major question concerns the localization and dynamic properties of the proteins. What conditions determine the length of the MreB filaments and how do length and movement relate to the various problems associated with cell shape determination, maintenance and repair?

There is a sense that the array of analytical methods we now possess, enabling us to localise proteins with increasing temporal and spatial resolution and to define the components of protein complexes and their stoichiometry, should allow details of the machinery and mechanisms to be resolved. Complexity may be the biggest barrier to progress.

FtsZ and the Cell Division Machinery

Most bacteria with a PG wall divide by directing the ingrowth of a sheet of wall material that eventually forms the new hemispherical poles of the daughter cells. In almost all bacteria, the key cytoskeletal protein involved in defining the site of division and then orchestrating the process is called FtsZ, which is structurally and biochemically homologous to tubulin (Löwe and Amos 1998). In bacteria where the process has been studied in detail, FtsZ appears to form a circumferential ring that defines the site of cell division (Bi and Lutkenhaus 1991). It also serves to recruit, directly or indirectly, multiple protein components of a division machine, sometimes called the "divisome" (Adams and Errington 2009; Egan and Vollmer 2013). Several divisome associated proteins might also be considered as cytoskeletal proteins (e.g. FtsA, DivIVA, MinD; see below), depending on the definition. *B. subtilis* is an interesting model for the study of bacterial cell division because it has two contrasting modes of division: a "conventional mode", carried out by vegetatively growing cells; and a modified, highly asymmetric division undertaken by sporulating cells.

Biochemical Properties of FtsZ

The presence of a tubulin GTP-binding signature motif in FtsZ (GGGTGTG) was first reported in the early 1990s (Raychaudhuri and Park 1992; De Boer et al. 1992; Mukherjee and Lutkenhaus 1994). Crystallographic studies confirmed the near congruence of the structures of FtsZ and tubulin proteins (Löwe and Amos 1998). Not surprisingly, the proteins also have similar biochemical properties. Like tubulin, FtsZ assembles in vitro in a head to tail fashion to form single stranded protofilaments, which can further assemble into bundles, sheets or rings. The protofilaments are also highly dynamic and go through cycles of turnover/polymerization, regulated by the binding and hydrolysis of GTP. See Chapter 5 for a detailed description of FtsZ polymerization dynamics.

FtsZ Visualization During Growth and Sporulation of B. subtilis

The ability of FtsZ to form tubulin-like protofilaments and protofilament bundles raised important questions about the abundance, assembly and dynamics of the protein in vivo. Estimations of protein abundance have suggested about 2000–6000 molecules per cell in *B. subtilis* (Feucht et al. 2001; Ishikawa et al. 2006; Muntel et al. 2014), giving a protein concentration of about 2–6 μM, well above the in vitro critical concentration for assembly (e.g. 0.72 μM for *E. coli* FtsZ; Chen et al. 2012). Also, given a protofilament subunit repeat length of 4.3 nm (Oliva et al. 2004), there is enough FtsZ to circumnavigate the cell about 3–10 times.

Several labs have investigated the localization of FtsZ in *B. subtilis*. Wang and Lutkenhaus (1993) used immunogold electron microscopy to demonstrate association of FtsZ with the leading edge of the invaginating cell division structure. The immunofluorescence studies of (Levin and Losick 1996) confirmed the presence of FtsZ bands (presumed to be rings) at the expected position near mid cell in vegetative cells but also unexpectedly, near both poles of early sporulating cells. This turns out to be a key feature of the mechanism used by *B. subtilis* to achieve asymmetric cell division during sporulation, which will be discussed in detail below. Several reports have highlighted the possible role of helical FtsZ structures as intermediates in assembly or constriction at cell division sites (Ben-Yehuda and Losick 2002; Feucht and Errington 2005; Peters et al. 2007; Strauss et al. 2012). Helical FtsZ structures are most prominent during the transition from vegetative growth to sporulation in *B. subtilis*, during which the site of division shifts from mid cell to near the pole. The mid cell Z ring appears to transform into a helix which grows length-wise, before breaking down into two short helices, one near each pole. Each helix then coalesces into a ring (Ben-Yehuda and Losick 2002). Therefore, each sporulating cell assembles two Z rings, one at each pole, but only one develops into a septum (see below). Several mutations in *ftsZ* have been described that promote a tendency to form spiral Z rings and similarly shaped division events (Feucht and Errington 2005; Michie et al. 2006), suggesting that the helical configuration has functional relevance. Peters et al. (2007) also described helical configurations in vegetative cells, based on modified immunofluorescence imaging methods. On the other hand, several higher resolution imaging methods, including super-resolution fluorescence imaging and cryo-EM of *B. subtilis* and other organisms have suggested that FtsZ rings may be more complex, and beaded or discontinuous (Jennings et al. 2011; Strauss et al. 2012; Li et al. 2007; Min et al. 2014).

Dynamic movement of FtsZ rings has been observed by time-lapse imaging (Strauss et al. 2012) but more quantitative and perhaps surprising information came from fluorescence recovery after photobleaching (FRAP) experiments. Erickson and colleagues established that FtsZ subunits in Z rings, either pre-constriction or during constriction, turn over with a half time of about 8 seconds (Anderson et al. 2004). This emphasises the likely importance of dynamics in FtsZ function but also, the difficulty in imaging such structures with the need for both high spatial and temporal resolution. Löwe's lab have recently described compelling evidence for the formation of regular circumferential bands of FtsZ in which the protofilaments are connected by regular lateral contacts for 2 Gram negative bacteria, *E. coli* and *Caulobacter crescentus*, as well as in an in vitro system. These observations support a model for constriction involving filament sliding (Szwedziak et al. 2014). It remains to be seen whether this model can be extended to *B. subtilis* and other Gram positive bacteria, but it seems unlikely that the fundamental features of FtsZ function in bacterial division are not well conserved.

The B. subtilis *Divisome*

The *B. subtilis* divisome has been studied in considerable detail: some properties of the proteins thought to contribute directly or indirectly to divisome function in this organism are described in Table 3.2. These proteins have been identified through homology to known division proteins in other organisms, by biochemical pull downs or through various genetic screens.

Imaging experiments suggest that the divisome assembles in at least two distinct steps (Gamba et al. 2009). In the first step, which seems to involve mainly cytosolic factors, a "ring" of FtsZ protein assembles, in parallel with the recruitment of "early" divisome proteins FtsA, SepF, ZapA and EzrA. After a delay representing about 20% of the cell cycle, the second step of assembly takes place, in which the "late" proteins are recruited. These are mainly proteins with major extracellular domains or integral membrane proteins. Various regulatory proteins, including GpsB, DivIVA, MinJ, MinD and MinC arrive at about the same time or slightly later, possibly being dependent on initiation of membrane or PG ingrowth. Ishikawa et al. (2006) detected interactions between the various early proteins in a series of biochemical pull-down experiments.

Three "early" cytosolic proteins appear to promote the formation of a functional Z ring in *B. subtilis* – FtsA, SepF and ZapA. FtsA was identified by its conserved location immediately upstream of and adjacent to FtsZ (Beall et al. 1988). Unlike *E. coli*, *ftsA* mutants of *B. subtilis* are viable, though they are substantially deficient in division (Beall and Lutkenhaus 1992). FtsZ still localizes at regular intervals but most of the Z rings are abnormal, often appearing as multiple diffuse bands rather than one clear, strong band (Jensen et al. 2005). Purified *B. subtilis* FtsA binds and hydrolyses ATP (Feucht et al. 2001) but little more work has been done on this protein so far. Using the *Thermotoga maritima* protein, Löwe and colleagues have demonstrated that even though FtsA has a different subdomain architecture to actin, the protein can form canonical actin-like protofilaments in vitro (Van Den Ent and Löwe 2000; Szwedziak et al. 2012). FtsA interacts specifically with the C-terminal domain of FtsZ. Despite a great deal of work over nearly 2 decades, little is known

Table 3.2 Proteins of the *B. subtilis* divisome and its regulators

Protein	MW (kDa)	Location[a]	Comments	Key references
FtsZ	40	C	Tubulin-like protein. Assembles into protofilaments and higher order structures to generate the "Z ring" at the division site. Recruits other divisome proteins to the ring.	Beall et al. (1988), Beall and Lutkenhaus (1991), and Wang and Lutkenhaus (1993)
FtsA	48	C	Actin / HSP70 superfamily ATPase. Dimerises and can form higher order structures. C-terminal amphipathic helix promotes membrane association. Direct interaction with FtsZ, which contributes to membrane association of the Z ring.	Beall and Lutkenhaus (1991), Feucht et al. (2001), Jensen et al. (2005) and Ishikawa et al. (2006)
SepF	17	C	Forms regular 50 nm diameter rings in vitro and interacts directly with FtsZ in vitro, promoting FtsZ bundling. Membrane targeting domain contributes to membrane association of the Z ring.	Hamoen et al. (2006) and Gündoğdu et al. (2011)
ZapA	9.0	C	Widely conserved protein that promotes Z ring formation by direct interaction with FtsZ.	Gueiros-Filho and Losick (2002)
EzrA	65	C	N-terminal transmembrane anchor. Cytosolic domain has a spectrin-like fold. Interacts with FtsZ, contributing to membrane association of the Z ring. Additional role in cell elongation via interactions with PBP 2B and GpsB.	Levin et al. (1999), Haeusser et al. (2004), Claessen et al. (2008), and Cleverley et al. (2014)
GpsB	11	C	DivIVA-related protein involved in both cell elongation and cell division. Interacts with the major PG synthase, PBP 1, and thought to be involved in shuttling of this protein between elongation and division complexes. Synthetic lethal in combination with *ftsA* mutation. Synthetic "sick" in combination with *ezrA*. EzrA-SepF interaction probably important for shuttling.	Claessen et al. (2008) and Tavares et al. (2008)

(continued)

Table 3.2 (continued)

Protein	MW (kDa)	Location[a]	Comments	Key references
FtsL	13	E	Bitopic membrane protein with short extracytoplasmic coiled-coil-like domain. Target of several cell division regulatory mechanisms. Unstable protein subject to degradation by a regulated intramembrane proteolysis (RIP) process involving YluC protease. Stability also regulated by interactions with DivIC and DivIB.	Daniel et al. (1998), Daniel and Errington (2000), Sievers and Errington (2000a, b), Kawai and Ogasawara (2006), Bramkamp et al. (2006) and Daniel et al. (2006)
DivIB	30	E	Bitopic membrane protein with large extracellular domain. Structural data from other organisms suggests two domains, one of which resembles the POTRA domain often involved in protein protein interactions. Complex pattern of interactions with FtsL and DivIC. Homologue called FtsQ in *E. coli*.	Beall and Lutkenhaus (1989), Harry and Wake (1989, 1997), Katis and Wake (1999), Katis et al. (2000), Daniel and Errington (2000) and Daniel et al. (2006)
DivIC	15	E	Bitopic membrane protein with short extracytoplasmic coiled-coil-like domain. Interacts with FtsL and DivIB. Likely homologue confusingly called FtsB in *E. coli*.	Katis et al. (1997), Katis and Wake (1999), Katis et al. (2000), Sievers and Errington (2000b), Robson et al. (2002) and Daniel and Errington (2000)
FtsW	44	I	Integral membrane protein closely related to RodA involved in cell elongation.	Lu et al. (2007)
Pbp2B	79	E	Penicillin binding protein. Monofunctional (class B) transpeptidase specifically required for cell division.	Yanouri et al. (1993), Daniel et al. (1996) and Daniel and Errington (2000)
DivIVA	19	C	Coiled coil protein with weak similarity to eukaryotic tropomyosins. Targeted to division sites and cell poles at least in part by sensing membrane curvature. Membrane interaction through conserved N-terminal domain containing essential tryptophan residue. Involved in a range of cell pole associated functions in Gram positive bacteria.	Cha and Stewart (1997), Edwards and Errington (1997), Hamoen and Errington (2003) and Lenarcic et al. (2009), Ramamurthi and Losick (2009) and Van Baarle et al. (2013)

(continued)

Table 3.2 (continued)

Protein	MW (kDa)	Location[a]	Comments	Key references
MinC	25	C	Widely conserved division inhibitor acting on FtsZ and possibly other steps in division.	Reeve et al. (1973), Levin et al. (1992), Marston and Errington (1999) and Gregory et al. (2008)
MinD	29	C	Widely conserved indirect division inhibitor that works by spatial regulation of MinC protein. Poorly characterised additional role in chromosome segregation during sporulation.	Reeve et al. (1973), Levin et al. (1992), Marston et al. (1998) and Marston and Errington (1999), Kloosterman et al. (2016)
MinJ	44	I / C	PDZ-domain protein targeted to cell poles by interaction with DivIVA (at least). Required for correct spatial localization of the MinCD complex and thus the regulation of cell division.	Patrick and Kearns (2008), Bramkamp et al. (2008) and Van Baarle and Bramkamp (2010)
Noc	33	C	Site-specific DNA binding protein. Inhibitor of division. Major factor effecting nucleoid occlusion.	Wu and Errington (2004), Wu et al. (2009) and Adams et al. (2015)
WhiA	36	C	Enigmatic nucleoid associated factor. *whiA* mutation causes severe filamentation when combined with *zapA*, *ezrA* or various regulatory proteins of cell division.	Surdova et al. (2013)
SpoIIE	92	C/I	Bifunctional sporulation-specific protein. C-terminal kinase domain regulates prespore-specific gene expression. C-terminal domain required for efficient switch in cell division position from mid cell to sub-polar position, probably via a direct interaction with FtsZ.	Arigoni et al. (1995), Feucht et al. (1996), Wu et al. (1998), Lucet et al. (2000), Carniol et al. (2005) and Bradshaw and Losick (2015)
MciZ	4.0	C	Mother cell-specific inhibitor of FtsZ assembly. Caps FtsZ protofilaments at the "minus" end.	Handler et al. (2008) and Bisson-Filho et al. (2015)
RefZ	24	C	Site-specific DNA-binding protein that contributes to precise relative positioning of chromosome and asymmetric division site during sporulation.	Wagner-Herman et al. (2012) and Miller et al. (2015)

[a]*C* cytosolic, *I* integral membrane, *E* extracytoplasmic

about the precise function of *ftsA* other than that it can form high MW dynamic complexes of various kinds with FtsZ (e.g., Loose and Mitchison 2014). Perhaps its best defined function lies in membrane association, which occurs through a C-terminal amphipathic helix (Pichoff and Lutkenhaus 2005) and enables the protein to anchor the Z ring to the membrane (Szwedziak et al. 2014). Interestingly, this interaction with the membrane is strongly dependent on the membrane potential (Strahl and Hamoen 2010).

Genetic and biochemical experiments suggest that SepF protein provides a second membrane anchor for the Z ring in Gram positive bacteria. *sepF* was discovered simultaneously in two labs by different methods. Ishikawa et al. (2006) identified the SepF protein in pull-down experiments using FtsZ, FtsA, EzrA and ZapA as bait. Yeast 2-hybrid experiments detected the formation of a SepF-SepF self interaction, as well as an interaction with FtsZ. Meanwhile, Hamoen et al. (2006) identified *sepF* as a candidate cell division gene from its conserved position (in Gram positive bacteria) between *ftsZ* and *divIVA*. Deletion of the gene gives a mild reduction in division frequency but the division septa formed are thick and morphologically abnormal. Mutation of *sepF* turned out to be lethal in the presence of mutations in *ftsA* or another division associated gene, *ezrA* (Ishikawa et al. 2006; Hamoen et al. 2006). In vitro, SepF protein assembles into large and regular protein rings with a diameter of about 50 nm: these rings are able to bundle FtsZ protofilaments into long tubular structures (Gündoğdu et al. 2011). Detailed structural analysis of the protein (Duman et al. 2013) suggests that the N-terminal region, like FtsA, contains a membrane associating amphipathic helix, whereas the C-terminal domain is globular and responsible for both the formation of SepF rings and association with FtsZ. Duman et al. (2013) suggest that the amphipathic helices of FtsA and SepF both serve to promote association of the Z ring with the leading edge of the septum, since this region contains positively curved (convex) membrane into which the helices can readily insert. Duman et al. (2013) proposed a model in which SepF polymers bind as arc onto the convex leading edge of the nascent division septum and maintain the Z ring in this position by bundling FtsZ protofilaments. However, this model is slightly unsatisfactory in leaving open the question of why SepF makes complete rings in vitro.

ZapA is a low MW (9 kDa) positive regulator of FtsZ assembly. It was identified in a screen for genes which, when overexpressed, could overcome the cell division block caused by overproduction of MinD (Gueiros-Filho and Losick 2002). Absence of ZapA gives no discernible phenotype under normal conditions but causes a severe division block in cells producing lower than normal levels of FtsZ, or lacking the *ezrA*, *divIVA* or *whiA* genes (Gueiros-Filho and Losick 2002; Surdova et al. 2013). ZapA interacts directly with FtsZ and, in vitro, it promotes FtsZ polymerisation as well as lateral association, yielding both single and bundled filaments (Gueiros-Filho and Losick 2002; Low et al. 2004). A temperature-sensitive mutant of FtsZ (FtsZ(Ts1)), defective in lateral association between FtsZ protofilaments at high temperatures, can be rescued by overexpressing ZapA. This supports the proposed function of ZapA as a promoter of FtsZ bunding (Monahan et al. 2009). ZapA of *Pseudomonas aeruginosa* forms dimers or tetramers in solution but oligomerizes at high concentrations; a property that could support its function as an effective FtsZ cross-linker (Low et al. 2004).

The *ezrA* gene, which is present only in Gram positive bacteria, was identified by mutations suppressing the division phenotype of a thermosensitive *ftsZ* allele (Levin et al. 1999). *ezrA* mutants can tolerate reduced levels of active FtsZ and the gene name derives from the observation that the *ezrA* single mutant makes extra Z-rings. The protein has an unusual topology with a single N-terminal transmembrane span followed by a major domain that is cytosolic. Curiously, one protein that shares this unusual topology is *E. coli* ZipA, which is an essential FtsZ-interacting protein in *E. coli* (Hale and De Boer 1997), but it seems that EzrA and ZipA are otherwise unrelated in sequence. *ezrA* mutants also have a slightly reduced cell diameter, indicating a mild defect in cell elongation. Genetic experiments suggest that this may be due to incorrect regulation of the activity of the major PBP (PBP 1) involved in synthesis of PG during both elongation and division (Claessen et al. 2008). The results of detailed mutational analysis of the gene suggest that different regions of the large protein contribute in different ways to the regulation of Z ring dynamics (Haeusser et al. 2007; Land et al. 2014). The crystal structure of the cytoplasmic domain of EzrA was recently solved (Cleverley et al. 2014) and shown to be similar to that of eukaryotic spectrins, comprising multiple, connected repeats of antiparallel α-helices, forming a complete semi-circle of 12 nm diameter. Spectrins are cytoskeletal proteins that can form two-dimensional polygonal networks lining the membrane, and they help maintain plasma membrane integrity and cytoskeletal structure in eukaryotic cells. The formation of a semi-circle could enable both the C-terminal four-helix bundle and the N-terminal transmembrane domain to interact with the membrane at the same time. Structural modelling indicates that an antiparallel dimer of EzrA molecules, as found in some crystal structure forms, could trap a paired FtsA-FtsZ protofilament inside the arch. In principle, this could serve to both anchor the protofilaments to the membrane and or locally prevent the formation of protofilament bundles.

Because FtsZ protein is thought to be indirectly associated with the cell membrane, through its interactions with FtsA, SepF and possible EzrA, it seems likely that most of the remaining divisome proteins, which are largely integral membrane or extracytoplasmic proteins (summarised in Table 3.2), do not interact directly with FtsZ. Their functions are probably concerned mainly with membrane dynamics, or peptidoglycan synthesis and turnover and will not be discussed in detail here.

Regulation of Z Ring Formation and Cell Division

Cell division needs to be tightly coordinated with other cell cycle events, particularly chromosome replication and segregation. Recently, the field of bacterial cell cycle regulation has been invigorated by the unexpected discovery that cell size homeostasis is achieved by an "adder" process in which new born cells grow by a relatively fixed length increment before dividing again, rather than by measuring a "division mass", according to a decades old dogma (Campos et al. 2014; taheri-araghi et al. 2015). The key questions now concern how the length increment is measured by the cell and used to regulate divisome function. In *B. subtilis* the intracellular concentration of FtsZ stays constant throughout the cell cycle and, although the frequency of Z ring formation varies with growth rate, the levels of FtsZ are unaffected. Artificially increasing the level of FtsZ in *B. subtilis* cells only leads to a small increase in Z ring frequency (Weart and Levin 2003).

One factor that could be involved in buffering the levels of available FtsZ is the two-component ATP-dependent protease, ClpXP. ClpX is a member of the AAA+ (ATPases associated with various cellular activities) family of ATPases. It recognizes and unfolds specific protein substrates and transfers the unfolded protein to the serine protease ClpP for degradation (Sauer et al. 2004). ClpXP is thought to participate in the regulation of FtsZ assembly by maintaining the pool of subunits available for ring formation (Weart et al. 2005; Camberg et al. 2009; Dziedzic et al. 2010). In *B. subtilis* ClpX inhibits FtsZ polymerization *in vivo* and *in vitro* in an ATP-and ClpP-independent manner but does not degrade it, though ATP hydrolysis has been shown to be required for maximum inhibition (Weart et al. 2005; Haeusser et al. 2009). This is in contrast to *E. coli* in which ClpX modulates the level of FtsZ by degrading FtsZ (Camberg et al. 2009).

Several spatial and or temporal regulators of divisome function have been identified, mainly acting by negative regulatory mechanisms.

Nucleoid Occlusion (NO)

The fact that cell division tends not ever to bisect the nucleoid, even in cells with perturbations in chromosome replication or organization, led Woldringh and colleagues to postulate a negative regulation exerted by the nucleoid, potentially by the DNA itself (Mulder and Woldringh 1989; Woldringh et al. 1991), which Rothfield later termed "nucleoid occlusion" (Cook et al. 1989). About 10 years ago, protein factors contributing to NO were identified almost simultaneously in *B. subtilis* (*noc*) and *E. coli* (*slmA*) (Wu and Errington 2004; Bernhardt and De Boer 2005): surprisingly, they were unrelated proteins that turned out to have different modes of division inhibition. *B. subtilis noc* was identified serendipitously as a factor that had a synthetic lethal division phenotype when combined with mutations affecting the Min system (which is described below). Three lines of evidence suggested that Noc protein was a NO factor: first, the protein had a classical helix-turn-helix motif and bound tightly to DNA; second, overexpression of Noc inhibited division; third, *noc* mutants had an increased frequency of nucleoid "guillotining" when DNA replication or segregation was perturbed (Wu and Errington 2004). Later work established that Noc is a site-specific DNA-binding protein with recognition sites (Noc binding sites; NBS) distributed all over the chromosome, except in the replication terminus region, where binding sites are scarce (Wu and Errington 2004; Wu et al. 2009). Noc appears to have been derived by gene duplication and divergence from the ParB (Spo0J) protein in Firmicutes, and like Spo0J it spreads from its primary binding sites to form arrays which are important for its function. While the N-terminal domain of Spo0J (ParB) is involved in interaction with its partner protein Soj, the N-terminal domain of Noc contains an amphipathic helix, which enables it to associate with the cell membrane (Adams et al. 2015). Both overexpression and deletion of Noc affect division at the level of FtsZ assembly (Wu and Errington 2004). It seems that the formation of Noc arrays around NBSs enhances membrane association and that recruitment of these DNA-Noc arrays to the membrane prevents the local formation of FtsZ ring assemblies (Adams et al. 2015). It is possible that Noc works by enhancing a natural NO system, akin to that originally described by Woldringh, in which the presence of chromosomal DNA excludes accumulation or formation of high MW divisome complexes.

The Min System

NO prevents division from occurring in the vicinity of the nucleoid, but it cannot protect the nucleoid free regions at the nascent and old cell poles. The system that prevents polar division is called Min and was first discovered *via E. coli* mutants that frequently divide near to the cell pole, giving small anucleate cells called minicells (Adler et al. 1967). The *B. subtilis* MinC and MinD proteins were identified by sequence homology to their *E. coli* counterparts (Levin et al. 1992). The *B. subtilis* Min system is now known to consist of at least four proteins: MinC, MinD, MinJ and DivIVA. MinC is an FtsZ inhibitor which interacts directly with FtsZ. In vitro studies have shown that like *E. coli* MinC, the *B. subtilis* protein inhibits FtsZ bundle formation by disrupting lateral interactions between protofilaments (Dajkovic et al. 2008; Scheffers 2008; Blasios et al. 2013), though *E. coli* MinC has been shown to also destabilize FtsZ protofilaments (Hu et al. 1999; Shen and Lutkenhaus 2010). MinD is a membrane-associated activator of MinC. Both MinC and MinD are relatively well conserved among bacteria. The two proteins form a heterotetrameric complex and in *B. subtilis* are recruited to the division site and the cell poles by MinJ, which in turn associates with the "topological specificity" determinant DivIVA (Edwards and Errington 1997; Marston et al. 1998; Bramkamp et al. 2008; Patrick and Kearns 2008). *E. coli* lacks counterparts of the MinJ and DivIVA proteins and instead uses an amazing oscillating MinCD mechanism to prevent division at the cell poles (Lutkenhaus 2007).

The key feature of DivIVA that enables it to spatially control the Min inhibitory effect lies in its targeted localization to division sites and cell poles. DivIVA oligomers have affinity for high negative membrane curvature, which normally occurs only at invaginating division septa or recently completed cell poles (Lenarcic et al. 2009; Ramamurthi and Losick 2009; Eswaramoorthy et al. 2011). It is probably recruited to the site of division as soon as membrane invagination begins, due to divisome constriction. Therefore, accumulation of DivIVA at the division site is dependent on the presence of a functional divisome but once curvature has been generated, the rings of DivIVA, one on each side of the growing septum, are no longer affected by contraction of the divisome (Eswaramoorthy et al. 2011). Upon completion of septation, the divisome disassembles and the septum splits to generate new cell poles for the two daughter cells. In cells that are not dividing, DivIVA-

GFP is concentrated at the hemispherical cell poles (Eswaramoorthy et al. 2011) but in dividing cells, DivIVA is remodelled and a portion of the DivIVA molecules remain at the pole, while some protein migrates to the new division site (Eswaramoorthy et al. 2011; Bach et al. 2014). The main structural feature of DivIVA is a parallel coiled coil, similar to the yeast tropomyosin Cdc8, a eukaryotic cytoskeletal protein involved in cytokinesis (Edwards et al. 2000; Oliva et al. 2010). This major C-terminal portion of DivIVA resembles the crescent shape of eukaryotic BAR domains normally found at the interface between the actin cytoskeleton and lipid membranes, which bind to curved membranes and also introduce curvature (Oliva et al. 2010). This raises the possibility that DivIVA senses membrane curvature using a mechanism similar to the Bar domain proteins. Structural and genetic evidence suggest that membrane interaction occurs via a hairpin structure with conserved exposed basic and hydrophobic residues in the N-terminal domain of the protein (Oliva et al. 2010).

MinJ is presently the least well characterised component of the Min system. It has 6 transmembrane helices with both N- and C-termini in the cytoplasm. The C-terminal globular portion of the protein comprises a classical PDZ domain; a protein fold often involved in protein-protein interactions (Van Baarle and Bramkamp 2010). MinJ can interact with both DivIVA and MinD, based on 2-hybrid experiments (Patrick and Kearns 2008; Bramkamp et al. 2008), suggesting that MinJ is the immediate polar target for recruitment of MinD, rather than DivIVA (Bramkamp et al. 2008).

As originally described by Marston et al. (1998) and reinforced by subsequent papers (Gregory et al. 2008; Bramkamp et al. 2008; Van Baarle and Bramkamp 2010), DivIVA and presumably now MinJ are recruited to mid cell soon after the initiation of division. MinJ, in turn, recruits the MinCD complex, which has no effect on the ongoing division but is poised to disassemble the divisome as division is completed, and or prevent the assembly of a new division complex. Although in general it appears that little, if any, of these proteins are retained at completed "old" cell poles, some activity is probably retained to prevent inappropriate minicell divisions from occurring there.

The mechanism of action of the MinCD inhibitor is not yet fully understood, despite over 20 years of study. MinD is a member of the Walker A cytoskeletal ATPase (WACA) family, a group of cytoskeletal proteins thought to be unique to bacteria (Löwe and Amos 2009; Shih and Rothfield 2006; Michie et al. 2006; Pilhofer and Jensen 2013). Characteristic of this family is a 'deviant' Walker A motif – KGGXXGKT' containing two conserved lysines, both important for binding and hydrolysis of ATP (Lutkenhaus 2012). As in *E. coli*, the ATPase activity of *B. subtilis* MinD is required for membrane binding and activation of MinC (Karoui and Errington 2001), but biochemical details of how the inhibitory activity of the MinCD complex is regulated remain elusive. One interesting recent development has been the report that *E. coli* MinC and MinD form alternating copolymeric cytomotive filaments with structural similarity to septins (Ghosal et al. 2014). Septins are a group of eukaryotic GTP-binding cytoskeletal proteins that polymerize into hetero-oligomeric protein complexes and play many important roles, including serving as membrane scaffolds for protein recruitment and as diffusion barriers for subcellular compartmentalization (Mostowy and Cossart 2012). It is not clear whether the *B. subtilis* homologues behave similarly or what the functional significance of the copolymer organization is.

Nutritional Regulation of Cell Division

In *B. subtilis* and probably many other bacteria, cell size is regulated according to the growth rate, such that fast growing cells are, on average, larger than slow growing cells (Sharpe et al. 1998). Weart et al. (2007) identified the UgtP protein as a key metabolic regulator of cell division in *B. subtilis*. UgtP is responsible for synthesis of glucolipids using UDP-glucose as a substrate. Mutations in the *ugtP* gene, or genes upstream in the UDP glucose synthetic pathway (*pgsC* or *gtaB*) had a small cell phenotype, in which both FtsZ ring formation and cell division occur at a smaller average cell size than in the wild type. UgtP turned out to interact directly with FtsZ in vitro and in vivo, and its inhibitory effect on FtsZ assembly is stimulated by UDP-Glucose (Weart et al. 2007). Under nutrient rich conditions UgtP levels are increased, as is the availability of its UDP-Glucose substrate, leading to an inhibition/delay in assembly of the FtsZ ring.

Monahan et al. (2014) identified a similar but quite distinct regulatory effect on cell division involving central carbon metabolism. They showed that a temperature sensitive *ftsZ* mutant could be rescued by mutations in genes encoding pyruvate kinase (*pyk*) or phosphoglycerate kinase (*pgk*) and established that these mutations work by limiting the supply of pyruvate from glycolysis. They identified the E1α subunit of pyruvate dehydrogenase, which uses pyruvate as a substrate in generating acetyl-CoA. Localization of E1α was found to shift between nucleoid associated and nucleoid excluded depending on the availability of nutrients (high vs low, respectively). Various genetic tests were consistent with a model in which E1α is a positive regulator of FtsZ ring formation helping to couple this to sensing of nutrient availability. Molecular details of the putative interaction between the various players in this process remain to be worked out.

Z Rings and Cell Division During Sporulation

Early in sporulation the cell division cycle is substantially modified to pave the way for generation of the distinct prespore and mother cell progeny and their subsequent differentiation. As mentioned above, this involves a repositioning of FtsZ rings away from the normal mid cell position to the cell poles (Errington 2003). The following section focuses mainly on events involving FtsZ. The key cell cycle changes, which are controlled by global changes in transcription in response to starvation, are as follows. First, medial division is blocked by a mechanism that is presently unclear, but a major change in chromosome configuration – the formation of a structure called the axial filament (Ryter 1965; Bylund et al. 1993) – probably contributes, through a nucleoid occlusion effect. Instead FtsZ rings are directed to sub-polar positions at each end of the cell (Levin and Losick 1996). Formation of these polar Z-rings requires both a small upregulation of FtsZ synthesis and the synthesis of a sporulation-specific protein SpoIIE (Ben-Yehuda and Losick 2002; Feucht et al. 1996) (see below). The upregulation of *ftsZ* occurs via a promoter controlled by the σ^H form of RNA polymerase, which is active only in stationary phase and sporulation (Gholamhoseinian et al. 1992; Gonzy-Tréboul et al. 1992).

Mutations in several key regulatory genes of sporulation give rise to an interesting phenotype called "disporic", in which prespore-like cells form at both poles of the cell (Piggot and Coote 1976). Lewis et al. (1994) showed that in these cells the asymmetric division events occurred sequentially, with the first preceding the second by about 20 min. This suggested that the Z rings at the two poles develop at different rates, ultimately contributing to the generation of asymmetry – whichever potential divisomes matures first defines the pole which the prespore cell forms (Lewis et al. 1994). The polar sporulation septum differs from a normal vegetative septum in having a much thinner layer of PG (Ryter 1965; Illing and Errington 1991; Tocheva et al. 2013). This thinning may be related to the fact that a little while later, the PG needs to be hydrolysed to enable the remarkable process of prespore engulfment to occur: the small prespore is engulfed by the mother cell to produce a cell within a cell, similar to eukaryotic phagocytosis (Illing and Errington 1991; Tocheva et al. 2013). In addition to FtsZ (Beall and Lutkenhaus 1991), FtsA is probably also required for sporulation division (Beall and Lutkenhaus 1992), though curiously, the latter protein only appears to accumulate at one of the polar potential division sites – presumably the one that goes on to support division (Feucht et al. 2001). DivIB, DivIC and FtsL, at least, are also required for formation of the sporulation septum (Levin and Losick 1994; Daniel et al. 1998; Feucht et al. 1999) but whether the other vegetative divisome proteins are also required for the sporulation septum has not been systematically studied. How the polar septum is formed despite continued presence of the Noc, MinCDJ and DivIVA proteins is also not clear.

The large (92 kDa) SpoIIE protein plays two distinct critical roles in the prespore developmental programme. In addition to its role in asymmetric septation, it is also essential for activation of the first compartment-specific transcription factor, σ^F, in the prespore (Duncan et al. 1995; Arigoni et al. 1996; Feucht et al. 1996). The N-terminal domain of SpoIIE contains 10 predicted transmembrane spans. This is followed by a central regulatory domain and a C-terminal PP2C phosphatase domain. SpoIIE is recruited to both polar Z rings sequentially (Arigoni et al. 1995; Levin et al. 1997; Wu et al. 1998), probably via a direct interaction with FtsZ (Lucet et al. 2000). The precise role of SpoIIE in FtsZ assembly is still not clear. Absence of SpoIIE causes a delayed and reduced frequency of polar divisions, as well as a vegetative-like thickening of the septa that are formed (Illing and Errington 1991; Barák and Youngman 1996, Feucht et al. 1996; Khvorova et al. 1998; Ben-Yehuda and Losick 2002; Carniol et al. 2005). Crucially, after septation, the SpoIIE protein from the septum is sequestered into the smaller prespore compartment (Wu et al. 1998; Campo et al. 2008; Bradshaw and Losick 2015), where the phosphatase domain helps to drive the activation of σ^F specifically in that compartment. Regulation of SpoIIE is highly complex and involves association with and release from the divisome, recruitment at the adjacent cell pole by interaction with DivIVA, oligomerization and controlled proteolysis (Bradshaw and Losick 2015).

At the transcriptional level, formation of a polar septum, through the localized action of SpoIIE phosphatase, triggers the prespore localized activation of σ^F, which then turns on the early prespore programme of gene expression. One of the newly expressed genes, *spoIIR*, encodes a factor that triggers the activation of a different sigma factor, σ^E, in the mother cell compartment. Among the proteins made as a result of transcription by σ^E-RNA polymerase is a specific inhibitor of *ftsZ* assembly, called MciZ (Handler et al. 2008) that appears to work as a protofilament capping protein (Bisson-Filho et al. 2015). MciZ helps to block the utilization of the second polar FtsZ ring located in the mother cell compartment. Three other σ^E-dependent proteins (SpoIID, SpoIIM and SpoIIP) also facilitate the formation of a second polar division, but apparently by working downstream on PG synthesis (Eichenberger et al. 2001).

FtsZ Inhibitors as Potential Antibiotics

FtsZ of *B. subtilis* and closely related Gram positive bacteria, including *Staphylococcus aureus*, is susceptible to inhibition by a family of related benzamide compounds, with potential for use as antibiotics. These compounds bind to an allosteric site in the protein and apparently trap the protein in the "open" state, which promotes protofilament assembly (Haydon et al. 2008; Tan et al. 2012). In vivo, this results in the formation of multiple discrete foci of FtsZ, which recruit all tested downstream divisome proteins (4 early and 4 late) (Haydon et al. 2008; Adams et al. 2011). However, productive Z rings are not formed, leading to a complete division block. The potential clinical use of these compounds has not yet been evaluated.

L-Form (Cell Wall Deficient) Bacteria

Despite the extraordinary complexity of the wall, its various important functions, and its role as the target for many powerful antibiotics, it is surprisingly easy for *B. subtilis* to lose its wall. Only one or two mutations are needed to enable *B. subtilis* (and many other organisms; Mercier et al. (2014)) to switch into a wall deficient mode called L-form (Leaver et al. 2009; Mercier et al. 2013; Kawai et al. 2015). L-forms require an osmoprotective medium to prevent them from incurring osmotic lysis and they have pleomorphic shapes, due to lack of the rigid cell wall. Remarkably, L-forms can tolerate the complete deletion of many genes that are normally essential for growth and division, including both FtsZ and the complete set of MreB homologues (Leaver et al. 2009; Mercier et al. 2012). They divide by a blebbing mechanism that requires only an increase in membrane synthesis (Mercier et al. 2013). Apart from interest in L-forms as potential biotechnological devices or possibly as agents responsible for certain infectious diseases, they are likely to be useful in basic studies of cell wall elongation and division through their ability to tolerate the disruption of many genes that are normally essential (Kawai et al. 2014).

The Future

Some of the open questions about the FtsZ system are similar to those of MreB. The advent of cryo-EM tomography is beginning to resolve the nature of the FtsZ ring (Szwedziak et al. 2014) but resolution of the detailed structure in vivo remains probably the biggest impediment to understand divisome function. Although various temporal and spatial regulators are now known and have been subjected to detailed study, again, understanding of their precise mechanism of action will probably await resolution of the Z-ring structure problem. Furthermore, even in the absence of the key negative regulators (MinCD and Noc), residual cell divisions still tend to occur between replicated chromosomes (Wu and Errington 2004; Rodrigues and Harry 2012), indicating the existence of as yet unidentified regulatory factors.

Once the structure of the Z ring has been resolved many details of the division process will need to be worked out, including the enigmatic roles of several membrane associated divisome proteins, such as DivIB, DivIC, FtsL and FtsW. Finally, it will be interesting to resolve how the division machinery is modified in order to bring about the various subtle changes associated with the asymmetric division of sporulating cells. How is mid cell division blocked? How is polar division promoted: in particular, how are the normal activities of the NO and Min systems overridden? Finally, how does the cell regulate the differential thickness of the vegetative vs sporulation septa?

Acknowledgements Work in the Errington lab is funded by a Wellcome Trust Senior Investigator Award (WT098374AIA) and a European Research Council Advanced Investigator Award (GA 670980 ELFBAD).

References

Abhayawardhane Y, Stewart GC (1995) *Bacillus subtilis* possesses a second determinant with extensive sequence similarity to the *Escherichia coli mreB* morphogene. J Bacteriol 177:765–773

Adams DW, Errington J (2009) Bacterial cell division: assembly, maintenance and disassembly of the Z ring. Nat Rev Microbiol 7:642–653

Adams DW, Wu LJ, Czaplewski LG, Errington J (2011) Multiple effects of benzamide antibiotics on FtsZ function. Mol Microbiol 80:68–84

Adams DW, Wu LJ, Errington J (2015) Nucleoid occlusion protein Noc recruits DNA to the bacterial cell membrane. EMBO J 34:491–501

Adler HI, Fisher WD, Cohen A, Hardigree AA (1967) Miniature *Escherichia coli* cells deficient in DNA. Proc Natl Acad Sci U S A 57:321–326

Anagnostopoulos C, Spizizen J (1961) Requirements for transformation in *Bacillus subtilis*. J Bacteriol 81:741–746

Anderson DE, Gueiros-Filho FJ, Erickson HP (2004) Assembly dynamics of FtsZ rings in *Bacillus subtilis* and *Escherichia coli* and effects of FtsZ-regulating proteins. J Bacteriol 186:5775–5781

Arigoni F, Pogliano K, Webb CD, Stragier P, Losick R (1995) Localization of protein implicated in establishment of cell type to sites of asymmetric division. Science 270:637–640

Arigoni F, Duncan L, Alper S, Losick R, Stragier P (1996) SpoIIE governs the phosphorylation state of a protein regulating transcription factor σ^F during sporulation in *Bacillus subtilis*. Proc Natl Acad Sci U S A 93:3238–3242

Bach JN, Albrecht N, Bramkamp M (2014) Imaging DivIVA dynamics using photo-convertible and activatable fluorophores in *Bacillus subtilis*. Front Microbiol 5:59

Barák I, Youngman P (1996) SpoIIE mutants of *Bacillus subtilis* comprise two distinct phenotypic classes consistent with a dual functional role for the SpoIIE protein. J Bacteriol 178:4984–4989

Beall B, Lutkenhaus J (1989) Nucleotide sequence and insertional inactivation of a *Bacillus subtilis* gene that affects cell division, sporulation, and temperature sensitivity. J Bacteriol 171:6821–6834

Beall B, Lutkenhaus J (1991) FtsZ in *Bacillus subtilis* is required for vegetative septation and for asymmetric septation during sporulation. Genes Dev 5:447–455

Beall B, Lutkenhaus J (1992) Impaired cell division and sporulation of a *Bacillus subtilis* strain with the *ftsA* gene deleted. J Bacteriol 174:2398–2403

Beall B, Lowe M, Lutkenhaus J (1988) Cloning and characterization of *Bacillus subtilis* homologs of *Escherichia coli* cell division genes *ftsZ* and *ftsA*. J Bacteriol 170:4855–4864

Ben-Yehuda S, Losick R (2002) Asymmetric cell division in *B. subtilis* involves a spiral-like intermediate of the cytokinetic protein FtsZ. Cell 109:257–266

Bernhardt TG, de Boer PAJ (2005) SlmA, a nucleoid-associated, FtsZ binding protein required for blocking septal ring assembly over Chromosomes in *E. coli*. Mol Cell 18:555–564

Bi EF, Lutkenhaus J (1991) FtsZ ring structure associated with division in *Escherichia coli*. Nature 354:161–164

Billings G, Ouzounov N, Ursell T, Desmarais SM, Shaevitz J, Gitai Z, Huang KC (2014) De novo morphogenesis in L-forms via geometric control of cell growth. Mol Microbiol 93:883–896

Bisson-Filho AW, Discola KF, Castellen P, Blasios V, Martins A, Sforca ML, Garcia W, Zeri AC, Erickson HP, Dessen A, Gueiros-Filho FJ (2015) FtsZ filament capping by MciZ, a developmental regulator of bacterial division. Proc Natl Acad Sci U S A 112:E2130–E2138

Blasios V, Bisson-Filho AW, Castellen P, Nogueira ML, Bettini J, Portugal RV, Zeri AC, Gueiros-Filho FJ (2013) Genetic and biochemical characterization of the MinC-FtsZ interaction in *Bacillus subtilis*. PLoS One 8:e60690

Bradshaw N, Losick R (2015) Asymmetric division triggers cell-specific gene expression through coupled capture and stabilization of a phosphatase. Elife 4 pii: e08145, doi: 10.7554/eLife.08145.

Bramkamp M, Weston L, Daniel RA, Errington J (2006) Regulated intramembrane proteolysis of FtsL protein and the control of cell division in *Bacillus subtilis*. Mol Microbiol 62:580–591

Bramkamp M, Emmins R, Weston L, Donovan C, Daniel RA, Errington J (2008) A novel component of the division-site selection system of *Bacillus subtilis* and a new mode of action for the division inhibitor MinCD. Mol Microbiol 70:1556–1569

Bylund JE, Haines MA, Piggot PJ, Higgins ML (1993) Axial filament formation in *Bacillus subtilis*: induction of nucleoids of increasing length after addition of chloramphenicol to exponential-phase cultures approaching stationary phase. J Bacteriol 175:1886–1890

Camberg JL, Hoskins JR, Wickner S (2009) ClpXP protease degrades the cytoskeletal protein, FtsZ, and modulates FtsZ polymer dynamics. Proc Natl Acad Sci U S A 106:10614–10619

Campo N, Marquis KA, Rudner DZ (2008) SpoIIQ anchors membrane proteins on both sides of the sporulation septum in *Bacillus subtilis*. J Biol Chem 283:4975–4982

Campos M, Surovtsev IV, Kato S, Paintdakhi A, Beltran B, Ebmeier SE, Jacobs-Wagner C (2014) A constant size extension drives bacterial cell size homeostasis. Cell 159:1433–1446

Carballido-López R, Errington J (2003) The bacterial cytoskeleton: in vivo dynamics of the actin-like protein Mbl of *Bacillus subtilis*. Dev Cell 4:19–28

Carballido-López R, Formstone A, Li Y, Ehrlich SD, Noirot P, Errington J (2006) Actin homolog MreBH governs cell morphogenesis by localization of the cell wall hydrolase LytE. Dev Cell 11:399–409

Carniol K, Ben-Yehuda S, King N, Losick R (2005) Genetic dissection of the sporulation protein SpoIIE and its role in asymmetric division in *Bacillus subtilis*. J Bacteriol 187:3511–3520

Cha J-H, Stewart GC (1997) The *divIVA* minicell locus of *Bacillus subtilis*. J Bacteriol 179:1671–1683

Chen Y, Milam SL, Erickson HP (2012) SulA inhibits assembly of FtsZ by a simple sequestration mechanism. Biochemistry 51:3100–3109

Claessen D, Emmins R, Hamoen LW, Daniel RA, Errington J, Edwards DH (2008) Control of the cell elongation-division cycle by shuttling of PBP1 protein in *Bacillus subtilis*. Mol Microbiol 68:1029–1046

Cleverley RM, Barrett JR, Basle A, Bui NK, Hewitt L, Solovyova A, Xu ZQ, Daniel RA, Dixon NE, Harry EJ, Oakley AJ, Vollmer W, Lewis RJ (2014) Structure and function of a spectrin-like regulator of bacterial cytokinesis. Nat Commun 5:5421

Cook WR, De Boer PA, Rothfield LI (1989) Differentiation of the bacterial cell division site. Int Rev Cytol 118:1–31

Dajkovic A, Lan G, Sun SX, Wirtz D, Lutkenhaus J (2008) MinC spatially controls bacterial cytokinesis by antagonizing the scaffolding function of FtsZ. Curr Biol 18:235–244

Daniel RA, Errington J (2000) Intrinsic instability of the essential cell division protein FtsL of *Bacillus subtilis* and a role for DivIB protein in FtsL turnover. Mol Microbiol 36:278–289

Daniel RA, Errington J (2003) Control of cell morphogenesis in bacteria. Cell 113:767–776

Daniel RA, Williams AM, Errington J (1996) A complex four-gene operon containing essential cell division gene *pbpB* in *Bacillus subtilis*. J Bacteriol 178:2343–2350

Daniel RA, Harry EJ, Katis VL, Wake RG, Errington J (1998) Characterization of the essential cell division gene *ftsL* (*yIID*) of *Bacillus subtilis* and its role in the assembly of the division apparatus. Mol Microbiol 29:593–604

Daniel RA, Noirot-Gros MF, Noirot P, Errington J (2006) Multiple interactions between the transmembrane division proteins of *Bacillus subtilis* and the role of FtsL instability in divisome assembly. J Bacteriol 188:7396–7404

De Boer P, Crossley R, Rothfield L (1992) The essential bacterial cell-division protein FtsZ is a GTPase. Nature 359:254–256

De Pedro MA, Cava F (2015) Structural constraints and dynamics of bacterial cell wall architecture. Front Microbiol 6:449

Defeu Soufo HJ, Graumann PL (2004) Dynamic movement of actin-like proteins within bacterial cells. EMBO Rep 5:789–794

Defeu Soufo HJ, Graumann PL (2006) Dynamic localization and interaction with other Bacillus subtilis actin-like proteins are important for the function of MreB. Mol Microbiol 62:1340–1356

Defeu Soufo HJ, Reimold C, Breddermann H, Mannherz HG, Graumann PL (2015) Translation elongation factor EF-Tu modulates filament formation of actin-like MreB protein in vitro. J Mol Biol 427:1715–1727

Dempwolff F, Reimold C, Reth M, Graumann PL (2011) *Bacillus subtilis* MreB orthologs self-organize into filamentous structures underneath the cell membrane in a heterologous cell system. PLoS One 6:e27035

Domínguez-Cuevas P, Porcelli I, Daniel RA, Errington J (2013) Differentiated roles for MreB-actin isologues and autolytic enzymes in *Bacillus subtilis* morphogenesis. Mol Microbiol 89:1084–1098

Domínguez-Escobar J, Chastanet A, Crevenna AH, Fromion V, Wedlich-Soldner R, Carballido-López R (2011) Processive movement of MreB-associated cell wall biosynthetic complexes in bacteria. Science 333:225–228

Duman R, Ishikawa S, Celik I, Strahl H, Ogasawara N, Troc P, Lowe J, Hamoen LW (2013) Structural and genetic analyses reveal the protein SepF as a new membrane anchor for the Z ring. Proc Natl Acad Sci U S A 110:E4601–E4610

Duncan L, Alper S, Arigoni F, Losick R, Stragier P (1995) Activation of cell-specific transciption by a serine phosphatase at the site of asymmetric division. Science 270:641–644

Dziedzic R, Kiran M, Plocinski P, Ziolkiewicz M, Brzostek A, Moomey M, Vadrevu IS, Dziadek J, Madiraju M, Rajagopalan M (2010) *Mycobacterium tuberculosis* ClpX interacts with FtsZ and interferes with FtsZ assembly. PLoS One 5:e11058

Edwards DH, Errington J (1997) The *Bacillus subtilis* DivIVA protein targets to the division septum and controls the site specificity of cell division. Mol Microbiol 24:905–915

Edwards DH, Thomaides HB, Errington J (2000) Promiscuous targeting of *Bacillus subtilis* cell division protein DivIVA to division sites in Escherichia coli and fission yeast

Egan AJ, Vollmer W (2013) The physiology of bacterial cell division. Ann N Y Acad Sci 1277:8–28

Eichenberger P, Fawcett P, Losick R (2001) A three-protein inhibitor of polar septation during sporulation in *Bacillus subtilis*. Mol Microbiol 42:1147–1162

Emami K et al (2017) RodA as the missing glycosyltransferase in Bacillus subtilis and antibiotic discovery for the peptidoglycan polymerase pathway. Nat Microbiol 2:16253. doi:10.1038/nmicrobiol.2016.253

Errington J (1993) *Bacillus subtilis* sporulation: regulation of gene expression and control of morphogenesis. Microbiol Rev 57:1–33

Errington J (2003) Regulation of endospore formation in *Bacillus subtilis*. Nat Rev Microbiol 1:117–126

Errington J (2015) Bacterial morphogenesis and the enigmatic MreB helix. Nat Rev Microbiol 13:241–248

Eswaramoorthy P, Erb ML, Gregory JA, Silverman J, Pogliano K, Pogliano J, Ramamurthi KS (2011) Cellular architecture mediates DivIVA ultrastructure and regulates min activity in *Bacillus subtilis*. MBio 2 (b). pii: e00257-11. doi: 10.1128/mBio.00257-11

Feucht A, Errington J (2005) *ftsZ* mutations affecting cell division frequency, placement and morphology in *Bacillus subtilis*. Microbiology 151:2053–2064

Feucht A, Magnin T, Yudkin MD, Errington J (1996) Bifunctional protein required for asymmetric cell division and cell-specific transcription in *Bacillus subtilis*. Genes Dev 10:794–803

Feucht A, Daniel RA, Errington J (1999) Characterization of a morphological checkpoint coupling cell-specific transcription to septation in *Bacillus subtilis*. Mol Microbiol 33:1015–1026

Feucht A, Lucet I, Yudkin MD, Errington J (2001) Cytological and biochemical characterization of the FtsA cell division protein of *Bacillus subtilis*. Mol Microbiol 40:115–125

Formstone A, Errington J (2005) A magnesium-dependent *mreB* null mutant: implications for the role of *mreB* in *Bacillus subtilis*. Mol Microbiol 55:1646–1657

Formstone A, Carballido-López R, Noirot P, Errington J, Scheffers DJ (2008) Localization and interactions of teichoic acid synthetic enzymes in *Bacillus subtilis*. J Bacteriol 190:1812–1821

Foulquier E, Pompeo F, Bernadac A, Espinosa L, Galinier A (2011) The YvcK protein is required for morphogenesis via localization of PBP1 under gluconeogenic growth conditions in *Bacillus subtilis*. Mol Microbiol 80:309–318

Gamba P, Veening JW, Saunders NJ, Hamoen LW, Daniel RA (2009) Two-step assembly dynamics of the *Bacillus subtilis* divisome. J Bacteriol 191:4186–4194

Garner EC, Bernard R, Wang W, Zhuang X, Rudner DZ, Mitchison T (2011) Coupled, circumferential motions of the cell wall synthesis machinery and MreB filaments in *B. subtilis*. Science 333:222–225

Gholamhoseinian A, Shen Z, Wu J-J, Piggot P (1992) Regulation of transcription of the cell division gene *ftsA* during sporulation of *Bacillus subtilis*. J Bacteriol 174:4647–4656

Ghosal D, Trambaiolo D, Amos LA, Lowe J (2014) MinCD cell division proteins form alternating copolymeric cytomotive filaments. Nat Commun 5:5341

Gonzy-Tréboul G, Karmazyn-Campelli C, Stragier P (1992) Developmental regulation of transcription of the *Bacillus subtilis ftsAZ* operon. J Mol Biol 224:967–979

Gregory JA, Becker EC, Pogliano K (2008) *Bacillus subtilis* MinC destabilizes FtsZ-rings at new cell poles and contributes to the timing of cell division. Genes Dev 22:3475–3488

Grundling A, Schneewind O (2007a) Genes required for glycolipid synthesis and lipoteichoic acid anchoring in *Staphylococcus aureus*. J Bacteriol 189:2521–2530

Grundling A, Schneewind O (2007b) Synthesis of glycerol phosphate lipoteichoic acid in *Staphylococcus aureus*. Proc Natl Acad Sci U S A 104:8478–8483

Gueiros-Filho FJ, Losick R (2002) A widely conserved bacterial cell division protein that promotes assembly of the tubulin-like protein FtsZ. Genes Dev 16:2544–2556

Gündoğdu ME, Kawai Y, Pavlendova N, Ogasawara N, Errington J, Scheffers DJ, Hamoen LW (2011) Large ring polymers align FtsZ polymers for normal septum formation. EMBO J 30:617–626

Haeusser DP, Schwartz RL, Smith AM, Oates ME, Levin PA (2004) EzrA prevents aberrant cell division by modulating assembly of the cytoskeletal protein FtsZ. Mol Microbiol 52:801–814

Haeusser DP, Garza AC, Buscher AZ, Levin PA (2007) The division inhibitor EzrA contains a seven-residue patch required for maintaining the dynamic nature of the medial FtsZ ring. J Bacteriol 189:9001–9010

Haeusser DP, Lee AH, Weart RB, Levin PA (2009) ClpX inhibits FtsZ assembly in a manner that does not require its ATP hydrolysis-dependent chaperone activity. J Bacteriol 191:1986–1991

Hale CA, De Boer PAJ (1997) Direct binding of FtsZ to ZipA, an essential component of the septal ring structure that mediates cell division in *E. coli*. Cell 88:175–185

Hamoen LW, Errington J (2003) Polar targeting of DivIVA in *Bacillus subtilis* is not directly dependent on FtsZ or PBP 2B. J Bacteriol 185

Hamoen LW, Meile JC, De Jong W, Noirot P, Errington J (2006) SepF, a novel FtsZ-interacting protein required for a late step in cell division. Mol Microbiol 59:989–999

Handler AA, Lim JE, Losick R (2008) Peptide inhibitor of cytokinesis during sporulation in *Bacillus subtilis*. Mol Microbiol 68:588–599

Harry EJ, Wake RG (1989) Cloning and expression of a *Bacillus subtilis* division initiation gene for which a homolog has not been identified in another organism. J Bacteriol 171:6835–6839

Harry EJ, Wake RG (1997) The membrane-bound cell division protein DivIB is localized to the division site in *Bacillus subtilis*. Mol Microbiol 25:275–283

Haydon DJ, Stokes NR, Ure R, Galbraith G, Bennett JM, Brown DR, Baker PJ, Barynin VV, Rice DW, Sedelnikova SE, Heal JR, Sheridan JM, Aiwale ST, Chauhan PK, Srivastava A, Taneja A, Collins I, Errington J, Czaplewski LG (2008) An inhibitor of FtsZ with potent and selective anti-staphylococcal activity. Science 321:1673–1675

Hu Z, Mukherjee A, Pichoff S, Lutkenhaus J (1999) The MinC component of the division site selection system in *Escherichia coli* interacts with FtsZ to prevent polymerization. Proc Natl Acad Sci U S A 96:14819–14824

Illing N, Errington J (1991) Genetic regulation of morphogenesis in *Bacillus subtilis*: roles of σ^E and σ^F in prespore engulfment. J Bacteriol 173:3159–3169

Ishikawa S, Kawai Y, Hiramatsu K, Kuwano M, Ogasawara N (2006) A new FtsZ-interacting protein, YlmF, complements the activity of FtsA during progression of cell division in *Bacillus subtilis*. Mol Microbiol 60:1364–1380

Jennings PC, Cox GC, Monahan LG, Harry EJ (2011) Super-resolution imaging of the bacterial cytokinetic protein FtsZ. Micron 42:336–341

Jensen SO, Thompson LS, Harry EJ (2005) Cell division in *Bacillus subtilis*: FtsZ and FtsA association is Z-ring independent, and FtsA is required for efficient midcell Z-Ring assembly. J Bacteriol 187:6536–6544

Jones LJF, Carballido-López R, Errington J (2001) Control of cell shape in bacteria: helical, actin-like filaments in *Bacillus subtilis*. Cell 104:913–922

Karoui ME, Errington J (2001) Isolation and characterization of topological specificity mutants of *minD* in *Bacillus subtilis*. Mol Microbiol 42:1211–1221

Katis VL, Wake RG (1999) Membrane-bound division proteins DivIB and DivIC of *Bacillus subtilis* function solely through their external domains in both vegetative and sporulation division. J Bacteriol 181:2710–2718

Katis VL, Harry EJ, Wake RG (1997) The *Bacillus subtilis* division protein DivIC is a highly abundant membrane-bound protein that localizes to the division site. Mol Microbiol 26:1047–1055

Katis VL, Wake RG, Harry EJ (2000) Septal localization of the membrane-bound division proteins of *Bacillus subtilis* DivIB and DivIC is codependent only at high temperatures and requires FtsZ. J Bacteriol 182:3607–3611

Kawai Y, Ogasawara N (2006) *Bacillus subtilis* EzrA and FtsL synergistically regulate FtsZ ring dynamics during cell division. Microbiology 152:1129–1141

Kawai Y, Asai K, Errington J (2009a) Partial functional redundancy of MreB isoforms, MreB, Mbl and MreBH, in cell morphogenesis of *Bacillus subtilis*. Mol Microbiol 73:719–731

Kawai Y, Daniel RA, Errington J (2009b) Regulation of cell wall morphogenesis in *Bacillus subtilis* by recruitment of PBP1 to the MreB helix. Mol Microbiol 71:1131–1144

Kawai Y, Marles-Wright J, Cleverley RM, Emmins R, Ishikawa S, Kuwano M, Heinz N, Bui NK, Hoyland CN, Ogasawara N, Lewis RJ, Vollmer W, Daniel RA, Errington J (2011) A widespread family of bacterial cell wall assembly proteins. EMBO J 30:4931–4941

Kawai Y, Mercier R, Errington J (2014) Bacterial cell morphogenesis does not require a preexisting template structure. Curr Biol 24:863–867

Kawai Y, Mercier R, Wu LJ, Dominguez-Cuevas P, Oshima T, Errington J (2015) Cell growth of wall-free L-form bacteria is limited by oxidative damage. Curr Biol 25:1613–1618

Khvorova A, Zhang L, Higgins ML, Piggot PJ (1998) The *spoIIE* locus is involved in the Spo0A-dependent switch in the location of FtsZ rings in *Bacillus subtilis*. J Bacteriol 180:1256–1260

Kloosterman TG, Lenarcic R, Willis CR, Roberts DM, Hamoen LW, Errington J, Wu LJ (2016) Complex polar machinery required for proper chromosome segregation in vegetative and sporulating cells of Bacillus subtilis. Mol Microbiol 101(2):333–350

Land AD, Luo Q, Levin PA (2014) Functional domain analysis of the cell division inhibitor EzrA. PLoS One 9:e102616

Leaver M, Errington J (2005) Roles for MreC and MreD proteins in helical growth of the cylindrical cell wall in *Bacillus subtilis*. Mol Microbiol 57:1196–1209

Leaver M, Dominguez-Cuevas P, Coxhead JM, Daniel RA, Errington J (2009) Life without a wall or division machine in *Bacillus subtilis*. Nature 457:849–853

Lenarcic R, Halbedel S, Visser L, Shaw M, Wu LJ, Errington J, Marenduzzo D, Hamoen LW (2009) Localisation of DivIVA by targeting to negatively curved membranes. EMBO J 28:2272–2282

Levin PA, Losick R (1994) Characterization of a cell division gene from *Bacillus subtilis* that is required for vegetative and sporulation septum formation. J Bacteriol 176:1451–1459

Levin PA, Losick R (1996) Transcription factor Spo0A switches the localization of the cell division protein FtsZ from a medial to a bipolar pattern in *Bacillus subtilis*. Genes Dev 10:478–488

Levin PA, Margolis PS, Setlow P, Losick R, Sun D (1992) Identification of *Bacillus subtilis* genes for septum placement and shape determination. J Bacteriol 174:6717–6728

Levin PA, Losick R, Stragier P, Arigoni F (1997) Localization of the sporulation protein SpoIIE in *Bacillus subtilis* is dependent upon the cell division protein FtsZ. Mol Microbiol 25:839–846

Levin PA, Kurtser IG, Grossman AD (1999) Identification and characterization of a negative regulator of FtsZ ring formation in *Bacillus subtilis*. Proc Natl Acad Sci U S A 96:9642–9647
Lewis PJ, Partridge SR, Errington J (1994) Sigma factors, asymmetry, and the determination of cell fate in *Bacillus subtilis*. Proc Natl Acad Sci U S A 91:3849–3853
Li Z, Trimble MJ, Brun YV, Jensen GJ (2007) The structure of FtsZ filaments *in vivo* suggests a force-generating role in cell division. EMBO J 26:4694–4708
Loose M, Mitchison TJ (2014) The bacterial cell division proteins FtsA and FtsZ self-organize into dynamic cytoskeletal patterns. Nat Cell Biol 16:38–46
Lovering AL, Safadi SS, Strynadka NC (2012) Structural perspective of peptidoglycan biosynthesis and assembly. Annu Rev Biochem 81:451–478
Low HH, Moncrieffe MC, Löwe J (2004) The crystal structure of ZapA and its modulation of FtsZ polymerisation. J Mol Biol 341:839–852
Löwe J, Amos LA (1998) Crystal structure of the bacterial cell-division protein FtsZ. Nature 391:203–206
Löwe J, Amos LA (2009) Evolution of cytomotive filaments: the cytoskeleton from prokaryotes to eukaryotes. Int J Biochem Cell Biol 41:323–329
Lu Z, Takeuchi M, Sato T (2007) The LysR-type transcriptional regulator YofA controls cell division through the regulation of expression of *ftsW* in *Bacillus subtilis*. J Bacteriol 189:5642–5651
Lucet I, Feucht A, Yudkin MD, Errington J (2000) Direct interaction between the cell division protein FtsZ and the cell differentiation protein SpoIIE. EMBO J 19:1467–1475
Lutkenhaus J (2007) Assembly dynamics of the bacterial MinCDE system and spatial regulation of the Z ring. Annu Rev Biochem 76:539–562
Lutkenhaus J (2012) The ParA/MinD family puts things in their place. Trends Microbiol 20:411–418
Marston AL, Errington J (1999) Selection of the midcell division site in *Bacillus subtilis* through MinD-dependent polar localization and activation of MinC. Mol Microbiol 33:84–96
Marston AL, Thomaides HB, Edwards DH, Sharpe ME, Errington J (1998) Polar localization of the MinD protein of *Bacillus subtilis* and its role in selection of the mid-cell division site. Genes Dev 12:3419–3430
Meeske AJ, Sham LT, Kimsey H, Koo BM, Gross CA, Bernhardt TG, Rudner DZ (2015) MurJ and a novel lipid II flippase are required for cell wall biogenesis in *Bacillus subtilis*. Proc Natl Acad Sci U S A 112:6437–6442
Meeske AJ et al (2016) SEDS proteins are a widespread family of bacterial cell wall polymerases. Nature. doi:10.1038/nature19331
Mercier R, Domínguez-Cuevas P, Errington J (2012) Crucial role for membrane fluidity in proliferation of primitive cells. Cell Rep 1:417–423
Mercier R, Kawai Y, Errington J (2013) Excess membrane synthesis drives a primitive mode of cell proliferation. Cell 152:997–1007
Mercier R, Kawai Y, Errington J (2014) General principles for the formation and proliferation of a wall-free (L-form) state in bacteria. Elife 3
Michie KA, Monahan LG, Beech PL, Harry EJ (2006) Trapping of a spiral-like intermediate of the bacterial cytokinetic protein FtsZ. J Bacteriol 188:1680–1690
Miller AK, Brown EE, Mercado BT, Herman JK (2015) A DNA-binding protein defines the precise region of chromosome capture during *Bacillus* sporulation. Mol Microbiol 99(1):111–122
Min J, Holden SJ, Carlini L, Unser M, Manley S, Ye JC (2014) 3D high-density localization microscopy using hybrid astigmatic/ biplane imaging and sparse image reconstruction. Biomed Opt Express 5:3935–3948
Monahan LG, Robinson A, Harry EJ (2009) Lateral FtsZ association and the assembly of the cytokinetic Z ring in bacteria. Mol Microbiol 74:1004–1017
Monahan LG, Liew AT, Bottomley AL, Harry EJ (2014) Division site positioning in bacteria: one size does not fit all. Front Microbiol 5:19

Morgenstein RM, Bratton BP, Nguyen JP, Ouzounov N, Shaevitz JW, Gitai Z (2015) RodZ links MreB to cell wall synthesis to mediate MreB rotation and robust morphogenesis. Proc Natl Acad Sci U S A 112:12510–12515

Mostowy S, Cossart P (2012) Septins: the fourth component of the cytoskeleton. Nat Rev Mol Cell Biol 13:183–194

Muchova K, Chromikova Z, Barak I (2013) Control of *Bacillus subtilis* cell shape by RodZ. Environ Microbiol 15:3259–3271

Mukherjee A, Lutkenhaus J (1994) Guanine nucleotide-dependent assembly of FtsZ into filaments. J Bacteriol 176:2754–2758

Mulder E, Woldringh CL (1989) Actively replicating nucleoids influence positioning of division sites in *Escherichia coli* filaments forming cells lacking DNA. J Bacteriol 171:4303–4314

Muntel J, Fromion V, Goelzer A, Maabeta S, Mader U, Buttner K, Hecker M, Becher D (2014) Comprehensive absolute quantification of the cytosolic proteome of *Bacillus subtilis* by data independent, parallel fragmentation in liquid chromatography/mass spectrometry (LC/MS(E)). Mol Cell Proteomics 13:1008–1019

Oliva MA, Cordell SC, Löwe J (2004) Structural insights into FtsZ protofilament formation. Nat Struct Mol Biol 11:1243–1250

Oliva MA, Halbedel S, Freund SM, Dutow P, Leonard TA, Veprintsev DB, Hamoen LW, Löwe J (2010) Features critical for membrane binding revealed by DivIVA crystal structure. EMBO J 29:1988–2001

Olshausen PV, Defeu Soufo HJ, Wicker K, Heintzmann R, Graumann PL, Rohrbach A (2013) Superresolution imaging of dynamic MreB filaments in *B. subtilis*--a multiple-motor-driven transport? Biophys J 105:1171–1181

Patrick JE, Kearns DB (2008) MinJ (YvjD) is a topological determinant of cell division in *Bacillus subtilis*. Mol Microbiol 70:1166–1179

Percy MG, Grundling A (2014) Lipoteichoic acid synthesis and function in gram-positive bacteria. Annu Rev Microbiol 68:81–100

Peters PC, Migocki MD, Thoni C, Harry EJ (2007) A new assembly pathway for the cytokinetic Z ring from a dynamic helical structure in vegetatively growing cells of *Bacillus subtilis*. Mol Microbiol 64:487–499

Pichoff S, Lutkenhaus J (2005) Tethering the Z ring to the membrane through a conserved membrane targeting sequence in FtsA. Mol Microbiol 55:1722–1734

Piggot PJ, Coote JG (1976) Genetic aspects of bacterial endospore formation. Bacteriol Rev 40:908–962

Pilhofer M, Jensen GJ (2013) The bacterial cytoskeleton: more than twisted filaments. Curr Opin Cell Biol 25:125–133

Pohl S, Harwood CR (2010) Heterologous protein secretion by *Bacillus* species. Adv Appl Microbiol 73:1–25

Popham DL, Setlow P (1996) Phenotypes of *Bacillus subtilis* mutants lacking multiple class a high-molecular-weight penicillin-binding proteins. J Bacteriol 178:2079–2085

Ramamurthi KS, Losick R (2009) Negative membrane curvature as a cue for subcellular localization of a bacterial protein. Proc Natl Acad Sci U S A 106:13541–13545

Raychaudhuri D, Park JT (1992) *Escherichia coli* cell-division gene *ftsZ* encodes a novel GTP-binding protein. Nature 359:251–254

Reeve JN, Mendelson NH, Coyne SI, Hallock LL, Cole RM (1973) Minicells of *Bacillus subtilis*. J Bacteriol 114:860–873

Reimold C, Defeu Soufo HJ, Dempwolff F, Graumann PL (2013) Motion of variable-length MreB filaments at the bacterial cell membrane influences cell morphology. Mol Biol Cell 24:2340–2349

Robson SA, Michie KA, Mackay JP, Harry E, King GF (2002) The *Bacillus subtilis* cell division proteins FtsL and DivIC are intrinsically unstable and do not interact with one another in the absence of other septasomal components. Mol Microbiol 44:663–674

Rodrigues CD, Harry EJ (2012) The Min system and nucleoid occlusion are not required for identifying the division site in *Bacillus subtilis* but ensure its efficient utilization. PLoS Genet 8:e1002561

Rueff AS, Chastanet A, Domínguez-Escobar J, Yao Z, Yates J, Prejean MV, Delumeau O, Noirot P, Wedlich-Soldner R, Filipe SR, Carballido-López R (2014) An early cytoplasmic step of peptidoglycan synthesis is associated to MreB in *Bacillus subtilis*. Mol Microbiol 91:348–362

Ryter A (1965) Etude morphologique de la sporulation de *Bacillus subtilis*. Annales de l'Institut Pasteur 108:40–60

Salje J, Van Den Ent F, De Boer P, Löwe J (2011) Direct membrane binding by bacterial actin MreB. Mol Cell 43:478–487

Sauer RT, Bolon DN, Burton BM, Burton RE, Flynn JM, Grant RA, Hersch GL, Joshi SA, kenniston JA, Levchenko I, Neher SB, Oakes ES, Siddiqui SM, Wah DA, Baker TA (2004) Sculpting the proteome with AAA(+) proteases and disassembly machines. Cell 119:9–18

Scheffers DJ (2008) The effect of MinC on FtsZ polymerization is pH dependent and can be counteracted by ZapA. FEBS Lett 582:2601–2608

Scheffers DJ, Tol MB (2015) LipidII: just another brick in the wall? PLoS Pathog 11:e1005213

Schirner K, Errington J (2009) The cell wall regulator σ^I specifically suppresses the lethal phenotype of *mbl* mutants in *Bacillus subtilis*. J Bacteriol 191:1404–1413

Schirner K, Marles-Wright J, Lewis RJ, Errington J (2009) Distinct and essential morphogenic functions for wall- and lipo-teichoic acids in *Bacillus subtilis*. EMBO J 28:830–842

Sewell EW, Brown ED (2014) Taking aim at wall teichoic acid synthesis: new biology and new leads for antibiotics. J Antibiot (Tokyo) 67:43–51

Shapiro L, Losick R (2000) Dynamic spatial regulation in the bacterial cell. Cell 100:89–98

Sharpe ME, Hauser PM, Sharpe RG, Errington J (1998) *Bacillus subtilis* cell cycle as studied by fluorescence microscopy: constancy of the cell length at initiation of DNA replication and evidence for active nucleoid partitioning. J Bacteriol 180:547–555

Shen B, Lutkenhaus J (2010) Examination of the interaction between FtsZ and $MinC_N$ in *E. coli* suggests how MinC disrupts Z rings. Mol Microbiol 75:1285–1298

Shih YL, Rothfield L (2006) The bacterial cytoskeleton. Microbiol Mol Biol Rev 70:729–754

Sievers J, Errington J (2000a) Analysis of the essential cell division gene *ftsL* of *Bacillus subtilis* by mutagenesis and heterologous complementation. J Bacteriol 182:5572–5579

Sievers J, Errington J (2000b) The *Bacillus subtilis* cell division protein FtsL localizes to sites of septation and interacts with DivIC. Mol Microbiol 36:846–855

Strahl H, Hamoen LW (2010) Membrane potential is important for bacterial cell division. Proc Natl Acad Sci U S A 107:12281–12286

Strauss MP, Liew AT, Turnbull L, Whitchurch CB, Monahan LG, Harry EJ (2012) 3D-SIM super resolution microscopy reveals a bead-like arrangement for FtsZ and the division machinery: implications for triggering cytokinesis. PLoS Biol 10:e1001389

Surdova K, Gamba P, Claessen D, Siersma T, Jonker MJ, Errington J, Hamoen LW (2013) The conserved DNA-binding protein WhiA is involved in cell division in *Bacillus subtilis*. J Bacteriol 195:5450–5460

Swulius MT, Jensen GJ (2012) The helical MreB cytoskeleton in *Escherichia coli* MC1000/pLE7 is an artifact of the N-Terminal yellow fluorescent protein tag. J Bacteriol 194:6382–6386

Swulius MT, Chen S, Jane Ding H, Li Z, Briegel A, Pilhofer M, Tocheva EI, Lybarger SR, Johnson TL, Sandkvist M, Jensen GJ (2011) Long helical filaments are not seen encircling cells in electron cryotomograms of rod-shaped bacteria. Biochem Biophys Res Commun 407:650–655

Szwedziak P, Wang Q, Freund SM, Löwe J (2012) FtsA forms actin-like protofilaments. EMBO J 31:2249–2260

Szwedziak P, Wang Q, Bharat TA, Tsim M, Löwe J (2014) Architecture of the ring formed by the tubulin homologue FtsZ in bacterial cell division. Elife 3:e04601

Taheri-Araghi S, Bradde S, Sauls JT, Hill NS, Levin PA, Paulsson J, Vergassola M, Jun S (2015) Cell-size control and homeostasis in bacteria. Curr Biol 25:385–391

Tan IS, Ramamurthi KS (2014) Spore formation in *Bacillus subtilis*. Environ Microbiol Rep 6:212–225

Tan CM, Therien AG, Lu J, Lee SH, Caron A, Gill CJ, Lebeau-Jacob C, Benton-Perdomo L, Monteiro JM, Pereira PM, Elsen NL, Wu J, Deschamps K, Petcu M, Wong S, Daigneault E, Kramer S, Liang L, Maxwell E, Claveau D, Vaillancourt J, Skorey K, Tam J, Wang H, Meredith TC, Sillaots S, Wang-Jarantow L, Ramtohul Y, Langlois E, Landry F, Reid JC, Parthasarathy G, Sharma S, Baryshnikova A, Lumb KJ, Pinho MG, Soisson SM, Roemer T (2012) Restoring methicillin-resistant *Staphylococcus aureus* susceptibility to β-lactam antibiotics. Sci Transl Med 4:126ra35

Tavares JR, De Souza RF, Meira GL, Gueiros-Filho FJ (2008) Cytological characterization of YpsB, a novel component of the *Bacillus subtilis* divisome. J Bacteriol 190:7096–7107

Tiyanont K, Doan T, Lazarus MB, Fang X, Rudner DZ, Walker S (2006) Imaging peptidoglycan biosynthesis in *Bacillus subtilis* with fluorescent antibiotics. Proc Natl Acad Sci U S A 103:11033–11038

Tocheva EI, Lopez-Garrido J, Hughes HV, Fredlund J, Kuru E, Vannieuwenhze MS, Brun YV, Pogliano K, Jensen GJ (2013) Peptidoglycan transformations during *Bacillus subtilis* sporulation. Mol Microbiol 88:673–686

Typas A, Banzhaf M, Gross CA, Vollmer W (2012) From the regulation of peptidoglycan synthesis to bacterial growth and morphology. Nat Rev Microbiol 10:123–136

Ursell TS, Nguyen J, Monds RD, Colavin A, Billings G, Ouzounov N, Gitai Z, Shaevitz JW, Huang KC (2014) Rod-like bacterial shape is maintained by feedback between cell curvature and cytoskeletal localization. Proc Natl Acad Sci U S A 111:E1025–E1034

Van Baarle S, Bramkamp M (2010) The MinCDJ system in *Bacillus subtilis* prevents minicell formation by promoting divisome disassembly. PLoS One 5:e9850

Van Baarle S, Celik IN, Kaval KG, Bramkamp M, Hamoen LW, Halbedel S (2013) Protein-protein interaction domains of *Bacillus subtilis* DivIVA. J Bacteriol 195:1012–1021

Van Den Ent F, Löwe J (2000) Crystal structure of the cell division protein FtsA from *Thermotoga maritima*. EMBO J 19:5300–5307

Van Den Ent F, Leaver M, Bendezu F, Errington J, De Boer P, Löwe J (2006) Dimeric structure of the cell shape protein MreC and its functional implications. Mol Microbiol 62:1631–1642

Van Teeffelen S, Gitai Z (2011) Rotate into shape: MreB and bacterial morphogenesis. EMBO J 30:4856–4857

Varley AW, Stewart GC (1992) The *divIVB* region of the *Bacillus subtilis* chromosome encodes homologs of *Escherichia coli* septum placement (MinCD) and cell shape (MreBCD) determinants. J Bacteriol 174:6729–6742

Vollmer W, Joris B, Charlier P, Foster S (2008) Bacterial peptidoglycan (murein) hydrolases. FEMS Microbiol Rev 32:259–286

Wachi M, Doi M, Tamaki S, Park W, Nakajima-Iijima S, Matsuhashi M (1987) Mutant isolation and molecular cloning of *mre* genes, which determine cell shape, sensitivity to mecillinam, and amount of penicillin-binding proteins in *Escherichia coli*. J Bacteriol 169:4935–4940

Wagner-Herman JK, Bernard R, Dunne R, Bisson-Filho AW, Kumar K, Nguyen T, Mulcahy L, Koullias J, Gueiros-Filho FJ, Rudner DZ (2012) RefZ facilitates the switch from medial to polar division during spore formation in *Bacillus subtilis*. J Bacteriol 194:4608–4618

Wang X, Lutkenhaus J (1993) The FtsZ protein of *Bacillus subtilis* is localized at the division site and has GTPase activity that is dependent upon FtsZ concentration. Mol Microbiol 9:435–442

Weart RB, Levin PA (2003) Growth rate-dependent regulation of medial FtsZ ring formation. J Bacteriol 185:2826–2834

Weart RB, Nakano S, Lane BE, Zuber P, Levin PA (2005) The ClpX chaperone modulates assembly of the tubulin-like protein FtsZ. Mol Microbiol 57:238–249

Weart RB, Lee AH, Chien AC, Haeusser DP, Hill NS, Levin PA (2007) A metabolic sensor governing cell size in bacteria. Cell 130:335–347

Woldringh CL, Mulder E, Huls PG, Vischer N (1991) Toporegulation of bacterial division according to the nucleoid occlusion model. Res Microbiol 142:309–320

Wu LJ, Errington J (2004) Coordination of cell division and chromosome segregation by a nucleoid occlusion protein in *Bacillus subtilis*. Cell 117:915–925

Wu LJ, Feucht A, Errington J (1998) Prespore-specific gene expression in *Bacillus subtilis* is driven by sequestration of SpoIIE phosphatase to the prespore side of the asymmetric septum. Genes Dev 12:1371–1380

Wu LJ, Ishikawa S, Kawai Y, Oshima T, Ogasawara N, Errington J (2009) Noc protein binds to specific DNA sequences to coordinate cell division with chromosome segregation. EMBO J 28:1940–1952

Yanouri A, Daniel RA, Errington J, Buchanan CE (1993) Cloning and sequencing of the cell division gene *pbpB*, which encodes penicillin-binding protein 2B in *Bacillus subtilis*. J Bacteriol 175:7604–7616

Chapter 4
Cytoskeletal Proteins in *Caulobacter crescentus*: Spatial Orchestrators of Cell Cycle Progression, Development, and Cell Shape

Kousik Sundararajan and Erin D. Goley

Abstract *Caulobacter crescentus*, an aquatic Gram-negative α-proteobacterium, is dimorphic, as a result of asymmetric cell divisions that give rise to a free-swimming swarmer daughter cell and a stationary stalked daughter. Cell polarity of vibrioid *C. crescentus* cells is marked by the presence of a stalk at one end in the stationary form and a polar flagellum in the motile form. Progression through the cell cycle and execution of the associated morphogenetic events are tightly controlled through regulation of the abundance and activity of key proteins. In synergy with the regulation of protein abundance or activity, cytoskeletal elements are key contributors to cell cycle progression through spatial regulation of developmental processes. These include: polarity establishment and maintenance, DNA segregation, cytokinesis, and cell elongation. Cytoskeletal proteins in *C. crescentus* are additionally required to maintain its rod shape, curvature, and pole morphology. In this chapter, we explore the mechanisms through which cytoskeletal proteins in *C. crescentus* orchestrate developmental processes by acting as scaffolds for protein recruitment, generating force, and/or restricting or directing the motion of molecular machines. We discuss each cytoskeletal element in turn, beginning with those important for organization of molecules at the cell poles and chromosome segregation, then cytokinesis, and finally cell shape.

The *Caulobacter crescentus* Life Cycle

Caulobacter crescentus is an aquatic, Gram-negative α-proteobacterium with a curved rod morphology that undergoes cell cycle-regulated morphogenesis and an obligate asymmetric cell division (Stove and Stanier 1962; Shapiro et al. 1971). Its dimorphic life cycle and overt cell polarity, along with readily available genetic tools and easy methods of synchronization, make *C. crescentus* an excellent model

---

K. Sundararajan • E.D. Goley (✉)
Department of Biological Chemistry, Johns Hopkins University School of Medicine, Baltimore, MD, USA
e-mail: egoley1@jhmi.edu

J. Löwe, L.A. Amos (eds.), *Prokaryotic Cytoskeletons*, Subcellular Biochemistry 84, DOI 10.1007/978-3-319-53047-5_4

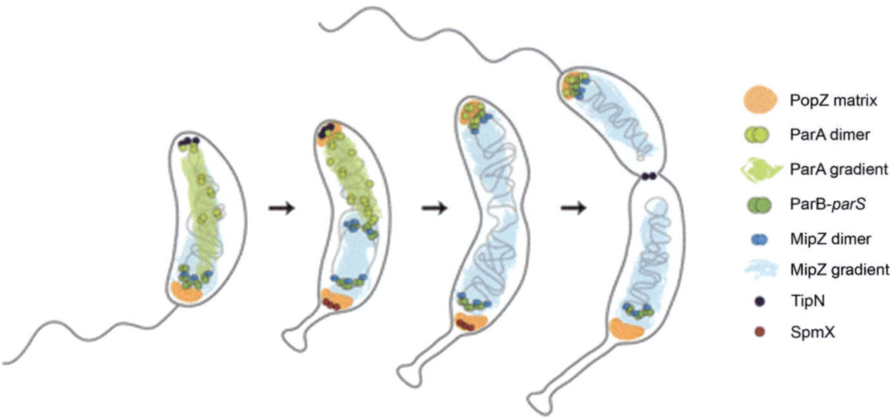

Fig. 4.1 Cytoskeletal elements involved in cell cycle progression. Swarmer cells (*left*, with polar flagellum) and newly divided stalked cells have a unipolar PopZ matrix and a single ParB focus (bound to *parS*) close to the old pole. ParA and MipZ bind DNA (*gray line*) non-specifically and localize as diffuse clouds enriched at the pole(s). In newborn cells, MipZ is concentrated at the old pole by ParB bound to PopZ whereas ParA is concentrated at the new pole by TipN. Following initiation of chromosome replication, a new PopZ matrix is assembled at the new pole. At the same time, PopZ at the old pole recruits SpmX to drive stalk biogenesis. As chromosome replication progresses, one of the ParB foci moves up the ParA concentration gradient towards the new pole, leading to an *ori-ter-ori* orientation of the chromosome. MipZ is segregated to the new pole with ParB-*parS*, leading to a MipZ concentration gradient with the minimum at approximately midcell (See text for more details)

organism for studying cell cycle-regulated processes and the role of the cytoskeleton in directing these events.

C. crescentus has two distinct cell types resulting from asymmetric cell division –a motile daughter or "swarmer" cell that possesses a polar flagellum and adhesive pili, and a sessile daughter or "stalked" cell that bears a slender, polar extension of the cell body called the stalk (Fig. 4.1) (Stove and Stanier 1962). An adhesive polysaccharide-rich matrix (holdfast) is secreted at the end of the stalk, and serves to attach the stalked cell to surfaces in the aquatic environments *C. crescentus* inhabits (Curtis and Brun 2010). Though the precise function of the stalk is debated, it likely serves to maximize nutrient access and/or uptake when *C. crescentus* is growing within a community in nutrient-poor aquatic environments (Klein et al. 2013). The flagellum and pili in swarmers and the stalk in stalked cells serve as morphological markers of cell polarity, and their polar locations are maintained from generation to generation.

Under laboratory growth conditions, the swarmer cell, which is non-replicating, differentiates into a stalked cell shortly after birth (Fig. 4.1). In nutrient-limited conditions in the wild, the swarmer cell stage is prolonged until a signal is received that there are enough nutrients available to produce progeny (Lesley and Shapiro 2008; England et al. 2010; Britos et al. 2011; Boutte et al. 2012). The swarmer-to-stalked cell differentiation is characterized by morphological changes, including elongation of the cell, ejection of the flagellum, retraction of the pili, and biogenesis of a stalk at the pole where the flagellum previously resided (Jenal 2000; Curtis and Brun 2010).

Accompanying the morphological changes associated with the swarmer-to-stalked transition, the stalked cell initiates replication and segregation of its single, circular chromosome. The chromosome is precisely positioned in *C. crescentus*, with each chromosomal locus localizing to a particular cellular address (Viollier et al. 2004). Notably, the chromosomal centromere (*parS*, a locus close to the origin of replication (*ori*)) is anchored at the old cell pole and the terminus (*ter*) is found near the new cell pole. Upon duplication, one copy of the centromere is rapidly segregated to the opposite pole and other loci follow into the opposite daughter compartment as they are replicated. The anchoring of the centromeres to opposite poles after segregation leads to an *ori-ter-ori* orientation of the chromosome in pre-divisional cells (Viollier et al. 2004). Following DNA segregation, a flagellum is built at the pole opposite the stalk, the envelope begins to constrict near midcell, and cytokinesis yields biochemically and morphologically distinct stalked and swarmer daughters (Fig. 4.1). The stalked cell immediately enters another round of division, while the swarmer first must differentiate to become a stalked cell.

Progression of *C. crescentus* through the cell cycle and execution of the associated morphogenetic events described above are tightly controlled through regulation of the abundance, activity, and localization of key proteins. Interconnected transcriptional circuits lead to sequential, cell cycle-dependent expression of over 500 genes (Laub et al. 2000; Zhou et al. 2015). Layered on top of transcriptional regulation, post-transcriptional and post-translational mechanisms, for example regulated proteolysis and phosphorylation, are in place to increase the robustness of the cell cycle program (Schrader et al. 2014; Zhou et al. 2015; Goley et al. 2007; Jenal 2009). In synergy with the regulation of protein abundance or activity, cytoskeletal elements are key contributors to cell cycle progression through spatial regulation of key developmental processes. These include: polarity establishment and maintenance, DNA segregation, cytokinesis, and cell elongation. Cytoskeletal proteins in *C. crescentus* are additionally required to maintain its rod shape, curvature, and pole morphology. In this chapter, we explore the mechanisms through which cytoskeletal proteins in *C. crescentus* orchestrate developmental processes by acting as scaffolds for protein recruitment, generating force, and/or restricting or directing the motion of molecular machines. We will discuss each cytoskeletal element in turn, beginning with those important for organization of molecules at the cell poles and chromosome segregation, then cytokinesis, and finally cell shape.

PopZ: Centromere Anchoring and Polar Organization

In *C. crescentus*, the cell poles serve as sites of assembly for molecules and activities involved in cell cycle progression and development, including the chromosomal centromere, proteins required for stalk and flagellar biosynthesis, and factors involved in the specification of stalked or swarmer cell fate. PopZ (Pole-organizing protein that affects FtsZ or Polar organizing protein Z) is a polar cytoskeletal protein that is conserved in nearly all α-proteobacteria and plays a key role in recruiting diverse proteins to the cell pole(s).

PopZ was first identified in *C. crescentus* as a polymerizing protein important for polar anchoring of the chromosomal centromere by two independent methods: a bioinformatic screen for proteins that are predicted to interact functionally with known polar proteins (Bowman et al. 2008) and a genetic screen for genes that cause cell division defects on overexpression (Ebersbach et al. 2008). It is a small acidic protein (177 amino acids in *C. crescentus*, pI = 3.88) that forms a scaffold for the recruitment of factors involved in stalk morphogenesis and centromere segregation and anchoring. While it is non-essential, deletion of *popZ* leads to cell elongation, minicell formation and prevention of stalk synthesis (Fig. 4.3B) (Bowman et al. 2008; Ebersbach et al. 2008).

PopZ Localization and Function in Cells

The first indication for the role of PopZ in polar organization came from investigating its localization over the cell cycle (Fig. 4.1 and 4.3I) (Bowman et al. 2008; Ebersbach et al. 2008). PopZ – visualized as a functional fluorescent fusion to YFP - forms a unipolar focus at the old pole in newly divided cells (Fig. 4.3I1). As chromosome replication is initiated and centromere segregation proceeds, PopZ transitions from a unipolar focus to bipolar foci (Fig. 4.3I2). PopZ oligomerizes and forms a matrix that excludes large macromolecules, including the nucleoid and ribosomes. This matrix can be seen by phase contrast microscopy as regions of low refractive index and by electron cryotomography (ECT) as ribosome-free zones (Fig. 4.4A) (Ebersbach et al. 2008; Bowman et al. 2010). The PopZ matrix has been proposed to function as a multivalent platform for the recruitment and/or retention of specific proteins at the cell pole(s) (Bowman et al. 2008, 2010; Ebersbach 2008).

A critical function of PopZ is in regulating the localization of the chromosomal centromere (*parS*) through its interactions with the ParA / ParB chromosome segregation machinery (Bowman et al. 2008, 2010; Ebersbach et al. 2008; Ptacin et al. 2014). At the old pole of newborn cells, the PopZ matrix binds to ParB, which binds to *parS*, thereby anchoring the *parS*-proximal regions of the chromosome (including the *ori*) close to the old pole. Following replication of the *parS* locus, one of the two ParB-*parS* complexes moves towards the new cell pole where it is captured by a newly assembled PopZ matrix (Fig. 4.1). Thus, the polar PopZ matrices anchor the duplicated centromeres at the poles and stabilize the *ori-ter-ori* orientation of the replicating chromosome. *In vitro* observations and heterologous expression studies indicate that PopZ can directly bind free ParB or ParB bound to *parS*, suggesting that PopZ and ParB are sufficient for centromere anchoring (Bowman et al. 2008; Ebersbach et al. 2008). In addition to anchoring the centromere, PopZ further contributes to chromosome organization by regulating the localization of ParA – the ATPase that directs migration of one copy of ParB-*parS* to the new pole (Bowman et al. 2008, 2010; Ebersbach et al. 2008; Ptacin et al. 2014). Regulation of ParA localization by PopZ is described in the section on ParA.

Upon the swarmer-to-stalked cell transition, PopZ at the old pole undergoes an apparent change in function (Ebersbach et al. 2008; Bowman et al. 2010). Roughly coincident with the initiation of DNA replication, PopZ releases the ParB-*parS* complex at the old pole and begins to recruit factors important for stalk morphogenesis and other developmental events at the old pole (Fig. 4.1). SpmX, a periplasmic muramidase that is the earliest recruit to the developing stalked pole, is localized to the old pole in a PopZ-dependent manner. SpmX recruitment is presumed to be indirect: the muramidase domain of SpmX, which is thought to reside in the periplasm, is sufficient for its localization, whereas PopZ is exclusively cytoplasmic (Bowman et al. 2010). Following localization of SpmX at the pole, PopZ recruits histidine kinases, DivJ and CckA, which regulate cell fate and stalk synthesis, and CpdR, ClpX, and RcdA, factors required for the regulated proteolysis of the master cell cycle regulator, CtrA (Ebersbach et al. 2008; Bowman et al. 2010). Co-immunoprecipitation experiments indicate that PopZ directly interacts with DivJ and CckA, further supporting the role for PopZ as a scaffold for the localization of these proteins (Ebersbach et al. 2008; Bowman et al. 2010). PopZ could recruit the other factors indirectly through its interaction with DivJ, CckA or other, as yet unknown, proteins.

Assembly Properties of PopZ

In vitro, PopZ forms large oligomers as evidenced by its behavior in native polyacrylamide gel electrophoresis (PAGE) and size exclusion chromatography. Although PopZ is a 17 kDa protein, it migrates in complexes as large as 650 kDa in a native PAGE (Bowman et al. 2008; Ebersbach et al. 2008). Size exclusion chromatography studies of PopZ mutants indicate that PopZ forms dimers of trimers, which then associate to form higher order structures (Bowman et al. 2013). Electron micrographs of purified PopZ show that PopZ polymers are 5 nm in diameter and vary widely in length – ranging from short filaments of 25–50 nm to long, branched filaments that are over 200 nm in length (Fig. 4.5A) (Bowman et al. 2008). Mutants of PopZ that are defective in forming higher order structures do not form a polar focus *in vivo* and are incapable of bipolar localization, suggesting that the ability to polymerize is required for PopZ localization (Laloux and Jacobs-Wagner 2013; Bowman et al. 2013).

PopZ is dissimilar in sequence and structure to other known polymerizing proteins, thus it establishes a new class of cytoskeletal proteins. Sequence comparison of PopZ homologs across α–proteobacterial species reveals that PopZ has three main regions:

1. An N-terminal region containing an α-helix (amino acids 1-26 in *C. crescentus* PopZ)
2. A proline-rich linker region (amino acids 26 – 106)
3. An α-helix-rich C-terminal region (amino acids 106 – 177) (Laloux and Jacobs-Wagner 2013; Bowman et al. 2013)

The N-terminal and C-terminal regions have distinct functions in regulating the assembly and polar localization of PopZ. The C-terminal α-helices are required for polar localization of PopZ, as mutants of PopZ that do not contain the C-terminal region fail to self-assemble and are diffuse throughout the cell (Laloux and Jacobs-Wagner 2013; Bowman et al. 2013). Size exclusion chromatography, multi-angle laser light scattering and native PAGE analyses indicate that the C-terminal region is sufficient for oligomerization (Bowman et al. 2013). This is further supported by observations in *C. crescentus* and upon heterologous expression in *E. coli* that fluorescent fusions to the C-terminal region of PopZ localize as a unipolar focus (Bowman et al. 2013).

While the C-terminal region is sufficient for polymerization of PopZ and its localization at the old pole in *C. crescentus*, PopZ requires its N-terminal α-helix for the transition from a unipolar focus to bipolar foci and for directing stalk biogenesis (Laloux and Jacobs-Wagner 2013). Mutations in the N-terminus lead to reduced frequency of bipolar PopZ foci, cell elongation and poor SpmX localization (Laloux and Jacobs-Wagner 2013). Moreover, isolation of N-terminal mutants of PopZ that prevent its interaction with ParA *in vitro,* but which do not affect PopZ bipolar localization *in vivo,* suggests that the N-terminus is also required for binding ParA (Ptacin et al. 2014). Cells expressing these mutants have delayed or incomplete migration of the ParB-*parS* complex, emphasizing the need for the PopZ-ParA interaction for efficient chromosome segregation (see ParA section for more detail).

PopZ: Unanswered Questions

While the details of the function of PopZ in regulating chromosome organization and polar development are beginning to become clear, how PopZ localizes to the poles in the first place has not yet been explained. PopZ does not rely on any other known polar protein for its localization and can localize to polar foci when expressed in *E. coli*, which lacks a PopZ homolog, suggesting that polar assembly is an intrinsic property of PopZ (Bowman et al. 2008; Ebersbach et al. 2008). One model argues that PopZ localizes to any nucleoid free region – which is the new pole in the case of swarmers – unless there is a pre-existing PopZ focus. The strongest evidence for this model comes from heterologous expression studies in *E. coli*. PopZ-YFP localizes as a single focus in *E. coli*, irrespective of expression levels (Bowman et al. 2008; Ebersbach et al. 2008; Laloux and Jacobs-Wagner 2013). Moreover, heterologous expression of *popZ-yfp* in *E. coli* following treatment with cephalexin, a cell division inhibitor, leads to PopZ foci between nucleoids (Laloux and

Jacobs-Wagner 2013). It therefore seems likely that the nucleoid physically excludes the large PopZ matrix, and vice versa, leading to PopZ assembly only in nucleoid-free regions. Whether these observations in *E. coli* are sufficient to explain polar PopZ localization in *C. crescentus* cells, which are smaller than *E. coli* and in which the nucleoid fills a larger portion of the cytoplasm, is unclear. Moreover, it is possible that *C. crescentus* has additional factors that assist in polar localization of the PopZ matrix. For example, either the inhibition of the MreB cytoskeleton or the loss of the polar marker TipN disrupts the normal distribution of PopZ (Bowman et al. 2008; Laloux and Jacobs-Wagner 2013). However, it has not been established if their effects on PopZ are direct.

Another unanswered question is how PopZ forms a new matrix at the new pole after DNA replication initiation. Unlike ParB and MipZ, which transition from unipolar to bipolar distribution by moving across the cell with the segregating centromere, PopZ forms a matrix at the new pole *de novo*, without disassembly or movement of the old matrix (Bowman et al. 2008; Ebersbach et al. 2008). PopZ overproduction in both *C. crescentus* and *E. coli* leads to expansion of a single PopZ matrix at one pole (Ebersbach et al. 2008; Laloux and Jacobs-Wagner 2013), therefore changes in PopZ levels cannot explain its appearance at the new pole. Initiation of DNA replication was originally suggested to be a signal for the formation of the new PopZ matrix (Bowman et al. 2008; Ebersbach et al. 2008). However, more recent evidence from cells treated with the replication inhibitor novobiocin suggests that it is the segregation of chromosome, specifically ParA-dependent migration of the ParB-*parS* complex and not replication itself, that is required for the bipolarization of PopZ (Laloux and Jacobs-Wagner 2013). It is estimated that 40% of total PopZ is not in a polar matrix, but rather is diffuse in the cytoplasm (Bowman et al. 2008; Laloux and Jacobs-Wagner 2013). It is possible that ParA, bound to TipN at the new pole, recruits cytoplasmic PopZ and increases its local concentration such that PopZ can form a new matrix (Laloux and Jacobs-Wagner 2013). How this series of interactions is triggered only upon centromere segregation remains to be determined.

ParA and the Chromosome Segregation Machinery

The *parABS* locus encodes a tripartite DNA partitioning system that functions in diverse bacteria to segregate plasmid or chromosomal DNA. It comprises a *cis*-acting DNA sequence, *parS*, and two proteins: ParB, which binds *parS* and *parS*-proximal sites along the DNA, and ParA, a Walker-type ATPase required for directed movement of ParB-*parS* complexes. As discussed below, there is conflicting data as to whether ParA assembles into cytoskeletal filaments or, alternatively, forms a nucleoprotein complex made of DNA-bound ParA dimers to promote DNA segregation (Howard and Gerdes 2010; Mierzejewska and Jagura-Burdzy 2012). We include ParA in our discussion of cytoskeletal proteins in *C. crescentus*, as one of the models proposed for chromosome segregation invokes ParA assembly into filaments.

ParABS: Localization and Function in Cells

ParA and ParB were first identified as proteins required for plasmid segregation in *E. coli* (van den Elzen et al. 1983). Since then, numerous studies have demonstrated their roles in chromosome segregation in diverse bacteria (Mohl and Gober 1997; H. J. Kim et al. 2000; Mohl et al. 2001; Godfrin-Estevenon et al. 2002; Fogel and Waldor 2006; Saint-Dic et al. 2006; Jakimowicz et al. 2007; Toro et al. 2008; Bartosik et al. 2009; Donovan et al. 2010; Iniesta 2014; Mierzejewska and Jagura-Burdzy 2012). *parA* and *parB* are essential in *C. crescentus* (Mohl and Gober 1997), and ParB depletion causes DNA segregation defects and cell elongation (Easter and Gober 2002).

ParB dimers bind specific AT-rich sites near the chromosomal origin of replication *(ori)*, known as *parS*, through a conserved helix-turn-helix motif (Toro et al. 2008; Figge et al. 2003). In pre-divisional cells, ParB bound to *parS* forms a polar focus that partially overlaps with PopZ at the old pole (Fig. 4.1) (Mohl and Gober 1997; Bowman et al. 2008; Ebersbach et al. 2008). Following the initiation of replication, *parS* is one of the first loci to be duplicated, as it lies only 8 kb away from *ori*. *parS* functions as the chromosomal centromere and is invariantly the first locus to be segregated (Toro et al. 2008). *parS* duplication leads to formation of two ParB-*parS* complexes. One ParB-*parS* focus migrates rapidly to the new pole, while the other focus is retained near the old pole (Mohl and Gober 1997; Viollier et al. 2004; Bowman et al. 2008; Toro et al. 2008; Shebelut et al. 2010). ParA is required for the completion of directed movement of the ParB-*parS* focus to the new pole (Mohl and Gober 1997; Toro et al. 2008).

ATP-bound ParA dimers bind sequence non-specifically to the nucleoid (Easter and Gober 2002) and may also polymerize (Ptacin et al. 2010). In swarmer cells, when imaged by widefield fluorescence microscopy, ParA forms a dynamic, cloud-like localization, with its highest concentration at the new pole opposite the ParB-*parS* complex (Fig. 4.3J) (Schofield et al. 2010). As DNA replication progresses, one of the two ParB-*parS* complexes comes in contact with the ParA cloud and is translocated rapidly across the cell to the new pole (Shebelut et al. 2010). ParB activates the ATPase activity of ParA and converts ParA-ATP – bound to DNA and/or assembled as a polymer - to free ParA-ADP monomers (Easter and Gober 2002; Figge et al. 2003). ParB, due to its affinity for ParA-ATP, moves up the concentration gradient of ParA towards the new pole (Toro et al. 2008; Ptacin et al. 2010; Shebelut et al. 2010). Thus, a tight coordination between ATPase activity, dimerization, and differential DNA binding affinities of the ParAB system drives chromosome segregation.

Mechanism of ParABS-Mediated Chromosome Segregation

While it is clear that ParA is required for directing the centromere to the new pole, whether ParA generates a pulling force on the centromere, like microtubules in a eukaryotic spindle, or if ParB follows the retracting ParA cloud through directed diffusion is unclear. Based on existing data, there are two models for ParA function and DNA translocation in *C. crescentus*. In the first model, ATP-bound ParA is proposed to form a linear, bundled polymer that is anchored at the new pole (Ptacin et al. 2010; Banigan et al. 2011). As the ParB-*parS* complex binds this filament at its unanchored end, it activates the ATPase activity of ParA and leads to depolymerization of the ParA filament. As ParA depolymerizes, it pulls ParB-*parS* along due to the affinity of ParB for ATP-bound ParA. Thus, ParABS system forms a "burning bridge" that helps translocate *parS*.

Evidence for this ParA depolymerization-based burning bridge model comes from the observation that ParA can form bundled filaments *in vitro* (Ptacin et al. 2010), and that ParA from other species have been observed to form filaments (Leonard et al. 2005; Ebersbach et al. 2006; Ringgaard et al. 2009). However, there is limited evidence for such a filament structure *in vivo*. While early super-resolution images of ParA-eYFP suggested a ~40 nm thick linear filament of ParA that stretches from the new pole to the ParB focus (Ptacin et al. 2010), more recent super-resolution microscopy of ParA-Dendra2 revealed a cloud-like localization that is distinct from a structured filament (Lim et al. 2014). Moreover, an estimation of ParA concentration in cells suggests that there are far fewer ParA molecules than would be required to form a linear polymer that can stretch across the cell from the new pole to where ParB first interacts with ParA (Lim et al. 2014). Furthermore, single molecule tracking of the ParB focus revealed that ParB-*parS* complex does not follow a linear path, as would be predicted for a burning bridge model (Lim et al. 2014).

For the above reasons, a diffusion-based "DNA-relay" model has been proposed wherein ParA-ATP dimers on the DNA form a cloud-like nucleoprotein complex that drives the motion of ParB-*parS* through directed diffusion coupled with a Brownian ratchet (Lim et al. 2014). Observations of the movement of chromosomal loci suggest that these loci are in constant oscillations about their equilibrium point. Thus, ParA dimers bound to DNA also undergo oscillation. When the ParB-*parS* complex binds a DNA-bound ParA dimer out of its equilibrium point, the elastic nature of the chromosome brings ParA/ParB-*parS* complex closer to the equilibrium point. Simultaneously, ParB stimulates ParA's ATPase activity leading to release of ParA monomers from the chromosome. As this cycle of ParB binding to ParA dimers and their subsequent removal repeats, the centromere moves to the new pole following the DNA bound ParA-dimer concentration gradient.

Ensuring Robust Chromosome Segregation: ParABS and Beyond

Maintaining a concentration gradient of DNA-bound ParA that increases towards the new pole is central to efficient segregation through the proposed DNA relay mechanism. Interaction of ParA with PopZ and the new pole marker TipN are required to establish this gradient (Ptacin et al. 2010, 2014; Schofield et al. 2010)

(Fig. 4.1). Heterologous expression of PopZ and ParA in *E. coli* suggests that PopZ interacts directly with ParA (Laloux and Jacobs-Wagner 2013). Moreover, in otherwise wild-type *C. crescentus* cells, a DNA binding mutant of ParA ($ParA_{G16V}$) localizes specifically as bipolar foci in complex with the PopZ matrix, whereas in *ΔpopZ* cells, this mutant is distributed throughout the cytoplasm with no apparent concentration gradient (Ptacin et al. 2014). These observations illustrate that binding to both PopZ and DNA are required to maintain the ParA gradient.

While necessary, the PopZ-ParA interaction is not sufficient to maintain the ParA gradient with its highest concentration near the new cell pole, as PopZ is enriched at the old pole or both poles depending on cell cycle status. For this, TipN, which is localized at the new cell pole, is required. In Δ*tipN* cells, the ParA concentration gradient redistributes to a cloud enriched at both poles, leading to inefficient ParB-parS migration (Ptacin et al. 2010; Schofield et al. 2010). Moreover, whereas the DNA-binding mutant $ParA_{G16V}$ forms bipolar foci in otherwise wild-type cells, it localizes to the old pole in Δ*tipN* cells. In addition, ParA and TipN have been shown to interact directly by FRET and *in vitro* pull down experiments (Schofield et al. 2010). From these observations, TipN has been suggested to restrict the ParA concentration maximum to the new pole, possibly by titrating away free ParA monomers that result from ParB-activated ATP hydrolysis by ParA and subsequent release of ParA from the DNA (Schofield et al. 2010). Simultaneous deletion of *popZ* and *tipN* is synthetic lethal, further supporting the cooperative contributions of these factors to regulating ParA localization and function.

While it is clear that ParA is required for efficient centromere migration, there appear to be other compensatory mechanisms for chromosome segregation in the absence of a functional ParA. These include MreB-driven cell elongation (Shebelut et al. 2009, 2010; Schofield et al. 2010) and bulk segregation mechanisms mediated by the motor activities associated with DNA replication and transcription (Shebelut et al. 2010). In fact, ParA is required only for the late stages of centromere translocation (Shebelut et al. 2010). Nevertheless, cells lacking a functional ParABS system are not viable and have severe cell division defects. The cytokinesis defects associated with *parABS* mutants are attributed to the mislocalization of MipZ, a ParB-associated factor that coordinates DNA segregation with assembly of the cytokinetic FtsZ ring (Thanbichler and Shapiro 2006).

FtsZ: The Orchestrator of Cytokinesis

Shortly after segregation of the chromosomal centromere, the multi-protein machinery that will ultimately divide the cell begins to assemble near midcell (Thanbichler and Shapiro 2006). At the core of this machinery is FtsZ, a tubulin-like GTPase that is conserved in most bacteria with a peptidoglycan cell wall. It polymerizes on binding GTP and forms a ring like structure at the division site (Z-ring) that serves to coordinate invagination of the cell envelope during cytokinesis (Erickson et al. 2010; Meier and Goley 2014). In the absence of FtsZ, cells fail to divide, elongate, and eventually lyse (Fig. 4.3D). FtsZ is proposed to act as a scaffold for the recruitment of the division machinery (divisome), and as a potential source of constrictive force for envelope invagination. In addition to its role in cell division, *C. crescentus* FtsZ also directs elongation of the cell prior to initiating constriction of the envelope for cytokinesis (Aaron et al. 2007). The conserved features of FtsZ assembly and function are discussed elsewhere in this book. Here, we focus on the characteristics of FtsZ that have been specifically observed for *C. crescentus*.

FtsZ Domain Architecture and Function

Based on sequence homology across species, FtsZ is divided into four regions – (i) a poorly conserved N-terminal peptide of unknown function (ii), a GTPase domain that adopts the tubulin fold, (iii) an intrinsically disordered C-terminal linker (CTL), and (iv) a C-terminal conserved (CTC) peptide (Vaughan et al. 2004). The GTPase domain is highly conserved and is sufficient for polymerization of FtsZ *in vitro* (X. Wang et al. 1997; Din et al. 1998; Vaughan et al. 2004). The CTC is the binding site for several essential regulators of FtsZ function, including membrane-anchoring proteins and the ClpXP protease (Ma et al. 1997; Din et al. 1998; Ma and Margolin 1999; Król et al. 2012; Duman et al. 2013; Williams et al. 2014). The disordered CTL is the least conserved region of FtsZ: it is highly variable in length and sequence across species and is particularly long (100-400 residues) in α–proteobacteria (Vaughan et al. 2004; Margolin 2005). Despite its poor conservation, the CTL plays an essential, length- and disorder-dependent, but sequence-independent function in *E. coli* and *B. subtilis* (Buske and Levin 2013; Gardner et al. 2013). In *C. crescentus*, however, the linker plays both sequence- and length-dependent roles in executing cytokinesis (Sundararajan et al. 2015). The mechanisms by which the length and sequence of the CTL contribute to efficient cytokinesis are not yet clear, although both of these features affect steady state FtsZ levels and Z-ring structure (Sundararajan et al. 2015).

Z-ring Structure In Vivo

Various studies have characterized the structure and dynamics of the Z-ring *in vivo* as a means to understand its function both as a scaffold and as a potential force generator. ECT and polarized fluorescence microscopy imaging of *C. crescentus* cells revealed relatively sparse, single-stranded FtsZ protofilaments oriented perpendicular to the long axis of the cell (Fig. 4.4B) (Z. Li et al. 2007; Si et al. 2013; Szwedziak et al. 2014). On average, protofilaments are positioned 16 nm from the inner membrane (Z. Li et al. 2007; Szwedziak et al. 2014). Whether these filaments assemble into a continuous ring, however, is currently subject to debate (Figs. 4.3K and 4.4B) (Z. Li et al. 2007; Holden et al. 2014; Szwedziak et al. 2014). An early

ECT study and later super-resolution light microscopy favor a model of FtsZ protofilaments organized into discontinuous clusters (Z. Li et al. 2007; Holden et al. 2014), but a more recent ECT study revealed protofilaments that appear to completely encircle the cell at its waist (Szwedziak et al. 2014). The diameter of the Z-ring gradually decreases with cell envelope invagination, starting at ~500 nm and decreasing to ~350 nm at a constant rate of 0.8 nm/min (Holden et al. 2014). This gradual constriction of the Z-ring then switches to a rapid constriction stage, during which the diameter decreases to ~150 nm towards the end of cytokinesis (Holden et al. 2014).

FtsZ Function Over the Cell Cycle

Following DNA replication initiation in stalked cells, FtsZ begins to assemble approximately at the middle of the cell into a circumferential Z-ring (Fig. 4.2) (Quardokus et al. 2001; Thanbichler and Shapiro 2006). The assembly of the Z-ring triggers the sequential recruitment of more than two dozen factors that function together to effect cytokinesis and other FtsZ-dependent events (Goley et al. 2011). These factors include (i) FtsZ-binding proteins that regulate Z-ring structure,

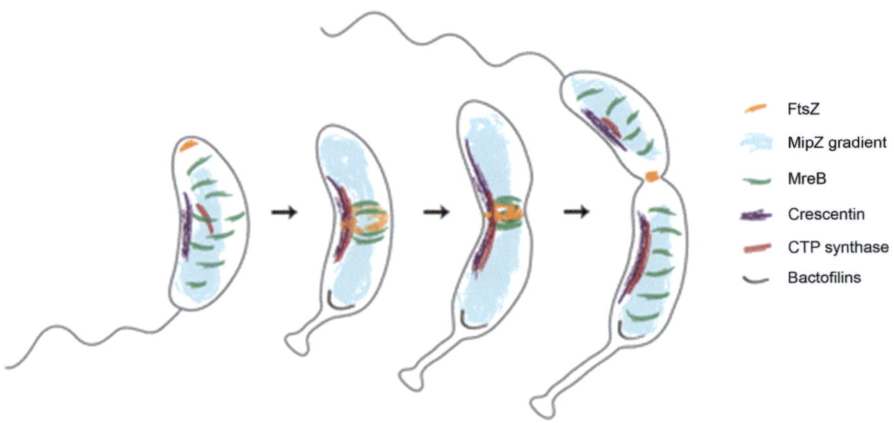

Fig. 4.2 Cytoskeletal regulation of cell shape and morphogenesis. In swarmer cells, dynamic clusters of MreB are distributed along the length of the cell and direct cell elongation. FtsZ is localized to the new cell pole in swarmers, but due to the dynamic relocation of the MipZ concentration maxima to the poles upon centromere segregation, FtsZ is displaced from the new pole and reassembled at midcell to form the Z-ring. Following Z-ring assembly, MreB is recruited to the midcell where it drives midcell elongation along with FtsZ. Finally, following the arrival of late division proteins, MreB is displaced from the midcell into dispersed clusters. Crescentin and CTP synthase work synergistically to direct cell curvature. Crescentin forms short structures at the inner cell curvature in swarmer cells, which elongate as the cell grows until these crescentin filaments reach the cell poles. CTP synthase forms shorter, predominantly cytoplasmic filaments in swarmer cells and longer, membrane-associated structures in stalked cells. Bactofilins assemble into a scaffold in early stalked cells at the base of the stalk and regulate pole morphogenesis

(ii) proteins that anchor FtsZ to the membrane, (iii) factors that stabilize the division machinery (divisome), (iv) peptidoglycan remodeling enzymes, (v) DNA translocation factors, (vi) outer membrane proteins, and (vii) pole morphogenesis factors. Although all of these proteins require FtsZ for their midcell localization, only a subset –FtsA, FzlA, FzlC, ZapA, FtsE, KidO, GdhZ, and ClpX – interact directly with FtsZ (Ma et al. 1997; X. Wang et al. 1997; Din et al. 1998; Goley et al. 2010; Radhakrishnan et al. 2010; Beaufay et al. 2015). FtsZ and the divisome together cause local remodeling of the cell envelope which results in cytokinesis (Meier and Goley 2014). While it is clear that FtsZ acts as a scaffold for the local assembly of the divisome, recent evidence suggests that FtsZ also contributes to the regulation of peptidoglycan enzyme activities through a CTL-dependent mechanism beyond its scaffolding function (Sundararajan et al. 2015).

Surprisingly, FtsZ directs a significant portion of cell wall synthesis for cell elongation in *C. crescentus* (Aaron et al. 2007; Kuru et al. 2012). Before the initiation of constriction, FtsZ recruits MreB, the actin homolog important for directing elongation-mode cell wall synthesis, and MurG, a critical component of the peptidoglycan precursor synthesis pathway, and directs longitudinal peptidoglycan synthesis at the midcell (Aaron et al. 2007; Goley et al. 2011). Midcell elongation continues until the arrival of the late divisome proteins FtsW and FtsB, after which peptidoglycan remodeling switches from an elongation mode to a division mode (Aaron et al. 2007; Goley et al. 2011).

Force Generation and FtsZ Filament Curvature

In vitro observations of *E. coli* FtsZ polymers reconstituted on liposomes have suggested that, in addition to acting as a scaffold, FtsZ is capable of generating force to deform membranes (Osawa et al. 2008, 2009). A curvature-based force generation mechanism has been proposed based on several observations. First, electron microscopy of FtsZ protofilaments indicates that FtsZ filaments become more curved upon GTP hydrolysis (Lu et al. 2000) and GTP hydrolysis mutants of FtsZ fail to complete cytokinesis (Y. Wang et al. 2001; Stricker and Erickson 2003). In addition, the geometry of membrane deformation by FtsZ *in vitro* depends on the orientation of the membrane anchor relative to the direction of filament curvature (Osawa et al. 2009; Osawa and Erickson 2011). Based on these observations and the presence of a discontinuous Z-ring, an "iterative pinching" model of FtsZ function has been proposed (Z. Li et al. 2007; Erickson et al. 2010; Y. Li et al. 2013; Holden et al. 2014). According to this model, membrane-anchored, curved FtsZ filaments cause local deformation of the membrane that is coupled with peptidoglycan remodeling. Repeated rounds of FtsZ filament assembly, membrane deformation, and peptidoglycan remodeling at the division site lead to coordinated cell envelope invagination.

Regulating the curvature of FtsZ filaments is central to the proposed "iterative pinching model". Interestingly, in *C. crescentus*, FzlA, a glutathione S-transferase family protein that is conserved in α-proteobacteria, binds directly to FtsZ and induces formation of highly curved, stable filament bundles (Goley et al. 2010). FzlA, an early recruit to the division site, is the only known FtsZ-binding factor that affects filament curvature and is essential for cytokinesis. Depletion of FzlA results in smooth elongated cells with no envelope invaginations, whereas overexpression of FzlA results in aberrant assembly of FtsZ as a tight focus instead of a ring-like structure (Goley et al. 2010). It has been proposed that FzlA-mediated curved filaments contribute to FtsZ-mediated force generation in *C. crescentus* and its relatives.

Cell Cycle Regulation of FtsZ Levels

In *C. crescentus*, assembly of the Z-ring is temporally regulated, in part, through transcriptional and post-translational regulation of FtsZ levels (Quardokus et al. 1996, 2001; Kelly et al. 1998; Williams et al. 2014). FtsZ levels are lowest in swarmer cells (Quardokus et al. 1996, 2001; Kelly et al. 1998). As swarmers mature to stalked cells, FtsZ is upregulated to approximately 1000 molecules/cell. This number further increases following DNA replication initiation to ~6400 molecules/cell and finally reaches a maximum of ~7600 molecules during active envelope constriction in pre-divisional cells. Just before the completion of cytokinesis, FtsZ levels drop back down to ~1700 molecules (Quardokus et al. 2001). This cycling of FtsZ levels is achieved by the cell cycle-controlled transcription of *ftsZ* through the regulated activity of three cell cycle master regulators: CtrA, which represses *ftsZ* transcription (Kelly et al. 1998); DnaA, which activates *ftsZ* transcription (Hottes et al. 2005); and CcrM, which regulates the methylation state and activity of the *ftsZ* promoter (Gonzalez and Collier 2013). In addition, FtsZ has a shorter half-life in swarmer cells than in stalked cells, and is degraded by the ClpXP and ClpAP protease complexes (Quardokus et al. 1996, 2001; Kelly et al. 1998; Williams et al. 2014). Notably, ClpX recognizes a degradation signal on the C-terminal helix of FtsZ and is recruited to the division plane towards the end of cytokinesis to promote FtsZ turnover (Williams et al. 2014).

Positioning of the Z-ring in Coordination with DNA Segregation

In addition to the regulation of cytokinesis through alteration of FtsZ levels, the timing and location of Z-ring assembly in *C. crescentus* is coordinated with segregation of the chromosomal centromere through the action of a negative regulator of FtsZ assembly, MipZ (Figs. 4.1 and 4.2) (Quardokus et al. 2001; Quardokus and Brun 2002; Thanbichler and Shapiro 2006; Kiekebusch et al. 2012). MipZ is a Walker-type ATPase that binds FtsZ and promotes disassembly of FtsZ filaments by increasing its GTP hydrolysis activity (Thanbichler and Shapiro 2006). MipZ interacts with the ParB-*parS* complex and with DNA, depending on its nucleotide-bound and dimerization state, forming a concentration gradient that is highest at the cell poles and lowest at midcell following *parS* segregation (Thanbichler and Shapiro 2006; Kiekebusch et al. 2012). This gradient establishes a region of low MipZ concentration at midcell where FtsZ can polymerize, thereby determining the position of the Z-ring and the future division site (Thanbichler and Shapiro 2006; Kiekebusch et al. 2012). MipZ is the only known spatial regulator of FtsZ localization in *C. crescentus*. Unlike *E. coli* and *B. subtilis*, nucleoid occlusion of Z-ring assembly has not been observed in *C. crescentus* and, in fact, Z-rings readily form over regions of high DNA concentration (Thanbichler and Shapiro 2006).

Another connection between FtsZ and DNA segregation exists through FtsK, a component of the divisome that translocates the chromosomal terminus (*ter*) (S. C. E. Wang et al. 2006). FtsK is required for *ter* partitioning and it is recruited to the Z-ring before the beginning of invagination (S. C. E. Wang et al. 2006; Goley et al. 2011). Depletion of FtsK results in formation of multiple Z-rings and disruption of division (S. C. E. Wang et al. 2006). Based on its association with the nucleoid, and its effect on the Z-ring, FtsK may play a role functionally analogous to nucleoid occlusion in *C. crescentus* (S. C. E. Wang et al. 2006).

Metabolic Regulation of FtsZ Function

Recent work in numerous organisms has demonstrated that cytokinesis must not only be regulated in a cell cycle-dependent fashion, but also in response to environmental and metabolic cues (Kirkpatrick and Viollier 2012). This is true in *C. crescentus* as well: KidO and GdhZ are metabolic enzymes implicated in the regulation of Z-ring assembly in response to metabolic state of the cell (Radhakrishnan et al. 2010; Beaufay et al. 2015). These two factors are proposed to work synergistically to prevent premature Z-ring assembly and promote Z-ring disassembly during constriction. On binding to its substrate NADH, KidO reduces lateral interaction between FtsZ filaments (Radhakrishnan et al. 2010; Beaufay et al. 2015). GdhZ, on the other hand, binds NAD^+ and glutamate, and increases the GTP hydrolysis rate of FtsZ, thereby favoring depolymerization (Beaufay et al. 2015). As both KidO and GdhZ are regulated by the metabolic coenzyme, NAD, it is hypothesized that they function as metabolic checkpoints for cell cycle progression through FtsZ. In addition, KidO and GdhZ are upregulated towards the end of cytokinesis, and are recruited to the division site. Hence, they could be involved in the disassembly of the cytokinetic ring at the end of the cell cycle, in parallel to proteolytic degradation of FtsZ (Beaufay et al. 2015; Williams et al. 2014). The significance of this mode of regulation remains to be fully elucidated (Radhakrishnan et al. 2010; Beaufay et al. 2015).

MreB: Regulating Cell Shape and Polarity

FtsZ contributes to cell shape maintenance primarily by maintaining consistent cell length. Uniform cell width, on the other hand, is maintained by MreB, a membrane-associated bacterial homologue of actin that is present in most rod-shaped bacteria (Daniel and Errington 2003; Figge et al. 2004; Shaevitz and Gitai 2010; Harris et al. 2014). *C. crescentus* cells depleted of MreB widen, become lemon-shaped, and eventually lyse (Fig. 4.3E) (Figge et al. 2004; Gitai et al. 2004). Beyond its role in defining rod shape, *C. crescentus* MreB has been implicated in the regulation of chromosome segregation, polar morphogenesis, and cell curvature (Gitai et al. 2004, 2005; Wagner et al. 2005; Divakaruni et al. 2007; Shebelut et al. 2009; Charbon et al. 2009; Dye et al. 2011). The assembly properties of MreB and its conserved role in maintaining rod shape are discussed elsewhere in this book. Here, we will focus on insights into MreB regulation and roles for MreB specific to cell cycle progression and development derived from studies in *C. crescentus*.

Regulation and Dynamics of MreB Localization

Like many proteins implicated in cell shape and polar development, the localization of MreB is dynamic over the cell cycle in *C. crescentus* (Fig. 4.2). In recently divided cells, MreB localizes in dynamic patches dispersed along the length of the cell that move circumferentially along the short cell axis (Figs. 4.2 and 4.3L) (Figge et al. 2004; Gitai et al. 2004, 2005; Wagner et al. 2005; S. Y. Kim et al. 2006; Harris et al. 2014). Following assembly of the cytokinetic FtsZ ring in early stalked cells, MreB is recruited to midcell in an FtsZ-dependent manner and is thought to help direct cell wall metabolism for elongation of the cell prior to cytokinesis (Figge et al. 2004; Gitai et al. 2004; Divakaruni et al. 2007; Goley et al. 2011). However,

Fig. 4.3 (continued) (Ausmees et al. 2003). (**G**) Phase contrast image of cells depleted of CTP synthase (Ingerson- Mahar et al. 2010). (**H**) Phase contrast image of cells overexpressing *bacA-venus* (Kühn et al. 2009). Figures (**I–O**) represent localization pattern of individual cytoskeletal proteins. (**I**) Epifluorescence microscopy showing polar localization of PopZ-YFP (in red overlaid on DIC images) in synchronized cells; (**I1**) – early swarmer with unipolar PopZ localization, (**I2**) – Pre-divisional cell with bipolar PopZ localization, (**I3**) – Bipolar PopZ localization in cell undergoing cytokinesis. Scale bar = 2μm (Ebersbach et al. 2008). (**J**) PALM image of ParA-Dendra2 expressed from native locus (**J1**) or under the control of P_{xyl} promoter (**J2**). Scale bar = 1 μm (Lim et al. 2014). (**K**) HT-PALM image of FtsZ-Dendra2 in a predivisional cell; (**K1**) and (**K2**) represent longitudinal and cross-sectional views of the Z-ring respectively. Scale bar = 500 nm (Holden et al. 2014). (**L**) PALM image of eYFP-MreB. Scale bar = 300 nm (Biteen et al. 2008). (**M**) Immunofluorescence microscopy image showing CreS-FLAG (*red*) and the nucleoid stained with DAPI (*blue*). Scale bar = 2 μm (Ausmees et al. 2003). (**N**) Epifluorescence microscopy of cells expressing mChy-CtpS (red overlaid on phase contrast image). Scale bar = 2 μm (Ingerson-Mahar et al. 2010). (**O**) Merged image showing localization of BacA-eCFP and BacB-Venus. Scale bar = 2 μm (Kühn et al. 2009)

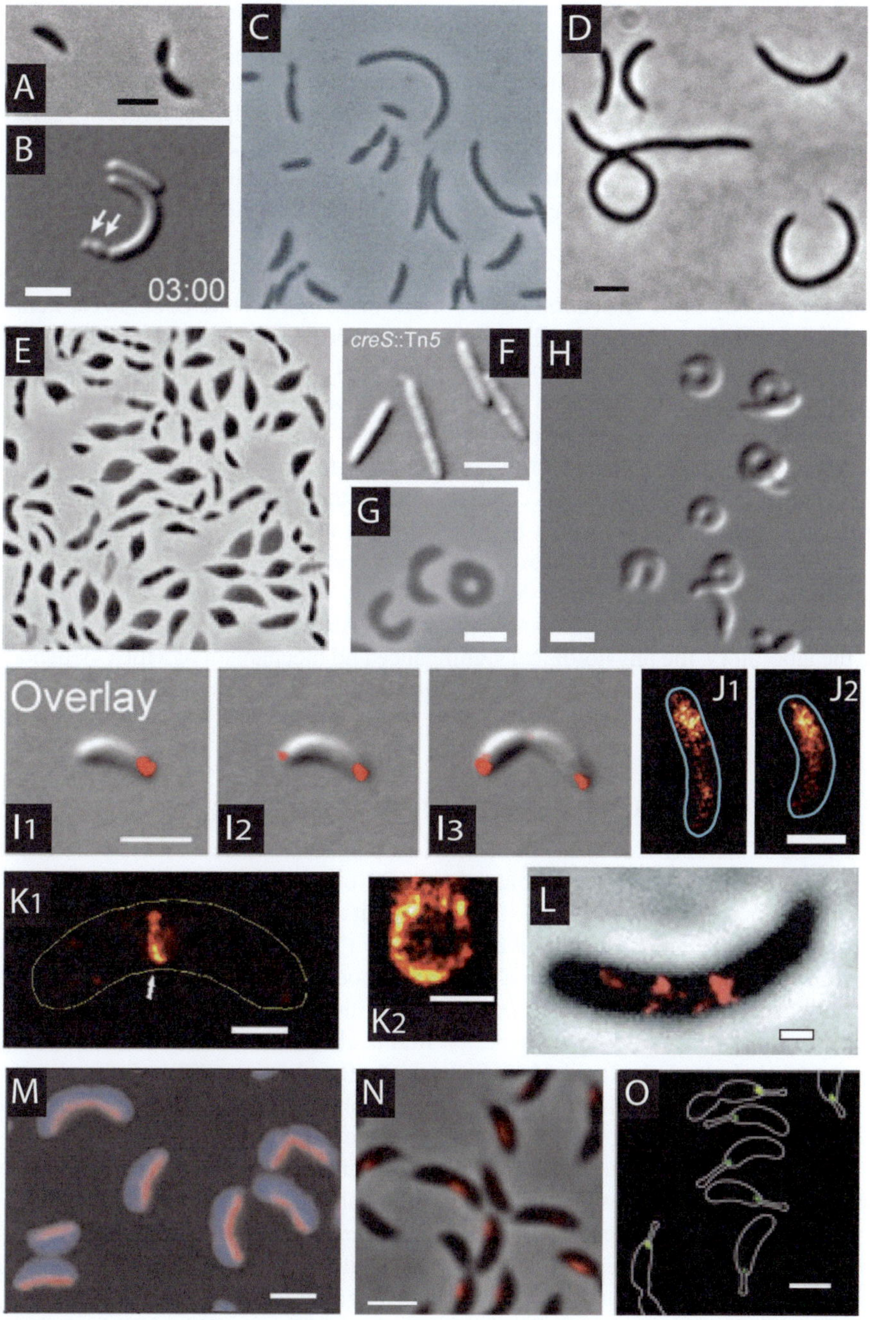

Fig. 4.3 Function and localization of *C. crescentus* cytoskeleton. (**A**) Phase contrast image of wild-type *C. crescentus* cells (Unpublished results). Figures (**B–H**) represent morphological changes upon deletion, depletion or overexpression of individual cytoskeletal elements. Scale bars = 2 μm. (**B**) DIC image of *ΔpopZ* cells. Arrows highlight minicell formation (Ebersbach et al. 2008). (**C**) Phase contrast image of cells overexpressing *parA* (Mohl and Gober 1997). (**D**) Phase contrast image of cells depleted of FtsZ for 4.5 h (Sundararajan et al. 2015). (**E**) Phase contrast image of cells depleted of MreB for 10 h (Figge et al. 2004). (**F**) DIC image of *ΔcreS* cells

the localization of MreB to midcell is not required for normal growth and morphology, as MreB mutants have been isolated that fail to accumulate at midcell, yet have apparently normal cell shape (Aaron et al. 2007; Dye et al. 2011). Roughly coincident with cytoplasmic compartmentalization in late predivisional cells, MreB is dispersed from midcell and once again adopts a dynamic patchy localization (Gitai et al. 2004; Goley et al. 2011).

While FtsZ is required for midcell recruitment of MreB, it is not clear if the two interact directly in *C. crescentus*, as reported for *E. coli* (White et al. 2010; Fenton and Gerdes 2013). However, MbiA, a factor identified in *C. crescentus* and conserved in a subset of α-proteobacteria, does bind MreB directly and regulate its localization (Yakhnina and Gitai 2012). While *mbiA* is not essential, overexpression of *mbiA* causes cell shape defects that resemble the effects of MreB depletion. Surprisingly, MbiA overproduction also leads to mislocalization of MreB: in the presence of excess MbiA, MreB condenses into a thick band at midcell. This occurs even in swarmer cells in which FtsZ has not yet assembled into a ring, implying FtsZ-independent midcell localization. Overexpression of *mbiA* also promotes midcell localization of a mutant of MreB - $MreB_{Q26P}$ - that is otherwise defective for midcell recruitment (Yakhnina and Gitai 2012). As *mbiA* is dispensable for growth in the laboratory, the physiological significance and mechanism of MbiA-dependent regulation of MreB localization and function remain to be elucidated.

MreB and C. crescentus *Morphogenesis*

Investigations into the physiological functions of MreB have been facilitated by identification of small molecule inhibitors of its assembly, including A22 (Gitai et al. 2004) and its less toxic derivative, MP265 (Takacs et al. 2010). A22 was first described as a molecule affecting *E. coli* cell shape (Iwai et al. 2002), but MreB was confirmed as its primary target through the identification of mutations in *C. crescentus* MreB that confer resistance to A22 (Gitai et al. 2005; Charbon et al. 2009; Dye et al. 2011). Indeed, A22 and MP265 cause MreB to become diffusely localized *in vivo* (Gitai et al. 2004; Takacs et al. 2010) and inhibit its assembly into antiparallel double filaments *in vitro* (Fig. 4.5D) (van den Ent et al. 2014).

While it is clear that the conserved function of MreB is the maintenance of cell width through the regulation of cell wall metabolism, in *C. crescentus,* MreB also influences cell curvature. A subset of A22-resistant MreB mutant strains have defects in cell curvature and are straight (Dye et al. 2011). This is attributed to a role for MreB in the membrane attachment of crescentin, the curvature-defining intermediate filament-like protein discussed later in this chapter (Charbon et al. 2009; Dye et al. 2011).

A second MreB-dependent morphogenetic process specific to *C. crescentus* and its relatives is the elaboration of a polar stalk. Stalk synthesis and elongation require specialized peptidoglycan metabolism to create a compartment with a smaller diameter than the cell body. Despite their different dimensions, elongation and synthesis of the stalk, like elongation of the cell body, requires MreB (Wagner et al. 2005). In *C. crescentus* cells depleted of MreB, stalks are either short, absent, or misshapen, appearing as polar "bulbs" much wider in diameter than wild-type stalks. When MreB is added back to these cells, stalks form at ectopic poles (i.e. emanating from the sides of cells) and are frequently branched (Wagner et al. 2005). These findings suggest that stalk elongation is a modified version of cell elongation, and that MreB is important for developing stalks of the appropriate diameter and for maintaining cell polarity so that stalks are assembled in the appropriate place.

MreB and Polar Development

In addition to its role in stalk placement, MreB is proposed to play a general role in cell polarity, as it influences the polar localization of the chromosomal centromere and of numerous proteins involved in cell cycle progression and development (Gitai et al. 2004, 2005; Bowman et al. 2008). Alterations in MreB levels cause severe defects in chromosome segregation (Gitai et al. 2004). Cells treated with A22 experience growth condition-specific defects in segregation of origin-proximal regions of the chromosome: cells grown in liquid media have delayed *parS* segregation, while those grown on solid media frequently fail to segregate *parS* entirely (Gitai et al. 2005; Shebelut et al. 2009). Moreover, chromatin immunoprecipitation experiments demonstrated physical association of MreB with multiple origin-proximal loci, including *parS* (Gitai et al. 2005). These observations initially led to the proposition that MreB forms a mitotic spindle-like segregation machine. However, the lack of a strict requirement for MreB for chromosome segregation under all tested growth conditions (Shebelut et al. 2009) and absence of a large-scale polar MreB structure *in vivo* (S. Y. Kim et al. 2006; Swulius et al. 2011; van den Ent et al. 2014) argue against this possibility.

In addition to defects in chromosome segregation, PopZ is completely delocalized in lemon-shaped cells depleted of MreB, and its assembly at the new cell pole is delayed in cells treated with A22 (Bowman et al. 2008; Laloux and Jacobs-Wagner 2013). MreB depletion also leads to mislocalization of polar developmental regulators including DivK, DivJ, and CckA (Gitai et al. 2004). However, since PopZ provides a platform for recruitment of these factors (Ebersbach et al. 2008; Bowman et al. 2010), the effects of MreB on their localization may be indirect. Indeed, as assembly of PopZ at the new cell pole requires chromosome segregation (Laloux and Jacobs-Wagner 2013), one possibility is that inhibition of MreB causes delayed or inhibited *parS* segregation, which prevents timely bipolarization of PopZ, which in turn prevents proper recruitment of other polar proteins. In general, MreB inhibition with A22 slows cell growth and cell cycle progression, providing a possible explanation for the observed delay in chromosome segregation and accumulation of PopZ at the new cell pole (Sliusarenko et al. 2011; Laloux and Jacobs-Wagner 2013).

MreB: Unanswered Questions

The use of small molecule inhibitors to assess the acute effects of MreB inhibition circumvents many of the caveats associated with the large-scale cell shape changes associated with long term MreB depletion. However, questions about the mechanisms by which MreB participates in polarity establishment and morphogenesis still remain. How is the dynamic localization of MreB to and its release from midcell achieved, and for what purpose? Does MreB directly interact with crescentin to anchor it to the membrane and promote cell curvature? How does MreB help to maintain both the width of the cell and the width of the stalk, when their diameters are vastly different? And does MreB play a specific role in defining cell polarity, or are its effects on polarity secondary to its effects on cell physiology and shape?

Crescentin and Cell Curvature

In addition to maintaining a rod shape through the action of MreB, *C. crescentus* cells adopt a gentle curvature as a result of another cytoskeletal protein, crescentin. Encoded by *creS*, crescentin is the first and best described bacterial intermediate filament (IF) protein. It was identified in a visual screen for cell shape mutants (Ausmees et al. 2003). *creS* is not essential: *creS* mutant cells grow as straight rods, but are otherwise indistinguishable from wild-type (Fig. 4.3F) (Ausmees et al. 2003) suggesting that this protein evolved specifically for determining cell shape. Emphasizing its role in defining cell shape, overexpression of *creS* causes hypercurvature of cells (Cabeen et al. 2009). While deleting *creS* has no effect on fitness in the laboratory, the vibrioid morphology of wild-type *C. crescentus* enhances its ability to colonize surfaces in conditions of fluid flow (Persat et al. 2014). This enhanced surface colonization is likely to confer a fitness advantage in the wild, where *C. crescentus* populates surfaces in aquatic environments such as streams and lakes.

Crescentin Localization and Assembly Properties

Crescentin forms a membrane-proximal, filamentous structure along the inner curvature of the cell when visualized by immunofluorescence microscopy or by imaging fluorescently-tagged crescentin variants (Figs. 4.2 and 4.3M) (Ausmees et al. 2003; Cabeen et al. 2009). As the cell cycle progresses, the crescentin structure elongates, then separates into two at the incipient division plane prior to completion of cytokinesis (Charbon et al. 2009). Several lines of evidence indicate that crescentin is membrane associated *in vivo*. In cells treated with inhibitors of the actin homolog, MreB, or with inhibitors of cell wall synthesis, the crescentin filament appears to dislodge from the inner curvature and recoil into a helical, cytoplasmic structure (Cabeen et al. 2009; Charbon et al. 2009). Similarly, an N-terminal truncation variant of crescentin ($CreS_{\Delta N27}$) assembles into a helical, cytoplasmic filament, indicating that membrane attachment is likely mediated by the N-terminus (Cabeen et al. 2009). Fluorescence microscopy of CreS-GFP in elongated cells revealed that crescentin helices have a uniform pitch of 1.6 ± 0.1 μm (Ausmees et al. 2003). This is likely a stretched conformation, as the $CreS_{\Delta N27}$ truncation mutant that is presumed to be dissociated from the membrane has a shorter pitch of 1.4 ± 0.15 μm (Cabeen et al. 2009).

Based on sequence identity and biochemical properties, crescentin is classified as an IF protein (Ausmees et al. 2003). Like eukaryotic IF proteins, crescentin contains 4 coiled-coil regions with seven-amino-acid (heptad) repeats separated by short linkers. Also like other IF proteins, crescentin bears a conserved stutter, an interruption in the heptad repeat, in the C-terminal coiled-coil region (Ausmees et al. 2003). Solubilized crescentin spontaneously assembles into filaments *in vitro* (Ausmees et al. 2003; Esue et al. 2010; Cabeen et al. 2011). While self-assembly is largely unaffected by salt concentrations, crescentin filaments form most readily at a low pH of 6.5 (Fig. 4.5E). At higher pH of up to 8.5, divalent cations (e.g. 5 mM Mg^{2+}) are required to stabilize polymers (Cabeen et al. 2011). At pH 6.5, crescentin structures are typically 8–10 nm (single filaments) or 17–20 nm (paired bundles) in width, but can form thicker bundles, particularly in the presence of divalent cations (Fig. 4.5E) (Ausmees et al. 2003; Cabeen et al. 2011). A study of the mechanical properties of crescentin *in vitro* determined that crescentin filamentous structures are elastic and solid-like, and can recover their network elasticity after shear stress (Esue et al. 2010). Both the self-assembly and mechanical properties of crescentin are consistent with its designation as an IF protein. It is worth noting that crescentin polymers have not been observed by ECT *in vivo* except when a membrane-dissociated form of crescentin was overproduced (Cabeen et al. 2010), so the physiologically relevant structure of crescentin awaits demonstration.

Crescentin-Mediated Cell Curvature: Mechanical Regulation of Peptidoglycan Synthesis

A growing body of evidence supports a model in which crescentin plays a mechanical role in regulating peptidoglycan synthesis to elicit a curved cell shape. First, crescentin-induced cell curvature depends on its polymerization and membrane attachment (Cabeen et al. 2009, 2010, 2011; Charbon et al. 2009). Crescentin variants that fail to polymerize under physiological pH and salt conditions *in vitro* (e.g. pH 7.5, 200 mM KCl) also fail to assemble into filaments *in vivo*, and cells producing these non-polymerizing variants are straight (Cabeen et al. 2011). Cells expressing membrane-dissociated $CreS_{\Delta N27}$ are also straight, even though the polymerization properties of this mutant are similar to wild-type (Cabeen et al. 2009, 2011).

A mutation in the lipopolysaccharide biosynthesis pathway (*wbqL*) also causes loss of cell curvature by disrupting membrane attachment of crescentin by a yet unknown mechanism (Cabeen et al. 2010).

Second, using a dominant negative variant of crescentin, Cabeen and colleagues demonstrated that loss of cell curvature upon crescentin inactivation requires cell growth (Cabeen et al. 2009). These data argue that crescentin does not bend the cell directly, but that it modulates synthesis of the peptidoglycan cell wall in a manner that leads to cell curvature. Consistent with that hypothesis, sacculi (isolated peptidoglycan) from wild-type, *ΔcreS*, or *creS*-overexpressing cells imaged by electron microscopy have the same relative degrees of curvature as the cells from which they are isolated (Cabeen et al. 2009). Moreover, when crescentin is present, a gradient in the rates of cell wall synthesis is present across the short axis of the cell: peptidoglycan is synthesized more slowly proximal to crescentin than it is distal to crescentin (Cabeen et al. 2009). These differential rates in peptidoglycan synthesis are thought to be sufficient to lead to the cell curvature observed. It is interesting to note that heterologous expression of *creS* in *E. coli* is sufficient to induce curvature (Cabeen et al. 2009), consistent with the idea that crescentin uses conserved pathways, i.e. those involved in cell wall metabolism, to alter cell shape.

Third, given the reduced helical pitch of membrane-dissociated crescentin, the physiologically relevant, membrane attached form is likely to be constrained and, therefore, exerts a strain on the cell envelope (Cabeen et al. 2009). This strain imposed on the inner curvature of the cell is proposed to locally compress the peptidoglycan, making bond hydrolysis less favorable and thereby locally altering the rates of cell wall hydrolysis and insertion (Cabeen et al. 2009). In support of the concept that peptidoglycan synthesis can be modulated by application of force, constraining growth of otherwise straight cells (*E. coli* or *C. crescentus* Δ*creS*) in a curved chamber induces curved morphology that is maintained when the cells are released (Cabeen et al. 2009). Collectively, these observations suggest that the physical constraint imposed on cells by crescentin is central to the mechanism of inducing curvature, and that it does so by modulating the kinetics of cell wall remodeling in a spatially regulated fashion. This hypothesis is upheld in a mechanochemical computational model of the influence of the crescentin filament on cell growth (Jiang and Sun 2012).

Crescentin: Unanswered Questions

How does crescentin assemble into functional structures *in vivo* and locally disassemble during cytokinesis? FRAP and FLIP experiments on fluorescently labeled crescentin suggest that these structures are nearly static (half-time of fluorescence recovery of 26 ± 2 min), and there is slow exchange between assembled and free crescentin in the cytoplasm (rate of loss of fluorescence recovery following photobleaching 0.5 ±0.18/h) (Esue et al. 2010). *De novo* assembly of crescentin structures in cells depleted of native crescentin shows biphasic growth: initial rapid assembly causing longitudinal extension, followed by a shift to a gradual incorporation of monomers and apparent thickening of structures as the filament reaches the cell poles (Charbon et al. 2009). Crescentin monomers are added laterally to a crescentin structure that extends bidirectionally, along the inner membrane, as the cell grows (Charbon et al. 2009). Curiously, as the structures extend to the poles, longitudinal elongation ceases despite continued lateral addition of monomers. This is true even in *creS*-overexpression conditions, arguing against limiting crescentin levels being responsible for its polar exclusion (Charbon et al. 2009). It is possible that crescentin assembly is negatively regulated by an unknown mechanism involving polar proteins.

The slow turnover of the crescentin structure poses a challenge for cell division. This is evident during heterologous expression of *creS* in *E. coli* and in overexpression of *creS* in *C. crescentus*. In the former, expression of *creS* causes curvature in *E. coli* but at the same time results in cell division defects. In the latter, overexpression of cytoplasmic $creS_{\Delta N27}$ disrupts cytokinesis, probably due to stable crescentin structures physically stalling constriction of cell envelope (Cabeen et al. 2010). Since such a disruption of cytokinesis is only evident in cases where crescentin may not have all of its interacting partners (as in *E. coli*) or is not attached to the membrane (as in the overexpression of $creS_{\Delta N27}$), it is proposed that an unidentified membrane-bound factor and/or activity is important for disassembly of crescentin structure during cell division.

CTP Synthase: Co-opting a Metabolic Enzyme to Define Cell Shape

CTP synthase (CtpS) is a universally conserved enzyme that generates CTP from ATP, UTP and glutamine (Long et al. 1970). Surprisingly, this essential metabolic enzyme was simultaneously discovered to form filaments in bacteria, yeasts, flies, and mammals (Ingerson-Mahar et al. 2010; Noree et al. 2010; Liu 2010; Carcamo et al. 2011). Polymerization of CtpS is required for feedback inhibition of its enzymatic activity in both eukaryotes and bacteria (Barry et al. 2014; Noree et al. 2014). However, CtpS fulfills an additional, apparently unique, cytoskeletal function in *C. crescentus* where it regulates cell curvature through a crescentin-dependent mechanism (Ingerson-Mahar et al. 2010).

CtpS Localization and Role in Cell Curvature

CtpS was first discovered to form filaments in *C. crescentus* through a microscopy-based screen for localized proteins where mCherry-CtpS localized to filament-like structures near the inner curvature of the cell (Figs. 4.2 and 4.3N) (Werner and Gitai

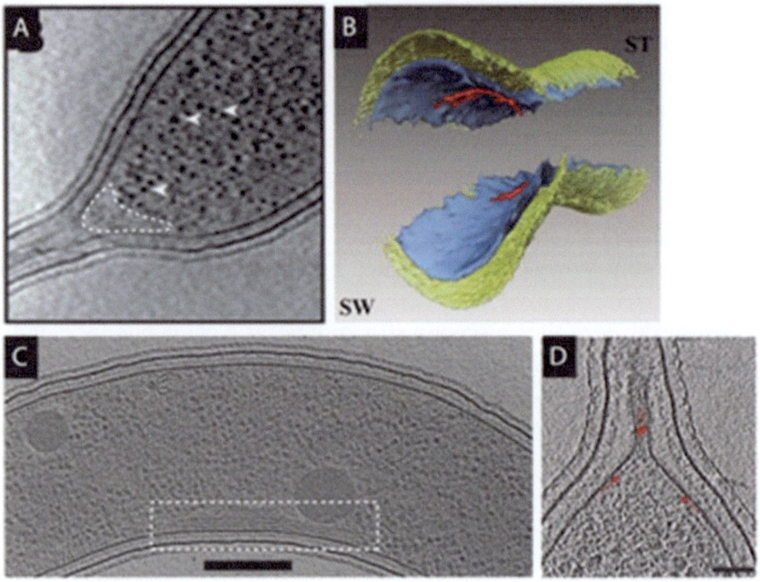

Fig. 4.4 Diverse cytoskeletal structures observed *in vivo*. A subset of the cytoskeletal elements described in *C. crescentus* has been imaged at high resolution *in vivo* using electron cryotomography (ECT). ECT images of *C. crescentus* cells showing (**A**) Ribosome-free zone associated with the PopZ matrix at the old cell pole (highlighted by *dotted line*) (Bowman et al. 2010), (**B**) FtsZ protofilaments (*red*) assembled into a ring proximal to the inner membrane (*blue*) at the site of division (Z. Li et al. 2007), (**C**) CTP synthase ribbons assembled at the inner curvature of the cell proximal to the inner membrane (scale bar = 200 nm) (Briegel et al. 2006; Ingerson-Mahar et al. 2010) (**D**) Bactofilin scaffold assembled at the base of the stalk (scale bar = 50 nm) (Kühn et al. 2009)

2010; Ingerson-Mahar et al. 2010). Early ECT images of wild-type *C. crescentus* revealed the presence of filament bundles (ECT filaments) in a similar location that were not composed of any of the then-known polymerizing proteins (Briegel et al. 2006). These ECT filaments were subsequently confirmed to be formed by CtpS using correlated fluorescence light microscopy and ECT (Fig. 4.4C) (Ingerson-Mahar et al. 2010).

ctpS is essential, and depletion of CtpS in *C. crescentus* causes slow growth and increased cell curvature (Fig. 4.3G) (Ingerson-Mahar et al. 2010). Conversely, overproduction of CtpS decreases cell curvature, resulting in nearly straight cells. When CtpS is depleted in a Δ*creS* background however, cells remain straight, indicating that CtpS exerts its influence on cell curvature through crescentin. Consistent with this hypothesis, in *ctpS*-overexpressing cells crescentin localization is aberrant, though crescentin protein levels are unaffected. Instead of assembling into filamentous structures, crescentin forms small foci in the presence of excess CtpS (Ingerson-Mahar et al. 2010). Heterologous expression in *E. coli* and bacterial two hybrid assays indicate that CtpS directly interacts with crescentin (Ingerson-Mahar et al. 2010). Based on these observations, it has been proposed that CtpS negatively regulates crescentin assembly, possibly by causing increased turnover of crescentin and/or fragmentation of polymers. This function of CtpS in regulating cell curvature does not require its ability to synthesize CTP, but rather relies on its glutamine amidotransferase (GAT) domain (Ingerson-Mahar et al. 2010).

Assembly Properties of CtpS

CtpS structures observed by ECT in cells typically comprise 4 parallel ribbons that are 10 nm apart (Fig. 4.4C). The three ribbons closest to the inner membrane are approximately 400–500 nm long, 7 nm thick, and 22 nm wide, while the ribbon farthest from the inner membrane is much shorter (Briegel et al. 2006). *C. crescentus* CtpS is refractory to purification, however CtpS from *E. coli* (EcCtpS) assembles into polymers *in vitro* that are similar to those observed for *C. crescentus* CtpS by ECT (Figs. 4.4D and 4.5F). They are 200–400 nm long and can assemble into bundles of 3–5 filaments spaced 8–9 nm apart (Ingerson-Mahar et al. 2010). EcCtpS polymerization is induced by its product, CTP, and inhibits the enzymatic activity of CtpS likely by preventing a critical conformational change (Barry et al. 2014).

Interestingly, the assembly and localization of CtpS is dynamic over the cell cycle (Figs. 4.2 and 4.3N) (Ingerson-Mahar et al. 2010). In new stalked cells, CtpS is present as a focus or short filament in the cytoplasm. The CtpS focus intensifies and elongates in late stalked cells and predivisional cells as it moves from the cytoplasm to the inner cell curvature, where it remains until the cell divides. The CtpS filament is often inherited asymmetrically, with stalked daughters receiving the CtpS filament and swarmers having little to no localized CtpS signal. Although CtpS polymers appear to be absent from swarmer cells, there is no change in the cellular levels of CtpS, suggesting that CtpS filament formation is triggered post-translationally during swarmer to stalk transition (Ingerson-Mahar et al. 2010). Since CTP synthesis feeds into multiple metabolic and cell cycle regulatory pathways, it may be that the assembly of CtpS structures is under the control of cell cycle-dependent metabolic regulation.

CtpS: Unanswered Questions

While it is clear that polymerization of CtpS is an ancient adaptation critical to the regulation of CTP metabolism, it is less obvious how and why *C. crescentus* co-opted these polymers for the determination of cell shape. It is interesting to note that EcCtpS can complement loss of *C. crescentus* CtpS both in promoting cell growth and maintaining wild-type cell curvature (Ingerson-Mahar et al. 2010), suggesting that *C. crescentus* CtpS has not acquired an additional activity. Instead, it is likely that crescentin evolved in such a way to allow interaction with and regulation by CtpS. How CtpS, specifically in the polymeric form, regulates the assembly and

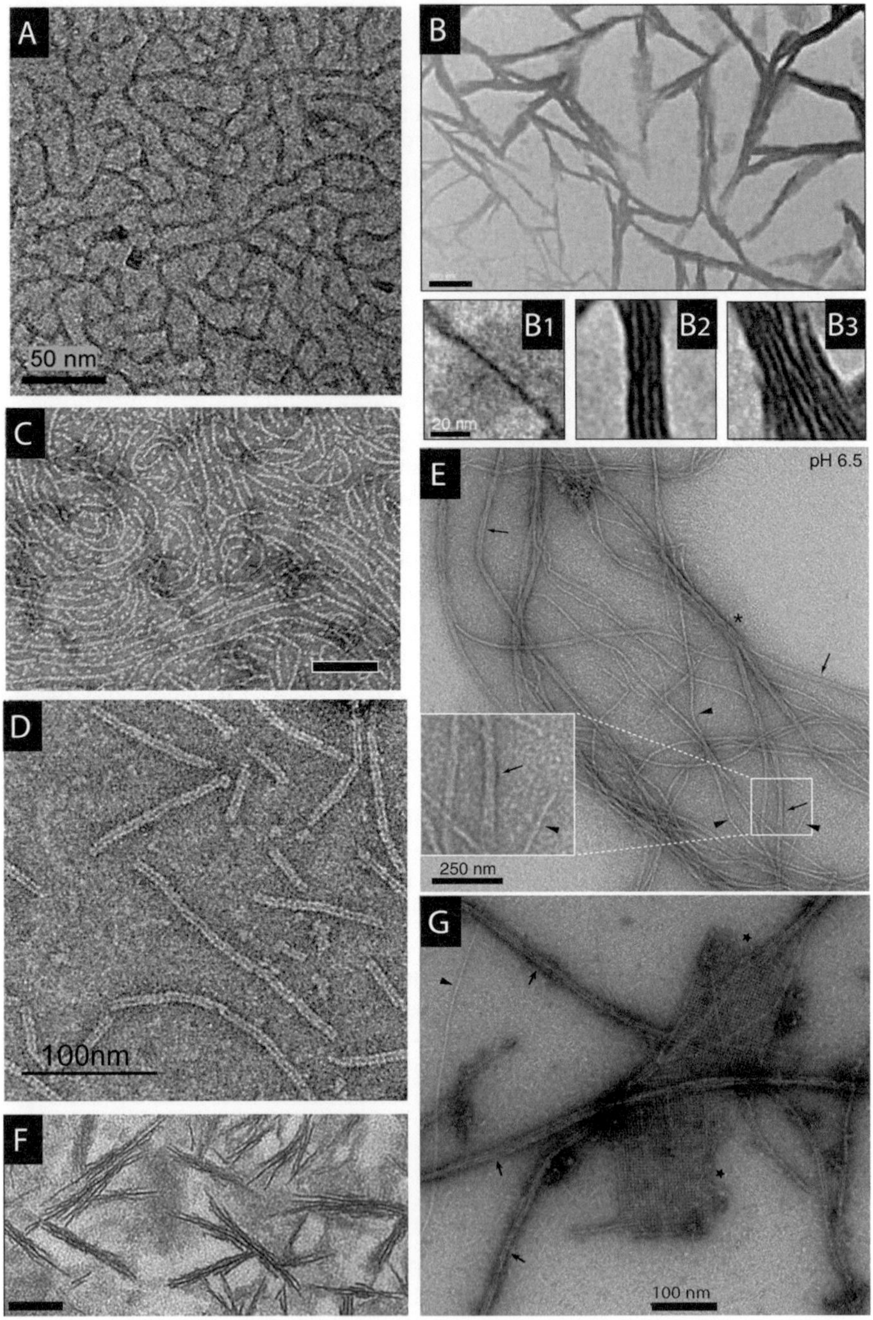

Fig. 4.5 Filaments and higher order structures formed by cytoskeletal proteins *in vitro*. Transmission electron microscopy (TEM) images of purified proteins reported to have been assembled into higher order structures *in vitro*. (**A**) PopZ filaments, scale bar = 50 nm (Bowman et al. 2008). (**B**) ParA filaments assembled in the presence of ATP, scale bar = 100 nm. (B1), (B2), (B3) show enlarged images of filament bundles, scale bar = 20 nm (Ptacin et al. 2010). (**C**) FtsZ protofilaments assembled in the presence of GTP, scale bar = 100 nm (Unpublished results). (**D**) Double protofilaments of MreB assembled on a lipid monolayer in the presence of ATP, scale bar = 100 nm (van den Ent et al. 2014). (**E**) Crescentin filaments (*arrowheads*) and bundles

activity of crescentin, however, is still unknown. Future work to characterize CtpS function and its interaction with crescentin at the molecular level will be required for a mechanistic understanding of CtpS and its role in cell shape. More generally, understanding how a polymerizing metabolic enzyme like CtpS has been co-opted for a cytoskeletal role may lend insight into how similar events occurred, for example in the evolution of the actin-hexokinase-hsp70 superfamily (Kabsch and Holmes 1995).

Bactofilins: Scaffolds for Stalk Morphogenesis

Bactofilins are a class of cytoskeletal proteins unique to bacteria. Representatives are found across all phyla of bacteria with sequenced genomes and function in various cellular processes, including morphogenesis, motility, and development (Kühn et al. 2009; Koch et al. 2011; Lin and Thanbichler 2013). The first bactofilins, BacA and BacB, were discovered in a microscopy-based screen for gene products involved in polar development during the swarmer-to-stalked cell transition in *C. crescentus*. BacA and BacB were demonstrated to function in stalk morphogenesis (Kühn et al. 2009) and are present throughout the cell cycle. However, their localization is cell cycle-regulated, as they are diffuse in swarmer cells, but assemble at the stalked pole at the swarmer-to-stalked cell transition (Figs. 4.2, 4.3O and 4.4E). They then remain at the stalked pole for the duration of the cell cycle and are inherited by the stalked daughter upon division.

Function and Localization of Bactofilins

Of the two bactofilins, BacA is apparently dominant to BacB, as *ΔbacB* cells have similar but less severe phenotypes than *ΔbacA* or *ΔbacAbacB* cells (Kühn et al. 2009; Hughes et al. 2013). Moreover, BacA is estimated to be in ten-fold excess of BacB (~200 molecules of BacA and only ~20 molecules of BacB per cell) (Kühn et al. 2009). Together they form an ordered sheet-like structure about 5 nm internal to the inner membrane at the base of the stalk in cells visualized by ECT (Fig. 4.4D)

Fig. 4.5 (continued) (*arrows* and *asterisks*) assembled at pH 6.5 in the absence of Mg^{2+}, scale bar = 250 nm (Cabeen et al. 2011). (**F**) Filaments formed by *E. coli* CTP synthase, scale bar = 100 nm (Ingerson-Mahar et al. 2010). (**G**) Filaments (*arrows*) and 2D-crystalline sheets (*asterisks*) formed by BacA, scale bar = 100 nm (Vasa et al. 2015). Note that the identities of structures seen in some TEM images published in papers on cytoskeletal filaments have been questioned (Ghosal et al. 2014, supplementary file; Griffith and Bonner 1973), since contaminating cellulose fibers and thin uranyl acetate crystals may each be mistaken for protein polymers; here, the FtsZ and MreB filaments in C and D appear white, as expected for protein embedded in negative stain, and are also unmistakably identifiable from longitudinal repeats corresponding to the sizes of the monomers, but the EM evidence for filament formation by other proteins is weaker

(Kühn et al. 2009). This structure serves as a scaffold for the recruitment of a stalk-specific peptidoglycan synthase, PbpC, that further recruits other factors required for stalk morphogenesis, such as StpX (Hughes et al. 2013). PbpC is a membrane-bound periplasmic enzyme, and it likely interacts with the bactofilin scaffold through its cytoplasmic tail (Kühn et al. 2009; Hughes et al. 2013). Consistent with the role of PbpC in stalk elongation, cells lacking BacA (or both BacA and BacB) fail to elongate their stalks as efficiently as wild-type (Kühn et al. 2009).

It is not clear how bactofilins are recruited to the base of the stalk and how this recruitment is coordinated with the cell cycle. Bactofilins are expressed throughout the cell cycle, however their transcript levels are upregulated during the swarmer to stalk transition (McGrath et al. 2007; Kühn et al. 2009). Although they are expressed in swarmers, bactofilins do not form a scaffold until the swarmer-to-stalked cell transition. In stalked cells, bactofilins can localize in the absence of proteins such as SpmX and DivJ that are known to be specifically recruited to the base of the stalk prior to stalk synthesis (Kühn et al. 2009). Due to their low physiological concentrations, it is possible that bactofilins require a nucleation factor to regulate their assembly temporally. Curiously, overexpression of either bactofilin causes morphological defects with cells becoming increasingly curved over time until they eventually lyse (Fig. 4.3H). In these experiments, both BacA and BacB were seen to localize to the inner curvature, suggesting a preference for positive curvature for bactofilin structures. This localization mechanism for bactofilins is reminiscent of the curvature-sensing mechanism of the BAR-domains of eukaryotic proteins (Kühn et al. 2009). It is important to note that the high curvature of bactofilin overexpressing cells is independent of the crescentin cytoskeleton that gives wild-type *C. crescentus* its curved morphology (Kühn et al. 2009). Whether the bactofilin structure assembles by sensing positively curved surfaces or if it induces positive curvature is yet to be determined.

Assembly Properties of Bactofilins

Bactofilins are spontaneously polymerizing proteins that can form linear polymers, bundles, sheets and crystalline arrays *in vitro* (Fig. 4.5G) (Kühn et al. 2009; Koch et al. 2011; Zuckerman et al. 2015). Based on sequence similarities and electron microscopy, BacA and BacB appear to behave similarly. BacA is easier to purify than BacB and has a lower critical concentration (250 nM), allowing for a more thorough characterization of the former. BacA assembles into polymers without a requirement for nucleotide and over a large range of pH and salt concentrations (Kühn et al. 2009). In this regard, bactofilins are similar to intermediate filaments. However, unlike intermediate filaments, bactofilins lack any coiled-coil regions. Like all bactofilins, BacA and BacB have a rigid core made of a DUF583 domain flanked by N- and C-terminal tails (Kühn et al. 2009; Vasa et al. 2015). Solid-state NMR and quantitative electron microscopy data led to a structural model in which each monomer of BacA adopts a β-helical structure comprising 6 turns made of 3 β-sheets each, thus making a triangular hydrophobic core. Head-to-tail assembly of DUF583 domains yields filaments that are 3 nm in diameter, which form loosely packed bundles at low salt concentrations. At physiological salt concentrations, these filaments assemble as ribbons and/or two-dimensional sheets. The physiological relevance of the different assemblies of bactofilins observed *in vitro* is currently unknown. The N- and C-terminal tails of BacA are flexible, and might serve as flanks that help position the structure relative to the inner membrane *in vivo* (Vasa et al. 2015). It is not clear whether the tails also function in interactions between bactofilins and other proteins such as PbpC.

C. crescentus Cytoskeletons: Future Perspectives

The past decade has seen the identification and characterization of numerous unique cytoskeletal elements in bacteria, and specifically in *C. crescentus*, owing largely to advances in microscopy techniques. However, a major question still remains for most of these factors – how do the structures observed *in vitro* correlate with the localization patterns and functions observed *in vivo*? Barring a few proteins such as PopZ and bactofilins, it is not clear if the filamentous structures formed by purified proteins *in vitro* are physiologically relevant for the function of these cytoskeletal proteins. In some cases such as ParA and MreB, it has not been definitively demonstrated that these proteins form filaments at all *in vivo* (Briegel et al. 2006; Swulius et al. 2011; Lim et al. 2014). While ECT and super-resolution microscopy have helped discern structures that had previously been resolution-limited, these techniques have their own challenges, making it difficult to identify non-canonical structures, dynamic assemblies, membrane-associated polymers, and/or heteropolymers.

Unlike eukaryotes, a change in cytoskeletal organization alone does not mean a change in bacterial cell shape; instead, cytoskeleton-driven cell shape changes require coordinated cell wall remodeling (Pichoff and Lutkenhaus 2007). It is not obvious how the differential assembly of cytoskeletal proteins affects cell wall remodeling enzymes to cause cell shape changes. Are MreB and FtsZ, for example, purely scaffolds that regulate assembly of cell wall synthetic complexes or do they more actively contribute to shape maintenance through mechanical means, as has been proposed for crescentin? A fascinating problem unique to *C. crescentus* and other stalked organisms is the development of a slender protrusion from a cylindrical cell of a much larger diameter (Wagner and Brun 2007). While it is apparent that many of the players involved in cell elongation are important for stalk biogenesis, it is not clear how these proteins are able to direct formation of a structure of predetermined dimensions, whether through as yet unknown forms of regulation of known players or *via* an unknown structural element.

Another question that deserves attention is, how do these cytoskeletal elements interact with each other? It is evident from individual characterizations of these proteins that there is considerable amount of crosstalk between cytoskeletons, directly and indirectly. For example, MreB interacts with FtsZ, and most probably with crescentin and affects PopZ localization as well. Tight spatiotemporal coordination between cytoskeletal structures is central to cell cycle progression while maintaining cell shape. A major goal for the future is to develop a holistic view of cytoskeletal regulation in the larger context of the cell cycle and developmental program of *C. crescentus*.

Acknowledgements We are grateful to members of the Goley laboratory for helpful discussions and for critical comments on this manuscript. Research in the Goley laboratory relevant to the subject of this chapter is supported by the National Institutes of Health under award number R01GM108640 (to E.D.G).

References

Aaron M et al (2007) The tubulin homologue FtsZ contributes to cell elongation by guiding cell wall precursor synthesis in Caulobacter crescentus. Mol Microbiol 64(4):938–952

Ausmees N, Kuhn JR, Jacobs-Wagner C (2003) The bacterial cytoskeleton: an intermediate filament-like function in cell shape. Cell 115(6):705–713

Banigan EJ et al (2011) Filament depolymerization can explain chromosome pulling during bacterial mitosis. PLoS Comput Biol 7(9):e1002145

Barry RM et al (2014) Large-scale filament formation inhibits the activity of CTP synthetase. eLife 3:e03638

Bartosik AA et al (2009) ParB deficiency in Pseudomonas aeruginosa destabilizes the partner protein ParA and affects a variety of physiological parameters. Microbiol (Reading, England) 155(Pt 4):1080–1092

Beaufay F et al (2015) A NAD-dependent glutamate dehydrogenase coordinates metabolism with cell division in Caulobacter crescentus. EMBO J 34(13):1786–1800

Biteen JS et al (2008) Super-resolution imaging in live Caulobacter crescentus cells using photoswitchable EYFP. Nat Methods 5(11):947–949

Boutte CC, Henry JT, Crosson S (2012) ppGpp and polyphosphate modulate cell cycle progression in Caulobacter crescentus. J Bacteriol 194(1):28–35

Bowman GR et al (2008) A polymeric protein anchors the chromosomal origin/ParB complex at a bacterial cell pole. Cell 134(6):945–955

Bowman GR et al (2010) CaulobacterPopZ forms a polar subdomain dictating sequential changes in pole composition and function. Mol Microbiol 76(1):173–189

Bowman GR et al (2013) Oligomerization and higher-order assembly contribute to sub-cellular localization of a bacterial scaffold. Mol Microbiol 90(*4*):*776*–795

Briegel A et al (2006) Multiple large filament bundles observed in Caulobacter crescentus by electron cryotomography. Mol Microbiol 62(1):5–14

Britos L et al (2011) Regulatory response to carbon starvation in Caulobacter crescentus. PLoS One 6(4):e18179

Buske PJ, Levin PA (2013) A flexible C-terminal linker is required for proper FtsZ assembly in vitro and cytokinetic ring formation in vivo. Mol Microbiol 89(2):249–263

Cabeen MT et al (2009) Bacterial cell curvature through mechanical control of cell growth. EMBO J 28:1–12

Cabeen MT et al (2010) Mutations in the Lipopolysaccharide biosynthesis pathway interfere with crescentin-mediated cell curvature in Caulobacter crescentus. J Bacteriol 192(13):3368–3378

Cabeen MT, Herrmann H, Jacobs-Wagner C (2011) The domain organization of the bacterial intermediate filament-like protein crescentin is important for assembly and function. Cytoskeleton 68(4):205–219

Carcamo WC et al (2011) Induction of cytoplasmic rods and rings structures by inhibition of the CTP and GTP synthetic pathway in mammalian cells. PLoS One 6(12):e29690

Charbon G, Cabeen MT, Jacobs-Wagner C (2009) Bacterial intermediate filaments: in vivo assembly, organization, and dynamics of crescentin. Genes Dev 23(9):1131–1144

Curtis PD, Brun YV (2010) Getting in the loop: regulation of development in Caulobacter crescentus. Microbiol Mol Biol Rev MMBR 74(1):13–41

Daniel RA, Errington J (2003) Control of cell morphogenesis in bacteria: two distinct ways to make a rod-shaped cell. Cell 113(6):767–776

Din N, Quardokus EM, Sackett MJ (1998) Dominant C-terminal deletions of FtsZ that affect its ability to localize in Caulobacter and its interaction with FtsA. Mol Microbiol 27(5):1051–1063

Divakaruni AV et al (2007) The cell shape proteins MreB and MreC control cell morphogenesis by positioning cell wall synthetic complexes. Mol Microbiol 66(1):174–188

Donovan C et al (2010) Subcellular localization and characterization of the ParAB system from Corynebacterium glutamicum. J Bacteriol 192(13):3441–3451

Duman R et al (2013) Structural and genetic analyses reveal the protein SepF as a new membrane anchor for the Z ring. Proc Natl Acad Sci 110(48):E4601–E4610

Dye NA et al (2011) Mutations in the nucleotide binding pocket of MreB can alter cell curvature and polar morphology in Caulobacter. Mol Microbiol 81(2):368–394

Easter J, Gober JW (2002) ParB-stimulated nucleotide exchange regulates a switch in functionally distinct ParA activities. Mol Cell 10(2):472–434

Ebersbach G et al (2006) Regular cellular distribution of plasmids by oscillating and filament-forming ParA ATPase of plasmid pB171. Mol Microbiol 61(6):1428–1442

Ebersbach G et al (2008) A self-associating protein critical for chromosome attachment, division, and polar organization in Caulobacter. Cell 134(6):956–968

England JC et al (2010) Global regulation of gene expression and cell differentiation in Caulobacter crescentus in response to nutrient availability. J Bacteriol 192(3):819–833

Erickson HP, Anderson DE, Osawa M (2010) FtsZ in bacterial cytokinesis: cytoskeleton and force generator all in one. Microbiol Mol Biol Rev (MMBR) 74(4):504–528

Esue O et al (2010) Dynamics of the bacterial intermediate filament crescentin in vitro and in vivo. PLoS One 5(1):e8855

Fenton AK, Gerdes K (2013) Direct interaction of FtsZ and MreB is required for septum synthesis and cell division in Escherichia coli. EMBO J 32(13):1953–1965

Figge RM, Easter J, Gober JW (2003) Productive interaction between the chromosome partitioning proteins, ParA and ParB, is required for the progression of the cell cycle in Caulobacter crescentus. Mol Microbiol 47(5):1225–1237

Figge RM, Divakaruni AV, Gober JW (2004) MreB, the cell shape-determining bacterial actin homologue, co-ordinates cell wall morphogenesis in Caulobacter crescentus. Mol Microbiol 51(5):1321–1332

Fogel MA, Waldor MK (2006) A dynamic, mitotic-like mechanism for bacterial chromosome segregation. Genes Dev 20(23):3269–3282

Gardner KAJA, Moore DA, Erickson HP (2013) The C-terminal linker of Escherichia coli FtsZ functions as an intrinsically disordered peptide. Mol Microbiol 89(2):264–275

Ghosal D, Trambaiolo D, Amos LA, Lowe J (2014) MinCD cell division proteins form alternating copolymeric cytomotive filaments. Nat Commun 5:5341

Gitai Z, Dye N, Shapiro L (2004) An actin-like gene can determine cell polarity in bacteria. Proc Natl Acad Sci U S A 101(23):8643–8648

Gitai Z et al (2005) MreB actin-mediated segregation of a specific region of a bacterial chromosome. Cell 120(3):329–341

Godfrin-Estevenon A-M, Pasta F, Lane D (2002) The parAB gene products of Pseudomonas putida exhibit partition activity in both P. putida and Escherichia coli. Mol Microbiol 43(1):39–49

Goley ED, Iniesta AA, Shapiro L (2007) Cell cycle regulation in Caulobacter: location, location, location. J Cell Sci 120(Pt 20):3501–3507

Goley ED et al (2010) Imaging-based identificationof a critical regulator of FtsZ protofilament curvature in Caulobacter. Mol Cell 39(6):975–987

Goley ED et al (2011) Assembly of the Caulobacter cell division machine. Mol Microbiol 80(6):1680–1698

Gonzalez D, Collier J (2013) DNA methylation by CcrM activates the transcription of two genes required for the division of Caulobacter crescentus. Mol Microbiol 88(1):203–218

Griffith JD, Bonner JF (1973) Chromatin-like aggregates of uranyl acetate. Nat New Biol 244:80–81

Harris LK, Dye NA, Theriot JA (2014) A Caulobacter MreB mutant with irregular cell shape exhibits compensatory widening to maintain a preferred surface area to volume ratio. Mol Microbiol 94(5):988–1005

Holden SJ et al (2014) High throughput 3D super-resolution microscopy reveals Caulobacter crescentus in vivo Z-ring organization. Proc Natl Acad Sci 111(12):4566–4571

Hottes AK, Shapiro L, McAdams HH (2005) DnaA coordinates replication initiation and cell cycle transcription in Caulobacter crescentus. Mol Microbiol 58(5):1340–1353

Howard M, Gerdes K (2010) What is the mechanism of ParA-mediated DNA movement? Mol Microbiol 78(1):9–12

Hughes HV et al (2013) Co-ordinate synthesis and protein localization in a bacterial organelle by the action of a penicillin-binding-protein. Mol Microbiol 90(6):1162–1177

Ingerson-Mahar M et al (2010) The metabolic enzyme CTP synthase forms cytoskeletal filaments. Nature 12(8):739–746

Iniesta AA (2014) ParABS system in chromosome partitioning in the bacterium Myxococcus xanthus. PLoS One 9(1):e86897

Iwai N, Nagai K, Wachi M (2002) Novel S-benzylisothiourea compound that induces spherical cells in Escherichia coli probably by acting on a rod-shape-determining protein(s) other than penicillin-binding protein 2. Biosci Biotechnol Biochem 66(12):2658–2662

Jakimowicz D et al (2007) Characterization of the mycobacterial chromosome segregation protein ParB and identification of its target in Mycobacterium smegmatis. Microbiol (Reading, England) 153(Pt 12):4050–4060

Jenal U (2000) Signal transduction mechanisms in Caulobacter crescentus development and cell cycle control. FEMS Microbiol Rev 24(2):177–191

Jenal U (2009) The role of proteolysis in the Caulobacter crescentus cell cycle and development. Res Microbiol 160(9):687–695

Jiang H, Sun SX (2012) Growth of curved and helical bacterial cells. Soft Matter 8(28):7446–7451

Kabsch W, Holmes KC (1995) The actin fold. FASEB J 9(2):167–174

Kelly AJ et al (1998) Cell cycle-dependent transcriptional and proteolytic regulation of FtsZ in Caulobacter. Genes Dev 12(6):880–893

Kiekebusch D et al (2012) Localized dimerization and nucleoid binding drive gradient formation by the bacterial cell division inhibitor MipZ. Mol Cell 46(3):245–259

Kim HJ et al (2000) Partitioning of the linear chromosome during sporulation of Streptomyces coelicolor A3(2) involves an oriC-linked parAB locus. J Bacteriol 182(5):1313–1320

Kim SY et al (2006) Single molecules of the bacterial actin MreB undergo directed treadmilling motion in Caulobacter crescentus. Proc Natl Acad Sci U S A 103(29):10929–10934

Kirkpatrick CL, Viollier PH (2012) Decoding Caulobacter development. FEMS Microbiol Rev 36(1):193–205

Klein EA et al (2013) Physiological role of stalk lengthening in Caulobacter crescentus. Commun Int Biol 6(4):e24561

Koch MK, McHugh CA, Hoiczyk E (2011) BacM, an N-terminally processed bactofilin of Myxococcus xanthus, is crucial for proper cell shape. Mol Microbiol 80(4):1031–1051

Król E et al (2012) Bacillus subtilis SepF binds to the C-terminus of FtsZ. PLoS One 7(8):e43293

Kühn J et al (2009) Bactofilins, a ubiquitous class of cytoskeletal proteins mediating polar localization of a cell wall synthase in Caulobacter crescentus. EMBO J 29(2):1–13

Kuru E et al (2012) In Situ probing of newly synthesized peptidoglycan in live bacteria with fluorescent D-amino acids. Angew Chem Int Ed Eng 51(50):12519–12523
Laloux G, Jacobs-Wagner C (2013) Spatiotemporal control of PopZ localization through cell cycle-coupled multimerization. J Cell Biol 201(6):827–841
Laub MT et al (2000) Global analysis of the genetic network controlling a bacterial cell cycle. Sci (New York, N.Y.) 290(5499):2144–2148
Leonard TA, Butler PJ, Löwe J (2005) Bacterial chromosome segregation: structure and DNA binding of the Soj dimer–a conserved biological switch. EMBO J 24(2):270–282
Lesley JA, Shapiro L (2008) SpoT regulates DnaA stability and initiation of DNA replication in carbon-starved Caulobacter crescentus. J Bacteriol 190(20):6867–6880
Li Z et al (2007) The structure of FtsZ filaments in vivo suggests a force-generating role in cell division. EMBO J 26(22):4694–4708
Li Y et al (2013) FtsZ protofilaments use a hinge-opening mechanism for constrictive force generation. Science (New York, N.Y.) 341(6144):392–395
Lim HC et al (2014) Evidence for a DNA-relay mechanism in ParABS-mediated chromosome segregation. eLife 3:e02758
Lin L, Thanbichler M (2013) Nucleotide-independent cytoskeletal scaffolds in bacteria. Cytoskeleton (Hoboken, NJ) 70(8):409–423
Liu J-L (2010) Intracellular compartmentation of CTP synthase in Drosophila. J Genet Genom Yi chuan xue bao 37(5):281–296
Long CW, Levitzki A, Koshland DE (1970) The subunit structure and subunit interactions of cytidine Triphosphate synthetase. J Biol Chem 245(1):80–87
Lu C, Reedy M, Erickson HP (2000) Straight and curved conformations of FtsZ are regulated by GTP hydrolysis. J Bacteriol 182(1):164–170
Ma X, Margolin W (1999) Genetic and functional analyses of the conserved C-terminal core domain of Escherichia coli FtsZ. J Bacteriol 181(24):7531–7544
Ma X et al (1997) Interactions between heterologous FtsA and FtsZ proteins at the FtsZ ring. J Bacteriol 179(21):6788–6797
Margolin W (2005) FtsZ and the division of prokaryotic cells and organelles. Nat Rev Mol Cell Biol 6(11):862–871
McGrath PT et al (2007) High-throughput identification of transcription start sites, conserved promoter motifs and predicted regulons. Nat Biotechnol 25(5):584–592
Meier EL, Goley ED (2014) Form and function of the bacterial cytokinetic ring. Curr Opin Cell Biol 26:19–27
Mierzejewska J, Jagura-Burdzy G (2012) Prokaryotic ParA-ParB-parS system links bacterial chromosome segregation with the cell cycle. Plasmid 67(1):1–14
Mohl DA, Gober JW (1997) Cell cycle-dependent polar localization of chromosome partitioning proteins in Caulobacter crescentus. Cell 88(5):675–684
Mohl DA, Easter J, Gober JW (2001) The chromosome partitioning protein, ParB, is required for cytokinesis in Caulobacter crescentus. Mol Microbiol 42(3):741–755
Noree C et al (2010) Identification of novel filament-forming proteins in Saccharomyces cerevisiae and Drosophila melanogaster. J Cell Biol 190(4):541–551
Noree C et al (2014) Common regulatory control of CTP synthase enzyme activity and filament formation. Mol Biol Cell 25(15):2282–2290
Osawa M, Erickson HP (2011) Inside-out Z rings--constriction with and without GTP hydrolysis. Mol Microbiol 81(2):571–579
Osawa M, Anderson DE, Erickson HP (2008) Reconstitution of contractile FtsZ rings in liposomes. Science (New York, NY) 320(5877):792–794
Osawa M, Anderson DE, Erickson HP (2009) Curved FtsZ protofilaments generate bending forces on liposome membranes. EMBO J 28(22):3476–3484
Persat A, Stone HA, Gitai Z (2014) The curved shape of Caulobacter crescentus enhances surface colonization in flow. Nat Commun 5:3824
Pichoff S, Lutkenhaus J (2007) Overview of cell shape: cytoskeletons shape bacterial cells. Curr Opin Microbiol 10(6):601–605

Ptacin JL et al (2010) A spindle-like apparatus guides bacterial chromosome segregation. Nature 12(8):791–798
Ptacin JL et al (2014) Bacterial scaffold directs pole-specific centromere segregation. Proc Natl Acad Sci 111(19):E2046–E2055
Quardokus EM, Brun YV (2002) DNA replication initiation is required for mid-cell positioning of FtsZ rings in Caulobacter crescentus. Mol Microbiol 45(3):605–616
Quardokus E, Din N, Brun YV (1996) Cell cycle regulation and cell type-specific localization of the FtsZ division initiation protein in Caulobacter. Proc Natl Acad Sci U S A 93(13):6314–6319
Quardokus EM, Din N, Brun YV (2001) Cell cycle and positional constraints on FtsZ localization and the initiation of cell division in Caulobacter crescentus. Mol Microbiol 39(4):949–959
Radhakrishnan SK, Pritchard S, Viollier PH (2010) Coupling prokaryotic cell fate and division control with a bifunctional and oscillating oxidoreductase homolog. Dev Cell 18(1):90–101
Ringgaard S et al (2009) Movement and equipositioning of plasmids by ParA filament disassembly. Proc Natl Acad Sci 106(46):19369–19374
Saint-Dic D et al (2006) A parA homolog selectively influences positioning of the large chromosome origin in Vibrio cholerae. J Bacteriol 188(15):5626–5631
Schofield WB, Lim HC, Jacobs-Wagner C (2010) Cell cycle coordination and regulation of bacterial chromosome segregation dynamics by polarly localized proteins. EMBO J 29(18):3068–3081
Schrader JM et al (2014) The coding and noncoding architecture of the Caulobacter crescentus genome. PLoS Genet 10(7):1004463
Shaevitz JW, Gitai Z (2010) The structure and function of bacterial actin homologs. Cold Spring Harb Perspect Biol 2(9):a000364
Shapiro L, Agabian-Keshishian N, Bendis I (1971) Bacterial differentiation. Science (New York, N.Y.) 173(4000):884–892
Shebelut CW, Jensen RB, Gitai Z (2009) Growth conditions regulate the requirements for Caulobacter chromosome segregation. J Bacteriol 191(3):1097–1100
Shebelut CW et al (2010) Caulobacter chromosome segregation is an ordered multistep process. Proc Natl Acad Sci 107(32):14194–14198
Si F et al (2013) Organization of FtsZ filaments in the bacterial division ring measured from polarized fluorescence microscopy. Biophys J 105(9):1976–1986
Sliusarenko O et al (2011) High-throughput, subpixel precision analysis of bacterial morphogenesis and intracellular spatio-temporal dynamics. Mol Microbiol 80(3):612–627
Stove JL, Stanier RY (1962) Cellular differentiation in stalked bacteria. Nature 196:1189–1192
Stricker J, Erickson HP (2003) In vivo characterization of Escherichia coli ftsZ mutants: effects on Z-ring structure and function. J Bacteriol 185(16):4796–4805
Sundararajan K et al (2015) The bacterial tubulin FtsZ requires its intrinsicallydisordered linker to direct robust cell wallconstruction. Nat Commun 6:7281
Swulius MT et al (2011) Biochemical and biophysical research communications. Biochem Biophys Res Commun 407(4):650–655
Szwedziak P et al (2014) Architecture of the ring formed by the tubulin homologue FtsZ in bacterial cell division. eLife 3:e04601
Takacs CN et al (2010) MreB Drives De Novo Rod Morphogenesis in Caulobacter crescentus via remodeling of the cell wall. J Bacteriol 192(6):1671–1684
Thanbichler M, Shapiro L (2006) MipZ, a spatial regulator coordinating chromosome segregation with cell division in Caulobacter. Cell 126(1):147–162
Toro E et al (2008) Caulobacter requires a dedicated mechanism to initiate chromosome segregation. Proc Natl Acad Sci 105(40):15435–15440
van den Elzen PJ et al (1983) Structure and regulation of gene expression of a Clo DF13 plasmid DNA region involved in plasmid segregation and incompatibility. Nucl Acids Res 11(24):8791–8808
van den Ent F et al (2014) Bacterial actin MreB forms antiparallel double filaments. eLife 3:e02634

Vasa S et al (2015) β-Helical architecture of cytoskeletal bactofilin filaments revealed by solid-state NMR. Proc Natl Acad Sci 112(2):E127–E136

Vaughan S et al (2004) Molecular evolution of FtsZ protein sequences encoded within the genomes of archaea, bacteria, and eukaryota. J Mol Evol 58(1):19–29

Viollier PH et al (2004) Rapid and sequential movement of individual chromosomal loci to specific subcellular locations during bacterial DNA replication. Proc Natl Acad Sci U S A 101(25):9257–9262

Wagner JK, Brun YV (2007) Out on a limb: how the Caulobacter stalk can boost the study of bacterial cell shape. Mol Microbiol 64(1):28–33

Wagner JK, Galvani CD, Brun YV (2005) Caulobacter crescentus requires RodA and MreB for stalk synthesis and prevention of ectopic pole formation. J Bacteriol 187(2):544–553

Wang X et al (1997) Analysis of the interaction of FtsZ with itself, GTP, and FtsA. J Bacteriol 179(17):5551–5559

Wang Y, Jones BD, Brun YV (2001) A set of ftsZ mutants blocked at different stages of cell division in Caulobacter. Mol Microbiol 40(2):347–360

Wang SCE, West L, Shapiro L (2006) The bifunctional FtsK protein mediates chromosome partitioning and cell division in Caulobacter. J Bacteriol 188(4):1497–1508

Werner JN, Gitai Z (2010) High-throughput screening of bacterial protein localization. Methods Enzymol 471:185–204

White CL, Kitich A, Gober JW (2010) Positioning cell wall synthetic complexes by the bacterial morphogenetic proteins MreB and MreD. Mol Microbiol 76(3):616–633

Williams B et al (2014) ClpXP and ClpAP proteolytic activity on divisome substrates is differentially regulated following the Caulobacter asymmetric cell division. Mol Microbiol 93(5):853–866

Yakhnina AA, Gitai Z (2012) The small protein MbiA interacts with MreB and modulates cell shape in Caulobacter crescentus. Mol Microbiol 85(6):1090–1104

Zhou B et al (2015) The global regulatory architecture of transcription during the Caulobacter cell cycle. PLoS Genet 11(1):e1004831

Zuckerman DM et al (2015) The bactofilin cytoskeleton protein BacM of Myxococcus xanthus forms an extended β-sheet structure likely mediated by hydrophobic interactions. PLoS One 10(3):e0121074

Chapter 5
FtsZ Constriction Force – Curved Protofilaments Bending Membranes

Harold P. Erickson and Masaki Osawa

Abstract FtsZ assembles in vitro into protofilaments (pfs) that are one subunit thick and ~50 subunits long. In vivo these pfs assemble further into the Z ring, which, along with accessory division proteins, constricts to divide the cell. We have reconstituted Z rings in liposomes in vitro, using pure FtsZ that was modified with a membrane targeting sequence to directly bind the membrane. This FtsZ-mts assembled Z rings and constricted the liposomes without any accessory proteins. We proposed that the force for constriction was generated by a conformational change from straight to curved pfs. Evidence supporting this mechanism came from switching the membrane tether to the opposite side of the pf. These switched-tether pfs assembled "inside-out" Z rings, and squeezed the liposomes from the outside, as expected for the bending model. We propose three steps for the full process of cytokinesis: (a) pf bending generates a constriction force on the inner membrane, but the rigid peptidoglycan wall initially prevents any invagination; (b) downstream proteins associate to the Z ring and remodel the peptidoglycan, permitting it to follow the constricting FtsZ to a diameter of ~250 nm; the final steps of closure of the septum and membrane fusion are achieved by excess membrane synthesis and membrane fluctuations.

Keywords *E. coli* • Z-ring constriction • FtsZ • Tubulin • Curved protofilaments • Intermediate curved pfs • Bacterial cell division • Constriction force • Liposomes • FtsZ-MTS • Reconstituted systems • FtsA • Substructure of Z ring • Final step of septum closure • Copy number of divisome proteins

Introduction

In 2010 we published an extensive review of FtsZ, covering in vitro assembly studies as well as mechanisms in vivo (Erickson et al. 2010). In the present article we focus on studies that have provided new insights, along with some repetition of issues that are still debated. The major question is, what generates the constriction

H.P. Erickson (✉) • M. Osawa
Department of Cell Biology, Duke University, Durham, NC 27710, USA
e-mail: h.erickson@cellbio.duke.edu; harold.erickson@duke.edu

J. Löwe, L.A. Amos (eds.), *Prokaryotic Cytoskeletons*, Subcellular Biochemistry 84, DOI 10.1007/978-3-319-53047-5_5

force for bacterial cytokinesis? We will review the evidence that the force is generated primarily by FtsZ, specifically by the conformational change of FtsZ protofilaments (pfs) from straight to curved, generating a bending force on the membrane.

Straight pfs and Minirings

When *Escherichia coli* FtsZ (EcFtsZ) is assembled to steady state with GTP, negative stain EM shows pfs that are one subunit wide and 100–200 nm (25–50 subunits) long (Fig. 5.1a). Erickson et al. (1996) found that when FtsZ was assembled onto cationic lipid monolayers, many pfs were straight, and others were highly curved to form minirings, with an outside diameter of 24 nm (Fig. 5.1b). In some cases one could see an abrupt transition from the straight pf to the highly curved conformation (arrow, Fig. 5.1b). Tubulin shows a similar transition from straight pfs in the microtubule wall, to spirals and rings that are released upon disassembly of tubulin-GDP (McIntosh et al. 2010).

pf curvature similar to EcFtsZ minirings has been documented in only a few species. *Caulobacter crescentus* FtsZ assembles rings or spirals of 34–43 nm diameter in GDP and GTP-Mg, respectively, when stabilized by the accessory protein FzlA (Goley et al. 2010). *Methanococcus jannaschii* FtsZ assembled spiral ribbons about 30 nm in diameter in ficoll plus GDP (Huecas and Andreu 2004) or in mant-GTP

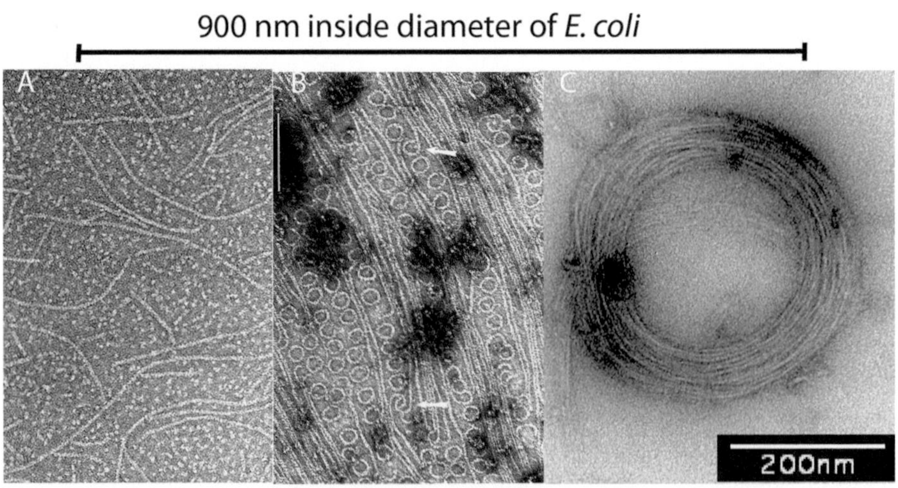

Fig. 5.1 Three conformations of FtsZ pfs. (**a**) pfs assembled in GTP are mostly straight, although some have intermediate curvature. (**b**) pfs assembled in GDP and adsorbed to a cationic lipid monolayer show the highly curved miniring conformation, mixed with straight pfs (Reprinted from Erickson et al. 1996 with permission of the author). (**c**) Toroids assembled in 1.6% methyl cellulose, where pfs are in the intermediate curved conformation (micrograph provided by Max Housman; these are somewhat larger than the 200 nm diameter reported previously (Popp et al. 2009)). The diameter of an undivided bacterium is indicated on *top*

(Huecas et al. 2015). A crystal form of *Mycobacterium tuberculosis* FtsZ showed antiparallel pfs in a tight spiral (Li et al. 2013). However, as discussed below, the curvature of these spirals is the opposite to that of EcFtsZ minirings.

The Intermediate Curved pf Conformation

In addition to the straight and highly curved minirings, there is an "intermediate curved conformation," with a diameter of ~100–300 nm. For EcFtsZ this is often seen as pf segments of uniform curvature, mixed with straight pfs (e.g., Fig. 1B, C in (Romberg et al. 2001); Fig. 3A, B in (Huecas et al. 2008); Fig. 4A–C in (Dajkovic et al. 2008). Under some conditions the pfs form closed circles, 160–250 nm in diameter that have been imaged by negative stain EM (Gonzalez et al. 2005; Chen et al. 2005). This curved conformation is highly favored when pfs are adsorbed to mica and imaged by AFM (Hamon et al. 2009; Mateos-Gil et al. 2012; Mingorance et al. 2005). The most reproducible production of intermediate curved pfs is assembly in polyvinyl alcohol or methyl cellulose, which are crowding reagents mimicking in the vivo condition (Popp et al. 2009). Here FtsZ assembles into a variety of toroids and helical pf bundles with diameters varying from 150 to 500 nm, depending on solution conditions. These are readily imaged by negative stain EM (Fig. 5.1c).

The intermediate curved conformation appears to be much more widespread than the miniring, and has been documented for FtsZ pfs from several species. *Bacillus subtilis* FtsZ (BsFtsZ) forms toroids that are 150–300 nm in diameter, mixed with straight pfs and bundles (Buske and Levin 2012; Michie et al. 2006). The curved pfs and toroids of BsFtsZ and *Staphylococcus aureus* FtsZ are enhanced by the drugs PC190723 and 8j, (Adams et al. 2011; Andreu et al. 2010). *M. tuberculosis* FtsZ assembled 200–250 nm toroids in methyl cellulose (Popp et al. 2010). *C. crescentus* FtsZ (CcFtsZ) assembled circles of ~75 nm diameter when stabilized by the protein FzlA in GMPCPP or in GTP plus EDTA (which blocks GTP hydrolysis) (Goley et al. 2010). Assembly of CcFtsZ alone in GTP gave uniformly curved pfs of this same curvature mixed with straight pfs (Goley et al. 2010; Milam and Erickson 2013). These probably correspond to the intermediate curved conformation. (As mentioned above, CcFtsZ also assembles circles of 34–43 nm diameter, which are probably the miniring conformation.) Co-assembly of EcFtsZ and FtsA* in vitro produced striking curved polymer bundles, often curling off the ends of straight bundles (Beuria et al. 2009). This is reminiscent of the toroids spooling off the ends of straight pf bundles in yeast (Srinivasan et al. 2008), mentioned below.

Some studies have imaged toroids by light microscopy. Srinivasan et al. (2008) expressed EcFtsZ-GFP in yeast and found a mixture of toroids and straight pf bundles. Many of the toroids were large enough (0.5 μm diameter) to resolve by conventional light microscopy, but there were also many smaller round patches that were probably toroids, but too small for light microscopy to resolve the hollow center. Similar mixtures of straight pf bundles and toroids were assembled by FtsZ from *M. tuberculosis* and *Xanthomonas*, attesting to the generality of the process. In one time series a yeast cell was initially filled with straight pf bundles, which later transitioned to toroids. Importantly, the toroids were often located on the end of a straight pf bundle, suggesting a transition where the straight pf bundle was spooled into the toroid (Srinivasan et al. 2008).

Toroids ~200 nm in diameter were imaged by light microscopy in *Arabidopsis*, under conditions where FtsZ2-GFP was overexpressed, or when Z-ring modulating factors ARC6 or MinE were deleted (Johnson et al. 2015). These toroids required super-resolution light microscopy for optimal resolution. Previous studies of chloroplast FtsZ had imaged small round patches under related conditions (Fujiwara et al. 2009; Zhang et al. 2013); these are likely toroids, but the conventional fluorescence microscopy used in these studies did not resolve the hollow center. In all of these studies the toroids were frequently attached to the end a linear filament, similar to the spooling of EcFtsZ seen in yeast (Srinivasan et al. 2008).

The preferred curvature of FtsZ was elegantly explored in an in vitro study where inside-out Z rings were assembled on lipid-coated capillaries of varying diameter (Arumugam et al. 2012). For capillary diameters of 500 nm or above, the Z rings were perpendicular to the axis, as expected if their preferred curvature was <500 nm. For capillary diameters 100–500 nm the Z rings were tilted. The tilt could be produced if the diameter of the capillary were smaller than the preferred curvature, or if the pfs had a helical pitch. Arumugam et al. obtained a good fit for a model with a native helix diameter of 55 nm and a pitch of 150 nm. A comparable fit could probably be obtained with a native diameter of 200 nm and a smaller pitch. Overall these measurements seem consistent with the intermediate curved pf, 100–300 nm in diameter, being the preferred conformation in the inside-out Z rings.

What Induces the Transition from Straight to Curved pfs?

The miniring conformation was discovered first (Erickson et al. 1996) and has received much emphasis. The simplest story of the mechanism was that GTP favored the straight conformation and GDP the miniring. This was supported by the observation of two forms of EcFtsZ polymers stabilized by DEAE dextran. In GMPCPP, or in GTP plus EDTA (which permits assembly but blocks GTP hydrolysis), the assembly formed sheets of straight pfs, while in GDP it formed helical tubes of the same diameter as minirings (Lu et al. 2000). However, this is an oversimplification. The mix of minirings and straight pfs shown in Fig. 5.1b was obtained in GDP. Polymers at steady state contain 20–50% GDP (Chen and Erickson 2009; Romberg and Mitchison 2004), but negative stain EM shows mostly straight pfs. The crystal structure of SaFtsZ shows it assembled into straight pfs, but the bound nucleotide is GDP (Matsui et al. 2012).

The intermediate curved conformation is observed more generally across species, and we think it is presently the best candidate for constriction-force generation. The intermediate curvature does not appear to be generated by GTP hydrolysis, since we have observed curved pfs of EcFtsZ in GMPCPP, and in pfs assembled by the mutant D212A, which is defective in hydrolysis (unpublished observations). The 75 nm circles of CcFtsZ were assembled in GMPCPP or in GTP plus EDTA, which blocks hydrolysis (Goley et al. 2011). Most important, Popp et al. (2010) reproducibly obtained toroids and spiral pf bundles in non-hydrolyzable GMPPNP and in GTP plus EDTA.

One thing that seems to favor the transition from straight to curved is contact with a surface. This is suggested by the strong preference for curved pfs when they are adsorbed to mica (Hamon et al. 2009; Mingorance et al. 2005). Tethering pfs to a phospholipid membrane, or contacts with neighboring filaments in pf bundles may similarly induce the transition. In contrast, single circles of intermediate curved pfs imaged by negative stain EM were presumably formed in solution by single pfs (Adams et al. 2011; Chen et al. 2005; Gonzalez et al. 2005). Overall, there is no consistent model explaining the transition from straight to curved pfs. The kinetics and thermodynamics of this important mechanism still need to be worked out.

Curved pfs Can Bend Membranes and Generate a Constriction Force

It was suggested in 1997 that FtsZ might generate the constriction force for cytokinesis by itself, without any motor protein, by using the conformational change from the straight to curved pf (Erickson 1997) (Fig. 5.2). For the next 10 years there was no way to test this hypothesis. In 2008 the reconstitution of Z rings in vitro provided the needed breakthrough (Osawa et al. 2008). This reconstitution was achieved by fusing an amphipathic helix (membrane targeting sequence, or mts) to the C terminus of FtsZ, so it could tether itself to the membrane. When this FtsZ-mts was incorporated inside tubular liposomes, it spontaneously assembled Z rings, and

A

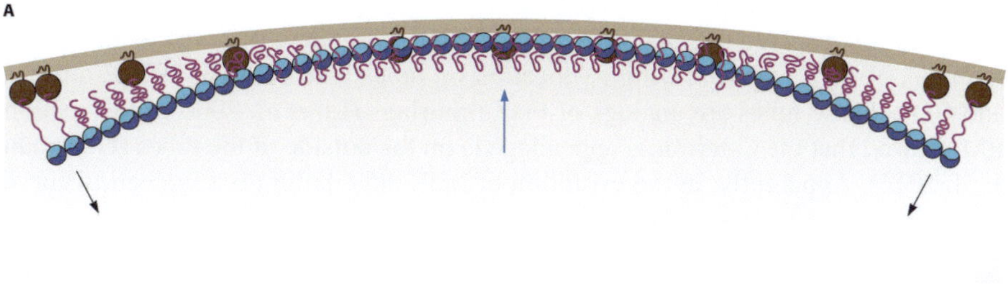

Fig. 5.2 Model for constriction force generated by pf curvature bending the membrane. The *brown arc* represents the inner membrane of the undivided *E. coli*. The *brown spheres* are FtsA, which has an amphipathic helix that inserts into the membrane. The *blue* and *cyan spheres* are the globular domains of FtsZ. The *magenta line* is the C terminus of FtsZ, which comprises a flexible linker and a short peptide that binds to FtsA. FtsZ is shown here as a pf that is straining to adopt a curved conformation. It pulls the membrane inward from its ends, and pushes outward at its middle, generating a bending force (Reprinted from Erickson et al. 2010 with permission of the author)

these Z rings constricted the liposome. Both Z-ring assembly and constriction were achieved with purified FtsZ-mts; no other protein was required. The constriction force generated by FtsZ was strong enough to invaginate multilamellar lipid vesicles that were up to 1 μm thick.

The constriction observed in this reconstituted system was consistent with the curvature hypothesis but did not prove it. Evidence in favor of pf curvature as the mechanism of constriction force was obtained by applying FtsZ-mts to the outside of large, unilamellar liposomes. There it formed patches that bent the membrane into concave depressions (Osawa et al. 2009). This direction of the bending was the same as inside liposomes, where it binds to the concave inner surface and constricts it to be more concave. These concave depressions on the outside of large liposomes were also observed by Arumugam et al. (2012).

The most definitive evidence that pf curvature generates the constriction force was obtained by switching the membrane tether to the opposite side of the pf. This was possible for FtsZ, because the N and C termini are flexible peptides emerging from the globular domain about 180° apart. The normal C-terminal attachment in FtsZ-mts was predicted to be on the outside of the curved pf. Switching the attachment to the N terminus gave mts-FtsZ, which would place the mts on the inside of the curved pf (Fig. 5.3). As predicted by the bending hypothesis, mts-FtsZ assembled Z rings on the outside of tubular liposomes (Osawa and Erickson 2011). Remarkably, these "inside-out Z rings" also generated a constriction force, squeezing the liposomes from the outside. These experiments provided strong support for the pf curvature hypothesis.

Tubulin pfs also have a curved conformation, which forms a bulge at the end of a disassembling MT as the pfs peel away. Curved tubulin pfs have been shown to generate force, since the bulge they create can drag an attached chromosome (McIntosh et al. 2010). There is, however, an apparent contradiction between the curvature of FtsZ and tubulin. It is well established for tubulin rings that the C terminus, corresponding to the outside of a MT, is on the inside of the ring (Moores and Milligan 2008; Nawrotek et al. 2011; Tan et al. 2008; Wang and Nogales 2005). In contrast, our model places the C terminus of FtsZ on the outside (Fig. 5.2). We have recently resolved this apparent contradiction by fusing a protein tag onto the C terminus of FtsZ and observing its location on tubes assembled in DEAE dextran and GDP. These tubes are analogs of FtsZ minirings (Lu et al. 2000). Thin section EM showed that the C-terminal appendage is on the outside of the tubes (Housman et al. 2016). Apparently, in the evolution of MTs the tubulin pfs switched to curve in the opposite direction to FtsZ.

Modeling studies have predicted that both FtsZ and tubulin pfs can bend in both directions (Grafmuller and Voth 2011; Hsin et al. 2012; Ramirez-Aportela et al. 2014; Theisen et al. 2012). A recent crystal structure of MtbFtsZ showed antiparallel pf pairs in a tight spiral, with the C terminus facing inward (Li et al. 2013). This curvature is the opposite of the EcFtsZ miniring/tube, and it is not yet clear if it is physiologically important. It has also not yet been determined what is the direction of the intermediate curved conformation. Since this form is likely the one generating the constriction force, we suggest that it has the C terminus on the outside facing the membrane, like the miniring.

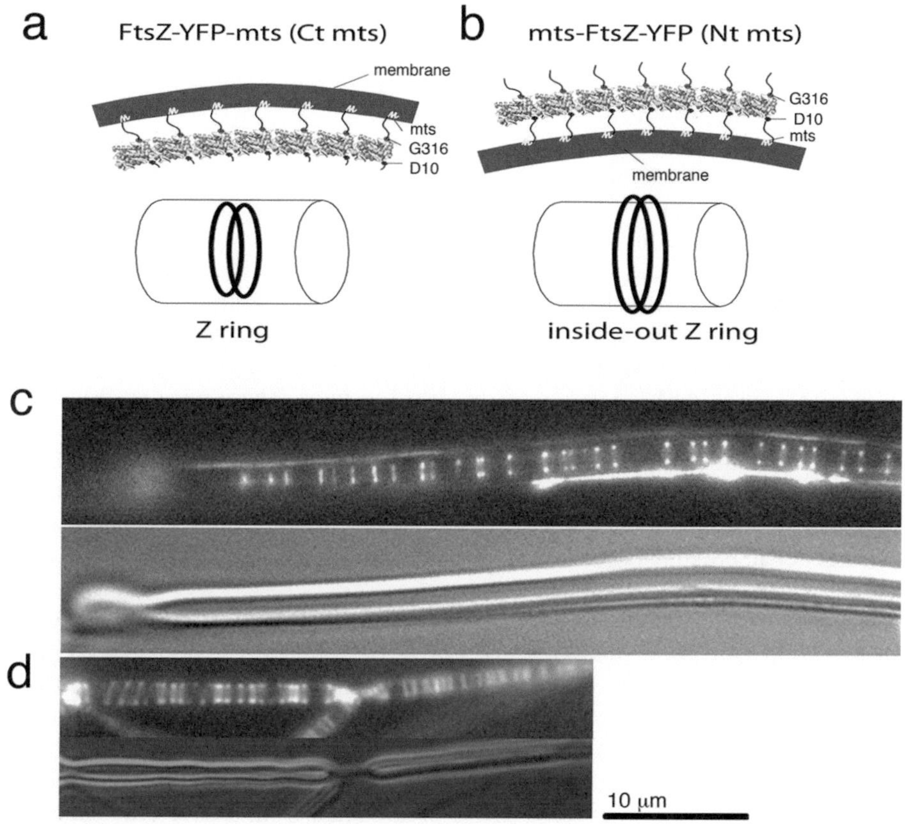

Fig. 5.3 (**a**) The normal tether on the C terminus is on the outside of the curved pf, and attaches to the concave membrane on the inside of a liposome or cell. (**b**) Switching the tether to the N terminus places it on the inside of the curved pf, for attachment to the convex surface on the outside of a liposome. (**c**) Z rings inside a tubular liposome reconstituted from FtsZ-mts. (**d**) Inside-out Z rings assembled on the outside of a liposome by mts-FtsZ. In (**c**) and (**d**) the *upper panel* shows FtsZ imaged by fluorescence, and the *lower panels* show the liposome imaged by DIC (Reprinted from Osawa et al. 2008; Osawa and Erickson 2011 with permission of the authors)

FtsZ pfs Are Mechanically Rigid

If the curved pfs are to generate a meaningful bending force on the membrane, they must be mechanically rigid. It is still not universally recognized how rigid the FtsZ pfs really are. For example, a recent review commented "*recent work has suggested that FtsZ filaments may be too flexible to bend membranes…*" (Eun et al. 2015). Ramirez-Aportela et al. (2014) commented on "*…the large disparity in measurements [of persistence length] found in the literature for GTP/GDP filaments of FtsZ from other species. For example, in a cryo-electron microscopy study of E. coli FtsZ (EcFtsZ), Turner et al. (*2012*) visualized relatively straight filaments with a persistence length of 1.4 μm. In contrast, we previously estimated protofilament* *persistence length values in the 100 nm range, also using cryo-electron microscopy of EcFtsZ but neglecting intrinsic filament curvature* (Huecas et al. 2008)." Dajkovic et al. (2008) also reported measuring a very low persistence length: Lp ~180 nm.

"…neglecting intrinsic filament curvature" is the key issue here. As we have noted previously (Erickson et al. 2010), in the two studies reporting very low persistence length (Dajkovic et al. 2008; Huecas et al. 2008) the authors assumed that the relaxed pf was straight, and that any curvature was due to random thermal bending. However, one can see in their micrographs (Fig. 3 in (Huecas et al. 2008)) that a significant fraction of the pfs have a fairly uniform curvature corresponding to the ~200 nm diameter intermediate curved conformation. The intermediate curved conformation was not well recognized at that time, so it is understandable that these curved pfs might have been attributed to thermal bending. In a separate study Hörger et al. (2008) used AFM to image pfs adsorbed to mica. These mica-adsorbed pfs seem to transition completely to the intermediate curved conformation, and this was recognized by the authors as the natural conformation of these pfs. They concluded that the pfs had a preferred curvature of 200 nm diameter, and a variation corresponding to a persistence length of ~4 μm. This is somewhat larger than the 1.4 μm of Turner et al, possibly because their pfs were adsorbed to mica, leaving them only the plane of the mica for thermal fluctuations.

The study by Turner et al. (2012) used cryoEM of samples where the pfs were suspended in aqueous medium over holes in the carbon film. The pfs had no contact with a carbon film, and they were widely dispersed to avoid any pf-pf contact. Their pfs were all straight (it is not clear why none were in the intermediate curved conformation). The small curvatures they measured were attributed to thermal bending, giving a persistence length of 1.4 μm. This is very close to the 2.9 μm estimated by Mickey and Howard (1995) for a single pf extracted from a MT. This study by Turner et al. should now be considered the definitive measurement of the mechanical rigidity of the FtsZ pfs.

This mechanical rigidity is not specific for FtsZ, but is a general property of globular proteins and their polymers. Twenty years ago Howard and colleagues measured the flexural rigidity of MTs and actin filaments suspended in aqueous solution (Gittes et al. 1993). They concluded that "If tubulin were homogeneous and isotropic, then the microtubule's Young's modulus would be ~1.2 GPa, similar to Plexiglas and rigid plastics". The generalization is that globular proteins and their polymers share this rigidity. Plexiglas is very rigid, and so are proteins.

In Vitro Systems Reconstituted with FtsZ and FtsA

In bacteria, FtsZ is tethered to the membrane by FtsA. FtsA has an amphipathic helix that binds the membrane, and FtsZ has a conserved C-terminal peptide that binds to the globular domain of FtsA (Pichoff and Lutkenhaus 2005) (Fig. 5.2). FtsA is an actin homolog, and it can assemble filaments when bound to a lipid bilayer (Szwedziak et al. 2012). Our original reconstitution in liposomes used

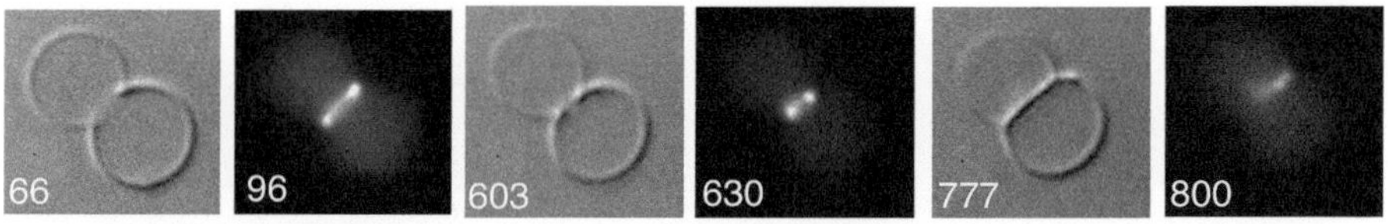

Fig. 5.4 Z rings reconstituted from EcFtsZ-YFP plus FtsA in a large unilamellar vesicle. Numbers on the *lower left* give time in s. At 66 s the vesicle is largely constricted, by a prominent Z ring seen by fluorescence at 96 s. At 603 s the constriction has advanced to where the invaginating membrane shows two dots about 1 μm apart. At 777 s the membranes appear to have fused into a broad, flat septum between the two vesicles, and the FtsZ-YFP is a diffuse smear at one end (800 s) (Adapted from Osawa and Erickson 2013 with permission of the authors)

FtsZ-mts, where the mts could tether FtsZ directly to the membrane, bypassing FtsA. More recently, three studies have achieved reconstitutions using the natural two-protein system, FtsZ plus FtsA, where the FtsZ pfs are tethered to the membrane by FtsA. Each study had to solve the problem that EcFtsA has been difficult to express and work with in vitro.

Osawa and Erickson (2013) reported a reconstitution using the *E. coli* mutant FtsA*, which is fully functional in vivo and gives much better results in vitro than wild type EcFtsA (Beuria et al. 2009; Geissler et al. 2007). When a 1:1 mixture of FtsZ-YFP:FtsA was incorporated inside large unilamellar vesicles, a small fraction of the liposomes developed fluorescent Z rings that constricted the liposome to the point of membrane contact (Fig. 5.4). Then the membranes abruptly formed a flat septum, apparently having fused at the point of tightest constriction. FtsA was essential to obtain constriction and septation of these large liposomes. Without FtsA, FtsZ-mts bound to and bent the membrane to form protrusions, and typically assumed the shape of small Z rings ~1 μm in diameter. Rarely a larger Z ring formed, but instead of constricting the liposome, the Z ring appeared to slide down the side as it constricted. Thus FtsA seems necessary to promote a stable Z ring around the center of the vesicle, but the mechanism is not yet known.

Szwedziak et al. (2014) recently presented beautiful cryoEM images of a similar reconstituted system of FtsZ plus FtsA inside membrane vesicles. Their proteins were from *Thermotoga maritima,* whose FtsA is well behaved in vitro. FtsZ was assembled into long pfs that associated in pairs and triplets (Fig. 5.5b). Favorable projections showed a dot between FtsZ and the membrane, which was interpreted as an FtsA filament in cross section. The FtsZ filaments formed arcs, with a diameter of ~100 nm, at various locations on the membranes. At sites of constriction the FtsZ pf pairs and triplets encircled the liposome, with a zone of overlap after completing the circuit. The maximum constriction observed was about 90 nm in diameter, consistent with the intermediate curved conformation.

Loose and Mitchison (2014) reconstituted FtsZ plus FtsA, not in liposomes but on planar lipid bilayers, using fluorescently tagged proteins that could be observed over time. Their proteins were from *E. coli,* and they obtained stable wild type FtsA by a SUMO fusion and expression at 18 °C. A 3:1 mixture of FtsZ:FtsA was placed over a planar lipid bilayer on a glass coverslip, and after ~5 min the adsorbed protein

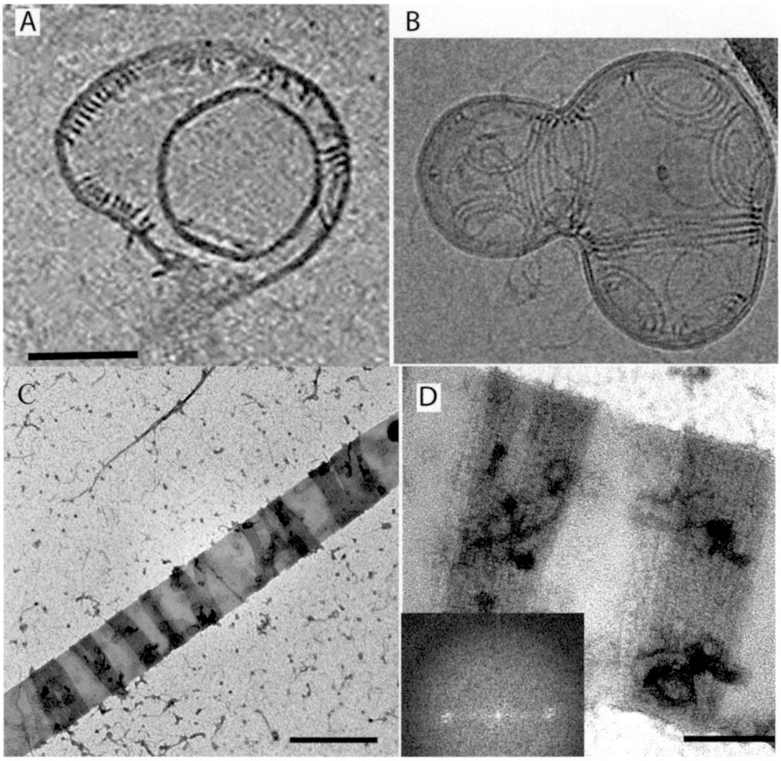

Fig. 5.5 *Panel A* shows a section through a cryoEM tomogram of Z rings reconstituted from FtsZ plus FtsA. The *left side* of the vesicle shows two Z rings in oblique section. They each comprise ribbons of pfs spaced ~7.5 nm apart. *Panel B* shows a projection of a an entire vesicle, with ribbons of 3 and 4 protofilaments forming arcs encircling sites of constriction (Reprinted from Szwedziak et al. 2014, with permission of the authors. **a** was extracted from their video 4, **b** is their Fig. 5.3c) *Panels C* and *D* show negative stain images of inside-out Z rings of mts-FtsZ, reconstituted on rigid lipid tubules. The dark stained Z rings comprise ribbons of parallel, close-packed pfs. The Fourier analysis (*inset* **d**) indicates a uniform 5 nm spacing of pfs (Reprinted from Milam et al. 2012 with permission of the authors). Scale bars are 100 nm (**a**, **b**) 500 nm (**c**) and 200 nm (**d**)

organized into rings, with a diameter of ~1.2 μm. The rings were ~0.5 μm thick, so they were clearly bundles or toroids of many pfs. Remarkably, the toroids showed rotational movement with a velocity of 110 nm/s = 25 subunits/s. (That is very fast, five times faster than the elongation rate of single pfs in solution at a steady state subunit concentration of 1 μM (Chen and Erickson 2005).) When single fluorescent molecules were sparsely doped they remained stationary for an average 7 s lifetime, which suggests that the mechanism driving the motion is subunit exchange, perhaps treadmilling, rather than sliding of pfs relative to each other. Because the bundle moves in one direction, rather than sliding apart, the pfs in the bundles or toroids must be arranged in a parallel fashion, all facing the same direction.

Arumugam et al. (2014) did a similar reconstitution on planar lipid bilayers, but they used FtsZ-mts without FtsA. The FtsZ-mts assembled small filament bundles, mostly straight, that grew longer and eventually anastomosed into a network. Although this network appeared stable, FRAP showed that it was rapidly exchanging subunits with a half time of 10 s, similar to the rate of exchange in the Z ring in vivo (Anderson et al. 2004) and single pfs in solution (Chen and Erickson 2005). Importantly, the exchange did not result in new pf bundles but took place within the existing bundles. This suggests that there are multiple pf ends within the bundles where exchange takes place. MinC was shown to disassemble the bundle network, apparently by first binding to these ends. The authors then reconstituted the MinCDE system and showed that the waves of peak Min resulted in disassembly of FtsZ, which reassembled at the Min trough.

The FtsZ polymers assembled on planar lipid bilayers are wider than the 250 nm resolution of the light microscope, whether tethered by FtsA or directly by the –mts. They have been termed "bundles," which implies a 3D structure, but they may actually be ribbons, one subunit thick, in which every pf is tethered to the membrane. It should be possible to resolve this question by negative stain or cryoEM. This may also address the discrepancy, discussed below, of the 7–9 nm pf spacing seen in reconstitution of FtsZ plus FtsA (Szwedziak et al. 2012), vs. the 5 nm spacing in ribbons of mts-FtsZ (Milam and Erickson 2013).

Sliding pfs – Do They Generate Constriction or Just Accommodate It?

The sliding-filament model of constriction force (Hörger et al. 2008; Lan et al. 2009) proposed that the Z ring is a long filamentous structure that can span the circumference of the cell, with the two ends meeting in an overlap zone where they can associate laterally. If the filaments could slide to increase the overlap, the increased lateral interaction should be thermodynamically favored, and this could generate constriction. This seems valid thermodynamically, but, as pointed out previously (Erickson 2009), this mechanism seems doomed by a kinetic trap. In order to slide, the filaments need first to break all lateral bonds, and as the number of lateral bonds increases, the time needed for complete rupture increases exponentially.

Although sliding itself is unlikely to generate a constriction force, sliding is probably essential to accommodate continued constriction. If the Z ring were only a partial circle, with a gap between the ends, constriction could continue until the gap closed. If a long pf pair made a complete circuit and the ends passed each other in an overlap zone (Szwedziak et al. 2014), continued constriction would require that the overlapping ends could slide. In this case it would be important that the overlap not have lateral bonds that would impose a brake on the sliding.

Consider next a model where a Z ring comprised four pfs, each of them making a complete circuit and then annealing to form a continuous circle. The complete circle would arrest constriction until a gap was created. We have suggested this model previously to explain the observation that Z rings reconstituted in GMPCPP generated an initial constriction, but then arrested (Osawa and Erickson 2011). This is in contrast to Z rings assembled in GTP, which continued constricting. Importantly, Z rings in GTP are constantly exchanging subunits between the Z ring and solution. Subunit exchange would necessarily create gaps, which would permit continued constriction until the gap is closed. Z rings assembled in GMPCPP do not exchange and would not create gaps, so constriction arrests.

Constriction Force by Partial Z Rings

Although FtsZ assembles complete circular Z rings in cells and in tubular liposomes, there are several examples of constriction force being generated without closure of the ring. The concave depressions generated on large liposomes probably contain only arcs of pfs, not a complete circle (Arumugam et al. 2012; Osawa et al. 2009). The *E. coli* mutant *ftsZ26* forms spiral Z rings, and scanning EM showed spiral constrictions, demonstrating force generation without closure into a ring (Addinall and Lutkenhaus 1996). An impressive example of partial Z rings generating arcs of constriction has been documented for a plastid (Sato et al. 2009). Force generation by these partial Z rings is difficult to explain by active filament sliding.

The pf Substructure of Z Rings – Ribbons or Scattered?

In *E. coli* there are ~6700 molecules of FtsZ in the cell (Li et al. 2014)[1], and fluorescence assays have found that 30–40% of the total FtsZ is in the Z ring (Geissler et al. 2007; Stricker et al. 2002). The 2000–2700 FtsZ molecules in the Z ring could make a pf 8600–11,600 nm long, enough to encircle the 900 nm inside diameter three to four times. For shorter and longer cells this could be 2–5 circuits, which we take to be the average number of pfs across the width of the Z ring. There are two competing models for how pfs are arranged in the Z ring. In the ribbon model the 2–5 pfs are thought to be parallel and in lateral contact, making a ribbon 10–25 nm wide. In the scattered model the pfs more widely spaced and not in contact, creating a band ~100 nm or more wide. In both models all pfs are tethered to the membrane.

The scattered model was first suggested by a cryoEM study of *C. crescentus*, which showed short pfs scattered around the Z ring and only rarely in contact with each other (Li et al. 2007). Several studies by super-resolution light microscopy have indirectly supported this model by reporting widths of the Z ring ~110 nm

[1] This study quantitates the number of molecules of almost all proteins in *E. coli* MG1655, under three growth conditions. Our Table 5.1 collects their data for cell division and cytoskeletal proteins.

Table 5.1 Copy number of cell division and cytoskeleton proteins, determined by Li et al. (2014)

Gene	MOPS complete	MOPS minimal	MOPS complete
			Without methionine
ftsA	984	575	1013
ftsB	487	140	173
ftsE	967	320	649
ftsI	349	144	226
ftsK	508	213	376
ftsL	416	201	423
ftsN	871	269	405
ftsP	905	196	410
ftsQ	336	147	172
ftsW	293	117	169
ftsX	838	244	436
ftsZ	6750	3335	5290
zipA	2128	501	874
minC	857	364	575
minD	5358	2444	3324
minE	3597	1970	2680
matP	601	360	347
slmA	2323	568	952
zapA	2275	738	2008
zapB	34,197	7797	18,090
ZapC/YcbW	155	155	155
ZapD/YacF	1983	373	538
ZapE/YhcM	255	214	262
mreB	11,304	2393	5570
mreC	738	176	333
mreD	367	71	148
rodZ	1309	576	669

This study used ribosome profiling to determine the rate of translation of 95% of the proteins in *E. coli.* The rates were converted to number of protein molecules synthesized in one cell cycle. Assuming no unusual degradation, the numbers can be taken as the average number of molecules per cell. The *E. coli* strain MG1655 was grown at 37 °C in three different media with different doubling times: 21.5 and 56.3 min in MOPS complete and incomplete media, and 26.5 min in MOPS complete without methionine. The latter condition required the cell to substantially upregulate the met-E enzyme, altering overall metabolism. We have extracted cell division and cytoskeleton proteins from their table S1. It seems quite valuable to have these numbers determined for all proteins in the identical cell and growth conditions. The 6750 molecules of FtsZ in a cell volume of 2×10^{-15} L is 5.6 μM

(Biteen et al. 2012; Fu et al. 2010; Holden et al. 2014); a ribbon of 2–5 pfs should be only 10–25 nm wide. A recent PALM study of the Z ring in *Streptococcus pneumoniae* measured the axial width of the Z ring to be 95 nm early in the cell cycle, increasing to 127 nm as the Z ring constricted (Jacq et al. 2015). It should be noted, however, that the most recent and comprehensive PALM studies (Coltharp et al. 2016; Holden et al. 2014; Jacq et al. 2015) also measured the thickness of the Z ring in the radial direction. The radial thickness was very close in magnitude to the axial width (71 nm axial width vs. 64 nm radial thickness (Holden et al. 2014); Fig. 7 in (Jacq et al. 2015); 99 nm axial vs. 59 nm radial (Coltharp et al. 2016)). This large radial width contradicts our expectation that all pfs are tethered to the membrane by FtsA and ZipA, necesitating a Z ring that is radially one subunit thick. It is important to recall that in *E. coli* the Z ring averages only 2–5 pfs thick (both axial and radial), so there is not enough material to build in the radial direction. This suggests that the actual radial thickness should be ~5 nm. The large radial thickness measured in the PALM images may therefore reflect the achieved resolution (which is worse than the resolution predicted theoretically in the studies). Since the measured axial width was only 7–40 nm more than the radial, the actual axial width of the Z ring might well be as small as 10–25 nm. Overall, the super-resolution light microscopy does not provide convincing support for the larger width of the scattered model.

A recent cryoEM study by Szwedziak et al. (2014) provides strong support for the ribbon model in both *C. crescentus* and *E. coli* . Their tomograms showed parallel arrays of pfs encircling the cells at the constriction site; these were clearly ribbons, one pf thick radially, and with variable axial widths. Szwedziak et al. (2014) also presented clear images of FtsZ rings reconstituted in liposomes with FtsA (Fig. 5.5). A typical Z ring comprised a ribbon of 2, 3 or 4 pfs that formed a continuous helical loop around the constriction, with a zone of overlap. In all cases the ribbons were one pf thick, with all pfs apparently linked to the membrane by FtsA. This supports the more important observation of Z rings in bacteria, which are ribbons, one pf thick.

The Spacing of pfs in the Ribbons

Curiously, in the cryo-EM study of Szwedziak et al. (2014) the pfs were spaced about 7–9 nm apart in both the in vitro reconstitution, and in cells of *C. crescentus* and *E. coli*. The earlier cryoEM study of Li et al. (2007) reported widely scattered pfs in wild type *C. crescentus*, but upon overexpression of FtsZ they also found ribbons of parallel pfs. Importantly, the pfs in their reconstructions were spaced 7–9 nm apart (7 nm by our measurements from their figures, 9.2 nm reported in their Fourier analysis).

In a separate study, Milam et al. (2012) succeeded in imaging reconstituted Z rings of mts-FtsZ by negative stain EM. These inside-out Z rings appeared as bands of uniform density and width, encircling lipid tubules 300 nm in diameter (Fig. 5.5).

The width of a given Z ring varied from ~15 nm to >100 nm, but was constant for a given Z ring, consistent with ribbons of three to tens of pfs. Fourier analysis indicated a pf spacing of 5 nm (Fig. 5.5D), similar to 2-D sheets assembled in DEAE dextran (Erickson et al. 1996). These images also support the ribbon model.

The cryoEM and the negative stain both support a ribbon model, but they differ in the pf spacing. The negative stain showed pfs in lateral contact, with a 5 nm spacing, while the cryoEM showed a spacing of 7–9 nm. The negative stain images should not be dismissed as artifactual (Szwedziak et al. 2014). Uranyl acetate is an extremely rapid protein fixative (Zhao and Craig 2003), and is generally considered as good as cryo for preserving protein structures at a 1–2 nm level (Ohi et al. 2004). A problem with the 7–9 nm spacing is that the FtsZ pf is maximally 5 nm wide, so the pfs in these cryoEM images are apparently not in contact with each other. No mechanism has been suggested for forces that would operate through 2–4 nm of solvent to bring the pfs together as parallel tracks, but keep them 2–4 nm apart. Nevertheless, the agreement of the two independent studies by cryoEM (Li et al. 2007; Szwedziak et al. 2014) makes a strong case that the pfs in vivo are indeed held at a separation of 7–9 nm. A potentially important clue is that the pfs in the cryoEM studies were tethered to the membrane by FtsA, whereas the pfs imaged by negative stain were directly attached to the membrane by the amphipathic helix on mts-FtsZ. This suggests that FtsA may enforce the larger separation of pfs. However, a mechanism for FtsA to bridge adjacent pfs is not obvious.

The pf Substructure of Z Rings – Continuous or Patchy?

A different point of controversy is whether the Z ring has a continuous or patchy structure. Both the cryoEM and the negative stain of reconstructed Z rings show continuous Z rings of mostly uniform width. CryoEM also showed a uniform width of Z rings in bacteria (Szwedziak et al. 2014). In contrast, most super-resolution light microscope studies have shown a patchy structure, where bright patches of FtsZ are separated by dim or dark zones apparently depleted of FtsZ. This was first reported for images obtained by structured illumination microscopy (SIM) (Strauss et al. 2012) and was confirmed in a more recent SIM study (Rowlett and Margolin 2014). A similar patchy structure was reported in two super-resolution studies using PALM, which is a completely different super-resolution technique (Holden et al. 2014; Jacq et al. 2015). It is not yet clear how to resolve the contradictory structures seen by EM and light microscopy.

Z-Ring Assembly and Initial Constriction

New Z-rings assemble in daughter cells as the mother cell is finishing septation. A recent PALM study followed the intensity of the Z ring through the cell cycle (Coltharp et al. 2016). With slow growth in M9 minimal medium, the Z-ring initially contained about 25% of the cell's total FtsZ, and this increased to 30% by 90 min. There was no obvious remodeling, but an earlier FRAP analysis showed that the Z ring is rapidly exchanging subunits with the cytoplasmic pool, with a half time of 12 s (Buss et al. 2015), very similar to the 8 s exchange observed under faster cell growth (Anderson et al. 2004). The Z ring then persisted without obvious change until ~145 min, when constriction began. The constriction comprised two phases: in the first 35 min, FtsZ remained present at a constant intensity (~35% of the total) as the Z ring constricted from 1000 nm to 250 nm; during the next 11 min FtsZ disassembled as the septum progressed to closure.

We suggest that FtsZ is generating a constriction force on the inner membrane as soon as the Z ring is assembled and throughout the first 145 min, but that attachment of the membrane to the rigid peptidoglycan wall prevents any actual constriction. During this time the Z ring serves as a scaffold for assembly of downstream division proteins, which will remodel the peptidoglycan wall. Constriction finally begins when this remodeling permits the peptidoglycan to follow the force that the FtsZ is exerting on the membrane. Invagination continues until the FtsZ reaches the 250 nm diameter of the intermediate curved conformation, which is probably the limit of pf bending. We suggest FtsZ as the driving force for constriction, primarily because it has been demonstrated to generate constriction in liposomes, with sufficient force to invaginate thick-walled, multilamellar liposomes.

This scenario, and the primacy of FtsZ in generating the constriction force, has been questioned by Coltharp et al. (2016). They examined how mutations in either FtsZ or peptidoglycan synthesis affected the timing of constriction onset and the rate of septum closure. They found that both constriction onset and rate of septum invagination were not affected by the FtsZ84 mutant, which has reduced GTPase and exchange dynamics. In contrast, alterations in peptidoglycan synthesis affected both constriction onset and the rate of closure. They concluded that "septum closure is likely driven by septum synthesis rather than Z-ring contraction."

The idea that peptidoglycan remodeling might provide the primary driving force for septation has a long history. In the extreme scenario, the ring of FtsZ is proposed to serve primarily as a docking site for the remodeling enzymes, and the constriction force is generated entirely by the inward remodeling of the peptidoglycan. This was largely discounted when FtsZ was discovered in mycoplasma and archaea, which have no peptidoglycan cell wall (Margolin et al. 1996; Wang and Lutkenhaus 1996a, b). Since FtsZ, but none of the other Fts proteins, are found in archaea, FtsZ was boosted as the prime candidate for generating the constriction. In bacteria with a peptidoglycan wall, peptidoglycan remodeling is still a candidate for contributing to the constriction force (Coltharp et al. 2016; Meier and Goley 2014), perhaps especially in the later stage. This remains an intriguing speculation; however, compelling evidence is lacking that peptidoglycan remodeling contributes to the constriction force.

The observations of Coltharp et al. (2016) are actually consistent with our model, where FtsZ is the primary source of constriction force. We suggest specifically that FtsZ84, although having reduced GTPase and dynamics, can still generate a constriction force that is more than sufficient to drive septum invagination. The onset of constriction is triggered by the onset of peptidoglycan remodeling, which is a chemical process that is probably independent of the force on the membrane. Likewise, the rate of septum invagination is probably limited, not by the force generated by FtsZ, but by the rate at which remodeling permits the peptidoglycan wall to follow the inner membrane. In the scenario we propose, septum invagination is driven by the constriction force of FtsZ, but its rate is limited by the peptidoglycan remodeling, which allows the wall to follow passively.

The Final Step of Septum Closure

The final step of septation must involve the advancing furrow constricting to the point where the curved membranes contact and fuse into separate membranes of the daughter cells. FtsZ apparently does not participate in this final step. The intermediate curved pfs could constrict the furrow to a diameter of ~250 nm, but something else is needed to finish septation. In a recent study Söderstrom et al. (2014) used a clever FRAP assay to distinguish daughter cells still connected by a narrow gap, versus cells separated by the completed septum. They found that FtsZ disappeared from the Z ring before the septum was closed. In contrast, FtsA and downstream proteins that remodel the peptidoglycan remained longer, and were typically still present when the septum was closed. This was confirmed in the recent PALM study of Coltrap et al. (2016), which found that the FtsZ intensity in the Z ring remained constant as it constricted to ~250 nm diameter, and then the FtsZ disassembled.

An attractive general mechanism for the final stage of septum closure is excess membrane synthesis. As a septum ingresses, new membrane is needed to provide the increased surface area. If the cell synthesizes membrane in excess of the minimum, the excess will be accommodated most easily at the highly curved invaginating septum. Thus, if the FtsZ ring generated constriction to a diameter of ~250 nm, and the remodeled peptidoglycan maintained this constriction after the FtsZ dissociated, a small excess of membrane synthesis could drive the constriction to septum closure and force membrane fusion. An important consideration is that the membrane at the septum has two points of high curvature: where the septum meets the cell membrane, and at the tip of the constriction. These points of curvature are more favorable for membrane expansion than starting a new bud. Membrane expansion would be essential to accommodate the invaginating septum, but once the curvature is established it should also be sufficient to drive the membranes together for the final closure and fusion.

The potential for excess membrane synthesis to drive cell division was demonstrated in a recent study of Mercier et al. (2013). In a previous study from the Errington lab, Leaver et al. (2009) had created *B. subtilis* L forms, which had no peptidoglycan wall. These cells propagated, but surprisingly they did not even need FtsZ. They divided by extruding pseudopods, which spontaneously resolved by breaking up into small spherical cells. The following study by Mercier et al. (2013) showed that for the L forms to propagate, it was essential to provide a mutation or manipulation that generated excess membrane synthesis. The authors concluded that the original life form may have relied on this simple mechanism for cell division – excess membrane synthesis and membrane fluctuations leading to vesicles pinching off. We suggest that this may still be the mechanism for the final step of septation in modern bacteria.

In summary, we suggest that FtsZ provides the primary constriction force for cytokinesis by exerting a bending force on the inner membrane. Actual constriction only begins when peptidoglycan remodeling permits the cell wall to follow the membrane. Bending pfs then constrict the inner membrane to the ~250 nm diameter of the intermediate curved pfs. The peptidoglycan wall follows by remodeling, and this remodeling limits the speed of septum invagination. The final steps of membrane closure extension and fusion of the two peptidoglycan cell walls completes the septation are driven by excess membrane synthesis, fluctuations and fusion.

References

Adams DW, Wu LJ, Czaplewski LG, Errington J (2011) Multiple effects of benzamide antibiotics on FtsZ function. Mol Microbiol 80:68–84

Addinall SG, Lutkenhaus J (1996) FtsZ-spirals and -arcs determine the shape of the invaginating septa in some mutants of *Escherichia coli*. Mol Microbiol 22:231–237

Anderson DE, Gueiros-Filho FJ, Erickson HP (2004) Assembly Dynamics of FtsZ rings in Bacillus subtilis and Escherichia coli and effects of FtsZ-regulating proteins. J Bacteriol 186:5775–5781

Andreu JM et al (2010) The antibacterial cell division inhibitor PC190723 is an FtsZ polymer-stabilizing agent that induces filament assembly and condensation. J Biol Chem 285:14239–14246

Arumugam S, Chwastek G, Fischer-Friedrich E, Ehrig C, Monch I, Schwille P (2012) Surface topology engineering of membranes for the mechanical investigation of the tubulin homologue FtsZ. Angew Chem Int Ed Eng 51:11858–11862

Arumugam S, Petrasek Z, Schwille P (2014) MinCDE exploits the dynamic nature of FtsZ filaments for its spatial regulation. Proc Natl Acad Sci U S A 111:E1192–E1200

Beuria TK, Mullapudi S, Mileykovskaya E, Sadasivam M, Dowhan W, Margolin W (2009) Adenine nucleotide-dependent regulation of assembly of bacterial tubulin-like FtsZ by a hypermorph of bacterial actin-like FtsA. J Biol Chem 284:14079–14086

Biteen JS, Goley ED, Shapiro L, Moerner WE (2012) Three-dimensional super-resolution imaging of the midplane protein FtsZ in live Caulobacter crescentus cells using astigmatism. ChemPhysChem 13:1007–1012

Buske PJ, Levin PA (2012) Extreme C terminus of bacterial cytoskeletal protein FtsZ plays fundamental role in assembly independent of modulatory proteins. J Biol Chem 287:10945–10957

Buss J, Coltharp C, Shtengel G, Yang X, Hess H, Xiao J (2015) A multi-layered protein network stabilizes the Escherichia coli FtsZ-ring and modulates constriction dynamics. PLoS Genet 11:e1005128

Chen Y, Erickson HP (2005) Rapid in vitro assembly dynamics and subunit turnover of FtsZ demonstrated by fluorescence resonance energy transfer. J Biol Chem 280:22549–22554

Chen Y, Erickson HP (2009) FtsZ filament dynamics at steady state: subunit exchange with and without nucleotide hydrolysis. Biochemistry 48:6664–6673

Chen Y, Bjornson K, Redick SD, Erickson HP (2005) A rapid fluorescence assay for FtsZ assembly indicates cooperative assembly with a dimer nucleus. Biophys J 88:505–514

Coltharp C, Buss J, Plumer TM, Xiao J (2016) Defining the rate-limiting processes of bacterial cytokinesis. Proc Natl Acad Sci U S A 113(8):E1044–E1053

Dajkovic A, Lan G, Sun SX, Wirtz D, Lutkenhaus J (2008) MinC spatially controls bacterial cytokinesis by antagonizing the scaffolding function of FtsZ. Curr Biol 18:235–244

Erickson HP (1997) FtsZ, a tubulin homolog, in prokaryote cell division. Trends Cell Biol 7:362–367

Erickson HP (2009) Modeling the physics of FtsZ assembly and force generation. Proc Natl Acad Sci U S A 106:9238–9243

Erickson HP, Taylor DW, Taylor KA, Bramhill D (1996) Bacterial cell division protein FtsZ assembles into protofilament sheets and minirings, structural homologs of tubulin polymers. Proc Natl Acad Sci U S A 93:519–523

Erickson HP, Anderson DE, Osawa M (2010) FtsZ in bacterial cytokinesis: cytoskeleton and force generator all in one. Microbiol Mol Biol Rev 74:504–528

Eun YJ, Kapoor M, Hussain S, Garner EC (2015) Bacterial filament systems: toward understanding their emergent behavior and cellular functions. J Biol Chem 290:17181–17189

Fu G, Huang T, Buss J, Coltharp C, Hensel Z, Xiao J (2010) In vivo structure of the E. coli FtsZ-ring revealed by photoactivated localization microscopy (PALM). PLoS ONE 5:e12682

Fujiwara MT, Sekine K, Yamamoto YY, Abe T, Sato N, Itoh RD (2009) Live imaging of chloroplast FtsZ1 filaments, rings, spirals, and motile dot structures in the AtMinE1 mutant and overexpressor of Arabidopsis thaliana. Plant Cell Physiol 50:1116–1126

Geissler B, Shiomi D, Margolin W (2007) The ftsA* gain-of-function allele of Escherichia coli and its effects on the stability and dynamics of the Z ring. Microbiology (Reading, England) 153:814–825

Gittes F, Mickey B, Nettleton J, Howard J (1993) Flexural rigidity of microtubules and actin filaments measured from thermal fluctuations in shape. J Cell Biol 120:923–934

Goley ED, Dye NA, Werner JN, Gitai Z, Shapiro L (2010) Imaging-based identification of a critical regulator of FtsZ protofilament curvature in Caulobacter. Mol Cell 39:975–987

Goley ED, Yeh YC, Hong SH, Fero MJ, Abeliuk E, McAdams HH, Shapiro L (2011) Assembly of the Caulobacter cell division machine. Mol Microbiol 80:1680–1698

Gonzalez JM et al (2005) Cooperative behavior of Escherichia coli cell-division protein FtsZ assembly involves the preferential cyclization of long single-stranded fibrils. Proc Natl Acad Sci U S A 102:1895–1900

Grafmuller A, Voth GA (2011) Intrinsic bending of microtubule protofilaments. Structure 19:409–417

Hamon L et al (2009) Mica surface promotes the assembly of cytoskeletal proteins. Langmuir 25:3331–3335

Holden SJ, Pengo T, Meibom KL, Fernandez Fernandez C, Collier J, Manley S (2014) High throughput 3D super-resolution microscopy reveals Caulobacter crescentus in vivo Z-ring organization. Proc Natl Acad Sci U S A 111:4566–4571

Hörger I, Velasco E, Mingorance J, Rivas G, Tarazona P, Velez M (2008) Langevin computer simulations of bacterial protein filaments and the force-generating mechanism during cell division. Phys Rev E Stat Nonlinear Soft Matter Phys 77:011902

Housman M, Milam SL, Moore DA, Osawa M, Erickson HP (2016) FtsZ protofilament curvature is the opposite of tubulin rings. Biochemistry 55:4085–4091

Hsin J, Gopinathan A, Huang KC (2012) Nucleotide-dependent conformations of FtsZ dimers and force generation observed through molecular dynamics simulations. Proc Natl Acad Sci U S A 109:9432–9437

Huecas S, Andreu JM (2004) Polymerization of nucleotide-free, GDP- and GTP-bound cell division protein FtsZ: GDP makes the difference. FEBS Lett 569:43–48

Huecas S, Llorca O, Boskovic J, Martin-Benito J, Valpuesta JM, Andreu JM (2008) Energetics and geometry of FtsZ polymers: nucleated self-assembly of single protofilaments. Biophys J 94:1796–1806
Huecas S et al (2015) Beyond a fluorescent probe: inhibition of cell division protein FtsZ by mant-GTP elucidated by NMR and biochemical approaches. ACS Chem Biol 10:2382–2392
Jacq M, Adam V, Bourgeois D, Moriscot C, Di Guilmi AM, Vernet T, Morlot C (2015) Remodeling of the Z-ring nanostructure during the Streptococcus pneumoniae cell cycle revealed by photo-activated localization microscopy. mBio 6:eo1108–1115
Johnson CBLZ, Luo Z, Shaik RS, Sung MW, Vitha S, Holzenburg A (2015) In situ structure of FtsZ mini-rings in Arabidopsis chloroplasts. Adv Struct Chem Imag 1:12
Lan G, Daniels BR, Dobrowsky TM, Wirtz D, Sun SX (2009) Condensation of FtsZ filaments can drive bacterial cell division. Proc Natl Acad Sci U S A 106:121–126
Leaver M, Dominguez-Cuevas P, Coxhead JM, Daniel RA, Errington J (2009) Life without a wall or division machine in Bacillus subtilis. Nature 457:849–853
Li Z, Trimble MJ, Brun YV, Jensen GJ (2007) The structure of FtsZ filaments in vivo suggests a force-generating role in cell division. EMBO J 26:4694–4708
Li Y et al (2013) FtsZ protofilaments use a hinge-opening mechanism for constrictive force generation. Science 341:392–395
Li GW, Burkhardt D, Gross C, Weissman JS (2014) Quantifying absolute protein synthesis rates reveals principles underlying allocation of cellular resources. Cell 157:624–635
Loose M, Mitchison TJ (2014) The bacterial cell division proteins FtsA and FtsZ self-organize into dynamic cytoskeletal patterns. Nat Cell Biol 16:38–46
Lu C, Reedy M, Erickson HP (2000) Straight and curved conformations of FtsZ are regulated by GTP hydrolysis. J Bacteriol 182:164–170
Margolin W, Wang R, Kumar M (1996) Isolation of an ftsZ homolog from the archaebacterium *Halobacterium salinarium*: implications for the evolution of FtsZ and tubulin. J Bacteriol 178:1320–1327
Mateos-Gil P et al (2012) FtsZ polymers bound to lipid bilayers through ZipA form dynamic two dimensional networks. Biochim Biophys Acta 1818:806–813
Matsui T, Yamane J, Mogi N, Yamaguchi H, Takemoto H, Yao M, Tanaka I (2012) Structural reorganization of the bacterial cell-division protein FtsZ from Staphylococcus aureus. Acta Crystallogr 68:1175–1188
McIntosh JR, Volkov V, Ataullakhanov FI, Grishchuk EL (2010) Tubulin depolymerization may be an ancient biological motor. J Cell Sci 123:3425–3434
Meier EL, Goley ED (2014) Form and function of the bacterial cytokinetic ring. Curr Opin Cell Biol 26:19–27
Mercier R, Kawai Y, Errington J (2013) Excess membrane synthesis drives a primitive mode of cell proliferation. Cell 152:997–1007
Michie KA, Monahan LG, Beech PL, Harry EJ (2006) Trapping of a spiral-like intermediate of the bacterial cytokinetic protein FtsZ. J Bacteriol 188:1680–1690
Mickey B, Howard J (1995) Rigidity of microtubules is increased by stabilizing agents. J Cell Biol 130:909–917
Milam SL, Erickson HP (2013) Rapid in vitro assembly of Caulobacter crescentus FtsZ protein at pH 6.5 and 7.2. J Biol Chem 288:23675–23679
Milam SL, Osawa M, Erickson HP (2012) Negative-stain electron microscopy of inside-out FtsZ rings reconstituted on artificial membrane tubules show ribbons of protofilaments. Biophys J 103:59–68
Mingorance J, Tadros M, Vicente M, Gonzalez JM, Rivas G, Velez M (2005) Visualization of single Escherichia coli FtsZ filament dynamics with atomic force microscopy. J Biol Chem 280:20909–20914
Moores CA, Milligan RA (2008) Visualisation of a kinesin-13 motor on microtubule end mimics. J Mol Biol 377:647–654

Nawrotek A, Knossow M, Gigant B (2011) The determinants that govern microtubule assembly from the atomic structure of GTP-tubulin. J Mol Biol 412:35–42
Ohi M, Li Y, Cheng Y, Walz T (2004) Negative staining and image classification – powerful tools in modern electron microscopy. Biol Proced Online 6:23–34
Osawa M, Erickson HP (2011) Inside-out Z rings – constriction with and without GTP hydrolysis. Mol Microbiol 81:571–579
Osawa M, Erickson HP (2013) Liposome division by a simple bacterial division machinery. Proc Natl Acad Sci U S A 110:11000–11004
Osawa M, Anderson DE, Erickson HP (2008) Reconstitution of contractile FtsZ rings in liposomes. Science 320:792–794
Osawa M, Anderson DE, Erickson HP (2009) Curved FtsZ protofilaments generate bending forces on liposome membranes. EMBO J 28:3476–3484
Pichoff S, Lutkenhaus J (2005) Tethering the Z ring to the membrane through a conserved membrane targeting sequence in FtsA. Mol Microbiol 55:1722–1734
Popp D, Iwasa M, Narita A, Erickson HP, Maeda Y (2009) FtsZ condensates: an in vitro electron microscopy study. Biopolymers 91:340–350
Popp D, Iwasa M, Erickson HP, Narita A, Maeda Y, Robinson RC (2010) Suprastructures and dynamic properties of Mycobacterium tuberculosis FtsZ. J Biol Chem 285:11281–11289
Ramirez-Aportela E, Lopez-Blanco JR, Andreu JM, Chacon P (2014) Understanding nucleotide-regulated FtsZ filament dynamics and the monomer assembly switch with large-scale atomistic simulations. Biophys J 107:2164–2176
Romberg L, Mitchison TJ (2004) Rate-limiting guanosine 5'-triphosphate hydrolysis during nucleotide turnover by FtsZ, a prokaryotic tubulin homologue involved in bacterial cell division. Biochemistry 43:282–288
Romberg L, Simon M, Erickson HP (2001) Polymerization of FtsZ, a bacterial homolog of tubulin. Is assembly cooperative? J Biol Chem 276:11743–11753
Rowlett VW, Margolin W (2014) 3D-SIM super-resolution of FtsZ and its membrane tethers in Escherichia coli cells. Biophys J 107:L17–L20
Sato M, Mogi Y, Nishikawa T, Miyamura S, Nagumo T, Kawano S (2009) The dynamic surface of dividing cyanelles and ultrastructure of the region directly below the surface in Cyanophora paradoxa. Planta 229:781–791
Soderstrom B, Skoog K, Blom H, Weiss DS, von Heijne G, Daley DO (2014) Disassembly of the divisome in Escherichia coli: evidence that FtsZ dissociates before compartmentalization. Mol Microbiol 92:1–9
Srinivasan R, Mishra M, Wu L, Yin Z, Balasubramanian MK (2008) The bacterial cell division protein FtsZ assembles into cytoplasmic rings in fission yeast. Genes Dev 22:1741–1746
Strauss MP, Liew AT, Turnbull L, Whitchurch CB, Monahan LG, Harry EJ (2012) 3D-SIM super resolution microscopy reveals a bead-like arrangement for FtsZ and the division machinery: implications for triggering cytokinesis. PLoS Biol 10:e1001389
Stricker J, Maddox P, Salmon ED, Erickson HP (2002) Rapid assembly dynamics of the Escherichia coli FtsZ-ring demonstrated by fluorescence recovery after photobleaching. Proc Natl Acad Sci U S A 99:3171–3175
Szwedziak P, Wang Q, Freund SM, Lowe J (2012) FtsA forms actin-like protofilaments. EMBO J 31:2249–2260
Szwedziak P, Wang Q, Bharat TA, Tsim M, Lowe J (2014) Architecture of the ring formed by the tubulin homologue FtsZ in bacterial cell division. eLife 3:e04601
Tan D, Rice WJ, Sosa H (2008) Structure of the kinesin13-microtubule ring complex. Structure 16:1732–1739
Theisen KE, Zhmurov A, Newberry ME, Barsegov V, Dima RI (2012) Multiscale modeling of the nanomechanics of microtubule protofilaments. J Phys Chem B 116:8545–8555
Turner DJ, Portman I, Dafforn TR, Rodger A, Roper DI, Smith CJ, Turner MS (2012) The mechanics of FtsZ fibers. Biophys J 102:731–738

Wang X, Lutkenhaus J (1996a) Characterization of FtsZ from *Mycoplasma pulmonis*, an organism lacking a cell wall. J Bacteriol 178:2314–2319
Wang X, Lutkenhaus J (1996b) FtsZ ring: the eubacterial division apparatus conserved in archaebacteria. Mol Microbiol 21:313–319
Wang HW, Nogales E (2005) Nucleotide-dependent bending flexibility of tubulin regulates microtubule assembly. Nature 435:911–915
Zhang M, Schmitz AJ, Kadirjan-Kalbach DK, Terbush AD, Osteryoung KW (2013) Chloroplast division protein ARC3 regulates chloroplast FtsZ-ring assembly and positioning in arabidopsis through interaction with FtsZ2. Plant Cell 25:1787–1802
Zhao FQ, Craig R (2003) Capturing time-resolved changes in molecular structure by negative staining. J Struct Biol 141:43–52

Chapter 6
Intermediate Filaments Supporting Cell Shape and Growth in Bacteria

Gabriella H. Kelemen

Abstract For years intermediate filaments (IF), belonging to the third class of filamentous cytoskeletal proteins alongside microtubules and actin filaments, were thought to be exclusive to metazoan cells. Structurally these eukaryote IFs are very well defined, consisting of globular head and tail domains, which flank the central rod-domain. This central domain is dominated by an α-helical secondary structure predisposed to form the characteristic coiled-coil, parallel homo-dimer. These elementary dimers can further associate, both laterally and longitudinally, generating a variety of filament-networks built from filaments in the range of 10 nm in diameter. The general role of these filaments with their characteristic mechano-elastic properties both in the cytoplasm and in the nucleus of eukaryote cells is to provide mechanical strength and a scaffold supporting diverse shapes and cellular functions.

Since 2003, after the first bacterial IF-like protein, crescentin was identified, it has been evident that bacteria also employ filamentous networks, other than those built from bacterial tubulin or actin homologues, in order to support their cell shape, growth and, in some cases, division. Intriguingly, compared to their eukaryote counterparts, the group of bacterial IF-like proteins shows much wider structural diversity. The sizes of both the head and tail domains are markedly reduced and there is great variation in the length of the central rod-domain. Furthermore, bacterial rod-domains often lack the sub-domain organisation of eukaryote IFs that is the defining feature of the IF-family. However, the fascinating display of filamentous assemblies, including rope, striated cables and hexagonal laces together with the conditions required for their formation both *in vitro* and *in vivo* strongly resemble that of eukaryote IFs suggesting that these bacterial proteins are deservedly classified as part of the IF-family and that the current definition should be relaxed slightly to allow their inclusion. The lack of extensive head and tail domains may well make the bacterial proteins more amenable for structural characterisation, which will be essential for establishing the mechanism for their association into filaments. What

---

Dedication This chapter is dedicated to the memory of Nora Ausmees, who was instrumental in the discovery of the first bacterial IF proteins but sadly passed away in autumn 2015.

G.H. Kelemen (✉)
School of Biological Sciences, University of East Anglia,
Norwich Research Park, Norwich NR4 7TJ, UK
e-mail: g.kelemen@uea.ac.uk

J. Löwe, L.A. Amos (eds.), *Prokaryotic Cytoskeletons*, Subcellular Biochemistry 84, DOI 10.1007/978-3-319-53047-5_6

is more, the well-developed tools for bacterial manipulations provide an excellent opportunity of studying the bacterial systems with the prospect of making significant progress in our understanding of the general underlying principles of intermediate filament assemblies.

Keywords Bacterial IF-like proteins • *Caulobacter* • Cell curvature • Crescentin • CreS filaments • Coiled-coil proteins • Unit length filaments • *Streptomyces coelicolor* • FilP • Scy

Introduction

What Are Eukaryote Intermediate Filaments?

In eukaryote cells there are three filament forming cytoskeletal assemblies: the microtubules are the largest, forming tubular filaments with ~25 nm diameter; actin filaments are the smallest, with diameters of ~6 nm and finally, as their name suggests, intermediate filaments are sized in between, with diameters around 10 nm. Both microtubules and actin filaments are built from a single, globular protein, tubulin and actin, respectively, whereas intermediate filaments can assemble from a variety of proteins with a common, tripartite molecular architecture, which includes a characteristic α-helical centre domain flanked by variable head and tail domains (Chernyatina et al. 2015; Herrmann and Aebi 2004; Parry and Steinert 1999).

Two defining biochemical characteristics of IF proteins are their notorious insolubility and the fact that, after denaturation in the presence of urea, IF proteins are capable of re-folding in physiological buffers in the absence of any co-factors. Therefore, unlike tubulin or actin that require GTP or ATP for their polymerisation, the higher-order assemblies of IF proteins do not require any charged nucleotides.

Eukaryote IF proteins are widespread in all metazoan cells and can be classified into five main families. Four families are found in the cytoplasm, including acidic and basic keratins (Types I and II) in epithelial cells such as in the skin or kidneys, vimentin (Type III) in mesenchymal cells and neurofilament (Type IV) in neurons. The fifth IF family (Type V) is represented by the lamins that are located in the nucleus in all nucleated cells. There are known to be ~70 different IF proteins in human and other mammalian genomes (Hesse et al. 2001).

IF proteins are structural components of the cells and generate diverse networks of filaments both in the cytoplasm and in the nucleus supporting the mechanical integrity of cells.

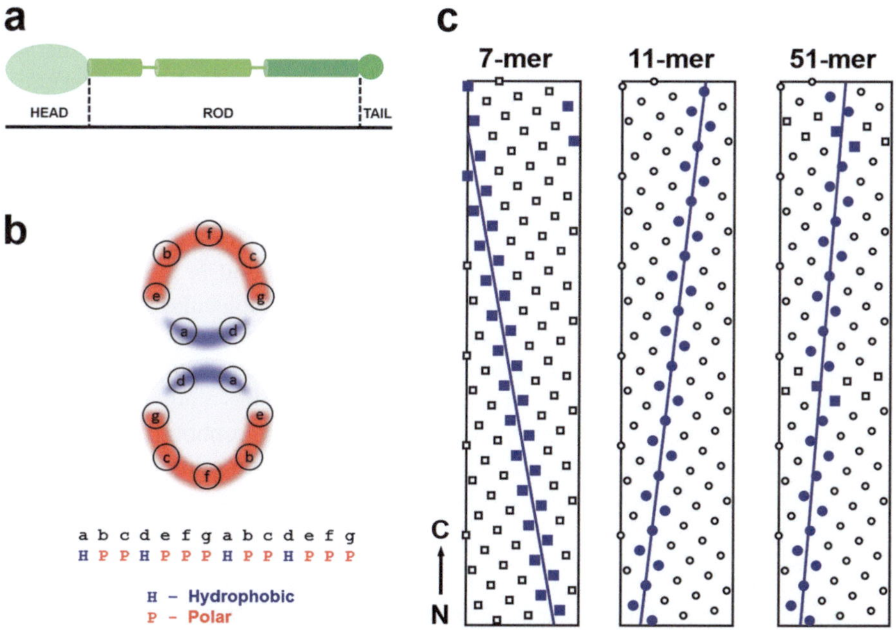

Fig. 6.1 Eukaryote intermediate filament proteins are dominated by α-helices forming coiled-coil assemblies. (**a**) The tripartite domain organisation of metazoan IFs. (**b**) The helical wheel representation of the classical arrangement of heptad repeats. The seven different amino acid positions of two helical turns are projected to a plane perpendicular to the axes of the α-helix. When the primary protein sequence displays a repeated occurrence of hydrophobic residues (*blue*) at positions (a) and (d) and polar residues (*red*) in all other positions, a hydrophobic streak is generated on one side of the α-helix. This streak promotes the assembly of two or more such α-helices. (**c**) Helical net diagram of the heptad (7-mer), hendecad (11-mer) and penindaenad (51-mer) repeat. *Circles* or *squares* represent C_α positions of each amino acids projected onto the α-helical cylinder that is "flattened" for representing the individual amino acid positions from the N-terminus (N) to the C-terminus (C) in 2D. The positions of hydrophobic amino acids in each repeats are marked *blue*. Squares and circles mark amino acid residues that are part of the heptad periodicity and hendecad periodicity, respectively. The *blue dashed lines* represent the hydrophobic streaks that create the interface between two α-helices producing left handed (7-mer) or right-handed (11-mer and 51-mer) coiled coils

Intermediate Filament Proteins Are Coiled-Coil Proteins

What are the defining characteristics of an intermediate filament protein? Intermediate filaments were first identified amongst metazoan species and were characterised in detail well before the first bacterial IF was identified. For a long time IFs were thought to be exclusive to metazoan species and the recent discoveries of IFs in insects and in bacteria extends the confines of the IFs although there is the apparent lack of IFs amongst plant and fungi. The eukaryote IF consists of variable, non α-helical, head and tail domains that flank the central α-helical rod domain (Fig. 6.1a). Heterogeneity of head and tail domains results in marked differences in the sizes of IF proteins, which range between 40 and 240 kDa, whereas the length of the rod domain is well conserved. This α-helical domain has a distinctive characteristic that enables the formation of a coiled-coil dimer, called the elementary dimer, where two IF strands associate in a parallel and in-register fashion. Apart from the keratins that form obligate hetero-dimers, for most intermediate filaments the elementary dimer is a homo-dimer. Further interactions between the elementary dimers generate complex higher order assemblies producing the 10 nm IF filaments.

Coiled-coil association is not exclusive to IF proteins and is found in a wide range of proteins including transcription factors, adhesins, molecular motors, receptors and signalling molecules, just to name a few (Parry et al. 2008). In general, coiled-coil assembly is promoted when the primary sequence of an α-helical protein exhibits a distinct pattern of regularly occurring hydrophobic- and charged amino acid residues (Fig. 6.1b). The repeated positioning of the hydrophobic residues generates a hydrophobic streak on one side of the α-helix, promoting the interaction between two, or sometimes three or more, such α-helical domains.

Both Heptad and Hendecad Repeats Are Found in Eukaryote IF Proteins

The most characterised α-helical sequence that promotes coiled-coil formation consists of the so-called canonical heptad repeat. However, the term "heptad repeat" does not refer to any specific seven amino acid long sequence that is repeated, instead, it refers to the repeated occurrence of hydrophobic amino acids. These residues such as Ala, Ile, Met, Leu or Val occupy the first (a) and fourth (d) positions within the consecutive seven amino acid sequences that are designated by the letters (a) to (g). When a sequence with such a heptad repeat forms an α-helix with the established 3.6 amino acid residues in every turn, every seven amino acids will correspond to approximately two turns on the helix with positions (a) and (d) found on the same side of the helix (Fig. 6.1b). Consecutive (a) and (d) positions of successive heptad sequences will produce a left handed hydrophobic streak on the α-helix that, in turn, will promote the association of two or more such α-helices coiled around each other in such a way that the hydrophobic streak is buried at the interface between the α-helices (Fig. 6.1c).

In addition to the classical heptad repeat, coiled-coil proteins with the less common 11-mer repeats (hendecad) are also found (Gruber and Lupas 2003; Kuhnel et al. 2004). In the hendecad repeat, 11 amino acids span 3 helical turns and their positions are marked by the letters (a) to (k). Hydrophobic amino acids are typically found at positions (a), (d) and (h) and generate a slightly right-handed hydrophobic streak (Fig. 6.1c). The shallow hydrophobic streak of the hendecad periodicity pattern results in an interface almost parallel to the axis of the α-helices (Nicolet et al. 2010; Nishimura et al. 2012; Stewart et al. 2012).

The rod domain of eukaryote IF proteins was predicted to consist of four sub-domains, Coil 1A, Coil 1B, Coil 2A and Coil 2B, linked by short, non α-helical linker sequences (Herrmann and Aebi 2004; Parry and Steinert 1999). However, this structural model was based on sequence analysis, with the assumption that the coiled-coil subdomains consist solely of heptad repeat patterns, which turns out not to be the case. Solving the crystal structure of overlapping fragments of Coil 2A and Coil 2B of human vimentin suggested a continuous, single Coil 2 subdomain (Nicolet et al. 2010; Strelkov et al. 2002). This in fact corroborates the very first sequence analysis of eukaryote IF proteins that identified three continuous α-helical segments of hamster vimentin (Quax-Jeuken et al. 1983). Therefore, the currently-recognised three sub-domains of the eukaryote IF rod domain are Coil 1A, Coil 1B and Coil 2 (Fig. 6.2). The amino acid sequence conservation amongst the rod domains of eukaryote IF proteins is very low, even though the length of the rod and the individual sub-domains are well conserved: 310 amino acid residues for the rod domain of the cytoplasmic IF proteins and 350 amino acid residues for the nuclear IFs (Chernyatina et al. 2015; Herrmann and Aebi 2004). The rod domain and its sub-domains are dominated by heptad periodicity: 6 heptads within the sub-domain Coil 1A and 13 heptads in both Coil 1B and Coil 2. In addition to the heptad repeats, at several distinct places of the rod domain a different periodicity, an 11-mer, hendecad pattern is observed. For example, a single hendecad is present towards the C terminal end of Coil 1B and four hendecads within Coil 2. Three of the latter 11-mers are at the beginning of the Coil 2 sub-domain while the fourth, single 11-mer register in the middle of Coil 2 is a highly conserved characteristic of the IF rod domain. This single discontinuity can be considered as a four-residue insert within the heptad periodicity pattern and is often described as a "stutter". Amongst eukaryote IF proteins, the first level of association of single IF protein strands generates a parallel, in-register, homo-dimer with an axial symmetry, but with a clear polarity (Fig. 6.2).

Structure of Eukaryote IF Proteins

The very first structural analysis of proteins using X-ray crystallography established a highly ordered organisation in an IF protein, keratin, by William Astbury in the 1930s (Astbury and Woods 1930; Astbury and Woods 1933). This was followed by Linus Pauling's discovery of the α-helical fold as a fundamental principle of protein secondary structure (Pauling et al. 1951) and the proposed interpretation of the coiled coil by Francis Crick (Crick 1953). However after this initial breakthrough, the structural characterisation of IF proteins proved to be very challenging because the "free" IF elementary dimer only exists under highly denaturing conditions (Herrmann et al. 1996) and the various higher order assemblies eluded any successful crystallisation attempts. To date, we still lack any X-ray crystallography data for a full length IF elementary dimer. Instead, the approach of building the full structure from successful crystallisation of overlapping short fragments, the so-termed

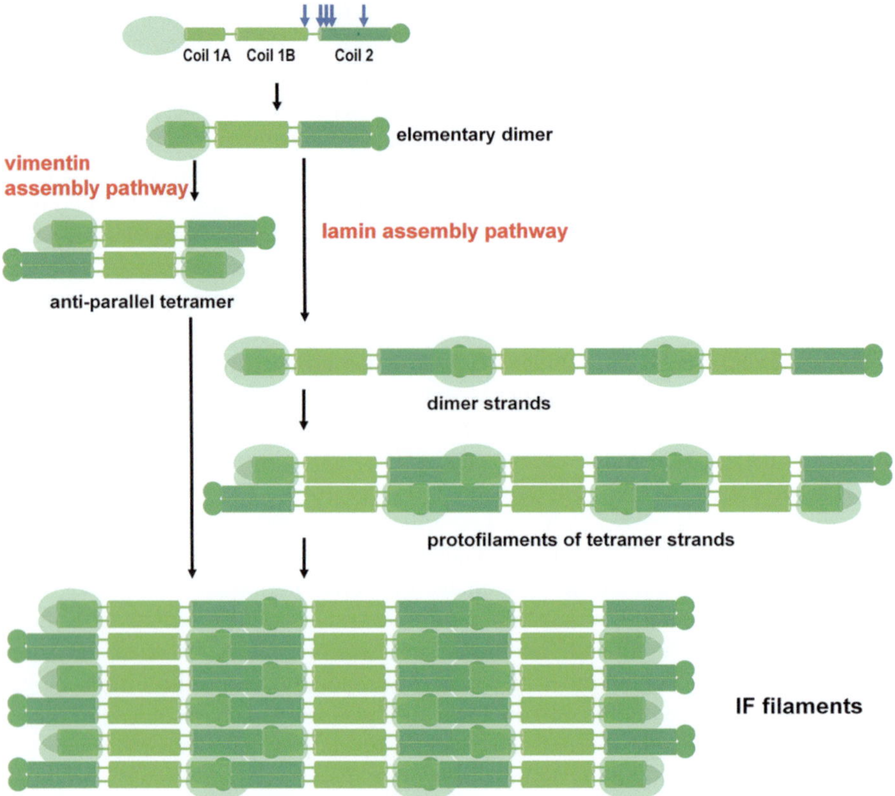

Fig. 6.2 Eukaryote IF assembly pathways. All metazoan IF proteins have the characteristic tripartite structure (TOP) with the head and tail domains flanking the α-helical rod domain. The sub-domain organisation of the rod domain includes Coil 1A, Coil 1B and Coil 2 connected via linker sequences. The Coil sub-domains are dominated by heptad periodicity and the scattered hendecad patterns are marked by vertical arrows. The elementary dimer is assembled by parallel, in-register homo-dimer formation. In the vimentin assembly pathway, two elementary dimers form an anti-parallel tetramer that lacks polarity. The lateral association of eight tetramers (16 elementary dimers) generates unit length filaments (ULFs, not shown), which further assemble both laterally and longitudinally to generate the fully assembled IF filaments. In the lamin assembly pathway elementary dimers associate longitudinally via head to tail associations. Two of these dimer strands form anti-parallel, tetramer strands called protofilaments that lack polarity. The lateral association of these protofilaments generates the final lamin IF network

"divide and conquer" approach, has generated a wealth of structural data (Aziz et al. 2012; Chernyatina et al. 2012; Chernyatina and Strelkov 2012; Meier et al. 2009; Nicolet et al. 2010; Strelkov et al. 2001, 2002). From these fragments a model of the full vimentin elementary dimer: a parallel, in-register homodimer, was proposed (Chernyatina et al. 2015), where each α-helical Coil 1A, 1B and Coil 2 sub-domain of one IF protein interacts with the same sub-domain of the partner protein. Structural data of peptide fragments confirmed that Coil 1B and Coil 2 form parallel homo-dimers, however, there are alternative interpretations about Coil 1A. Besides the "closed" coiled-coil configuration of Coil 1A dimers, which is the interpretation that is largely accepted and reviewed, an alternative model was also proposed. X-ray crystallography data of a fragment consisting of Coil 1A, linker L1 and part of Coil 1B suggests that the α-helices of Coil 1A are spread apart and this arrangement might be important for the further, longitudinal assembly of the elementary dimers, where the N-terminus of Coil 1A of an elementary dimer was shown to interact with the C-terminus of Coil 2 of an adjacent, second elementary dimer (Chernyatina et al. 2012). Moreover, both configurations, the "closed" coiled-coil and the "open" α-helices might exist at different stages of the dynamic IF assembly (Meier et al. 2009; Smith et al. 2002).

Interestingly, the IF rod domain is dominated by heptad periodicity patterns with the occasional hendecad pattern promoting coiled-coil dimerisation. As discussed above, α-helices made either of exclusively heptad or of hendecad periodicity patterns generate distinct hydrophobic streaks that are either left-handed or slightly right-handed, respectively (Fig. 6.1c). This, in turn, promotes the formation of coiled-coil dimers where the coil itself is left handed for heptads or slightly right handed for hendecads. In Coil 2, where hendecad and heptad periodicities are interspersed (Fig. 6.2), the characteristic left handed coil of the heptad pattern will locally unwind at the positions of the hendecads, so that the two α-helices become very nearly parallel, generating a so-called parallel bundle rather than a coil, per se (Chernyatina et al. 2015). The role of the local unwinding at the position of the hallmark stutter of Coil 2 was investigated by introducing a stutterless vimentin mutation, which did affect the higher order assembly pathway for filament formation (see below) but not the elementary dimer formation (Herrmann et al. 1999).

Very little is known about the structures of the head and tail domains that appear to be intrinsically disordered, however, they have important roles in the assembly of the elementary dimers into IF filaments. Interestingly, chemical crosslinking and site-directed spin labelling (SDSL) electron paramagnetic resonance (EPR) spectroscopy suggested that the head domain of vimentin folds back onto Coil 1A and comes in close proximity of Coil 1B (Aziz et al. 2009; Aziz et al. 2010; Herrmann and Aebi 2004).

Higher Order Assemblies of Eukaryote IF Proteins

If our understanding of a full-length elementary dimer is incomplete, the higher order assemblies built from these homo-dimers are even further away from any high-resolution structural analysis. Transmission electron-microscopy of negatively stained protein samples, atomic force microscopy and more recently, small-angle X-ray scattering (SAXS) has been used to establish further stages of IF assemblies, even though the assembly pathway of the different IF groups might vary and the precise arrangement of the elementary dimers within the filamentous IF assembly is not fully understood (Herrmann and Aebi 2004; Block et al. 2015). The assembly

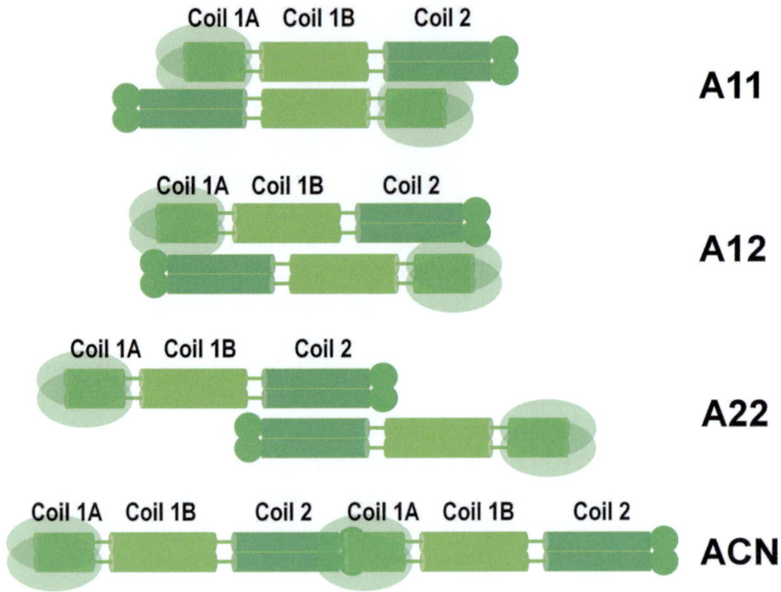

Fig. 6.3 Initial associations of the elementary dimers. The majority of "dimers of dimers" detected during the assembly of vimentin consists of A11 tetramers, where the Coil 1B sub-domains of each dimers are aligned in an anti-parallel fashion. Alternative lateral associations also exist, including the A12 tetramers, where two elementary dimers are aligned end-to-end, in an anti-parallel arrangement or the A22 tetramers, where the Coil 2 subdomains of each elementary dimers are aligned in an anti-parallel fashion. Finally head-to-tail associations of two elementary dimers generate the ACN arrangement

pathway for filament formation of all IFs is hierarchical and it involves both lateral and longitudinal associations of the rod-like and polar elementary dimers (Fig. 6.2).

Perhaps the most understood assembly pathway concerns the Type III IF protein, vimentin. Initially, two elementary dimers of vimentin form an anti-parallel, slightly staggered tetramer (Fig. 6.2) in buffers with low ionic strengths *in vitro* and these tetramers can also be observed *in vivo* (Herrmann et al. 1996; Soellner et al. 1985). Based on the structure of the elementary dimer, the model of the vimentin tetramer was proposed where the Coil 1B sub-domains of two elementary dimers are aligned in an anti-parallel fashion (Figs. 6.2 and 6.3) and polar or charged residues on the outer surface of either coiled coils make contact to produce the lateral association of two elementary dimers (Chernyatina et al. 2012). Further lateral association of the vimentin tetramers can be induced by increasing the ionic strength of the sample buffer *in vitro* (Herrmann et al. 1999). Typically, the lateral association of eight tetramers (16 elementary dimers) generates rod-like structures that are termed unit-length filaments (ULFs) with an average length of ~60–65 nm and diameter of ~16 nm (Herrmann et al. 1999; Strelkov et al. 2003). These ULFs subsequently associate longitudinally, in a head-to-tail fashion producing the characteristic IF

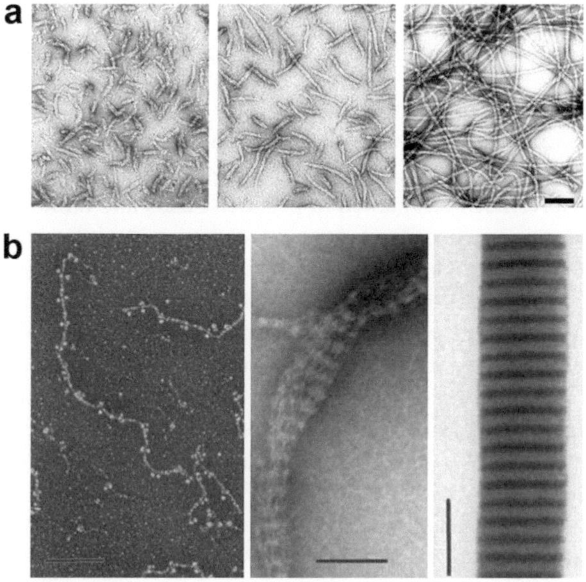

Fig. 6.4 Intermediate Filament formation *in vitro*. (**a**) Filament assembly of recombinant human vimentin was monitored 10 sec (*left*), 1 min (*middle*) and 1h (*right*) after the addition of filament buffer to fully denatured vimentin. Scale bar represents 100 nm (Images were taken from Herrmann and Aebi 2004). (**b**) Lamin filament formation *in vitro*. Chicken lamin B_2 was expressed in *E. coli* and the purified protein was dialysed against 25 mM MES, 150 mM NaCl, 1 mM EGTA pH = 6.5 (*left*), the same buffer in the presence of 25 mM $CaCl_2$ (*middle*) and finally 25 mM Tris, 25 mM $CaCl_2$, pH = 8.5 (*right*). Protein *assemblies* were monitored using transmission electron microscopy with rotary shadowing (*left*) or with negative staining (*middle* and *right*). Scale bars represent 100 nm (Images were taken from ©1991 Heitlinger et al. Journal of Cell Biology. 113:485–495. DOI: 10.1083/jcb.113.3.485)

filaments (Figs. 6.2 and 6.4). During the extension of the ULFs the filament diameter decreases to ~11 nm, suggesting some internal rearrangement of the subunit-association (Strelkov et al. 2003).

While the soluble vimentin tetramers correspond to the so-called A11, anti-parallel arrangement of the elementary dimers, where Coil 1B sub-domains are aligned (Fig. 6.3), chemical crosslinking and electron paramagnetic resonance spectroscopy with site-directed spin labelling have suggested several possible alternative lateral alignments within fully-formed IF filaments. All of the proposed lateral alignments are anti-parallel but they differ in the individual sub-domains involved in the intermolecular, dimer-dimer interactions. Besides the non-staggered lateral contact between two elementary dimers (A12 association), the A22 association may also be generated when Coil 2 sub-domains are aligned, creating a more elongated, staggered tetramer unit (Fig. 6.3).

The assembly pathway of the Type V IF family of nuclear lamins is less understood. Although the filaments of nuclear lamins, just as those of vimentin, are built by the lateral and longitudinal associations of the elementary dimers, the proposed key steps of the association of these dimers differ from those found during vimentin assembly. In vimentin the lateral interactions between elementary dimers and the production of ULFs precede the longitudinal associations and filament extension. Nuclear lamins have not been shown to produce ULFs and unlike the cytoplasmic IFs, lamins produce complex filamentous networks *in vivo* rather than the typical IF filaments with 10 nm diameter (Aebi et al. 1986). The nuclear lamina is a meshwork of intermediate-type filaments (Aebi et al. 1986; Gerace and Huber 2012; Goldberg et al. 2008). The first stage of association in nuclear lamins involves the longitudinal, head-to-tail assembly of the lamin elementary dimers into dimer strands of variable length (Fig. 6.2). This is followed by the lateral association of two of these dimer strands in an anti-parallel and staggered fashion, producing lamin protofilaments of tetramer strands (Ben-Harush et al. 2009; Dittmer and Misteli 2011; Stuurman et al. 1998). These protofilaments (Figs. 6.2 and 6.4) can further associate via lateral interactions to produce fibers with varying width and a characteristic striated pattern when viewed using transmission electron-microscopy of negatively stained samples (Aebi et al. 1986; Heitlinger et al. 1991, 1992). These highly ordered assemblies, termed paracrystalline fibers, were originally considered as *in vitro* artefacts. However, they are the dominant structures formed in physiological buffers (Zwerger et al. 2013) and paracrystalline arrays were also detected in cells expressing lamins, albeit at a higher-than-usual cellular level, confirming that this assembly can exist *in vivo* as well (Klapper et al. 1997).

The head-to-tail contact between elementary dimers relies on the head and tail domains and, importantly, the extremities of the coiled-coil rod domain. Remarkably, in spite of the very low amino acid sequence conservation amongst IF proteins in general, there is a highly conserved box of 26 and 32 amino acids at the N-terminus of Coil 1A and the C-terminus of Coil 2, respectively. Structural studies of "minilamins" carrying the N-terminal or C-terminal fragments of lamin with a truncated rod domain demonstrated hetero-coil formation between the ends of Coil 1A and Coil 2, suggesting that these conserved stretches might play an important role in the head-to tail contacts between elementary dimers and consequently, the longitudinal extension during filament formation of all IFs (Kapinos et al. 2010; Koster et al. 2015).

For the lateral interaction of elementary dimers and the head-to-tail dimer strands, polar and charged amino acid residues located on the outer surface of the coiled-coil rod domain are important. Therefore, these lateral interactions are sensitive to conditions that affect the surface charge of an elementary dimer or dimer strand. Depending on the surface charge distribution, individual elementary dimers have distinct electrostatic properties. Overall, the surface of elementary dimers is negatively charged and therefore their further association is highly dependent on the absence or presence of cations and the pH of the environment. Accordingly, the presence of cations, such as Mg^{2+}, Ca^{2+} can readily mediate lateral interactions between IF filaments promoting the paracrystalline assemblies of nuclear lamins or cytoplasmic IF bundle formation (Block et al. 2015). In addition, whilst low pH promotes the lateral assembly of elementary dimers or dimer strands *in vitro*, at high pH, in the absence of metal cations, IF assembly is prevented, keeping the elementary dimers or dimer strands, with their negative surface charge, apart.

Whether the assembly of a eukaryote IF follows the vimentin pathway, the lamin pathway (Fig. 6.2), or some other pathway not yet established, it is a fundamental characteristic of the native IF filaments that they are non-polar (Fig. 6.2). While the elementary dimer and the dimer strands of lamins have polarity with distinct N and C terminal ends, the assembled tetramers or the tetramer protofilaments are non-polar because they form in an anti-parallel arrangement. The lack of polarity of the IF filaments and the fact that the assembly pathway does not rely on any co-factors constitute the characteristic distinction between intermediate filaments and microtubules or actin filaments, which in turn are polar, with plus (+) and minus (−) ends and require nucleotide co-factors for their assembly. Unlike microtubules and actin filaments which often extend at one end and disassemble at the other, IFs usually extend at both ends during assembly; they do not spontaneously disassemble from either end. Moreover, subunit exchange can take place throughout IF filaments and not only at ends. Newly synthesised vimentin is incorporated in a random fashion alongside existing filaments and subunit exchange occurs evenly along the filaments *in vivo*; this was shown using fluorescence recovery after photo bleaching (FRAP) and photoactivation (Colakoglu and Brown 2009; Vikstrom et al. 1992).

Eukaryote IF Filaments in the Cellular Environment

The basic building blocks, the elementary dimers and their lateral and longitudinal higher order assemblies, are organised by the same fundamental principles. However, the eukaryote IF filaments of the various classes (Type I-V) form very diverse cellular fibers within the different cells they preside. This is not surprising, as both the head-to-tail and, more importantly, the lateral associations of the elementary dimers of the diverse IF family of proteins depend upon the individual amino acid sequences that control the outer surface charges both along the length and at the ends of the elementary dimers, and the primary amino acid sequences of IFs are not conserved. This explains the striking differences in the assembly pathway and the final cellular structures of different IF filaments together with their mechanical characteristics and individual biological function. The cytoplasmic IFs form filamentous networks between the plasma membrane and the outer nuclear membrane, generating a cytoplasmic network that stabilises the intracellular space and its contents, such as the nucleus or organelles. Keratins consists of bundles and fibers that stabilise the cell shape and mechanical stiffness of keratinocytes as cells that lack keratin can be deformed more easily than wild-type cells (Ramms et al. 2013; Seltmann et al. 2013). On the other hand, vimentin and desmin forms highly connected finer filament-networks with a smaller mesh size organising internal mechanics of the cells (Guo et al. 2013). Nuclear lamins form filaments under the nuclear membrane and inside the nucleus, organising the chromatin and affecting nuclear shape and even gene expression.

It is important to emphasise that IFs are part of the cytoskeletal cellular network that also includes microtubules, microfilaments and proteins that interconnect these three filament networks. Therefore, in order to understand the cellular function of IFs one must also consider the interaction between IFs and other cellular proteins controlling cell division, growth and cellular integrity (Chang and Goldman 2004).

Single point mutations within genes encoding specific IFs are the causes for a wide range of human diseases. Keratin mutations are implicated in a number of keratinopathies causing skin or liver disorders, cataract development is associated with mutations in vimentin or neurological diseases caused by mutations in neurofilaments (Toivola et al. 2015; Lowery et al. 2015).

Mechanical Characteristics of IF Filaments

IFs form more flexible biopolymers than those of microtubules or actin microfilaments. Generally, the flexibility of a filament is characterised by the so-called persistence length that is defined by the length along which the filament is straight and does not change its direction. The persistence lengths of IFs are a magnitude shorter than that of microfilaments, and are in the range between hundreds of nm and a few μm, depending on the Type of IFs (see review Block et al. 2015).

IFs are also stretchable and can be elongated. Quite remarkably, desmin was stretched to 3.6 times its original length (Kreplak et al. 2005). There are several proposed structural changes of the filament-assembly during stretching. One of these is the axial sliding of either the tetramers or tetramer protofilaments, or the elementary dimers within the tetramers or tetramer protofilament (Figs. 6.2 and 6.3). The mechanical force is proposed to disrupt the existing lateral interactions between elementary dimers, tetramers or protofilaments and promote sliding of these units in the longitudinal direction. Detachment is followed by a different "lock in", establishing a new alignment of the elementary dimers, tetramers or dimer strands and tetramer filaments. In addition to axial sliding, individual elementary dimers can also be stretched to double their length, during which there is a transition of the α-helices of the rod-domain to β-sheet conformation (Kreplak et al. 2005).

Dynamics of Eukaryote IFs Is Controlled by Post-translational Modifications

Microtubules or microfilaments rely on a pool of soluble intracellular subunits and nucleoside triphosphate co-factors to incorporate new subunits at their (+) ends and conversely, depolymerisation at the (−) ends of the filaments will release soluble subunits into the cytoplasm. Assembly of IFs does not require co-factors, and overall there is a very low intracellular pool of IF elementary dimers, if any, in the cytoplasm because of the strong tendency of association of these elementary dimers into fibers of different kinds via the different assembly pathways we discussed above. Newly synthesised IF proteins are immediately incorporated into the existing IF network (Coleman and Lazarides 1992; Ngai et al. 1990). However, soluble tetrameric vimentin has been observed *in vivo* (Soellner et al. 1985) suggesting that IFs assemble and disassemble in a dynamic fashion inside cells. Newly synthesised IF proteins were shown to incorporate in random fashion along the existing IF filaments and FRAP experiments confirmed that subunits of existing filaments can rearrange (Goldman et al. 2012; Vikstrom et al. 1992).

The dynamics of IFs are regulated by their synthesis, degradation and post-translational modification. The most widespread post-translational modification is phosphorylation, but sumoylation, lysine acetylation and farnesylation of IFs are all well documented (Snider and Omary 2014). Phosphorylation plays a key role in the disassembly of IFs. For example, phosphorylation of vimentin induces the disassembly of vimentin IFs at the leading edge lamellopodia during the cell motility of fibroblasts. Whereas local disassembly of vimentin is required for cell polarisation and lamellopodia formation during fibroblast motility, complete lack of vimentin by gene silencing causes lack of cell polarity and indiscriminate lamellopodia formation around the cell, affecting cell migration (Helfand et al. 2011). In turn, specific kinases and phosphatases, that are themselves highly regulated in response to mechanical stresses, control the assembly dynamics of vimentin (Chou et al. 1990; Eriksson et al. 1992a, b, 2004). Changes in the phosphorylation pattern of IFs have been also indicated in certain human diseases, including Alzheimer disease, myopathies and liver diseases (Snider and Omary 2014).

Bacterial Intermediate Filament-Like Proteins

Crescentin, the First Identified Bacterial IF-Like Protein

The first attempts to characterise the structure of eukaryote IFs such as keratin, took place over 90 years ago. Since then, especially in the last 30 years, a plethora of research has focussed on deciphering the structure, assembly and cellular function of eukaryote IFs. By comparison, the first bacterial protein with possible intermediate filament function was only identified relatively recently in the crescent-shaped bacterium, *Caulobacter crescentus* (Ausmees et al. 2003). This bacterium has a curved shape that is maintained throughout its complex life-cycle. In stationary phase cells, where the level of the cell division protein, FtsZ, is lower, *C. crescentus* forms elongated, helical cells. Cell curvature both in actively growing and stationary phase cells depends on the presence of the protein called crescentin, encoded by the gene *creS* (Ausmees et al. 2003).

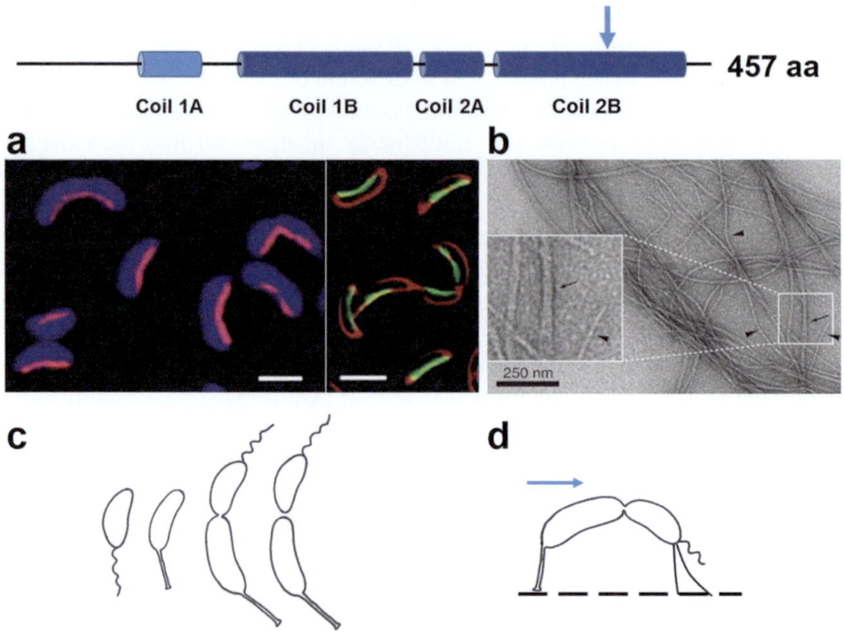

Fig. 6.5 Crescentin, the first bacterial IF-like protein. The domain organisation of crescentin resembles that of eukaryote IF proteins (TOP). Vertical arrow marks the position of the stutter in the last Coil sub-domain. (**a**) Immuno-localisation of FLAG-tagged crescentin (*left*) and monitoring GFP-tagged crescentin in the presence of the native crescentin (*right*). Scale bar represents 2 μm (Taken from Ausmees et al. 2003). (**b**) The *in vitro* polymerisation of crescentin was monitored using negatively stained protein samples by transmission electron-microscopy (Taken from Cabeen et al. 2011). (**c**) The life cycle of *C. crescentus* includes the transformation of swarmer cells to stalked cells that can initiate DNA replication and cell division. The pre-divisional cells develop flagella at the pole opposite to the stalked pole. Completion of the asymmetric cell division generates the smaller, motile swarmer cells and the larger, immobile stalked cells. (**d**) The curved, crescent shape of *C. crescentus* promotes the attachment of its Type IV pili, located at the flagellated poles of the pre-divisional cells, to solid surfaces during fluid flow (*horizontal arrow*) in its natural environment of freshwater lakes. The swarmer cells, that otherwise would swim away after completion of division, will be immobilised to the solid surface, promoting biofilm formation (Persat et al. 2014)

Both immunolocalisation and CreS-GFP fusion demonstrated that CreS forms a continuous, pole to pole filament at the inner curvature of the crescent shaped *C. crescentus* cells (Fig. 6.5a, Ausmees et al. 2003). For visualisation of CreS using fluorescent protein fusion it was necessary to generate a merodiploid strain carrying both *creS* and *creS-gfp* because CreS-GFP is not functional and cells carrying an only copy of *creS-gfp* lose their crescent shape and are straight. Interestingly, although CreS-GFP is not functional, it can still form filaments inside these straight cells. In older cultures, where long, straight cells are produced, CreS-GFP generates curved helical filaments that are very similar to the helical CreS filaments in old wild-type cells. This could suggest that it is not filament formation *per se* that is affected in cells carrying CreS-GFP. Instead, some special feature of the produced filaments is affected and/or the CreS-GFP filaments have lost their ability to interact with the cell membrane, cell wall or some hirtherto uncharacterised partner that links crescentin filaments and cell shape. Indeed, mecillinam, an inhibitor of cell wall cross-linking, or phosphomycin, which inhibits an early step of cell wall synthesis, both induce the detachment of crescentin from its static position from the inner curvature of the cell membrane and, instead, allow the now freely floating crescentin filaments to collapse into helical shapes, which are particularly discernible in elongated cells (Cabeen et al. 2009). The free-floating crescentin cables are helical but extended when attached, directly or indirectly, to the cell wall in wild-type cells. This suggests that attachment of the helical crescentin cables to the cell wall generates a mechanical tension. This mechanical force is proposed to affect the rate of peptidoglycan synthesis rather than the structure or crosslinking of peptidoglycan at the site of crescentin localisation (Cabeen et al. 2009). Accordingly, the rate of cell wall synthesis was shown to be lower at the inner curvature next to the crescentin cables and higher at opposite sites in wild-type *C. crescentus* cells.

The Domain Organisation of Crescentin and Eukaryote IFs Are Very Similar

Crescentin shows the domain organisation characteristic of eukaryote IF proteins with their tripartite structure, including the α-helical rod domain flanked by head and tail domains, although sequence identity between crescentin and eukaryote IFs is very low (25% identity; Ausmees et al. 2003). This is not unexpected, as even amongst eukaryote IF proteins sequence conservation is very low. Crescentin consists of 430 amino acids, where the α-helical rod extends between positions 80 and 444, leaving the first 80 and the last 13 amino acids as head and tail domains (Fig. 6.5). The length of the rod domain (365 amino acid) is comparable, if somewhat longer, than that of the conserved length of cytoplasmic (310 amino acid) or nuclear (350 amino acids) eukaryote IFs. The four sub-domains (1A, 1B, 2A and 2B) of the crescentin rod domain (Fig. 6.5) were originally identified in comparison with the eukaryote sub-domain organisation proposed at that time, however, as we discussed above, instead of four, three distinct sub-domains, Coil 1A, 1B and 2 represent the structure of vimentin, and eukaryote IFs in general, more accurately. In the absence of any data about the structure of crescentin, nevertheless taken into consideration the very similar biochemical and biophysical characteristics of crescentin and eukaryote IFs, we can only speculate that the crescentin rod domain might also be built from 3 sub-domains, where 2A and 2B might be part of a single sub-domain, Coil 2. Within the last coil sub-domain of crescentin, a stutter, that is a four residue insertion creating an 11 residue discontinuity amongst the heptad repeat periodicity, has been proposed at a similar position to that in eukaryote IFs (Figs. 6.2 and 6.5). However, high resolution structural analysis of crescentin will be required not only for establishing the sub-domain organisation of the crescentin rod domain but also for addressing questions whether, similarly to eukaryote IFs, the crescentin's heptad-based coiled coil is interrupted by 11-mer periodicity not only at the position of the stutter, but also at other positions such as the end of Coil 1B and the beginning of Coil 2 and, most importantly, to confirm the assumption that the first association of the crescentin protein is indeed a parallel, in-register elementary homo-dimer.

Filament Formation of Crescentin *in vitro* and *in vivo*

CreS assembles *in vitro* into long filaments with ~10–20 nm diameter in dilute buffers (5–50 mM Tris, HEPES or PIPES) with pH $\leq$ 7.0 in the absence of any co-factors (Fig. 6.5b) but fails to form filaments in buffers with pH $\geq$ 8.4 (Ausmees et al. 2003; Cabeen et al. 2011). This is consistent with the characteristic electrolyte properties of IFs and the overall negative surface charge of the elementary dimer, which, similarly to eukaryote IFs, is assumed to be the building block of the CreS filaments. Although conditions were tested for *in vitro* and *in vivo* filament formation of CreS (Ausmees et al. 2003; Cabeen et al. 2011) and the formed filaments were characterised using transmission electron-microscopy, there is no structural information about the soluble CreS elementary dimer and nothing is known about the assembly pathway for CreS filament formation *in vitro*. Although the mechanical properties and filamentation of CreS is compared to the eukaryote cytoplasmic IF, vimentin, for which the assembly pathway includes the anti-parallel tetramerisation of the elementary dimers followed by their lateral assembly into ULFs (Herrmann et al. 1999, Strelkov et al. 2003, see above) and their further longitudinal assembly into the final, 10–11 nm widths filaments, no soluble oligomers of CreS or ULFs have been visualised after negative staining using TEM, as yet. Alternatively, the assembly pathway for the filamentation of crescentin could follow that of nuclear lamins, where first the head-to-tail association of the elementary dimers produces a dimer strand, followed by the anti-parallel, lateral association of two dimer strands creating tetramer protofilaments and the further lateral association of these protofilaments generates the final lamin filaments (Fig. 6.2).

The role of the domain organisation of crescentin in its *in vitro* assembly was tested by generating crescentin fragments lacking distinct domains, such as the head, tail and rod-domain, or the linker sequences between Coil 1A and 1B (Linker 1) and the stutter within the Coil 2 sub-domain and monitoring filament formation by pelleting assays and transmission electron-microscopy of negatively stained samples (Cabeen et al. 2011). Interestingly, apart from the mutant lacking the entire rod domain, all mutants were able to produce filaments *in vitro* albeit with marked features affected by different buffer conditions. Mutants lacking either the first 27 amino acids of the head or the entire tail domains behaved very similarly to wild-type crescentin *in vitro*. In contrast, the absence of Linker 1 increased lateral associations generating thicker filaments with 20–30 nm diameter even under conditions where crescentin would not readily filament. However, in the presence of 200 mM KCl, a buffer condition that is perhaps the closest to physiological conditions, a linker-less crescentin formed aggregates rather than filaments. This suggests that Linker 1 is important for the correct lateral assembly, perhaps for the correct alignment of the Coil sub-domains, during filament formation. Finally, crescentin that lacked its stutter sequence failed to filament in the physiological buffer containing 200 mM KCl, although readily formed wide bundles at low pH and in the presence of Mg^{2+}, producing filaments with striated patterns when viewed with transmission electron microscopy (Cabeen et al. 2011). These striated patterns resemble the para-crystalline association observed in nuclear lamins, and might represent a unique lateral assembly. Just as in the case of eukaryote IFs, lowering the pH of the buffers and the presence of divalent cations, such as Mg^{2+}, strongly affect filament formation of crescentin. Both conditions promote lateral bundling presumably by generating a bridge between the negatively charged outer surfaces of the crescentin elementary dimers as it is proposed for eukaryote IFs (see above).

The same mutants were also tested for their biological function and filament formation *in vivo*. Remarkably, whilst most mutants, apart from the rod-less one, were able to form filaments of some sort *in vitro*, correct filament formation and localisation *in vivo* was dependent on all domains tested. As these crescentin mutants were introduced into the straight-shaped *creS* null mutant, in addition to filament formation, their ability to complement the knockout mutation and to promote cell curvature could also be monitored. The mildest defect was observed with the tail-less mutant, which partially complemented the *creS* mutation, generating some, 25%, curved cells. Whereas no filament formation, only diffuse localisation was observed for the mutants that lacked the entire head and rod domains, consequently, producing straight cells (Cabeen et al. 2011). Crescentin with a deletion of only the first 27 amino acids of its head domain, still failed to complement the *creS* mutant, but produced helical filaments that were detached from the cell envelope. This suggests that the positively charged first 27 amino acids are important for attachment of crescentin to the cell membrane (Cabeen et al. 2009). Interestingly, lack of Linker 1 produced polar aggregates and straight cells (Cabeen et al. 2009, 2011), which is consistent with the *in vitro* behaviour of this mutant forming aggregates rather than filaments in buffers resembling physiological conditions. Finally, stutterless crescentin produced straight cells with either short filaments or diffused localisation in approximately equal proportion. The effect of the stutter, which represents an 11-mer periodicity instead of the heptad periodicity, on the coiled coil assembly of the elementary dimer was discussed above. Similarly to the crescentin mutation, a stutterless vimentin mutation affected the higher order assembly pathway for filament formation although not the elementary dimer formation (Herrmann et al. 1999).

Interestingly, most of these crescentin mutants, except the tail-less mutant, had a negative dominant effect when they were introduced into wild-type cells, producing non-curved cells even in the presence of wild-type crescentin. This suggests that these crescentin mutants do not only alter their own assembly but they also obstruct the assembly pathway of the wild-type crescentin (Cabeen et al. 2011). This is not entirely surprising and has been also demonstrated for several eukaryote IFs including nuclear lamins or the cytoplasmic vimentin, where introduction of a dominant-negative mutant into the cells disrupted the *in vivo* organization of the native filaments (Helfand et al. 2011; Schirmer et al. 2001).

Dynamics of Crescentin in Its Cellular Environment

Transcription of *creS* might be regulated throughout the cell cycle of *C. crescentus* with the highest level expression in the reproductive stalked cells, which can initiate cell division (Laub et al. 2002; Mcgrath et al. 2007). However, the protein levels of crescentin do not change significantly (Charbon et al. 2009). Although, overall, eukaryote IF filaments are more stable than filaments formed from actin or tubulin, they are dynamic structures with reasonable turnover (Vikstrom et al. 1992; Yoon et al. 2001). The rate of this turnover varies between IFs, for example, the turnover of vimentin is almost 10 times faster than that of keratin when tested using FRAP (Yoon et al. 1998, 2001). Crescentin is fairly stable with very little sign of a soluble crescentin pool that, when produced, is immediately incorporated at random places alongside the existing crescentin filaments (Charbon et al. 2009). There is very slow, if any, subunit exchange within an existing filament as was assessed using FRAP or FLIM either in the presence or absence of protein synthesis (Charbon et al. 2009; Esue et al. 2010). *De novo* crescentin assembly was shown to be biphasic, where first a thin filament extending the entire length of the cell is formed, followed by the lateral insertion of more crescentin to produce an increase in the diameter of the filament (Charbon et al. 2009). As mentioned above, we do not have any structural information on the assembly pathway of crescentin, which would be required to fully interpret the biphasic *de novo* crescentin assembly observed in *C. crescentus*. However, it is attractive to speculate that, perhaps, crescentin assembly resembles more the lamin assembly pathway where the elementary dimers first associate longitudinally into protofilaments and this longitudinal association precedes the further lateral association of these protofilaments into the final IFs.

Even though there is increasing structural information about some of the assembly pathways for eukaryote IFs, there is a lack of understanding of how IFs are nucleated *in vivo* in their cellular environment. One possible clue for IF nucleation could be provided by their potential interactions with the other two main filamentous cytoskeletons, the microtubules and actin microfilaments, sometimes via other proteins called, cytolinkers (Chang and Goldman 2004). Interestingly, a functional crescentin-FLAG fusion pulled down MreB, the bacterial actin homologue, from cell extracts of *C. crescentus* or even from that of *E. coli*. Further supporting evidence for a link between MreB and crescentin came from finding that the introduction of crescentin into *E. coli* cells made them curved, whereas introduction of crescentin into *Agrobacterium tumefaciens,* a bacterium that lacks MreB, did not cause any change of cell curvature even though *A. tumefaciens* is more closely related to *C. crescentus*, belonging to the same class of alpha-proteobacteria (Charbon et al. 2009). Further dissection of the link between crescentin and MreB will establish whether this interaction is direct or indirect, and whether MreB acts as a nucleator for crescentin.

Crescentin turnover has not been studied in detail. One of the intriguing questions to address is the fate of the static crescentin filament during cell division, when there must be a localised disassembly of crescentin at the division site (Fig. 6.5c). During cell migration of fibroblasts, there is a dynamic disassembly of vimentin filaments into smaller assemblies at the front of the migrating cells, where lamellipodia are formed, whereas the fully assembled vimentin filament network extends at the rear of the cell (Helfand et al. 2011). The mechanism for the disassembly of vimentin includes the post-translational modification, phosphorylation. During mitosis and cytokinesis of fibroblasts the filamentous network of vimentin undergoes depolymerisation in order to distribute soluble vimentin into the daughter cells. This depolymerisation is achieved by transient phosphorylation that promotes the dramatic re-organisation of the vimentin filaments into soluble particles, which, after their distribution into the daughter cells, are de-phosphorylated to promote re-assembly of vimentin filaments (Chou et al. 1990; Yamaguchi et al. 2005). As yet, there is no information on covalent modification of crescentin. However, it is intriguing to consider that localised phosphorylation or other post-translational modification might play a role in its localised disassembly at the division sites during cell division. Interestingly overproduction of an essential tyrosine phosphatase, CtpA caused straightening of *C. crescentus* cells, although the effect of CtpA on cell shape was proposed to involve the cell wall synthetic enzymes rather than crescentin itself (Shapland et al. 2011).

A new technology, cryo electron-tomography opened up exciting possibilities of visualising cellular filaments at high resolution (Swulius et al. 2011; Szwedziak et al. 2012); however, filaments detected at the inner curvature of *C. crescentus* using this technique were shown to be not crescentin but CtpS, a metabolic enzyme of CTP synthase, as they persisted in the *creS* null mutant (Ingerson-Mahar et al. 2010).

Mechanical Properties of Crescentin

In the bacterial cells, crescentin cables are attached to the cell envelope and only show a slight curvature. However, when attachment is affected either by inhibitors of peptidoglycan synthesis (Cabeen et al. 2009) or by removing the first 27 amino acids of the head-domain that is important for attachment to the cell envelope, the crescentin cables collapse into helical filaments (Cabeen et al. 2009). This supports the notion that crescentin structures are flexible, stretchable and elastic, mimicking the intrinsic properties of eukaryote IF filaments. Indeed, quantitative rheology studies *in vitro* confirmed that crescentin behaves as a visco-elastic solid that can recover a considerable fraction of its elasticity after shearing (Esue et al. 2010). Stretching of eukaryote IFs can involve gliding of the dimer strands along their long axis, which in turn depends on the lateral interactions determined by the surface charge of these dimer strands. Crescentin showed increased stiffness in the presence of divalent cations such as Mg^{2+} or Ca^{2+}, which was also shown to promote filament bundling, confirming that these cations affect the lateral interactions between IF strands and filaments (Esue et al. 2010).

What Is the Biological Function of Crescentin Dependent Cell Curvature?

Lack of crescentin only affected the ability of *C. crescentus* cells to produce their characteristic crescent cell shape, but no other detectable defects have been found. Growth, cell division and chromosome segregation were all unaffected in the *creS* mutant under laboratory conditions. Originally, it was proposed that the crescent shape promotes mobility in the aquatic environment, where this bacterium naturally resides. Indeed, corkscrew shape of *Helicobacter* is advantageous for swimming as it promotes faster speed in viscous medium such as the gastric fluid (Martinez et al. 2016).

Recent studies propose a remarkable link between the crescent shape of *Caulobacter* and its fitness to colonise in its native environment (Persat et al. 2014). The natural environment for *Caulobacter* species is in freshwater lakes where natural fitness is influenced by the bacterium's ability to attach to solid surfaces to form micro-colonies leading to biofilm formation even in the presence of a naturally fluctuating fluid (O'Toole et al. 2000). For adhesion to solid surfaces *Caulobacter* predominantly uses its stalk, which is a compartmentalised cytoplasmic extension with a strong adhesive tip exclusive to the stalked cells (Schlimpert et al. 2012; Tsang et al. 2006). Beside its stalk, *Caulobacter* swarmer cells have Type IV pili which have also been shown to attach swarmer cells to solid surfaces. The stalk and pili reside in different cell types, the stalked cells and the swarmer cells, respectively. These cells are the products of an asymmetric cell division initiated exclusively in the stalked cells and are kept together at the predivisional stage (Fig. 6.5c). Recent studies suggested that for effective colonisation of solid surfaces in the presence of fluid flow, the crescent shape of the wild-type predivisional cell is advantageous, because only this crescent shape allows the attachment of the pili in the swarmer part of the predivisional cell at the same time when the stalk forms strong attachment in the stalked part of the predivisional cell (Fig. 6.5d, Persat et al. 2014). The surface attachment of the swarmer cell compartment in the predivisional cell is important for biofilm formation in the wild-type strain, because this maintains the association of the swarmer cells to the solid surface even after division is completed. In contrast, the straight geometry of the predivisional cells of the *creS* mutant does not allow the attachment of the pili of the swarmer cell compartment in the predivisional cell to the solid surface, consequently these swarmer cells will be dispersed in the fluid flow when division is completed.

An Extended IF Family

The discovery of the first example of a bacterial intermediate filament-like protein, crescentin, was soon followed by the characterisation of further bacterial proteins that form filamentous structures both *in vitro* and *in vivo.* These are also built from coiled-coil proteins and originate from diverse organisms, such as *Streptomyces coelicolor* or *Helicobacter pylori* (Bagchi et al. 2008; Fuchino et al. 2013; Holmes et al. 2013; Walshaw et al. 2010; Specht et al. 2011; Waidner et al. 2009). Strikingly, these bacterial proteins show more structural diversity than their eukaryote counterparts. All share the tripartite architecture including short head and tail domains flanking a central rod domain. However, there are marked differences in the length of the rod domains, which often lack the characteristic three sub-domain organisation raising the question whether these proteins are *bona fide* intermediate filaments. The current view on the structural features that define the IF family was based exclusively on members of the eukaryote IF family where there is a remarkable conservation of the rod domain (see above). Accordingly, all members of the IF family must have a rod domain of 310 or 340 amino acids, for cytoplasmic or nuclear IFs, respectively, and the three sub-domains, Coil 1A, Coil 1B and Coil 2. On the other hand, bacterial IF-like proteins share similar biochemical and biophysical properties and cellular localisation pattern with those of eukaryote IFs. This suggests they are part of the wider IF-family, a group which, in the light of its extended members, might be calling for a new definition. In addition to the bacterial proteins, there are eukaryote proteins of non-metazoan origins that do not fully comply with the strict IF definition and so far have been excluded from the IF family due to different length and sub-domain organisation of their rod domains. These include the lamin-like nuclear proteins from the simple protozoan *Dyctiostelium discoid*eum and *Trypanosome brucei* (Batsios et al. 2012; Dubois et al. 2012; Kruger et al. 2012) or the plant nuclear matrix constituent proteins, NMCPs (Ciska and Moreno Diaz de la Espina 2014).

Novel Structural Characteristics of IF-Like Proteins: FilP and Scy from *Streptomyces*

The two IF-like proteins, FilP and Scy of the filamentous bacterium, *Streptomyces coelicolor,* are markedly different in size, 310 amino acids and 1326 amino acids, respectively. Despite of this, analysis of their sequence suggests that they are strikingly similar (Walshaw et al. 2010). Both proteins have tripartite domain organisation, consisting of very short head and tail domains that extend 17 and 27 amino acids for Scy and 16 and 18 amino acids for FilP (Fig. 6.6a). Unlike the eukaryote IF proteins, the α-helical rod domains of Scy and FilP lack the three subdomain-organisation, instead, they have two sub-domains, CC7 and CC51 connected by a short, 7–9 amino acids linker (Walshaw et al. 2010). The CC7 subdomain is dominated by a canonical heptad periodicity, hence the name (coiled coil 7) and extends

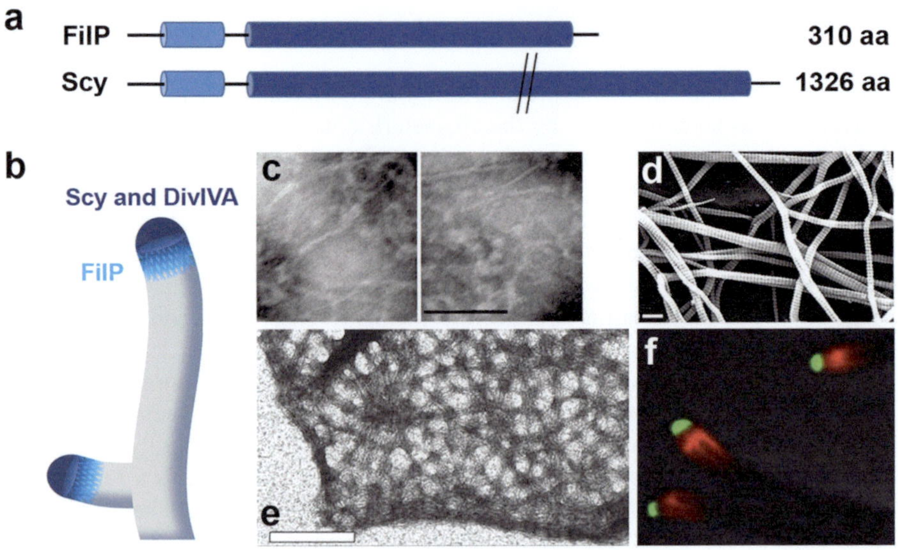

Fig. 6.6 Two IF-like proteins, FilP and Scy, from *S. coelicolor*. (**a**) The rod-domains of FilP and Scy each consist of two sub-domains, CC7 and CC51. The CC7 sub-domain is organised very similarly to the Coil 1A sub-domain of eukaryote IFs and is characterised by heptad periodicity. The CC51 domain is built from periodically modified hendecad repeats, that is, one heptad for every four hendecads: 7,11,11,11,11 (Walshaw et al. 2010). (**b**) Both Scy and FilP are part of the TIPOC. Scy and the polarity determinant DivIVA form a polar scaffold at the hyphal tips whereas FilP forms a filamentous network at sub-apical positions. It is conceivable that, similarly to that in *Mycobacterium*, polar growth in *Streptomyces* is restricted to the zone between the tip-dome and the sub-apical FilP network (Meniche et al. 2014). (**c**) Transmission electron microscopy of negatively stained Scy protein samples in Tris pH = 8.0. Scale bar represents 100 nm. (**d**) Scanning electron microscopy of FilP in 50 mM Tris, pH = 7.0, taken from (Bagchi et al. 2008). Scale bar represents 200 nm. (**e**) Transmission electron microscopy of the negatively stained lace-like networks of FilP in 50 mM Tris, pH = 7.0, taken from (Fuchino et al. 2013). Scale bar represents 200 nm. (**f**) DivIVA-EGFP (*green*) localises to hyphal tips. Immuno-localisation of FilP (*red*) indicates a filamentous network at sub-apical locations (Taken from Fuchino et al. 2013)

46 and 48 amino acids for Scy and FilP, respectively. This corresponds to 6 full heptads, which is very similar to the organisation of 6 heptads in Coil 1A of eukaryote IFs (Chernyatina et al. 2015; Walshaw et al. 2010). Remarkably, bioinformatic analysis of the CC51 sub-domains of Scy and FilP sequences using Fourier transformation to search for the periodicity-pattern of the positions of hydrophobic amino acids within the α-helical rod domain revealed a novel, highly regular 51 amino acids (penindaenad) repeat periodicity (Holmes et al. 2013; Walshaw et al. 2010). One likely interpretation of the 51-residue repeat is a periodically modified hendecad repeat that is one heptad for every four hendecads: 7,11,11,11,11. The occurrence of hendecads within Coil 1A and Coil 2 of vimentin and eukaryote IFs is well documented (Chernyatina et al. 2015) and we have discussed above how deletion of the hendecad in Coil 2, also known as a stutter, affected higher order assembly of

both the eukaryote vimentin and the bacterial IF-like crescentin (Herrmann et al. 1999; Cabeen et al. 2011). However, unlike eukaryote IFs and crescentin, the CC51 sub-domain of Scy and FilP is dominated by hendecad repeats. Comparisons of the hydrophobic streak of the heptad, hendecad and penindaenad repeats can be displayed using helical net diagrams (Fig. 6.1c). Unlike the left-handed streak of the heptad repeat, the 51-residue motif produces a right-handed hydrophobic streak on the surface of the α-helix, a streak that is more similar to that of the uninterrupted hendecad (Fig. 6.1c). The hydrophobic streak of the 51-repeat is so shallow, with a very long pitch, that burying these streaks at the interface between two such α-helices will generate an almost parallel bundle of the two strands rather than a coiled-coil *per se*. Interestingly, recent structural studies of hendecad coiled-coils demonstrated either shallow right-handed coiled coils (Lee et al. 2010) or even "straight" hydrophobic streaks between the two interacting strands forming a nearly parallel bundle (Nicolet et al. 2010; Nishimura et al. 2012; Stewart et al. 2012).

High resolution structural information is much needed to characterise the novel, 51-mer repeat of Scy and FilP and the effect of this periodically interrupted hendecad repeat on the interface of their protein assemblies. The predicted elementary unit for Scy and FilP is a parallel, in-register homo-dimer, based on the packing considerations of the residues at the interface positions of (a) and (h) of the hendecad-motifs. In the stalk of the bacterial F_0F_1 ATP synthase that forms homodimers, the most common amino acid at positions (a) and (h) is alanine (78% and 42% respectively; Del Rizzo et al. 2002; Walshaw et al. 2010). Whereas, tetrabrachion, another filament-forming protein from the archaea *Staphylothermus marinus*, is also dominated by hendecad-repeats but assembles into parallel tetramers. In the structural organisation of tetrabrachion, larger amino acids, isoleucine (58%) and leucine (48%), can be afforded at positions (a) and (h) (Peters et al. 1996; Stetefeld et al. 2000). In the CC51 sub-domain of Scy the most common amino acid at positions (a) and (h) is alanine (63% and 64%, respectively), which favours the hypothesis that the minimal assembly unit of Scy, and also FilP, is a dimer. Although the oligomerisation states of Scy and FilP have not been established experimentally, biochemical characterisation of AbpS, a FilP orthologue with 93% sequence identity originating from another *Streptomyces* species, *Streptomyces reticuli,* have demonstrated the presence of dimers in all overproduced and purified AbpS samples, regardless whether the full length or truncated AbpS were tested (Walter et al. 1998, 1999; Walter and Schrempf 2003). The *in vitro* chemical cross-linking experiments also detected AbpS tetramers that were dependent on the presence of the N-terminal region, suggesting that the AbpS elementary dimers can further assemble into dimer of dimers. In the absence of further experimental data we can only speculate whether dimerisation of the AbpS dimers is the result of a lateral or a longitudinal interaction between two elementary dimers. The elementary dimers of eukaryote vimentin form an anti-parallel, slightly staggered tetramer via lateral interaction between two dimers (Fig. 6.2), which could be one possible arrangement for the AbpS dimers and, conceivably, FilP or Scy dimers. Alternatively, longitudinal assembly of the two dimers, where interaction is established between the N-terminal region of one of the dimers and either the N- or C-terminal regions of the second dimer, is also plausible and consistent with the finding that tetramerisation of AbpS is dependent on its N-terminal region (Walter and Schrempf 2003).

Assemblies of FilP and Scy *in vitro*

The likely parallel, in-register FilP and Scy homo-dimers further assemble into filamentous structures *in vitro* that can be visualised by negative staining using transmission electron microscopy (Bagchi et al. 2008; Fuchino et al. 2013; Holmes et al. 2013). Assuming that both the CC7 and CC51 domains are fully extended, the coiled-coil dimers of FilP and Scy are expected to have an elongated rod shape with a predicted length of ~48 nm and ~198 nm, respectively. Purified Scy forms a network of rope-like filaments where long (>100 nm) filaments can be readily detected (Fig. 6.6c; Holmes et al. 2013). However, it is likely that the Scy homo-dimers are further assembled in these samples as the width of the Scy filaments is larger than 2 nm, which is the width of a coiled-coil homo-dimer. Moreover, no single Scy filaments of ~190 nm have so far been detected. Remarkably, FilP, is able to form a fascinating array of assemblies (Fig. 6.6d, e) including an interconnected lace-like network (Fuchino et al. 2013) or a compact, striated filament (Bagchi et al. 2008) that is reminiscent of the paracrystalline arrays observed in preparations of nuclear lamins (Heitlinger et al. 1992). In the lace-like assembly filaments are interconnected via “nodes” where the average distance between the “nodes” was around ~60 nm (Fuchino et al. 2013) , which is slightly longer than the predicted length of a single FilP dimer of ~48 nm. Interestingly, the nodes were often positioned in a hexagonal arrangement, built from tessellation of equilateral triangle patterns with “nodes” at the vertices linked by filaments at the sides. It has also been shown that the lace-like assembly can merge into the striated pattern (Fuchino et al. 2013), which suggests that both of these structures are organised by common principles, though what these principles are is not known at present. Organisation of eukaryote IF filaments *in vitro* depends markedly on the buffer conditions, including the pH and the presence or absence of cations, which can affect the lateral and longitudinal associations of the elementary dimers. The buffer conditions are also critical for the filamentous structures that are formed *in vitro* in protein preparations of FilP and Scy. Similarly to the investigations carried out for crescentin (Cabeen et al. 2011), it is paramount that the *in vitro* filament formation of both FilP and Scy is further investigated using systematic analysis of the effect of buffer conditions on the higher order assembly not only of the full length FilP and Scy, but also of truncated derivatives that lack the N-terminal CC7 or part of the C-terminal CC51 domains. This latter will address the question of which part of FilP, the N- or C-terminus, is essential for the formation of the nodes observed in the lace-like or striated FilP networks. In spite of the striking difference in size (Scy is four times longer), the coiled-coil domain organisation of Scy and FilP is remarkably similar (Walshaw et al. 2010). Therefore, it is highly likely that each of these proteins will affect filament formation of the other, either by acting as a mimetic inhibitor and preventing higher order associations of the other or by promoting co-polymer formation.

Both Scy and FilP Support Filamentous Growth in *Streptomyces*

The genes encoding Scy and FilP are adjacent in the *Streptomyces* chromosome, which makes it likely that they share a common ancestor and were generated by gene duplication during evolution. The phenotypes of the knockout mutants of *filP* and *scy* are distinct, suggesting that FilP and Scy are paralogues with separate biological functions that emerged after gene duplication. Although simple BLAST searches are not reliable for finding true homologues of coiled-coil proteins, both FilP and Scy have firm homologues amongst actinobacteria. FilP is more widely spread even amongst the morphologically simple, rod-shaped organisms, although it is absent from *Corynebacterium*, while Scy is found almost exclusively amongst the filamentous actinobacteria (Bagchi et al. 2008; Chandra and Chater 2014; Holmes et al. 2013). Lack of Scy and FilP have a profound, albeit very different, effect on filamentous growth, while Scy also has a long range effect on cell division. In order to discuss the role of these bacterial IF-like proteins, first the morphological complexity of *Streptomyces* will be introduced.

Streptomyces – Polar Growth with Suspended Cell Division

The life-cycle of *Streptomyces* (Fig. 6.7) has two striking features that distinguish these bacteria from the most studied, model bacteria, such as *E. coli*, *B. subtilis* or *C. crescentus*. First, polar growth is employed for extending the hyphal filaments at their tip ends, as opposed to lateral cell wall extension, a mechanism wide-spread in most rod-shaped bacteria. The second remarkable feature of *Streptomyces* bacteria concerns the link between two fundamental processes: growth and division. Unlike in uni-genomic bacteria, where growth and division are essentially interlinked, in *Streptomyces* growth can take place in the absence cell division and the two phases, cellular growth and fully completed cell division are separated both in time and space.

Cellular growth (Fig. 6.7) is initiated when a single ovoid spore carrying a single chromosome produces one or two germ tubes (Flardh and Buttner 2009; Jyothikumar et al. 2008; McCormick and Flardh 2012). These hyphal filaments extend exclusively at their apical tips and as a result of chromosome replication and segregation they carry increasing numbers of chromosomes. At several, irregularly spaced positions, hyphal septa are formed generating multi-genomic cellular compartments. However, this infrequent septum formation is not followed by cell-cell separation, hence cell division is "suspended" during this stage of filamentous growth. In uni-genomic bacteria with polar growth, such as the actinomycete *Mycobacterium* or *Corynebacterium*, cell division generates new poles for growth as the divisional septum sites become the new poles of the daughter cells. In *Streptomyces* one of the implications of the suspended cell division is that new poles are not generated at the hyphal septa. Instead, new poles, called tips, are generated at lateral hyphal locations, usually far behind an existing growing tip, which in turn extend as branches of the original filament (Fig. 6.7a). During the first stage of rapid growth of *S.*

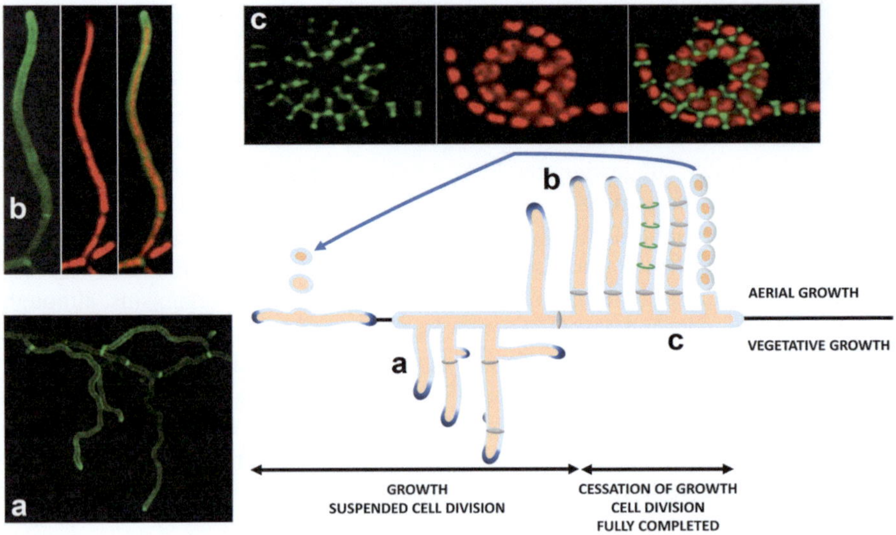

Fig. 6.7 The life cycle of *Streptomyces coelicolor*. (*Middle*) Spore germination produces two germ tubes that extend by polar growth orchestrated by the TIPOC (*dark blue*). During filamentous growth, cell division is "suspended" and this generates a network of branching, multi-genomic filaments, producing first the vegetative mycelium with occasional septum formation but no cell division (**a**) then the aerial mycelium with less frequent branching (**b**). The switch to sporulation begins with the cessation of polar growth. This is followed by chromosome segregation and the polymerisation of the key cell division protein, FtsZ into regularly spaced rings (*green*). The FtsZ-mediated synchronous septation (**c**) generates the cylindrical prespore-compartments, each carrying a single chromosome. The spores, after maturation, will disperse and, under favourable conditions, can initiate growth again. Confocal microscopy was used to monitor selected stages of the life cycle. Fluorescent wheat germ agglutinin (WGA, *green*, **a**, *left panels* in **b** and **c**) marks the cell wall and propidium iodide (*red*, *middle panels* in **b** and **c**) stains the nucleoid. Overlayered images (*right panels* in **b** and **c**) are also shown

coelicolor, a dense network of branching filaments, the vegetative mycelium, is produced capitalising on the nutrient resources provided by the medium. This is followed by a second stage of growth on semisolid medium, where branches emerge into the air producing the aerial mycelium (Fig. 6.7b). Perhaps because less nutrients are available, there is much less hyphal branching of the aerial filaments and the extension of their tips finally comes to a halt. Importantly, the cessation of polar growth coincides with the switch to a now fully completed cell division, where 50–100 unigenomic pre-spore compartments are generated by a synchronous septation event in the tip-proximal, unbranched aerial compartments (Fig. 6.7c).

Growth at One Pole Is Prevalent in the Orders *Actinomycetales* and *Rhizobiales*

Polar growth, where growth takes place at a "single", polar location, provides an alternative mechanism to lateral growth for building rod-shaped or filamentous bacteria. Lateral cell wall extension is mediated by the bacterial actin-homologues, the MreB-family of cytoskeletal proteins (Chapter 8; Cabeen and Jacobs-Wagner 2010; Dominguez-Escobar et al. 2011; Garner et al. 2011; van Teeffelen et al. 2011). However, polar growth provides an alternative, MreB-independent, mechanism in the Gram-positive order *Actinomycetales,* for generating either rod-shaped cells of the pathogenic bacteria, *Mycobacterium* and *Corynebacterium* (Thanky et al. 2007; Letek et al. 2008a, b) or long, branching filaments amongst *Streptomyces* species (Brana et al. 1982; Miguelez et al. 1992; Flardh and Buttner 2009, 2012). Remarkably, unidirectional polar growth, at the sites generated during cell division, was also demonstrated for an entirely different group of bacteria, the alpha-proteobacteria, including *Agrobacterium tumefaciens* and other rod-shaped members of the order *Rhizobiales* (Brown et al. 2012). In these latter bacteria, the cytoskeletal cell division protein, FtsZ localises to the actively growing pole for a large part of the cell cycle, and then re-locates to the middle of the cells at the time of division (Zupan et al. 2013; Cameron et al. 2015). In addition, the pole-organizing protein PopZ and polar organelle development protein PodJ are important for growth at the new pole and the transformation of new poles to old poles (Cameron et al. 2014, 2015; Grangeon et al. 2015).

The Essential Polarity Determinant, DivIVA

Amongst the *Actinomycetales* the coiled-coil protein DivIVA functions as an essential polarity determinant for localised cell wall synthesis at the poles for the uni-genomic *Mycobacterium* and *Corynebacterium* (Nguyen et al. 2007; Kang et al. 2008; Letek et al. 2008b) or at the growing tips and at the sites of new tip formation, hence branching, in *S. coelicolor* (Flardh 2003; Hempel et al. 2008). DivIVA is essential in all these three genera and changes in the cellular DivIVA levels have profound effects on cell shape. Lowering the level of DivIVA in *Mycobacterium* and *Corynebacterium* leads to coccoidal cells, where cell wall synthesis becomes exclusive to the division septum. On the other hand, elevated DivIVA levels generate club-shaped *Mycobacterium* or *Corynebacterium* cells (Kang et al. 2008; Letek et al. 2008b) that are reminiscent of the "tennis-racket" tips of *S. coelicolor* observed following DivIVA overproduction (Flardh 2003). Excess DivIVA not only accumulates at existing poles triggering their expansion, but it also assembles at ectopic locations, where new polarity centres are established, producing branching both in *Streptomyces* and *Mycobacterium* (Flardh 2003; Kang et al. 2008).

Interestingly, DivIVA is not restricted to bacteria that exhibit polar growth and is widespread in Gram-positive bacteria with rod or spherical shape, albeit implicated in cell division rather than growth. DivIVA was first identified and characterised in

B. subtilis where it assembles at the divisional septum and the poles (Edwards and Errington 1997) and recruits the FtsZ-antagonist MinCD complex via MinJ (Bramkamp et al. 2008; Patrick and Kearns 2008). Originally it was proposed that DivIVA tethered the MinCD complex at the poles, allowing FtsZ ring formation at midcell position, a mechanism resembling that found in *E. coli*, where there is no DivIVA, MinCD oscillates between the two poles (Lutkenhaus 2012). However, subsequently DivIVA has been shown to assemble primarily into rings at each side of the developing septal membrane, a post-FtsZ event, where its role is to prevent the formation of further septation (Eswaramoorthy et al. 2011). As the septum becomes a future pole, the septal DivIVA rings change into polar DivIVA patches (Eswaramoorthy et al. 2011). DivIVA has several partners beside MinJ, icluding RacA that binds to the chromosomes tethering them to the poles during *Bacillus* sporulation (Ben-Yehuda et al. 2003; Wu and Errington 2003) or the competence proteins, Maf and ComN (Briley et al. 2011; dos Santos et al. 2012).

It is intriguing how DivIVA seems to possess diverse cellular function in different bacteria, guiding cell division in *B. subtilis* and organising polar growth in the *Actinomycetales*. These functional differences are unlikely to be the result of major structural differences amongst the DivIVA proteins, all of which have an N- and C-terminal coiled-coil domain joined by a linker of varied length, although, DivIVA of *S. coelicolor* has an unusually long (132 aa) linker and an extra 86 aa at the C-terminus that might affect partner-binding. Perhaps the common "core" DivIVA function is to generate a molecular scaffold that assembles on negatively curved membranes (see below) such as the developing septal membrane or the poles, where it recruits its partners. Hence, it might, in fact, be the partner proteins that determine the cellular process that DivIVA controls. Interestingly, homologues of the *B. subtilis* DivIVA partners RacA, MinJ, ComN and Maf are all absent from the *Actinomycetales*. Similarly, DivIVA in *S. coelicolor* interacts with the IF-like cytoskeletal proteins FilP and Scy together with the cellulose synthase, CslA (Holmes et al. 2013; Xu et al. 2008), proteins with no homologues in *B. subtilis*. Alternatively, the role of DivIVA in the *Actinomycetales* might not as yet have been fully explored. In addition to polar locations, DivIVA also assembles at the divisional septum in *Mycobacterium* and *Corynebacterium* and at occasional vegetative septa in *Streptomyces*.

The localisation of *B. subtilis* DivIVA to the septum and the poles even when expressed in heterologous hosts like *E. coli* or *Schizosaccharomyces pombe* (Edwards et al. 2000) suggests that DivIVA does not require any additional proteins for its localisation. Instead, negative membrane curvature, a geometrical cue, alone is sufficient for DivIVA recruitment (Lenarcic et al. 2009; Ramamurthi and Losick 2009; Strahl and Hamoen 2012) suggesting that cellular architecture controls DivIVA localisation and function. Mathematical modelling of the DivIVA-membrane interaction led to the so-called "molecular bridging" model (Lenarcic et al. 2009; Strahl and Hamoen 2012). This model is built on two assumptions: the association of the DivIVA tetramers into higher order assemblies and the affinity of DivIVA for the cell membrane via the hydrophobic Phe residue (F17) at its N-terminus (Oliva et al. 2010). According to this model, DivIVA assemblies bind to negatively curved membranes as this geometry maximises both the DivIVA-DivIVA and the DivIVA-membrane interfaces (Lenarcic et al. 2009; Strahl and Hamoen 2012). *Bacillus*, *Mycobacterium* or *Corynebacterium,* all share similar cellular geometry with negatively curved membranes at their poles and at the divisional septum. In contrast, meandering hyphal filaments of *S. coelicolor* generate negatively curved, concave membrane surfaces in abundance (Fig. 6.7a). Whereas new branches tend to form at concave hyphal curvatures (Hempel et al. 2008), not all concave curvatures give rise to new branches. Furthermore, DivIVA can only be detected at a few, selected sites, in spite of the fact that the geometry of the meandering hyphae presents ample negatively curved membrane surfaces for DivIVA recruitment.

Sustained Polar Growth in the Filamentous *Streptomyces* Is Orchestrated by a Multi-protein Assembly, the Tip Organising Centre (TIPOC), Including the IF-Like Proteins Scy and FilP

Why does DivIVA only localise to selected sites at the lateral wall of the *Streptomyces* hyphae? Unlike that of *B. subtilis* DivIVA (Oliva et al. 2010), no solved structure is available for *S. coelicolor* DivIVA. The phenylalanine residue, F17, implicated in membrane association in *B. subtilis* (Oliva et al. 2010) is replaced by a leucine residue, L18, in *S. coelicolor* DivIVA, which was shown to purify predominantly in the cytoplasmic fraction of cell extracts with very little of it detected in the membrane fraction (Wang et al. 2009), suggesting that the affinity of DivIVA to the membrane is weak. Compositional cues in the membrane itself might affect DivIVA recruitment in *Streptomyces*. Certain membrane components, such as cardiolipins, were shown to localize at hyphal tips and at branch points in *S. coelicolor* and changes in the expression of the essential cardiolipin synthase gene (*clsA*) produced hyphal filaments with altered branching patterns (Jyothikumar et al. 2012). In contrast, lack of cardiolipin or phosphatidyl-ethanolamine in the cell membrane of *B. subtilis* does not affect DivIVA localisation (Nishibori et al. 2005; Lenarcic et al. 2009) even though these membrane components are enriched at the poles and at the division site in *Bacillus* (Kawai et al. 2004). Finally, and perhaps most importantly, in *Streptomyces* polar growth does not rely on a single polarity determinant alone. Instead, DivIVA, together with the IF-like proteins, Scy and FilP, form a polar, multi-protein assembly, the Tip Organising Centre (TIPOC) at the growing hyphal tips and at emerging branching sites (Fuchino et al. 2013; Holmes et al. 2013). Direct, pair-wise interactions amongst DivIVA, FilP and Scy were demonstrated both *in vitro* and *in vivo* in the heterologous host *E. coli* using bacterial two hybrid assays (Fuchino et al. 2013; Holmes et al. 2013). At present we do not have any molecular details of the complex assembly of these three proteins, however, the affinity of DivIVA alone to negatively curved membranes will undoubtedly be altered by its association with further TIPOC components, Scy and FilP. Furthermore, when new branches emerge in the lateral wall of the *Streptomyces* hyphae some bending of the cell membrane might be required. DivIVA was proposed to have similarities to eukaryotic BAR-domain proteins that are known not only to assemble at curved membranes but also to deform membranes for organelle biogenesis (Oliva et al.; Frost et al. 2009). However, *in vitro* experiments failed to demonstrate membrane bending properties for DivIVA of *B. subtilis* (Lenarcic et al. 2009). Future work is required to establish whether and/or how Scy or some other, as yet unknown, TIPOC components might induce membrane deformation at the sites of tip-emergence in *Streptomyces*.

Cellular localisation of the TIPOC components *in vivo* suggests a complex and dynamic interaction between these three proteins. DivIVA and Scy co-localise both at fully developed hyphal tips and also at lateral hyphal locations leading to branching (Holmes et al. 2013). DivIVA-EGFP and mCherry-Scy foci start with overlapping, single point-like foci at the sites of the emergence of new branches. This is followed by an extended, dome-shaped DivIVA-EGFP and an mCherry-Scy patch spreading at the fully developed hyphal tips (Flardh 2003; Hempel et al. 2008; Holmes et al. 2013). Localisation of DivIVA-EGFP and mCherry-Scy coincides with the sites marked by a fluorescent conjugate of the antibiotic vancomycin, Van-Fl, which binds to the yet unincorporated or incorporated but not crosslinked peptidoglycan precursors, presumably marking the sites of new cell wall synthesis (Daniel and Errington 2003; Flardh 2003; Hempel et al. 2008; Holmes et al. 2013). This suggested that both DivIVA and Scy mark the sites of new peptidoglycan synthesis. Intriguingly, the immuno-localisation of FilP showed no clear overlap with DivIVA localisation *in vivo*, instead, FilP formed extended sub-apical gradients adjacent to the DivIVA domes at hyphal tips of *S. coelicolor* (Fig. 6.6b, f; Fuchino et al. 2013). In addition to the sub-apical FilP gradients, FilP cables were also observed adjacent to branch-formation during induced expression of DivIVA, raising the possibility, that the different FilP assemblies, the cables and the lace-like networks that were observed *in vitro,* might in fact reflect distinct FilP assemblies *in vivo*, albeit the FilP cables were observed when DivIVA was induced artificially (Fuchino et al. 2013). Interestingly, DivIVA-EGFP and FilP did not co-localise at all in the heterologous host *E. coli*, suggesting that despite the fact that FilP and DivIVA proteins are able interact *in vitro*, they might require an additional protein for their adjacent localisation at the hyphal tips in *Streptomyces*. Some of the localisation studies used fluorescent protein fusions (Bagchi et al. 2008; Hempel et al. 2008; Holmes et al. 2013), others immune-localisation (Fuchino et al. 2013) and both methods have some shortcomings. Fluorescent protein fusions are seldom fully functional and immune-localisation require fixed samples, where fixing itself can abolish fine structures that exists *in vivo*. Therefore, further experimental data are needed to clarify the dynamic localisation of the three TIPOC components not only at the hyphal tips but also at other hyphal sites, where fluorescent signals were detected for all three proteins, but these signals have not yet been fully interpreted. Furthermore, localisation studies testing the *in vivo* interaction between FilP and Scy, the two IF-like proteins with highly similar structural composition but with markedly different sizes, will establish whether these paralogous proteins evolved independent functions or more likely, whether their functions are linked.

When the concept of polar growth was introduced above, a "single" polar site was proposed as the site where cell wall extension takes place. However, there is no experimental evidence that this is indeed the case. The polar cap at the hyphal tip of *Streptomyces* that is marked by the fluorescent conjugate, Van-Fl, might label the sites where the yet unincorporated peptidoglycan precursors accumulate rather than those of already incorporated but not crosslinked sites, where nascent cell wall synthesis takes place. Fascinatingly, new cell wall synthesis in *Mycobacterium* was recently shown to be confined to a sub-polar zone suggesting that the polar caps are in fact inert (Meniche et al. 2014). If this was the case for cellular growth of the *Streptomyces* hyphae, the growth zone might be located at the interface between the sub-apical FilP filaments and the apical tip-cap where DivIVA and Scy are located (Fig. 6.6b).

FilP Supports Hyphal Mechanics During Growth

The absence of FilP in a knockout mutant of *S. coelicolor* rendered the hyphal filaments more flexible and meandering with much reduced rigidity as was measured using atomic force microscopy (Bagchi et al. 2008). However, the *filP* mutants did not have thinner cell walls and did not show any branching defects or sensitivity to cell wall damaging agents, such as lysozyme or vancomycin (Fuchino et al. 2013), suggesting that cell wall itself is not affected in the *filP* knockout mutant. The intriguing lace-like filament formation of FilP *in vitro* together with the *in vivo* localisation of FilP to a sub-apical filamentous network are consistent with a mechanical role for FilP in strengthening the hyphal wall next to the active growth zone. This role is in fact is very similar to the role of eukaryote IFs supporting cellular shape or in the case of lamins, supporting the shape of the nucleus.

Interestingly, *filP* expression is controlled during the cell cycle of *Streptomyces* by two transcription factors, WhiA and WhiB, which have a common regulon in *Streptomyces* (Bush et al. 2013, 2015, 2016). The activities of both WhiA and WhiB positively regulate the expression of *ftsZ* and repress transcription of *filP* after hyphal growth comes to a halt and a synchronous cell division is initiated leading to spore formation (Fig. 6.8e). Hence, WhiA and WhiB restrict *filP* expression to the stages of active polar growth and switch off *de novo* FilP production at the time of sporulation. In addition to regulation at the level of transcription, FilP filamentation might also be controlled dynamically by post-translational modifications. This is similar to eukaryote IFs that often undergo post translational modification including phosphorylation, acetylation and other covalent modifications (Snider and Omary 2014). FilP phosphorylation was documented at the late stages of *Streptomyces* development, coinciding with aerial development and sporulation (Manteca et al. 2011). This suggests that at the time of cessation of hyphal growth, in addition to no further transcription of *filP* mediated via repression by WhiA and WhiB, the existing FilP filaments might also disassemble in response to phosphorylation, similarly to vimentin disassembly during cellular movement at the leading edge of fibroblasts

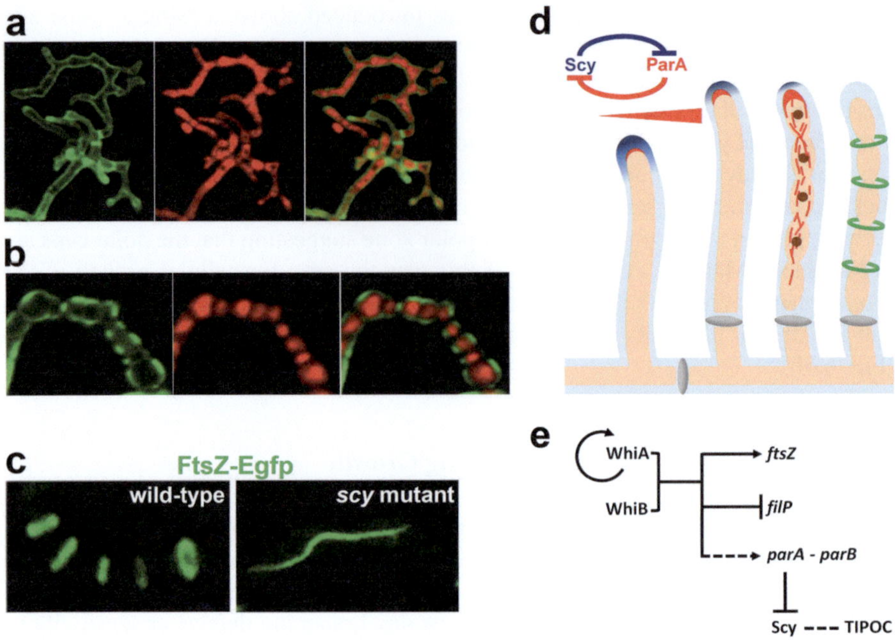

Fig. 6.8 Scy, the molecular scaffold component of the Tip Organising Centre (TIPOC), has a long-range effect on cell division during the transition from aerial growth to sporulation. (**a**) Lack of Scy affects polar growth and branching by the dispersal of DivIVA, producing uneven hyphal diameter and ectopic branches. (**b**) In the *scy* null mutant chromosome segregation and cell division are affected, producing spores with irregular DNA content and septation. (**a** and **b**) Fluorescent wheat germ agglutinin (WGA, *green*, *left*) marks the cell wall and propidium iodide (*red*, *middle*) stains the nucleoid. Overlayed images are also shown (*right*). For comparison, see wild-type development in Fig. 6.7. (**c**) FtsZ-EGFP rings forming a ladder in the aerial hyphae of wild-type *Streptomyces* (*left*) and a prominent FtsZ-EGFP cable runs alongside the lateral hyphal wall in the *scy* mutant (*right*). (**d**) The molecular switch from growth to division relies on the interaction between Scy and ParA. During filamentous growth, Scy (*blue*) orchestrates polar growth and recruits ParA (*red*) to the hyphal tips. Increasing ParA levels in the aerial hyphae offset the switch and promotes disassembly of Scy, hence polar growth will come to a halt. In the absence of Scy, ParA filamentation is initiated, which, in turn, organises the *ori* bound ParB (*brown*) and the alignment of chromosomes along the hyphae. (**e**) Regulatory network controlling both FilP and Scy. Transcription of *filP* is switched off by the transcriptional regulators, WhiA and WhiB when hyphal growth stops (Bush et al. 2013, 2016). At the same time, both *ftsZ* and *parA* transcriptions are upregulated in readiness for chromosome segregation and division. The elevated ParA levels de-stabilise the Scy assemblies, which lead to cessation of growth. Direct interactions (*solid lines*) and indirect effects (*dashed lines*) are shown

(Helfand et al. 2011). FilP was also shown to be N-acetylated, although nothing is known about the role of this modification in FilP dynamics (Hesketh et al. 2002).

Scy Is a Molecular Organiser at the Hyphal Tip

Although Scy is not essential for growth, lack of Scy severely affects polar growth and branching in *Streptomyces*. Scy and DivIVA interact directly, building a multi-protein scaffold, the TIPOC, for controlling tip extension in *Streptomyces* (Holmes et al. 2013). As both Scy and DivIVA are key components of this polar scaffold, changing their cellular levels dramatically affects the establishment of new polarity centres. Both lack of Scy and depletion of DivIVA lead to the increase in new cell wall synthesis occurring at ectopic lateral sites promoting aberrant branching, suggesting that both Scy and DivIVA are important for the integrity of the polar assembly. When any of its structural components is removed, the complex "disperses", producing ectopic new polarity centres. Intriguingly, dispersal of the tip-complex in wild-type *S. coelicolor* was proposed to provide a mechanism for new tip emergence during branching (Richards et al. 2012). Live monitoring of actively growing hyphal tips showed that from the large DivIVA-EGFP patches at the hyphal tip small DivIVA-EGFP spots were "left behind" at the hyphal wall, which later on initiated branching at the sites marked by the spots. This is an attractive proposition for the mechanism of new branch formation, but it does not exclude the possibility of an additional, alternative mechanism for *de novo* assembly of new polarity centres. Interestingly, DivIVA phosphorylation by the kinase AfsK was proposed to contribute to the disassembly and re-cycling of DivIVA in *Streptomyces* (Hempel et al. 2012; Saalbach et al. 2013). Lack of AfsK reduces new branch formation while its overexpression induces DivIVA disassembly from existing tips and the re-location of DivIVA, promoting new, multiple DivIVA assemblies at lateral hyphal sites (Hempel et al. 2012).

Two growth phases precede sporulation, the first generates the vegetative mycelium and is followed by a second growth phase when aerial hyphae emerge into the air (Fig. 6.7). Polar growth governs both phases, however aerial hyphae tend to branch much less than vegetative hyphae, and the tip-proximal part of the aerial filament that undergoes sporulation septation is always un-branched in wild-type *S. coelicolor*. Although nothing is known about the TIPOC components, DivIVA or Scy, during aerial growth, the fact that branching patterns of the *vegetative* filaments are so sensitive to changes of both DivIVA and Scy levels suggest that the reduced branching of the *aerial* hyphae could reflect changes in the levels of DivIVA and Scy, or indeed changes in not yet identified TIPOC or associated protein components.

Predictably, increasing the levels of DivIVA or Scy promotes new branch formation, presumably by the assembly of new polarity centres where the respective partner is also recruited (Flardh 2003; Holmes et al. 2013). Interestingly, while Scy deletion or DivIVA depletion have very similar effects on growth, there are differences when these two TIPOC components are overproduced. Elevated DivIVA levels first lead to the "tennis-racket-like" expansion of existing tips followed by ectopic tip emergence (Flardh 2003; Hempel et al. 2008), whereas Scy overproduction initially generates multiple, new, lateral tips followed by a "hook on hook" growth pattern (Holmes et al. 2013). In both cases, although multiple new tips are generated, most of them are not capable of prolonged growth due to the limiting availability of components of cell wall synthetic machinery and, possibly, the TIPOC partner. One possible explanation for the two phenotypes involves differences in the way in which DivIVA and Scy forms supramolecular self-assemblies, based on their different shape, size and physico-chemical characteristics. Excess DivIVA might be primarily attracted to existing assemblies leading to the tennis-racket tips. In contrast, an upsurge in the production of the long, rope-like Scy in the cytoplasm of the hyphal filament might instead nucleate multiple assemblies close to the sites of Scy production, rather than travelling to existing tips. In addition, DivIVA-mediated tennis-racket tips might indicate that it is DivIVA, not Scy, that directly interacts with cell wall synthetic or lytic enzymes and their recruitment in abundance to the tips leads to isotropic tip expansion (Hempel et al. 2008).

Is Synchronous Cell Division Orchestrated from the TIPOC?

Intriguingly, Scy, the molecular scaffold located at hyphal tips seems to have a long range effect along the aerial filament, which is committed to fully completed cell division during sporulation. Fewer aerial hyphae transform into spore chains in the *scy* knockout mutant and those that do, undergo irregular septation producing spore compartments with uneven DNA contents (Fig. 6.8a and b; Holmes et al. 2013). This suggests, that lack of the TIPOC component, Scy, affects, directly or indirectly, chromosome organisation and division. In the following, we will discuss the proposed mechanism for the long range influence of the TIPOC on the synchronous division event along the hyphal filaments. Accordingly, on the one hand the integrity of TIPOC during the growth phase is not only important for polar growth but also for the recruitment of cellular components that later control the synchronous division. On the other hand, the disassembly of TIPOC during the switch from polar growth to division is a molecular cue for initiating changes in chromosome organisation, and in turn, prompts the correct positioning of synchronous division.

Transition of the multi-genomic hyphae into uni-genomic pre-spore compartments is a highly-orchestrated but poorly-understood process that begins with the cessation of growth (Flardh and Buttner 2009; Jakimowicz and van Wezel 2012; McCormick and Flardh 2012). This is followed by the re-organisation of the chromosomes and the appearance of regularly-positioned FtsZ rings (Fig. 6.7), generating the so-called FtsZ "ladder" that marks the positions of the division machinery, the divisome, where cross-walls are built. Once septation is completed, each pre-spore compartment carries a single chromosome. During filamentous growth cell division is suspended, therefore, it is not surprising that components of the divisome, including the key division protein, FtsZ, are not essential for growth in *Streptomyces*. Knockout mutants of divisome components affect sporulation-

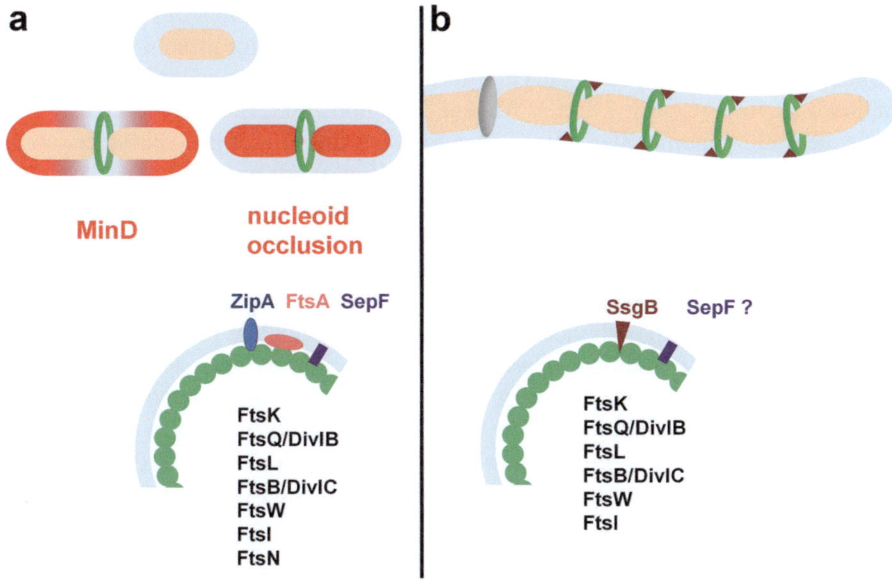

Fig. 6.9 Controlling FtsZ-ring formation. (**a**) Polymers of FtsZ at mid-cell of the rod-shaped *E. coli* or *B. subtilis* rely on both nucleoid occlusion where proteins that are associated with the chromosomes (*red*) block FtsZ (*green*) assembly at sites that are occupied by DNA and the negative regulation of FtsZ assembly by MinD (*red*), which interacts with the division inhibitor MinC and sequesters it to the cell poles, allowing FtsZ polymerisation at the mid-cell position (Wu and Errington 2003; Lutkenhaus 2012). (**b**) In the multi-genomic filaments of *Streptomyces*, the co-ordinated polymerisation of FtsZ at regular intervals during sporulation requires a complex organisation that is not yet fully understood. *S. coelicolor* has no obvious MinC homologue nor a nucleoid occlusion mechanism, in the strictest sense. Instead, FtsZ positioning is proposed to be under positive control by an FtsZ partner protein SsgB (Willemse et al. 2011). Comparison of the divisome compositions demonstrates that *Streptomyces* lacks some of the key proteins known for FtsZ stabilisation on the membrane (Jakimowicz and van Wezel 2012; McCormick and Flardh 2012)

septation but not hyphal growth (McCormick et al. 1994; Schwedock et al. 1997; Jakimowicz and van Wezel 2012; McCormick and Flardh 2012). This is in stark contrast to uni-genomic bacteria, where deletion of components of the divisome is lethal. Composition of the divisome itself is also different in *Streptomyces* (Fig. 6.9), lacking either FtsA or ZipA that are important for the membrane association of the FtsZ ring in *E. coli* or *B. subtilis* (Hale and de Boer 1997; Raychaudhuri 1999; Donachie et al. 1979; Pichoff and Lutkenhaus 2002; Szwedziak et al. 2012; van den Ent and Löwe 2000).

The key question in bacterial cell division is how the GTP-induced polymerisation of the bacterial tubulin-like protein, FtsZ is orchestrated in the three dimensional space of a bacterial cell. As summarised in Fig. 6.9a, in rod-shaped *E. coli* or *B. subtilis*, the formation of a ring of FtsZ polymers at mid-cell relies on two mechanisms, each involving negative regulation of FtsZ assembly: (1) nucleoid occlusion, where proteins associated with the chromosomes block FtsZ assembly at sites that are occupied by DNA (see Chap. 9), or (2) the negative regulation of FtsZ assembly by the ParA-like ATPase, MinD, in complex with the inhibitor protein MinC; these proteins, when sequestered to the cell poles, allow FtsZ polymerisation only at the mid-cell position (Chaps. 7 and 9; Wu and Errington 2003; Lutkenhaus 2012).

In the multi-genomic filaments of *Streptomyces*, the co-ordinated polymerisation of FtsZ at regular intervals during sporulation (Fig. 6.9b) must require a complex organisation that is not yet fully understood. Importantly, a significant increase in cellular FtsZ levels is required for the synchronous septation of aerial hyphae, which is achieved via the activation of one of the *ftsZ* promoters by the transcription factors, WhiA and WhiB (Bush et al. 2013, 2015, 2016). *S. coelicolor* has no obvious MinC homologues although there are several ParA/MinD like proteins that are not yet characterised. In addition, sporulation septa formation precedes full chromosome segregation (Flardh and Buttner 2009), so a nucleoid occlusion mechanism, in the strictest sense, is unlikely. On the other hand, FtsZ positioning in *S. coelicolor* is proposed to be under positive control by an FtsZ partner protein SsgB (Fig. 6.9b), which in turn depends on SsgA (Willemse et al. 2011). SsgA and SsgB are small, 130–145 amino acid long, proteins that are exclusive to the morphologically more complex Actinomycetales (Noens et al. 2005; Traag and van Wezel 2008), however, it is not understood what controls the cellular localisation of SsgA and SsgB. Another actinomycete specific protein, CrgA, also affects FtsZ assembly but its link to cell division is not yet fully addressed (Del Sol et al. 2006; Plocinski et al. 2011).

The lack of explicit nucleoid occlusion does not exclude the possibility that nucleoid organisation could provide positional cues for the FtsZ-ladders in *Streptomyces*. Indeed, mutations that affect chromosome organisation and placement in hyphal filaments also affect septum positioning during sporulation (Jakimowicz and van Wezel 2012). For example, knockout mutations of *parA* not only affect chromosome segregation but also septum positioning. ParA belongs to the MinD/ParA family of ATPases and its ATP-dependent polymerisation into filaments is essential for the correct positioning of its partner protein ParB, which, in turn, associates with the *ori* proximal *parS* sites on the chromosome to organise chromosome segregation during sporulation (Jakimowicz et al. 2002, 2007).

Remarkably, the chromosome partitioning protein, ParA interacts with the IF-like TIPOC component, Scy and the dynamic nature of this interaction was proposed to contribute to the switch from polar growth to the completion of cell division in the aerial hyphae of *S. coelicolor* (Fig. 6.8d; Ditkowski et al. 2013). Both Scy and ParA form assemblies for their respective roles, Scy assembles at the hyphal tips and organises the TIPOC for polar growth (Holmes et al. 2013), whereas ParA forms filamentous structures nucleated at the hyphal tips prior to sporulation (Jakimowicz et al. 2002, 2007). Interestingly, the Scy-ParA interaction destabilises the assemblies of both Scy and ParA (Fig. 6.8d), and this mutually negative regulation of their respective assemblies generates a possible molecular switch from growth to cell division. Such a molecular switch that relies on two components, each negatively regulating the other, is widely established in nature with some of the best examples being the switch from the lytic to lysogenic cycle after lambda phage infection (Ptashne 2011) and the synthetic toggle switch (Gardner et al. 2000).

Here, instead of a transcriptional switch, Scy and ParA negatively regulate each other's assembly, hence, the activity of their partner protein.

Depending on the protein levels of Scy and ParA, either polar growth or the initiation of chromosome organisation followed by sporulation ensues. During filamentous growth, Scy is actively produced and becomes part of TIPOC, while ParA levels are very low due to transcription of *parA* from a weak promoter. At this stage, ParA is recruited to the tips by Scy, however, the low levels of ParA are not sufficient to affect the integrity of TIPOC and polar growth continues. In addition, high Scy levels in TIPOC block ParA polymerisation into filaments, thereby blocking concerted chromosome segregation during the growth phase. In contrast, in the aerial hyphae ParA expression is markedly upregulated due to elevated transcription from a second promoter, parABp(2) (Jakimowicz et al. 2006) which offsets the switch. Increasing ParA levels will promote disassembly of Scy and the destabilisation of TIPOC, hence polar growth will come to a halt. Concomitantly, disintegration of the TIPOC will also abolish the negative effect of Scy on ParA polymerisation. Hence, nucleated from the hyphal tip, ParA filamentation is initiated, which, in turn, organises the *ori* bound ParB and the alignment of chromosomes along the hyphae (Fig. 6.8d).

Interestingly, two non-sporulating developmental mutants of *S. coelicolor*, *whiA* and *whiB,* generate unusually long aerial hyphae, suggesting that the TIPOC destabilising morphogen is dependent on *whiA* and *whiB*. Indeed, the *parABp(2)* promoter is dependent, albeit indirectly, on both *whiA* and *whiB* genes (Jakimowicz et al. 2006; Bush et al. 2013, 2015, 2016). Intriguingly *filP* transcription is derepressed in the *whiA* and *whiB* mutants (Bush et al. 2013, 2015, 2016) suggesting that the transcription factors WhiA and WhiB affect polar growth and the switch from growth to division by affecting two components of the TIPOC, Scy and FilP (Fig. 6.8e). Accordingly, WhiA and WhiB indirectly activate *parA* transcription in the aerial hyphae, which in turn, leads to the ParA-mediated destabilisation of Scy and the TIPOC, leading to the cessation of polar growth. Simultaneously, *filP* expression is directly repressed by WhiA and WhiB in the aerial hyphae, as presumably, no further mechanical support is required in the absence of cell wall extension.

Recently ParA has been shown to interact with DivIVA in *Mycobacterium* (Ginda et al. 2013) and actinomycete DivIVA-ParB complexes were demonstrated when expressed in the heterologous host, *E. coli* (Donovan et al. 2012). In addition, the dynamic interaction between the polar coiled-coil protein TipN and ParA in *C. crescentus* (Huitema et al. 2006; Lam et al. 2006; Ptacin et al. 2010), an example outside of the *Actinomycetales*, suggests that chromosome organisation by polar scaffolds is widespread amongst bacteria.

ParA filamentation in the aerial hyphae of *S. coelicolor* is followed by FtsZ ring formation at regular intervals leading to evenly spaced pre-spore compartments with a single chromosome. In addition to the connection between polar scaffolds and chromosome organisation via the dynamic ParA and Scy interaction, further components of the division machinery might also be organised from the tips in *Streptomyces*. This is likely because the irregular septation and DNA segregation defects in the pre-spores of the *scy* null mutant are more severe than those in the *parA* mutant (Jakimowicz et al. 2007; Holmes et al. 2013). Moreover, many of the aerial hyphae of the *scy* mutant failed to transform into spore chains and in these hyphae, instead of the regular FtsZ rings, FtsZ cables, often parallel to the lateral wall of the aerial hyphae, were observed (Fig. 6.8c; Holmes et al. 2013). This strongly suggests that further components of both chromosome organisation and FtsZ assembly are dependent, directly or indirectly, on the integrity of the TIPOC.

Is There a Case for an Extended IF-Like Family?

The previous examples of bacterial coiled-coil proteins, such as crescentin, FilP and Scy, provided us with sufficient experimental evidence for their *in vitro* and *in vivo* assemblies into filaments and scaffolds. Their functions to support bacterial cell shape and the mechanical properties of bacterial cells together with integrating the cytoskeleton for growth and division are all functions associated with intermediate filaments. Although further investigation into the structure and pathway of their assembly is paramount, based on their biochemical and biophysical characteristics and their cellular role, the term, bacterial IF-like proteins, is appropriate.

However, it is important to emphasize that all coiled-coil proteins are likely to have an innate propensity to form ordered assemblies, so this feature, on its own, does not identify a filament forming coiled-coil protein as being IF-like. Indeed, the 81-amino acid long, bacterial coiled-coil protein ZapB forms filaments both *in vivo* and *in vitro* and, remarkably, the *in vitro* assembly of ZapB resembles the paracrystalline assemblies of lamin and the striated filaments of FilP (Ebersbach et al., 2008). Nevertheless, ZapB is not an IF-like protein. Furthermore, synthetic, self-assembling fibers (SAFs) can be generated from short, 24–28 amino acid long, oligopeptides (Woolfson 2005) producing filaments of ~40 μm in length and ~80 nm in diameter (Sharp et al. 2012). Precisely because not all coiled-coil proteins are IFs, the strict definition of the sub-domain organisation of the IF rod domains has been established (Herrmann et al. 2007; Parry et al. 2007). In this review, it is argued that the discovery of IF-like proteins in bacteria makes a case for a revision of this definition. This concept is also supported by the IF-like proteins identified in the simple protozoan *Dyctiostelium discoideum* and *Trypanosome brucei* (Batsios et al. 2012; Dubois et al. 2012; Kruger et al. 2012) and even in plants (Ciska and Moreno Diaz de la Espina 2014). However, the proposed changes for the extended IF-like family including the relaxation of the length and subdomain organisation of the rod-domain comes with a major challenge that is how to classify proteins of the IF-like family in the absence of the straightforward sub-domain based classification. The definition of the IF-like family will have to be based on both structural and functional characteristics that allow the inclusion of proteins with *bona fide* IF functions and, the same time, places a constraint on the large number of proteins with the propensity for coiled-coil assemblies.

Bacterial Coiled-Coil Rich Proteins

In the last section of this chapter, bacterial examples, which currently lack sufficient experimental evidence to fully support their classification into the IF-like family, are described. Some of these are unlikely to be IF-like proteins and others might be classified as such at a later stage, when further experimental evidence will be available.

The helical shape of *Helicobacter* species, residents of the human gut flora, depends on numerous filament-forming, Coiled-coil rich proteins that prompted the designation of CCRP for the group of these molecular organisers (Waidner et al. 2009; Specht et al. 2011). The CCRP proteins are highly variable amongst the different *Helicobacter* species. Deletion of any of the four CCRPs (Ccrp58, Ccrp59, Ccrp1042 and Ccrp1043) of *Helicobacter pylori* resulted in cells that lost their helical shape to a varying degree, but this phenotype was highly dependent on the particular genetic background of the strains used. All four proteins were shown to form filaments *in vitro* although their *in vivo* localisation has yet to be fully explored. Regular, patchy Ccrp59 foci were observed along the helical cells prompting the assumption that Ccrp59 forms helical filaments *in vivo* (Waidner et al. 2009). Similarly to crescentin in *C. crescentus,* the helical Ccrp59 filaments are proposed to locally affect the rate of cell wall synthesis. Although all four proteins contain extended α-helical sub-domains likely to form coiled-coil oligomers, the number and length of these sub-domains vary and show no resemblance to any of the previously discussed bacterial IF-like proteins. The helical shape of *Helicobacter* is important for its colonisation of the gastric mucosa (Sycuro et al. 2010) and the Ccrp proteins were indicated in the virulence of this human pathogenic bacterium (Schatzle et al. 2015).

Just as Scy is specific to *Streptomyces*, the essential coiled-coil protein RsmP (Rod-shape morphology protein) is specific to *Corynebacterium* (Fiuza et al. 2010). RsmP is 291 amino acids long but its rod-domain (116 amino acids) is very short and consists of two sub-domains of 41 and 52 amino acids. *Corynebacterium*, a unigenomic member of the *Actinomycetales*, exhibits polar growth that is governed by the polarity determinant, DivIVA (Letek et al. 2008a, b). RsmP was originally identified as a highly expressed protein during DivIVA depletion. Reduced RsmP levels lead to changes in cell shape from club-shape to spherical cells, whilst overproduction generates extenuated club-shaped cells. When expressed from a strong, but not its native, promoter, RsmP-GFP was shown to form curved filamentous structures *in vivo* along the long axis of *Corynebacterum* cells (Fiuza et al. 2010). Interestingly, RsmP activity is regulated by phosphorylation, a posttranslational modification often employed by controlling the dynamics of eukaryote IFs. RsmP is phosphorylated by PknA and PknL and this phosphorylation affects its localisation. A GFP fusion to an RsmP mutant that mimics the non-phosphorylated form localizes as long filaments, similar to the wild-type RsmP-GFP, whereas, the phosphomimetic RsmP-GFP is restricted to the poles, where DivIVA is located (Fiuza et al. 2010).

This suggest that RsmP and DivIVA might form a dynamic polar complex in *Corynebacterium*, similarly to the TIPOC in *Streptomyces*.

As the curved crescent shape of *C. crescentus* is supported by the IF-like protein crescentin, the spiral shaped spirochetes contain remarkable ribbon-like internal filaments that can be visualised using cryo-electron-tomography (Izard et al. 2004) or transmission electron microscopy (You et al. 1996). However, neither CfpA of *Treponema* nor Scc of *Leptospira* are likely to belong to the IF-like proteins and they are both implicated in nucleoid organisation as they interact with DNA (Izard 2006; Mazouni et al. 2006).

Finally, both AglZ and FrzS are key proteins for gliding motility of *Myxococcus xanthus*. Both proteins contain extensive α-helical sub-domains and, presumably via coiled-coil association, generate striated assemblies when overproduced in the heterologous host *E. coli* (Mauriello et al. 2009; Mignot et al. 2007; Ward et al. 2000; Yang et al. 2004). AglZ has been shown to interact with the MreB filaments and form a motility complex that stays fixed compared to the external surface propelling the cells forward in a corkscrew-like movement (Mignot et al. 2007). However, neither AglZ nor FrzS have been documented to have IF characteristics, apart from the striated assemblies, that are reminiscent of those exhibited by IF lamins. However, as discussed above, striated assemblies can be the result of a wide variety of coiled-coil associations, and therefore not specific to IFs.

Conclusions

The identification of the first bacterial IF-like protein in 2003 represented a major breakthrough, completing the list of bacterial orthologues of all three main cytoplasmic components, microtubules, actin filaments and intermediate filaments, of the eukaryote filamentous cytoskeleton. IF-like proteins are widespread in bacteria, but only extensively studied in a handful of organisms, including the crescent-shaped, aquatic bacterium *C. crescentus,* the filamentous soil-dwelling *Streptomyces* (best known for its propensity to produce a wide range of secondary metabolites with anti-bacterial, anti-cancer or anti-fungal properties), the club-shaped *Corynebacterium* and the helical, often pathogenic, *Helicobacter*, which can colonise the human gut flora.

There are several advantages of studying bacterial IF-like proteins over their eukaryote counterparts. Overall, bacteria are more tractable and there are excellent genetic tools for their manipulation. Moreover, high resolution microscopy technologies, such as 3D-SIM, PALM, STORM together with cryo-electron tomography, have been developed and are available for the characterisation of mutant phenotypes and assessing the cellular localisation of these filamentous proteins. Comparing IF-like proteins and their cellular function in model systems of bacteria and eukaryotes will enable us to identify the core characteristics of their structure, assemblies, dynamics and cellular function.

The following questions must be addressed through future research:

1. What is the role of the sub-domain organisation within IF rod domains? Evidently, some of the bacterial proteins, FilP and Scy of *Streptomyces*, RsmP of *Corynebacterium* and the Ccrp proteins of *Helicobacter* do not comply with the strict sub-domain organisation defined for metazoan IFs, which includes Coil 1A, Coil 1B and Coil 2 separated by linkers. Deletion of some of these Coil IF sub-domains or linkers did not affect elementary dimer formation but abolished their ability to form higher order assemblies. This confirms that the sub-domain organisation is essential for the further assemblies of elementary dimers. However, the bacterial protein, FilP can readily assemble into filaments that are very similar to those of eukaryote IFs in spite of the fact that it has markedly different sub-domain organisation. In order to assess the effect of the sub-domain organisation on the overall protein structure, it is paramount that in the near future we acquire high resolution structural information of the various bacterial elementary dimers and their assembly pathway. It is conceivable, that the role of the sub-domain organisation amongst eukaryote IFs is to act as "cogwheels" for securing a "lock-in" mechanism for the correct *lateral* alignment between elementary dimers. For this lock-in mechanism, the principle of cogwheels are important but not the actual location or the number of cogwheels. Therefore, it is plausible that whilst the bacterial IF-like proteins use different sub-domain organisations, they employ the same lock-in mechanism for their assembly.
2. Further supporting evidence is needed to confirm the nature of the *in vivo* assemblies in their cellular environment in bacteria and to corroborate congruence between the different assemblies, such as filaments or lace-like networks found *in vitro* and the cables or the extended filamentous network detected *in vivo*.
3. Beside structural information, further investigations are needed into the cellular dynamics of bacterial IF-like proteins, addressing questions such as what is responsible for the nucleation of the emerging assemblies and conversely, what controls the disassembly of the filamentous networks. Does this latter involve post-translational modifications, specifically protein phosphorylation?
4. The evolutionary link between bacterial and eukaryote IF-like proteins is currently not understood. Bacterial and eukaryote IFs could originate from a common coiled-coil rich ancestor protein via convergent evolution. Alternatively eukaryote IFs may have evolved from their bacterial counterparts. In addition, it is also conceivable that the bacterial IF-like proteins arose by horizontal gene transfer from metazoan IFs. *Helicobacter* species colonise the gastrointestinal tracts of animals and decomposing animals could allow the transfer of genetic material to the soil dwelling *Streptomyces* or aquatic *Caulobacter* species. There is evidence for ongoing evolution in bacteria. Both in *Streptomyces* and *Helicobacter* two adjacent genes encode IF-like proteins, which are likely to have arisen via gene-duplication. In *Streptomyces*, the two paralogous proteins, FilP and Scy, have distinct cellular functions suggesting that they derived via convergent evolution.
5. How can we refine the definition of the IF-like protein family? The discovery of bacterial IF-like proteins is a compelling reason to consider extending the meta-

zoan IF family into a wider, IF-like protein family. This extended family is proposed to be defined by a set of shared biochemical, biophysical and mechanical properties with similar *in vitro* and *in vivo* higher order assemblies. These assemblies are built from a basic, parallel, coiled-coil elementary dimer and rely on the same principles for their further lateral and longitudinal associations. Beyond bacterial and metazoan members, the newly defined IF-like family should also encompass proteins that share the above defining characteristics from unicellular eukaryotes, such as the protozoa *Dictyostelium* and *Trypanosoma*, the insect lamins and the lamin-like, plant nuclear matrix constituent proteins.

References

Aebi U, Cohn J, Buhle L, Gerace L (1986) The nuclear lamina is a meshwork of intermediate-type filaments. Nature 323:560–564

Astbury WT, Woods HJ (1930) The X-ray interpretation of the structure and elastic properties of hair keratin. Nature 126:913–914

Astbury WT, Woods HJ (1933) X-ray studies on the structure of hair, wool and related fibres II: the molecular structure and elastic properties of hair keratin. Phil Trans Roy Soc Lond A 232:333–394

Ausmees N, Kuhn JR, Jacobs-Wagner C (2003) The bacterial cytoskeleton: an intermediate filament-like function in cell shape. Cell 115:705–713

Aziz A, Hess JF, Budamagunta MS, Fitzgerald PG, Voss JC (2009) Head and rod 1 interactions in vimentin: identification of contact sites, structure, and changes with phosphorylation using site-directed spin labeling and electron paramagnetic resonance. J Biol Chem 284:7330–7338

Aziz A, Hess JF, Budamagunta MS, Voss JC, Fitzgerald PG (2010) Site-directed spin labeling and electron paramagnetic resonance determination of vimentin head domain structure. J Biol Chem 285:15278–15285

Aziz A, Hess JF, Budamagunta MS, Voss JC, Kuzin AP, Huang YJ, Xiao R, Montelione GT, Fitzgerald PG, Hunt JF (2012) The structure of vimentin linker 1 and rod 1B domains characterized by site-directed spin-labeling electron paramagnetic resonance (SDSL-EPR) and X-ray crystallography. J Biol Chem 287:28349–28361

Bagchi S, Tomenius H, Belova LM, Ausmees N (2008) Intermediate filament-like proteins in bacteria and a cytoskeletal function in *Streptomyces*. Mol Microbiol 70:1037–1050

Batsios P, Peter T, Baumann O, Stick R, Meyer I, Graf R (2012) A lamin in lower eukaryotes? Nucleus 3:237–243

Ben-Harush K, Wiesel N, Frenkiel-Krispin D, Moeller D, Soreq E, Aebi U, Herrmann H, Gruenbaum Y, Medalia O (2009) The supramolecular organization of the *C. elegans* nuclear lamin filament. J Mol Biol 386:1392–1402

Ben-Yehuda S, Rudner DZ, Losick R (2003) RacA, a bacterial protein that anchors chromosomes to the cell poles. Science 299:532–536

Block J, Schroeder V, Pawelzyk P, Willenbacher N, Koster S (2015) Physical properties of cytoplasmic intermediate filaments. Biochim Biophys Acta 1853:3053–3064

Bramkamp M, Emmins R, Weston L, Donovan C, Daniel RA, Errington J (2008) A novel component of the division-site selection system of *Bacillus subtilis* and a new mode of action for the division inhibitor MinCD. Mol Microbiol 70:1556–1569

Brana AF, Manzanal MB, Hardisson C (1982) Mode of cell-wall growth of *Streptomycesantibioticus*. FEMS Microbiol Lett 13:231–235

Briley K Jr, Prepiak P, Dias MJ, Hahn J, Dubnau D (2011) Maf acts downstream of ComGA to arrest cell division in competent cells of *B. subtilis*. Mol Microbiol 81:23–39

Brown PJ, De Pedro MA, Kysela DT, Van der Henst C, Kim J, de Bolle X, Fuqua C, Brun YV (2012) Polar growth in the Alphaproteobacterial order *Rhizobiales*. Proc Natl Acad Sci U S A 109:1697–1701

Bush MJ, Bibb MJ, Chandra G, Findlay KC, Buttner MJ (2013) Genes required for aerial growth, cell division, and chromosome segregation are targets of WhiA before sporulation in *Streptomyces venezuelae*. MBio 4:e00684–e00613

Bush MJ, Tschowri N, Schlimpert S, Flardh K, Buttner MJ (2015) c-di-GMP signalling and the regulation of developmental transitions in streptomycetes. Nat Rev Microbiol 13:749–760

Bush MJ, Chandra G, Bibb MJ, Findlay KC, Buttner MJ (2016) Genome-wide ChIP-seq analysis shows that WhiB is a transcription factor that co-controls its regulon with WhiA to initiate developmental cell division in *Streptomyces*. MBio 7:e00523–e00516

Cabeen MT, Charbon G, Vollmer W, Born P, Ausmees N, Weibel DB, Jacobs-Wagner C (2009) Bacterial cell curvature through mechanical control of cell growth. EMBO J 28:1208–1219

Cabeen MT, Herrmann H, Jacobs-Wagner C (2011) The domain organization of the bacterial intermediate filament-like protein crescentin is important for assembly and function. Cytoskeleton (Hoboken) 68:205–219

Cabeen MT, Jacobs-Wagner C (2010) The bacterial cytoskeleton. Annu Rev Genet 44:365–392

Cameron TA, Anderson-Furgeson J, Zupan JR, Zik JJ, Zambryski PC (2014) Peptidoglycan synthesis machinery in *Agrobacterium tumefaciens* during unipolar growth and cell division. MBio 5:e01219–e01214

Cameron TA, Zupan JR, Zambryski PC (2015) The essential features and modes of bacterial polar growth. Trends Microbiol 23:347–353

Chandra G, Chater KF (2014) Developmental biology of *Streptomyces* from the perspective of 100 actinobacterial genome sequences. FEMS Microbiol Rev 38:345–379

Chang L, Goldman RD (2004) Intermediate filaments mediate cytoskeletal crosstalk. Nat Rev Mol Cell Biol 5:601–613

Charbon G, Cabeen MT, Jacobs-Wagner C (2009) Bacterial intermediate filaments: *in vivo* assembly, organization, and dynamics of crescentin. Genes Dev 23:1131–1144

Chernyatina AA, Guzenko D, Strelkov SV (2015) Intermediate filament structure: the bottom-up approach. Curr Opin Cell Biol 32:65–72

Chernyatina AA, Nicolet S, Aebi U, Herrmann H, Strelkov SV (2012) Atomic structure of the vimentin central alpha-helical domain and its implications for intermediate filament assembly. Proc Natl Acad Sci U S A 109:13620–13625

Chernyatina AA, Strelkov SV (2012) Stabilization of vimentin coil2 fragment via an engineered disulfide. J Struct Biol 177:46–53

Chou YH, Bischoff JR, Beach D, Goldman RD (1990) Intermediate filament reorganization during mitosis is mediated by p34cdc2 phosphorylation of vimentin. Cell 62:1063–1071

Ciska M, Moreno Diaz de la Espina S (2014) The intriguing plant nuclear lamina. Front Plant Sci 5:166

Colakoglu G, Brown A (2009) Intermediate filaments exchange subunits along their length and elongate by end-to-end annealing. J Cell Biol 185:769–777

Coleman TR, Lazarides E (1992) Continuous growth of vimentin filaments in mouse fibroblasts. J Cell Sci 103(Pt 3):689–698

Crick FHC (1953) The packing of alpha-helices – simple coiled-coils. Acta Crystallogr 6:689–697

Daniel RA, Errington J (2003) Control of cell morphogenesis in bacteria: two distinct ways to make a rod-shaped cell. Cell 113:767–776

Del Rizzo PA, Bi Y, Dunn SD, Shilton BH (2002) The "second stalk" of *Escherichia coli* ATP synthase: structure of the isolated dimerization domain. Biochemistry 41:6875–6884

Del Sol R, Mullins JG, Grantcharova N, Flardh K, Dyson P (2006) Influence of CrgA on assembly of the cell division protein FtsZ during development of *Streptomyces coelicolor*. J Bacteriol 188:1540–1550

Ditkowski B, Holmes N, Rydzak J, Donczew M, Bezulska M, Ginda K, Kedzierski P, Zakrzewska-Czerwinska J, Kelemen GH, Jakimowicz D (2013) Dynamic interplay of ParA with the polarity protein, Scy, coordinates the growth with chromosome segregation in *Streptomyces coelicolor*. Open Biol 3:130006

Dittmer TA, Misteli T (2011) The lamin protein family. Genome Biol 12:222

Dominguez-Escobar J, Chastanet A, Crevenna AH, Fromion V, Wedlich-Soldner R, Carballido-Lopez R (2011) Processive movement of MreB-associated cell wall biosynthetic complexes in bacteria. Science 333:225–228

Donachie WD, Begg KJ, Lutkenhaus JF, Salmond GP, Martinez-Salas E, Vincente M (1979) Role of the ftsA gene product in control of *Escherichia coli* cell division. J Bacteriol 140:388–394

Donovan C, Sieger B, Kramer R, Bramkamp M (2012) A synthetic *Escherichia coli* system identifies a conserved origin tethering factor in Actinobacteria. Mol Microbiol 84:105–116

dos Santos VT, Bisson-Filho AW, Gueiros-Filho FJ (2012) DivIVA-mediated polar localization of ComN, a posttranscriptional regulator of *Bacillus subtilis*. J Bacteriol 194:3661–3669

Dubois KN, Alsford S, Holden JM, Buisson J, Swiderski M, Bart JM, Ratushny AV, Wan Y, Bastin P, Barry JD, Navarro M, Horn D, Aitchison JD, Rout MP, Field MC (2012) NUP-1 Is a large coiled-coil nucleoskeletal protein in trypanosomes with lamin-like functions. PLoS Biol 10:e1001287

Ebersbach G, Galli E, Moller-Jensen J, Lowe J, Gerdes K (2008) Novel coiled-coil cell division factor ZapB stimulates Z ring assembly and cell division. Mol Microbiol 68:720–35

Edwards DH, Errington J (1997) The *Bacillus subtilis* DivIVA protein targets to the division septum and controls the site specificity of cell division. Mol Microbiol 24:905–915

Edwards DH, Thomaides HB, Errington J (2000) Promiscuous targeting of *Bacillus subtilis* cell division protein DivIVA to division sites in *Escherichia coli* and fission yeast. EMBO J 19:2719–2727

Eriksson JE, Brautigan DL, Vallee R, Olmsted J, Fujiki H, Goldman RD (1992a) Cytoskeletal integrity in interphase cells requires protein phosphatase activity. Proc Natl Acad Sci U S A 89:11093–11097

Eriksson JE, He T, Trejo-Skalli AV, Harmala-Brasken AS, Hellman J, Chou YH, Goldman RD (2004) Specific *in vivo* phosphorylation sites determine the assembly dynamics of vimentin intermediate filaments. J Cell Sci 117:919–932

Eriksson JE, Opal P, Goldman RD (1992b) Intermediate filament dynamics. Curr Opin Cell Biol 4:99–104

Esue O, Rupprecht L, Sun SX, Wirtz D (2010) Dynamics of the bacterial intermediate filament crescentin *in vitro* and *in vivo*. PLoS One 5:e8855

Eswaramoorthy P, Erb ML, Gregory JA, Silverman J, Pogliano K, Pogliano J, Ramamurthi KS (2011) Cellular architecture mediates DivIVA ultrastructure and regulates *min* activity in *Bacillus subtilis*. MBio 2

Fiuza M, Letek M, Leiba J, Villadangos AF, Vaquera J, Zanella-Cleon I, Mateos LM, Molle V, Gil JA (2010) Phosphorylation of a novel cytoskeletal protein (RsmP) regulates rod-shaped morphology in *Corynebacterium glutamicum*. J Biol Chem 285:29387–29397

Flardh K (2003) Essential role of DivIVA in polar growth and morphogenesis in *Streptomyces coelicolor* A3(2). Mol Microbiol 49:1523–1536

Flardh K, Buttner MJ (2009) *Streptomyces* morphogenetics: dissecting differentiation in a filamentous bacterium. Nat Rev Microbiol 7:36–49

Flardh K, Richards DM, Hempel AM, Howard M, Buttner MJ (2012) Regulation of apical growth and hyphal branching in *Streptomyces*. Curr Opin Microbiol 15:737–743

Frost A, Unger VM, de Camilli P (2009) The BAR domain superfamily: membrane-molding macromolecules. Cell 137:191–196

Fuchino K, Bagchi S, Cantlay S, Sandblad L, Wu D, Bergman J, Kamali-Moghaddam M, Flardh K, Ausmees N (2013) Dynamic gradients of an intermediate filament-like cytoskeleton are recruited by a polarity landmark during apical growth. Proc Natl Acad Sci U S A 110:E1889–E1897

Gardner TS, Cantor CR, Collins JJ (2000) Construction of a genetic toggle switch in *Escherichia coli*. Nature 403:339–342

Garner EC, Bernard R, Wang W, Zhuang X, Rudner DZ, Mitchison T (2011) Coupled, circumferential motions of the cell wall synthesis machinery and MreB filaments in *B. subtilis*. Science 333:222–225

Gerace L, Huber MD (2012) Nuclear lamina at the crossroads of the cytoplasm and nucleus. J Struct Biol 177:24–31

Ginda K, Bezulska M, Ziolkiewicz M, Dziadek J, Zakrzewska-Czerwinska J, Jakimowicz D (2013) ParA of *Mycobacterium smegmatis* co-ordinates chromosome segregation with the cell cycle and interacts with the polar growth determinant DivIVA. Mol Microbiol 87:998–1012

Goldberg MW, Fiserova J, Huttenlauch I, Stick R (2008) A new model for nuclear lamina organization. Biochem Soc Trans 36:1339–1343

Goldman RD, Cleland MM, Murthy SN, Mahammad S, Kuczmarski ER (2012) Inroads into the structure and function of intermediate filament networks. J Struct Biol 177:14–23

Grangeon R, Zupan JR, Anderson-Furgeson J, Zambryski PC (2015) PopZ identifies the new pole, and PodJ identifies the old pole during polar growth in *Agrobacterium tumefaciens*. Proc Natl Acad Sci U S A 112:11666–11671

Gruber M, Lupas AN (2003) Historical review: another 50th anniversary--new periodicities in coiled coils. Trends Biochem Sci 28:679–685

Guo M, Ehrlicher AJ, Mahammad S, Fabich H, Jensen MH, Moore JR, Fredberg JJ, Goldman RD, Weitz DA (2013) The role of vimentin intermediate filaments in cortical and cytoplasmic mechanics. Biophys J 105:1562–1568

Hale CA, de Boer PA (1997) Direct binding of FtsZ to ZipA, an essential component of the septal ring structure that mediates cell division in *E. coli*. Cell 88:175–185

Heitlinger E, Peter M, Haner M, Lustig A, Aebi U, Nigg EA (1991) Expression of chicken lamin B2 in *Escherichia coli*: characterization of its structure, assembly, and molecular interactions. J Cell Biol 113:485–495

Heitlinger E, Peter M, Lustig A, Villiger W, Nigg EA, Aebi U (1992) The role of the head and tail domain in lamin structure and assembly: analysis of bacterially expressed chicken lamin A and truncated B2 lamins. J Struct Biol 108:74–89

Helfand BT, Mendez MG, Murthy SN, Shumaker DK, Grin B, Mahammad S, AEBI U, Wedig T, Wu YI, Hahn KM, Inagaki M, Herrmann H, Goldman RD (2011) Vimentin organization modulates the formation of lamellipodia. Mol Biol Cell 22:1274–1289

Hempel AM, Cantlay S, Molle V, Wang SB, Naldrett MJ, Parker JL, Richards DM, Jung YG, Buttner MJ, Flardh K (2012) The Ser/Thr protein kinase AfsK regulates polar growth and hyphal branching in the filamentous bacteria *Streptomyces*. Proc Natl Acad Sci U S A 109:E2371–E2379

Hempel AM, Wang SB, Letek M, Gil JA, Flardh K (2008) Assemblies of DivIVA mark sites for hyphal branching and can establish new zones of cell wall growth in *Streptomyces coelicolor*. J Bacteriol 190:7579–7583

Herrmann H, Aebi U (2004) Intermediate filaments: molecular structure, assembly mechanism, and integration into functionally distinct intracellular Scaffolds. Annu Rev Biochem 73:749–789

Herrmann H, Bar H, Kreplak L, Strelkov SV, Aebi U (2007) Intermediate filaments: from cell architecture to nanomechanics. Nat Rev Mol Cell Biol 8:562–573

Herrmann H, Haner M, Brettel M, Ku NO, Aebi U (1999) Characterization of distinct early assembly units of different intermediate filament proteins. J Mol Biol 286:1403–1420

Herrmann H, Haner M, Brettel M, Muller SA, Goldie KN, Fedtke B, Lustig A, Franke WW, Aebi U (1996) Structure and assembly properties of the intermediate filament protein vimentin: the role of its head, rod and tail domains. J Mol Biol 264:933–953

Hesketh AR, Chandra G, Shaw AD, Rowland JJ, Kell DB, Bibb MJ, Chater KF (2002) Primary and secondary metabolism, and post-translational protein modifications, as portrayed by proteomic analysis of Streptomyces coelicolor. Mol Microbiol 46:917–932

Hesse M, Magin TM, Weber K (2001) Genes for intermediate filament proteins and the draft sequence of the human genome: novel keratin genes and a surprisingly high number of pseudogenes related to keratin genes 8 and 18. J Cell Sci 114:2569–2575

Holmes NA, Walshaw J, Leggett RM, Thibessard A, Dalton KA, Gillespie MD, Hemmings AM, Gust B, Kelemen GH (2013) Coiled-coil protein Scy is a key component of a multiprotein assembly controlling polarized growth in *Streptomyces*. Proc Natl Acad Sci U S A 110:E397–E406

Huitema E, Pritchard S, Matteson D, Radhakrishnan SK, Viollier PH (2006) Bacterial birth scar proteins mark future flagellum assembly site. Cell 124:1025–1037

Ingerson-Mahar M, Briegel A, Werner JN, Jensen GJ, GITAI Z (2010) The metabolic enzyme CTP synthase forms cytoskeletal filaments. Nat Cell Biol 12:739–746

Izard J et al (2004) Tomographic reconstruction of treponemal cytoplasmic filaments reveals novel bridging and anchoring components. Mol Microbiol 51:609–618

Izard J (2006) Cytoskeletal cytoplasmic filament ribbon of *Treponema*: a member of an intermediate-like filament protein family. J Mol Microbiol Biotechnol 11:159–66

Jakimowicz D, Chater K, Zakrzewska-Czerwinska J (2002) The ParB protein of *Streptomyces coelicolor* A3(2) recognizes a cluster of *parS* sequences within the origin-proximal region of the linear chromosome. Mol Microbiol 45:1365–1377

Jakimowicz D, Mouz S, Zakrzewska-Czerwinska J, Chater KF (2006) Developmental control of a *parAB* promoter leads to formation of sporulation-associated ParB complexes in *Streptomyces coelicolor*. J Bacteriol 188:1710–1720

Jakimowicz D, van Wezel GP (2012) Cell division and DNA segregation in *Streptomyces*: how to build a septum in the middle of nowhere? Mol Microbiol 85:393–404

Jakimowicz D, Zydek P, Kois A, Zakrzewska-Czerwinska J, Chater KF (2007) Alignment of multiple chromosomes along helical ParA scaffolding in sporulating *Streptomyces* hyphae. Mol Microbiol 65:625–641

Jyothikumar V, Klanbut K, Tiong J, Roxburgh JS, Hunter IS, Smith TK, Herron PR (2012) Cardiolipin synthase is required for Streptomyces coelicolor morphogenesis. Mol Microbiol 84:181–197

Jyothikumar V, Tilley EJ, Wali R, Herron PR (2008) Time-lapse microscopy of *Streptomyces coelicolor* growth and sporulation. Appl Environ Microbiol 74:6774–6781

Kang CM, Nyayapathy S, Lee JY, Suh JW, Husson RN (2008) Wag31, a homologue of the cell division protein DivIVA, regulates growth, morphology and polar cell wall synthesis in mycobacteria. Microbiology 154:725–735

Kapinos LE, Schumacher J, Mucke N, Machaidze G, Burkhard P, Aebi U, Strelkov SV, Herrmann H (2010) Characterization of the head-to-tail overlap complexes formed by human lamin A, B1 and B2 "half-minilamin" dimers. J Mol Biol 396:719–731

Kawai F, Shoda M, Harashima R, Sadaie Y, Hara H, Matsumoto K (2004) Cardiolipin domains in *Bacillus subtilis marburg* membranes. J Bacteriol 186:1475–1483

Klapper M, Exner K, Kempf A, Gehrig C, Stuurman N, Fisher PA, Krohne G (1997) Assembly of A- and B-type lamins studied *in vivo* with the baculovirus system. J Cell Sci 110(Pt 20):2519–2532

Koster S, Weitz DA, Goldman RD, Aebi U, Herrmann H (2015) Intermediate filament mechanics *in vitro* and in the cell: from coiled coils to filaments, fibers and networks. Curr Opin Cell Biol 32:82–91

Kreplak L, Bar H, Leterrier JF, Herrmann H, AEBI U (2005) Exploring the mechanical behavior of single intermediate filaments. J Mol Biol 354:569–577

Kruger A, Batsios P, Baumann O, Luckert E, Schwarz H, Stick R, Meyer I, Graf R (2012) Characterization of NE81, the first lamin-like nucleoskeleton protein in a unicellular organism. Mol Biol Cell 23:360–370

Kuhnel K, Jarchau T, Wolf E, Schlichting I, Walter U, Wittinghofer A, Strelkov SV (2004) The VASP tetramerization domain is a right-handed coiled coil based on a 15-residue repeat. Proc Natl Acad Sci U S A 101:17027–17032

Lam H, Schofield WB, Jacobs-Wagner C (2006) A landmark protein essential for establishing and perpetuating the polarity of a bacterial cell. Cell 124:1011–1023

Laub MT, Chen SL, Shapiro L, Mcadams HH (2002) Genes directly controlled by CtrA, a master regulator of the *Caulobacter* cell cycle. Proc Natl Acad Sci U S A 99:4632–4637

Lee LK, Stewart AG, Donohoe M, Bernal RA, Stock D (2010) The structure of the peripheral stalk of *Thermus thermophilus* H + −ATPase/synthase. Nat Struct Mol Biol 17:373–378

Lenarcic R, Halbedel S, Visser L, Shaw M, Wu LJ, Errington J, Marenduzzo D, Hamoen LW (2009) Localisation of DivIVA by targeting to negatively curved membranes. EMBO J 28:2272–2282

Letek M, Fiuza M, Ordonez E, Villadangos AF, Ramos A, Mateos LM, Gil JA (2008a) Cell growth and cell division in the rod-shaped actinomycete *Corynebacterium glutamicum*. Antonie Van Leeuwenhoek 94:99–109

Letek M, Ordonez E, Vaquera J, Margolin W, Flardh K, Mateos LM, Gil JA (2008b) DivIVA is required for polar growth in the MreB-lacking rod-shaped actinomycete *Corynebacterium glutamicum*. J Bacteriol 190:3283–3292

Lowery J, Kuczmarski ER, Herrmann H, Goldman RD (2015) Intermediate filaments play a pivotal role in regulating cell architecture and function. J Biol Chem 290:17145–17153

Lutkenhaus J (2012) The ParA/MinD family puts things in their place. Trends Microbiol 20:411–418

Manteca A, Ye J, Sanchez J, Jensen ON (2011) Phosphoproteome analysis of *Streptomyces* development reveals extensive protein phosphorylation accompanying bacterial differentiation. J Proteome Res 10:5481–5492

Martinez LE, Hardcastle JM, Wang J, Pincus Z, Tsang J, Hoover TR, Bansil R, Salama NR (2016) *Helicobacter pylori* strains vary cell shape and flagellum number to maintain robust motility in viscous environments. Mol Microbiol 99:88–110

Mauriello EM, Nan B, Zusman DR (2009) AglZ regulates adventurous (A-) motility in *Myxococcus xanthus* through its interaction with the cytoplasmic receptor, FrzCD. Mol Microbiol 72:964–977

Mazouni K, Pehau-Arnaudet G, England P, Bourhy P, Saint Girons I, Picardeau M (2006) The scc spirochetal coiled-coil protein forms helix-like filaments and binds to nucleic acids generating nucleoprotein structures. J Bacteriol 188:469–76

McCormick JR, Flardh K (2012) Signals and regulators that govern *Streptomyces* development. FEMS Microbiol Rev 36:206–231

McCormick JR, Su EP, Driks A, Losick R (1994) Growth and viability of *Streptomyces coelicolor* mutant for the cell division gene ftsZ. Mol Microbiol 14:243–254

Mcgrath PT, Lee H, Zhang L, Iniesta AA, Hottes AK, Tan MH, Hillson NJ, Hu P, Shapiro L, Mcadams HH (2007) High-throughput identification of transcription start sites, conserved promoter motifs and predicted regulons. Nat Biotechnol 25:584–592

Meier M, Padilla GP, Herrmann H, Wedig T, Hergt M, Patel TR, Stetefeld J, Aebi U, Burkhard P (2009) Vimentin coil 1A-A molecular switch involved in the initiation of filament elongation. J Mol Biol 390:245–261

Meniche X, Otten R, Siegrist MS, Baer CE, Murphy KC, Bertozzi CR, Sassetti CM (2014) Subpolar addition of new cell wall is directed by DivIVA in mycobacteria. Proc Natl Acad Sci U S A 111:E3243–E3251

Mignot T, Shaevitz JW, Hartzell PL, Zusman DR (2007) Evidence that focal adhesion complexes power bacterial gliding motility. Science 315:853–856

Miguelez EM, Martin C, Manzanal MB, Hardisson C (1992) Growth and morphogenesis in *Streptomyces*. FEMS Microbiol Lett 100:351–359

Ngai J, Coleman TR, Lazarides E (1990) Localization of newly synthesized vimentin subunits reveals a novel mechanism of intermediate filament assembly. Cell 60:415–427

Nguyen L, Scherr N, Gatfield J, Walburger A, Pieters J, Thompson CJ (2007) Antigen 84, an effector of pleiomorphism in *Mycobacterium smegmatis*. J Bacteriol 189:7896–7910

Nicolet S, Herrmann H, Aebi U, Strelkov SV (2010) Atomic structure of vimentin coil 2. J Struct Biol 170:369–376

Nishibori A, Kusaka J, Hara H, Umeda M, Matsumoto K (2005) Phosphatidylethanolamine domains and localization of phospholipid synthases in *Bacillus subtilis* membranes. J Bacteriol 187:2163–2174

Nishimura T, Molinard G, Petty TJ, Broger L, Gabus C, Halazonetis TD, Thore S, Paszkowski J (2012) Structural basis of transcriptional gene silencing mediated by *Arabidopsis* MOM1. PLoS Genet 8:e1002484

Noens EE, Mersinias V, Traag BA, Smith CP, Koerten HK, Van Wezel GP (2005) SsgA-like proteins determine the fate of peptidoglycan during sporulation of *Streptomyces coelicolor*. Mol Microbiol 58:929–944

O'Toole G, Kaplan HB, Kolter R (2000) Biofilm formation as microbial development. Annu Rev Microbiol 54:49–79

Oliva MA, Halbedel S, Freund SM, Dutow P, Leonard TA, Veprintsev DB, Hamoen LW, Löwe J (2010) Features critical for membrane binding revealed by DivIVA crystal structure. EMBO J 29:1988–2001

Parry DA, Fraser RD, Squire JM (2008) Fifty years of coiled-coils and alpha-helical bundles: a close relationship between sequence and structure. J Struct Biol 163:258–269

Parry DA, Steinert PM (1999) Intermediate filaments: molecular architecture, assembly, dynamics and polymorphism. Q Rev Biophys 32:99–187

Parry DA, Strelkov SV, Burkhard P, Aebi U, Herrmann H (2007) Towards a molecular description of intermediate filament structure and assembly. Exp Cell Res 313:2204–2216

Patrick JE, Kearns DB (2008) MinJ (YvjD) is a topological determinant of cell division in *Bacillus subtilis*. Mol Microbiol 70:1166–1179

Pauling L, Corey RB, Branson HR (1951) The structure of proteins; two hydrogen-bonded helical configurations of the polypeptide chain. Proc Natl Acad Sci U S A 37:205–211

Persat A, Stone HA, Gitai Z (2014) The curved shape of *Caulobacter crescentus* enhances surface colonization in flow. Nat Commun 5:3824

Peters J, Baumeister W, Lupas A (1996) Hyperthermostable surface layer protein tetrabrachion from the archaebacterium *Staphylothermus marinus*: evidence for the presence of a right-handed coiled coil derived from the primary structure. J Mol Biol 257:1031–1041

Pichoff S, Lutkenhaus J (2002) Unique and overlapping roles for ZipA and FtsA in septal ring assembly in *Escherichia coli*. EMBO J 21:685–693

Plocinski P, Ziolkiewicz M, Kiran M, Vadrevu SI, Nguyen HB, Hugonnet J, Veckerle C, Arthur M, Dziadek J, Cross TA, Madiraju M, Rajagopalan M (2011) Characterization of CrgA, a new partner of the *Mycobacterium tuberculosis* peptidoglycan polymerization complexes. J Bacteriol 193:3246–3256

Ptacin JL, Lee SF, Garner EC, Toro E, Eckart M, Comolli LR, Moerner WE, Shapiro L (2010) A spindle-like apparatus guides bacterial chromosome segregation. Nat Cell Biol 12:791–798

Ptashne M (2011) Principles of a switch. Nat Chem Biol 7:484–487

Quax-Jeuken YE, Quax WJ, Bloemendal H (1983) Primary and secondary structure of hamster vimentin predicted from the nucleotide sequence. Proc Natl Acad Sci U S A 80:3548–3552

Ramamurthi KS, Losick R (2009) Negative membrane curvature as a cue for subcellular localization of a bacterial protein. Proc Natl Acad Sci U S A 106:13541–13545

Ramms L, Fabris G, Windoffer R, Schwarz N, Springer R, Zhou C, Lazar J, Stiefel S, Hersch N, Schnakenberg U, Magin TM, Leube RE, Merkel R, Hoffmann B (2013) Keratins as the main

component for the mechanical integrity of keratinocytes. Proc Natl Acad Sci U S A 110:18513–18518
Raychaudhuri D (1999) ZipA is a MAP-Tau homolog and is essential for structural integrity of the cytokinetic FtsZ ring during bacterial cell division. EMBO J 18:2372–2383
Richards DM, Hempel AM, Flardh K, Buttner MJ, Howard M (2012) Mechanistic basis of branch-site selection in filamentous bacteria. PLoS Comput Biol 8:e1002423
Saalbach G, Hempel AM, Vigouroux M, FLARDH K, Buttner MJ, Naldrett MJ (2013) Determination of phosphorylation sites in the DivIVA cytoskeletal protein of *Streptomyces coelicolor* by targeted LC-MS/MS. J Proteome Res 12:4187–4192
Schatzle S, Specht M, Waidner B (2015) Coiled coil rich proteins (Ccrp) influence molecular pathogenicity of *Helicobacter pylori*. PLoS One 10:e0121463
Schirmer EC, Guan T, Gerace L (2001) Involvement of the lamin rod domain in heterotypic lamin interactions important for nuclear organization. J Cell Biol 153:479–489
Schlimpert S, Klein EA, Briegel A, Hughes V, Kahnt J, Bolte K, Maier UG, Brun YV, Jensen GJ, Gitai Z, Thanbichler M (2012) General protein diffusion barriers create compartments within bacterial cells. Cell 151:1270–1282
Schwedock J, McCormick JR, Angert ER, Nodwell JR, Losick R (1997) Assembly of the cell division protein FtsZ into ladder-like structures in the aerial hyphae of *Streptomyces coelicolor*. Mol Microbiol 25:847–858
Seltmann K, Fritsch AW, Kas JA, Magin TM (2013) Keratins significantly contribute to cell stiffness and impact invasive behavior. Proc Natl Acad Sci U S A 110:18507–18512
Shapland EB, Reisinger SJ, Bajwa AK, Ryan KR (2011) An essential tyrosine phosphatase homolog regulates cell separation, outer membrane integrity, and morphology in *Caulobacter crescentus*. J Bacteriol 193:4361–4370
Sharp TH, Bruning M, Mantell J, Sessions RB, Thomson AR, Zaccai NR, Brady RL, Verkade P, Woolfson DN (2012) Cryo-transmission electron microscopy structure of a gigadalton peptide fiber of *de novo* design. Proc Natl Acad Sci U S A 109:13266–13271
Smith TA, Strelkov SV, Burkhard P, Aebi U, Parry DA (2002) Sequence comparisons of intermediate filament chains: evidence of a unique functional/structural role for coiled-coil segment 1A and linker L1. J Struct Biol 137:128–145
Snider NT, Omary MB (2014) Post-translational modifications of intermediate filament proteins: mechanisms and functions. Nat Rev Mol Cell Biol 15:163–177
Soellner P, Quinlan RA, Franke WW (1985) Identification of a distinct soluble subunit of an intermediate filament protein: tetrameric vimentin from living cells. Proc Natl Acad Sci U S A 82:7929–7933
Specht M, Schatzle S, Graumann PL, Waidner B (2011) *Helicobacter pylori* possesses four coiled-coil-rich proteins that form extended filamentous structures and control cell shape and motility. J Bacteriol 193:4523–4530
Stetefeld J, Jenny M, Schulthess T, Landwehr R, Engel J, Kammerer RA (2000) Crystal structure of a naturally occurring parallel right-handed coiled coil tetramer. Nat Struct Biol 7:772–776
Stewart AG, Lee LK, Donohoe M, Chaston JJ, Stock D (2012) The dynamic stator stalk of rotary ATPases. Nat Commun 3:687
Strahl H, Hamoen LW (2012) Finding the corners in a cell. Curr Opin Microbiol 15:731–736
Strelkov SV, Herrmann H, Aebi U (2003) Molecular architecture of intermediate filaments. Bioessays 25:243–251
Strelkov SV, Herrmann H, Geisler N, Lustig A, Ivaninskii S, Zimbelmann R, Burkhard P, Aebi U (2001) Divide-and-conquer crystallographic approach towards an atomic structure of intermediate filaments. J Mol Biol 306:773–781
Strelkov SV, Herrmann H, Geisler N, Wedig T, Zimbelmann R, Aebi U, Burkhard P (2002) Conserved segments 1A and 2B of the intermediate filament dimer: their atomic structures and role in filament assembly. EMBO J 21:1255–1266
Stuurman N, Heins S, Aebi U (1998) Nuclear lamins: their structure, assembly, and interactions. J Struct Biol 122:42–66

Swulius MT, Chen S, Jane Ding H, Li Z, Briegel A, Pilhofer M, Tocheva EI, Lybarger SR, Johnson TL, Sandkvist M, Jensen GJ (2011) Long helical filaments are not seen encircling cells in electron cryotomograms of rod-shaped bacteria. Biochem Biophys Res Commun 407:650–655

Sycuro LK, Pincus Z, Gutierrez KD, Biboy J, Stern CA, Vollmer W, Salama NR (2010) Peptidoglycan crosslinking relaxation promotes *Helicobacter pylori*'s helical shape and stomach colonization. Cell 141:822–833

Szwedziak P, Wang Q, Freund SM, Löwe J (2012) FtsA forms actin-like protofilaments. EMBO J 31:2249–2260

Thanky NR, Young DB, Robertson BD (2007) Unusual features of the cell cycle in mycobacteria: polar-restricted growth and the snapping-model of cell division. Tuberculosis (Edinb) 87:231–236

Toivola DM, Boor P, Alam C, Strnad P (2015) Keratins in health and disease. Curr Opin Cell Biol 32:73–81

Traag BA, van Wezel GP (2008) The SsgA-like proteins in actinomycetes: small proteins up to a big task. Antonie Van Leeuwenhoek 94:85–97

Tsang PH, Li G, Brun YV, Freund LB, Tang JX (2006) Adhesion of single bacterial cells in the micronewton range. Proc Natl Acad Sci U S A 103:5764–5768

van den Ent F, Löwe J (2000) Crystal structure of the cell division protein FtsA from *Thermotoga maritima*. EMBO J 19:5300–5307

van Teeffelen S, Wang S, Furchtgott L, Huang KC, Wingreen NS, Shaevitz JW, Gitai Z (2011) The bacterial actin MreB rotates, and rotation depends on cell-wall assembly. Proc Natl Acad Sci U S A 108:15822–15827

Vikstrom KL, Lim SS, Goldman RD, Borisy GG (1992) Steady state dynamics of intermediate filament networks. J Cell Biol 118:121–129

Waidner B, Specht M, Dempwolff F, Haeberer K, Schaetzle S, Speth V, Kist M, Graumann PL (2009) A novel system of cytoskeletal elements in the human pathogen *Helicobacter pylori*. PLoS Pathog 5:e1000669

Walshaw J, Gillespie MD, Kelemen GH (2010) A novel coiled-coil repeat variant in a class of bacterial cytoskeletal proteins. J Struct Biol 170:202–215

Walter S, Rohde M, Machner M, Schrempf H (1999) Electron microscopy studies of cell-wall-anchored cellulose (Avicel)-binding protein (AbpS) from *Streptomyces reticuli*. Appl Environ Microbiol 65:886–892

Walter S, Schrempf H (2003) Oligomerization, membrane anchoring, and cellulose-binding characteristics of AbpS, a receptor-like Streptomyces protein. J Biol Chem 278:26639–26647

Walter S, Wellmann E, Schrempf H (1998) The cell wall-anchored *Streptomyces reticuli* avicel-binding protein (AbpS) and its gene. J Bacteriol 180:1647–1654

Wang SB, Cantlay S, Nordberg N, Letek M, Gil JA, Flardh K (2009) Domains involved in the *in vivo* function and oligomerization of apical growth determinant DivIVA in *Streptomyces coelicolor*. FEMS Microbiol Lett 297:101–109

Ward MJ, Lew H, Zusman DR (2000) Social motility in *Myxococcus xanthus* requires FrzS, a protein with an extensive coiled-coil domain. Mol Microbiol 37:1357–1371

Willemse J, Borst JW, de Waal E, Bisseling T, van Wezel GP (2011) Positive control of cell division: FtsZ is recruited by SsgB during sporulation of *Streptomyces*. Genes Dev 25:89–99

Woolfson DN (2005) The design of coiled-coil structures and assemblies. Adv Protein Chem 70:79–112

Wu LJ, Errington J (2003) RacA and the Soj-Spo0J system combine to effect polar chromosome segregation in sporulating *Bacillus subtilis*. Mol Microbiol 49:1463–1475

Xu H, Chater KF, Deng Z, Tao M (2008) A cellulose synthase-like protein involved in hyphal tip growth and morphological differentiation in *Streptomyces*. J Bacteriol 190:4971–4978

Yamaguchi T, Goto H, Yokoyama T, Sillje H, Hanisch A, Uldschmid A, Takai Y, Oguri T, Nigg EA, Inagaki M (2005) Phosphorylation by Cdk1 induces Plk1-mediated vimentin phosphorylation during mitosis. J Cell Biol 171:431–436

Yang R, Bartle S, Otto R, Stassinopoulos A, Rogers M, Plamann L, Hartzell P (2004) AglZ is a filament-forming coiled-coil protein required for adventurous gliding motility of *Myxococcus xanthus*. J Bacteriol 186:6168–6178

Yoon KH, Yoon M, Moir RD, Khuon S, Flitney FW, Goldman RD (2001) Insights into the dynamic properties of keratin intermediate filaments in living epithelial cells. J Cell Biol 153:503–516

Yoon M, Moir RD, Prahlad V, Goldman RD (1998) Motile properties of vimentin intermediate filament networks in living cells. J Cell Biol 143:147–157

You Y et al (1996) Characterization of the cytoplasmic filament protein gene (cfpA) of *Treponema pallidum subsp. pallidum*. J Bacteriol 178:3177–3187

Zupan JR, Cameron TA, Anderson-Furgeson J, Zambryski PC (2013) Dynamic FtsA and FtsZ localization and outer membrane alterations during polar growth and cell division in *Agrobacterium tumefaciens*. Proc Natl Acad Sci U S A 110:9060–9065

Zwerger M, Jaalouk DE, Lombardi ML, Isermann P, Mauermann M, Dialynas G, Herrmann H, Wallrath LL, Lammerding J (2013) Myopathic lamin mutations impair nuclear stability in cells and tissue and disrupt nucleo-cytoskeletal coupling. Hum Mol Genet 22:2335–2349

Chapter 7
FtsZ-ring Architecture and Its Control by MinCD

Piotr Szwedziak and Debnath Ghosal

Abstract In bacteria and archaea, the most widespread cell division system is based on the tubulin homologue FtsZ protein, whose filaments form the cytokinetic Z-ring. FtsZ filaments are tethered to the membrane by anchors such as FtsA and SepF and are regulated by accessory proteins. One such set of proteins is responsible for Z-ring's spatiotemporal regulation, essential for the production of two equal-sized daughter cells. Here, we describe how our still partial understanding of the FtsZ-based cell division process has been progressed by visualising near-atomic structures of Z-rings and complexes that control Z-ring positioning in cells, most notably the MinCDE and Noc systems that act by negatively regulating FtsZ filaments. We summarise available data and how they inform mechanistic models for the cell division process.

Keywords FtsZ • FtsA • Bacterial cell division • Z-ring structure • MinCD copolymers • Cytomotive filaments • Cell constriction • Liposome constriction • Sliding filaments • CryoET • Tomography • Collaborative filaments

---

P. Szwedziak (✉)
Department of Biology, Institute of Molecular Biology & Biophysics, ETH Zürich, Zürich, Switzerland
e-mail: piotr.szwedziak@mol.biol.ethz.ch

D. Ghosal
Broad Center for the Biological Sciences, California Institute of Technology, Pasadena, CA, USA
e-mail: dghosal@caltech.edu

J. Löwe, L.A. Amos (eds.), *Prokaryotic Cytoskeletons*, Subcellular Biochemistry 84, DOI 10.1007/978-3-319-53047-5_7

Introduction

The Divisome

In most bacteria and archaea, cell division is facilitated by a putative multiprotein complex called the divisome (Adams and Errington 2009). Its assembly starts with the polymerisation of FtsZ into the Z-ring, which subsequently recruits other protein components to the division site. In *Escherichia coli*, this takes place in a highly linear process that can be subdivided into two stages. First, early (inner) divisome proteins (in *E. coli* most prominently FtsA) tether FtsZ filaments to the membrane and coordinate ring assembly. Then late (outer) divisome proteins, largely responsible for cell wall remodelling and synthesis are recruited. Currently, the divisome is thought to consist of more than 20 protein components and, as mentioned, can be divided into the inner and outer divisome (Vicente and Rico 2006). This distinction reflects the fact the divisome needs to remodel not only the inner membrane (Z-ring and associated proteins) but also the stress-bearing cell wall (PBPs and associated proteins) and outer membrane (Tol-Pal and associated proteins).

Fts Proteins

In *E. coli*, the vast majority of proteins participating in bacterial cell division were first identified through a screen to detect thermosensitive defects that cause cells to filament, thus their genes were termed *fts* (Hirota et al. 1968). One of these proteins, FtsZ, is the endogenous homolog of eukaryotic tubulin (Löwe and Amos 1998). Biochemical assays and electron microscopy of purified FtsZ unveiled its polymerisation (Bramhill and Thompson 1994; Erickson et al. 1996; Mukherjee and Lutkenhaus 1994). FtsZ's GTP binding and hydrolysis, as well as filament formation, are relatively well-understood and it is known how nucleotide is hydrolysed efficiently only upon polymerisation through a mechanism that is conserved between tubulin and FtsZ (Oliva et al. 2004). Because of its ubiquity in bacteria and the fact that cell division in some archaea and all eukaryotes involves ESCRT-III (Guizetti et al. 2011; Katzmann et al. 2001; Lindas et al. 2008; Samson et al. 2008), it seems likely that FtsZ and FtsZ-based cell division are ancient (Davis et al. 2002; Faguy and Doolittle 1998). It seems reasonable to assume that a mechanism for cytokinesis would be required at a fairly early stage of evolution of cellular life, although it is known that L-forms, bacteria without cell wall, produce progeny without ordered cell division through the production of large amounts of membrane and spontaneous blebbing and abscission (Mercier et al. 2013). FtsZ protein consists of a globular core domain, which is responsible for GTP-binding and polymerisation and an extended C-terminal tail, of which the last 15–20 residues are a particularly busy region, since it is known to bind many proteins including FtsA, SepF, ZipA and MinC. In between the tail and globular body, a flexible, unstructured linker connects the two parts (Erickson et al. 2010).

FtsA Protein

FtsA is a prominent FtsZ binding partner and is an essential membrane anchor of the Z-ring in *E. coli* (Pichoff and Lutkenhaus 2005). In many other organisms, FtsA and its membrane anchoring function seems to have been replaced by the unrelated protein SepF (Duman et al. 2013). FtsA was first predicted to be an actin-like protein from structure-based sequence alignments involving the then newly solved actin crystal structure (Bork et al. 1992). The crystal structure of FtsA with an ATP molecule bound in the active site later revealed that the protein is indeed a member of the actin/HSP70 family, although its fold was shown to considerably deviate from the canonical actin fold (van den Ent and Löwe 2000). Three out of the four subdomains characteristic of the actin fold (1A, 2A, 2B) are conserved in FtsA but the fourth (1C) is located on the opposite side of subdomain 1A, compared with actin's 1B subdomain.

FtsZ Protein

The first hint that FtsZ might be central to cell constriction came in the 1980s, when it was reported that an increase in the level of FtsZ induces minicell formation in *E. coli* (Ward and Lutkenhaus 1985). Subsequently, after a report (Dai and Lutkenhaus 1992) that a proper ratio of FtsZ to FtsA is critical for cell division, it was shown that a simultaneous increase in both FtsZ and FtsA concentration in cells increases the frequency of central divisions (Begg et al. 1998). But the discovery that FtsZ arranges into a ring whose assembly coincides with division in bacteria (Bi and Lutkenhaus 1991) was the most important landmark discovery, focussing the efforts to understand cell division and also starting a series of discoveries that led to the understanding that bacteria have intracellular organisation, harbour filamentous proteins and a cytoskeleton. Defects in FtsZ-ring assembly and in controlling its position lead to opposite effects. Whereas the former blocks septation and yields long, filamentous cells (Hirota et al. 1968), the latter results in near-polar divisions, producing anucleate minicells (Adler et al. 1967). Since precise spatial positioning of the Z-ring decides the fate of the daughter cells, bacteria use several regulatory mechanisms to guide the Z-ring to midcell. Two very widely used negative regulatory mechanisms are the Min and nucleoid occlusion (NO) systems. An increasing number of studies have suggested other negative regulators as well as positive signals, and cell cycle-chromosome coordination also plays critical roles in spatiotemporal regulation of the Z-ring (see Chap. 9).

Z-ring Structure and Constriction Mechanism

Although progress in understanding FtsZ ring structure, its mechanics and spatial positioning have steadily advanced over the past 25 years, many fundamental questions remain. Here, we discuss structural and biochemical studies of Z-ring architecture and its regulatory proteins and how these inform models of the constriction process that FtsZ facilitates during cell division. Possible sources of energy in the system are: binding energies of proteins between each other, nucleotide hydrolysis by FtsA (ATP) and FtsZ (GTP) (both inner divisome), as well as cell wall synthesis from precursors in the periplasm (PBPs, outer divisome).

Since understanding Z-ring architecture and function is required to unravel the mechanism of cell division, the cytokinetic ring in bacteria has been the subject of extensive investigations. Various methods have been employed, ranging from fluorescence light microscopy (FLM), more recently including super-resolution techniques, to electron microscopy (EM), including negative-stain EM, electron cryomicroscopy (cryoEM) and cryotomography (cryoET); experiments ranging from *in vivo* investigations of fixed or live cells to *in vitro* studies of purified components under well-defined conditions (Fig. 7.1a) are discussed in more detail later.

Based on these studies, three major models for generating constriction force by the divisome have been proposed:

(a) The scaffold model, where FtsZ's role is purely structural and the ring is required to correctly position the constriction machinery, including the peptidoglycan-synthetic machinery. Then, peptidoglycan (PG) synthesis would exert force by inserting new PG strands and remodelling cell wall displacing the (inner) cell membrane inwards to produce constriction.
(b) The iterative pinching model based on *in vitro* observations that FtsZ filaments seem to adopt two main conformations – straight or bent – dependent on the nucleotide state (Erickson et al. 1996; Lu et al. 2000). Repeated hydrolysis-induced bending of FtsZ filaments in a very loosely assembled ring could produce an inward force on the membrane (Fig. 7.1b**, bottom**).
(c) The sliding-filament model (Lan et al. 2009; Szwedziak et al. 2014), where FtsZ filaments in a continuous and dense ring would slide past each other, driven by the increase in lateral filament overlap (Fig. 7.1b**, top**).

Spatiotemporal Regulation of Z-ring Assembly

Positioning of the Z-ring at the right time and in the right place is critical to ensure the integrity of genetic information transmitted to the daughter cells. *E. coli* and *B. subtilis* use the combined effect of two negative mechanisms to inhibit Z-ring assembly everywhere except at midcell. These negative regulators are nucleoid occlusion (NO) system (Bernhardt and de Boer 2005; Woldringh et al. 1991; Wu et al. 2009) and the Min system (Adler et al. 1967).

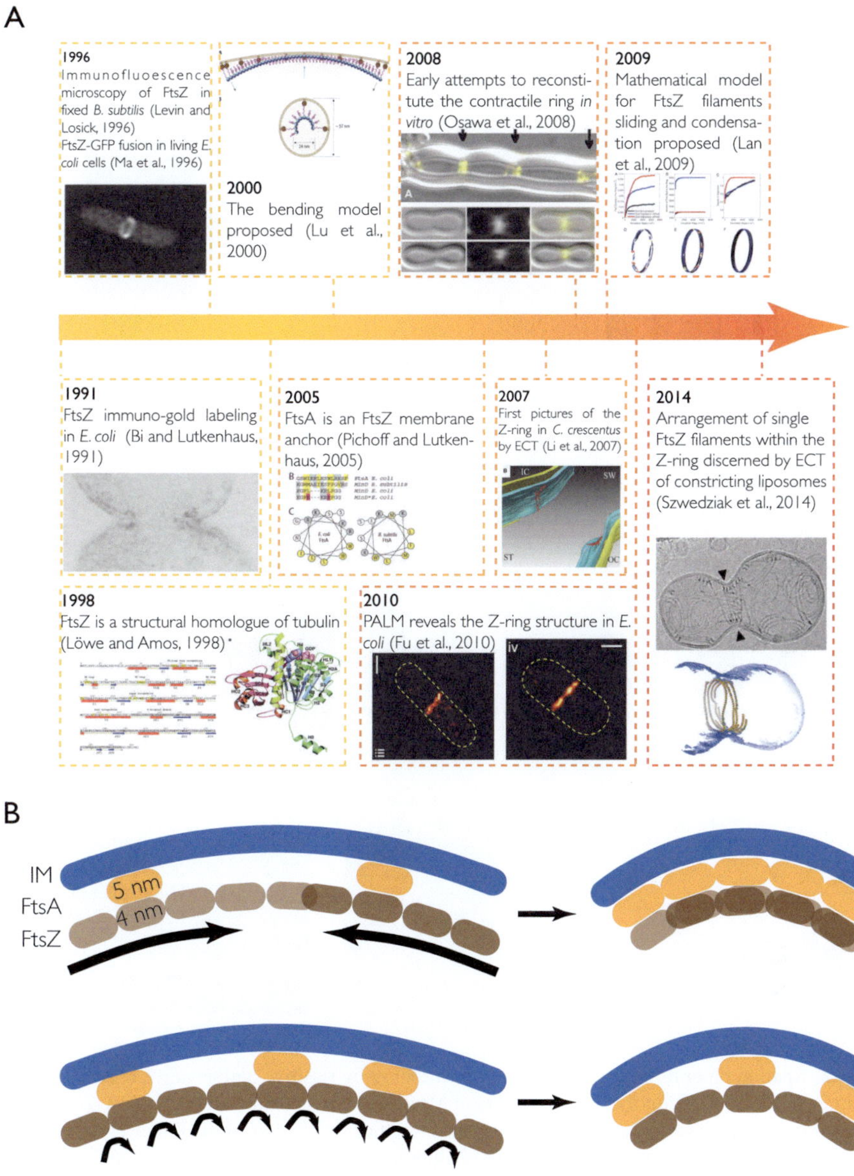

Fig. 7.1 Two models for constriction force generation and a time lapse of key discoveries in the field. (**a**) An overview of crucial discoveries in the field of bacterial cell biology related to the Z-ring architecture represented as a time arrow. This exemplifies the plethora of techniques and approaches that have been employed over the past 25 years to dissect the structure and mechanics of the Z-ring. (**b**) Two prevailing models for constriction force generation. *Upper panel* depicts the sliding model. At early stages FtsZ filaments overlap only partially and are attached to the inner membrane (IM) by sparsely positioned FtsA as there is not enough protein to make a continuous A-ring. As the FtsZ filaments slide past each other the overlap increases and more and more binding energy is produced. This is accompanied by increased membrane curvature, which enables the addition of more FtsA according to the repeat length mismatch between the two proteins. FtsZ filament shortening via GTP hydrolysis ensures that the filaments are not stalled by avidity. *Lower panel* depicts the bending model. Here, membrane-associated FtsZ filaments undergo a conformational change upon GTP hydrolysis. This straight to bent rearrangement between every monomer, given the number of subunits encircling the cell, could provide sufficient energy to perform constriction

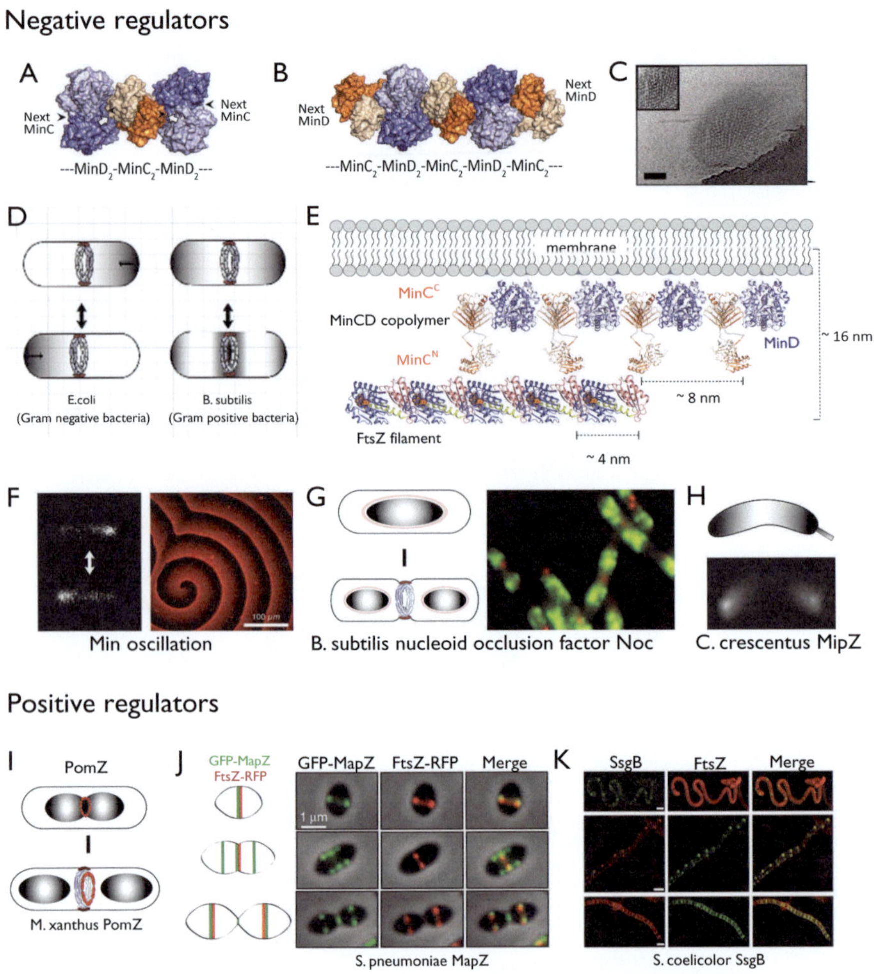

Fig. 7.2 Spatiotemporal regulation of Z-ring positioning by positive and negative regulators. (**a**, **b**) Composite model of copolymeric filaments of MinD and C-terminal domain of MinC (MinC^C), built by combining *A. aeolicus* MinC^{C}D cocrystal structure and *E. coli* MinD dimer crystal structure (Modified from Ghosal et al. 2014). MinD binding sites on a dimer of the MinC^C domain (*wheat* and *orange*) are exactly on the opposite sides (indicated by *white arrows*). Similarly, MinC^C binding site on a MinD dimer (*slate* and *light blue*) are also at the opposite sides (shown by *red arrow-heads*). This means, MinC and MinD will always form an open MinC$_2$-MinD$_2$ complex with available binding sites at both side of the complex. Thus, it could translationally extend on the membrane (e.g. –MinC$_2$-MinD$_2$-MinC$_2$-MinD$_2$–) and form an extended alternating copolymeric filament. (**c**) Electron cryo-tomography of MinCD cofilament decorated liposome. Shown is surface view of a tomographic reconstruction (Modified from Ghosal et al. 2014). *Inset* showing low-pass filtered view. Scale bar 100 nm. (**d**) Schematic representation of the Min protein localisation in *E. coli* and in *B. subtilis*. In *E. coli*, (*left*) the Min system continuously oscillates between the two cell poles. In *B. subtilis* (*right*), however, the Min system remains static, anchored at the cell poles in a newborn cell. Later, when replication of DNA is completed and cell division is continuing, the Min system is recruited at the division site. This ensures the Min system is present at both cell poles of newborn daughter cells. (**e**) Plausible mechanism of FtsZ inhibition by the Min system. Complete MinCD copolymers are tethered to the membrane through MinD's membrane

Nucleoid Occlusion – Protecting from Incision

Nucleoid occlusion inhibits Z-ring assembly in the vicinity of the nucleoid and prevents lethal bisection of nucleoids (Fig. 7.2g). The DNA binding proteins Noc in *B. subtilis* and SlmA in *E. coli* have been identified as the factors mediating nucleoid occlusion (Bernhardt and de Boer 2005; Wu et al. 2009). A null mutation of the nucleoid occlusion factors Noc or SlmA leads to the formation of the Z-ring over the unsegregated nucleoids and results in bisection of the chromosome, whereas overproduction of these proteins generates longer cells (Bernhardt and de Boer 2005; Wu et al. 2009). (see Chap. 9 for more details on NO)

Min Systems in Gram Negative and Gram Positive Bacteria

Later in this Chapter, we will focus on the second negative regulatory mechanism, the Min system. In the Gram positive bacterium *B. subtilis*, the system is statically targeted to membrane around the cell poles by the polar membrane-anchor DivIVA (Marston et al. 1998) (Fig. 7.2d**, right**). In contrast, in the Gram negative bacterium *E. coli* the Min system continuously oscillates between the two poles (Raskin and de Boer 1999) (Fig. 7.2d, f). In either case, the (time-averaged) concentration of Min proteins is sufficiently high at both cell poles for the inhibition of cell division. Taken together, a combination of the nucleoid occlusion and the Min systems positions the Z-ring correctly in time and space, thereby synchronising cell growth, DNA duplication and septation.

Fig. 7.2 (continued) targeting sequence (MTS), and interact with the FtsZ filaments through N-terminal domains of MinC ($MinC^N$). The MinCD copolymers cooperatively bind to FtsZ filaments and could inhibit Z-ring formation by disrupting lateral interaction between FtsZ filaments or by antagonising their scaffolding function (Modified from Ghosal et al. 2014). (**f**) *Left*: GFP-MinC oscillation in *E. coli*, frames are separated by 20 s (Ghosal et al., *Encyclopedia of Cell Biology*). *Right*: *in vitro* reconstitution of the Min waves (Image modified from Loose et al. 2008). (**g**) *Left*: Schematic of Nucleoid occlusion (NO). NO proteins (SlmA in *E coli*, Noc in *B subtilis*: coloured in *salmon*) coat the nucleoid and protect it from bisectioning. *Right*: Localisation of Noc-YFP in *Bacillus subtilis*. Noc is shown in *green*, replication terminus region (Ter) is shown in *red* (Wu and Errington 2012). (**h**) Schematic of MipZ gradient formation in *C. crescentus* (*top*). Subcellular localisation of MipZ-YFP (*below*) (Kiekebusch et al. 2012). (**i**) Schematic of PomZ localisation in *Myxococcus xanthus*. PomZ localises at the midcell before FtsZ and acts as a positive signal (*top*), this is then followed by other divisome assembly. (**j**) *Left*: schematic of MapZ localisation during different stages of *S. pneumoniae* cell cycle. In a newborn cell, MapZ (*green*) and FtsZ (*red*) colocalise at the midcell. As the cell grows, MapZ ring splits into two rings that move in opposite directions. Subsequently, a third MapZ ring appears at the midcell and during this time, FtsZ ring splits into two rings that colocalise with the other MapZ rings to define future division sites. *Right*: GFP-MapZ and FtsZ-RFP localisation in *S. pneumoniae* during different stages of cell cycle (Modified from Fleurie et al. 2014). Scale bar 1 μm. (**k**) Localisation of SsgB-eGFP and FtsZ-mCherry in *S. coelicolor* cells during sporulation-specific cell division. SsgB localises into foci much before FtsZ's arrival. Later stages, they colocalise. *Top panel*: early aerial hyphae stage (stage IIA), *middle panel*: predivision localisation in aerial hyphae (stage IIC), *bottom panel*: Z-rings (stage III) (Modified from Willemse et al. 2011). Scale bar 1 μm

Regulation of Z-ring Position in Spherical Bacteria

Compared to the rod-shaped model organisms, much less is known about the mechanisms of Z-ring positioning in spherical bacteria. A significant difference to rod-shaped bacteria is that cocci have an infinite number of possible division planes yet many of them (such as the Gram negative *Neisseria gonorrhoeae* and the Gram positive *Staphylococcus aureus*) divide in alternating perpendicular planes (Ramirez-Arcos et al. 2001). In the Gram negative cocci, the Min system has been shown to play an important role in Z-ring positioning, whereas the Gram positive cocci altogether lack this system (Ramirez-Arcos et al. 2001) and rely more on the nucleoid occlusion mechanism for correct positioning of the divisome complex (Veiga et al. 2011). Recent studies revealed that the deletion of the nucleoid occlusion factor Noc in *Staphylococcus aureus* results in the formation of the divisome complex near the nucleoid, causing DNA breaks. This suggests that in the Gram positive cocci a nucleoid occlusion factor is essential in defining the Z-ring positioning (Veiga et al. 2011). It has been reported that spherical mutants of *E. coli* behave like cocci and also divide in alternating perpendicular planes (Begg and Donachie 1998). Investigation of the mechanism of division site selection in these spherical *E. coli* mutants may provide hints about the underlying regulatory mechanism of divisome assembly in cocci.

Alternative Regulators of Z-ring Position

Although numerous bacteria solve the problem of Z-ring positioning by using one of the above or by combining effects of both the negative regulatory mechanisms, increasing evidence suggests these are not universal or even sufficient solutions. Several bacteria use other mechanisms to guide the Z-ring to midcell, and in these bacteria the Min system and/or NO system are either absent or redundant (Harry et al. 2006; Margolin 2005). These alternative regulators can be both negative and positive.

MipZ

One such example is the Gram negative bacterium *Caulobacter crescentus*. *Caulobacter* uses a protein called MipZ (Fig. 7.2h), which is conserved across all α-proteobacteria that lack the Min system (Kiekebusch et al. 2012; Thanbichler and Shapiro 2006). MipZ is a weak FtsZ polymerisation inhibitor that interacts with the nucleoid in a ParB-dependent manner and generates a bipolar gradient from the cell poles towards the midcell. This gradient of MipZ inhibits Z-ring assembly near the cell poles and on the nucleoid, thereby localising the Z-ring at midcell (Kiekebusch et al. 2012). Interestingly, MipZ is also a deviant WACA ATPase, like MinD, that acts as a negative regulator of the Z-ring assembly but by a different gradient-generating mechanism.

PomZ

In parallel to the negative regulators already mentioned, evidence suggests that several bacteria use different positive regulatory mechanisms to regulate Z-ring positioning. For example, the δ-proteobacterium *Myxococcus xanthus* lacks any of the negative regulators discussed so far – instead a protein called PomZ is involved (Treuner-Lange et al. 2013). PomZ independently localises at the division site prior to arrival of FtsZ and acts as a positive signal for proper Z-ring positioning (Fig. 7.2i). Division site localisation of PomZ has been shown to correlate with cell-length and cell cycle progression and it has been suggested that PomZ recruits and stabilises the Z-ring at the division site. Intriguingly, PomZ is also a WACA family protein, like MinD (Treuner-Lange et al. 2013), suggesting WACA family members harbour functionalities that are well suited to the task of intracellular site selection and positioning of molecules.

MapZ

Another example of a positive regulator, recently reported in *Streptococcaceae* and most other *Lactobacillales*is, is MapZ (midcell-anchored protein Z) (Fleurie et al. 2014; Holečková et al. 2015) (Fig. 7.2j). The extracellular domain of MapZ interacts with the newly-synthesised peptidoglycan at midcell and remains permanently associated at the equator of the cell, whereas the intracellular domain physically anchors the Z-ring through direct protein-protein interaction, thereby positioning the Z-ring at the division plane (Fleurie et al. 2014). Thus, MapZ also plays a dual role, reminiscent of PomZ. Additionally, MapZ also undergoes phosphorylation through the endogenous Ser/Thr kinase StkP and plays an essential role in Z-ring constriction.

SsgAB

One more example of positive regulation of the Z-ring localisation is that played by the SsgAB system of *Streptomyces coelicolor* (Monahan and Harry 2013; Willemse et al. 2011) (Fig. 7.2k). In the absence of canonical negative regulators, sporulating cells of *Streptomyces coelicolor* use a two-protein system SsgA/SsgB to spatially regulate Z-ring positioning. In *Streptomyces*, FtsZ is only required for sporulation as vegetative growth is through branching hyphae. SsgA promotes the localisation of SsgB at midcell, whereas SsgB stimulates Z-ring assembly by facilitating FtsZ polymerisation (Willemse et al. 2011). The SsgAB system was the first positive guiding signal for the Z-ring to be discovered, but it seems to be present only in *Actinomycetes*.

The Min System in Rod-Shaped Bacteria

The term Min is derived from the fact that modifications of this system either results in the formation of anucleate minicells or multinucleate filaments (Adler et al. 1967; Reeve et al. 1973). Recent studies have established that, in addition to its presence in bacteria and archaea, the Min system is widely conserved in eukaryotic plastids and mitochondria, in line with their bacterial ancestry (Leger et al. 2015). In *E. coli*, the Min system is encoded by three genes *minC*, *minD* and *minE*, all of which are products of the *minB* operon (de Boer et al. 1988, 1989;). *minC* and *minD* are widely conserved amongst bacteria, whereas *minE* and *DivIVA* are found among the Gram negative and the Gram positive bacteria, respectively (Harry et al. 2006; Lutkenhaus 2007; Rothfield et al. 2005).

MinC Dimers

MinC is a dimeric protein with two distinct functional domains connected by a flexible linker (Cordell et al. 2001; Hu and Lutkenhaus 2000). MinC's N-terminal domain ($MinC^N$) interacts with FtsZ directly and interferes with its ability to form the Z-ring, thus inhibiting division (Dajkovic et al. 2008; Hu and Lutkenhaus 2000). The C-terminal domain ($MinC^C$), a small beta-helix fold, forms a constitutive homodimer and interacts with MinD (Hu and Lutkenhaus 2000). Recent studies have revealed that $MinC^C$ competes with FtsA and ZipA for the conserved C-terminal tail-domain of FtsZ and alters lateral association of FtsZ filaments (Dajkovic et al. 2008; Shen and Lutkenhaus 2009).

MinD Dimers

MinD is a deviant Walker A type ATPase and belongs to the broad superfamily of P-loop GTPases (de Boer et al. 1991; Lutkenhaus 2012). Members of the deviant Walker A type ATPases are characterised by three distinct motifs critical for nucleotide catalysis: switch I, switch II, and a deviant Walker A motif (XKGGXXK[T/S]), as compared to the classic P-loop signature (GXXGXGK[T/S]) (Koonin 1993). Some of the closely related members of the deviant Walker A type ATPases that are involved in cytoskeletal roles such as intracellular positioning, have been further grouped into a new class, known as the Walker A cytoskeletal ATPases (WACAs) (Michie and Löwe 2006). MinD is a member of this WACA family, along with Soj/ParA, SopA and ParF (Michie and Löwe 2006) and the ones already mentioned. Probably all deviant Walker A type ATPases and hence all WACAs form ATP-sandwich dimers since the deviant P-loop sequence is required for this feature because of steric considerations. Two ATP molecules are sandwiched between two protein monomers such that each ATP molecule is contacting surfaces from the two proteins. It was believed that the WACA family proteins form filaments on their own, without surfaces or other support. However, recent studies have challenged this notion and demonstrated that most, if not all, WACA family proteins only form collaborative filaments, meaning matrix-assisted (e.g. on DNA or, membrane or other surfaces) (Ghosal and Löwe 2015). The MinD sandwich dimer was solved by crystallography (Hu et al. 2003; Wu et al. 2011) and MinD dimerisation leads to the exposure of a short (10–12 amino acids) C-terminal amphipathic helix (also known as membrane targeting helix: MTS) that targets MinD to the cytoplasmic membrane (Szeto et al. 2003). Membrane bound MinD recruits MinC to the cytoplasmic membrane and forms an "active" inhibitor complex that inhibits Z-ring assembly anywhere but at midcell.

MinE Activates MinD ATPase

MinE is a small dimeric protein whose monomers have two separate functional domains (Ghasriani et al. 2010; Kang et al. 2010; Park et al. 2011). The N-terminal domain restricts MinCD's inhibitory activity by disassembling the complex from the membrane, while the C-terminal domain is responsible for MinE's topological specificity, that is, it regulates MinE's localisation (Ramos et al. 2006). Structural and biochemical evidence suggest that MinE switches between two very different conformations. In its freely diffusible cytoplasmic form, the MinD interacting residues and the membrane targeting sequence (MTS) are sequestered within a six-stranded $\beta -$ sheet of the MinE dimer. However, on sensing the membrane-bound MinD, MinE undergoes drastic conformational changes and collapses into a four-stranded $\beta -$ sheet MinE dimer, exposing two contact helices that interact with the membrane bound MinD (Park et al. 2011). First, MinE displaces MinC from the MinCD inhibitory-complex, followed by stimulation of the ATPase activity of MinD. This results in dissociation of the MinCD complex from the membrane since MinD membrane binding requires the dimer and dimerisation is ATP dependent. Free MinC and MinD diffuse away and rebind to the cytoplasmic membrane at other locations, only to repeat the process. Starting from small random fluctuations this results in a biological oscillation or pattern formation reaction following principles as described theoretically by Alan Turing (Gierer and Meinhardt 1972; Turing 1952).

Structure and Activity of MinCD

The dynamic behavior of the Min system was first discovered by imaging fusions of MinD protein to GFP in *E. coli* (Raskin and de Boer 1999). GFP-MinD was found to continuously oscillate between the two poles of a cell. Subsequently, MinC was also found to oscillate, in a pattern very similar to MinD (Hu and Lutkenhaus 1999). The MinDE component of the Min system is suggested to constitute a reaction-diffusion device that drives oscillation (Kruse et al. 2007), and MinC is a cargo in this oscillation process. *In vitro* reconstitution studies provided direct evidence that MinD and MinE only self-organise to produce regular surface waves/patterns on a planar lipid bilayer in the presence of ATP (Ivanov and Mizuuchi 2010; Loose et al. 2008, 2011), although the wavelengths observed did not fit the *in vivo* situation (Fig. 7.2f).

Molecular Interactions

Studies with *E. coli* Min proteins have revealed important insights into the molecular basis of Min oscillation. However, the molecular basis of FtsZ inhibition by the Min system has remained somewhat elusive. Two major questions that remain are: 1) How does MinD activate MinC *in vivo* or, what is the molecular architecture of the MinCD inhibitory complex? 2) What is the mechanism of Z-ring inhibition by the MinCD inhibitor complex? In cells, the MinC concentration has been estimated to be ~40 times lower than that of FtsZ. MinD recruits MinC to the membrane and increases its inhibitory activity 25–50 fold, thus making it an efficient inhibitor. It has been suggested that the recruitment of MinC to membrane by MinD constitutes the activation process. However, Joe Lutkenhaus's group identified a mutation within the switch II region providing a MinD molecule that could recruit MinC to the division site but did not activate it, suggesting that activation of MinC requires more than recruitment of MinCD complex to the membrane.

MinC-MinD form Bipolar Copolymers in vitro *and in cells*

The MinC-MinD cocrystal structure (Ghosal et al. 2014) revealed an interesting feature of their interaction. The MinC binding sites on the MinD dimer are on opposite sides and similarly, the MinD binding sites on the MinC dimer are equally on opposite sides of the MinC dimers (Fig. 7.2a). This means that MinC and MinD cannot form a closed symmetrical (2 + 2 dimer) complex but, instead, must form an open $MinC_2$-$MinD_2$ complex with further binding sites available at both ends of the structure. Thus, the $MinC_2$-$MinD_2$ complex will further extend linearly, to form –$MinC_2$-$MinD_2$-$MinC_2$-$MinD_2$–, and indefinitely, given enough free monomers. In short, the two proteins are capable of forming an alternating copolymeric filament (Fig. 7.2b). Indeed, using biochemical studies, electron microscopy and *in vivo* analyses Ghosal and colleagues (Ghosal et al. 2014) showed that, at physiological concentrations, MinCD forms a new class of alternating copolymeric filaments on membrane bilayers (Fig. 7.2c). Reconstitution of the entire MinCD-FtsZ super-complex on liposome membranes then revealed that MinCD filaments bind membranes and also bind to FtsZ filaments with high avidity but, surprisingly, do not disassemble them. Based on these observations, Ghosal and colleagues proposed that MinD optimally activates MinC by forming copolymers and that their periodic structure, following the 4 nm repeat of FtsZ, seems ideal to help them to inhibit FtsZ filaments from forming a Z-ring, by either altering the structural integrity of the filaments or by disrupting lateral interactions (discussed later) (Fig. 7.2e). Through avidity, MinCD copolymers will thus target FtsZ filaments for inhibition over FtsZ monomers and this could provide an explanation for how substoichiometric amounts (~ 40 times less) of MinC effectively inhibit Z-ring assembly *in vivo* (Fig. 7.2e). Recently, Park and colleagues (Park et al. 2015) extended the above study using genetics with mutants that interfere with the copolymer-forming interfaces but not dimerisation and proposed that membrane recruitment of individual MinCD complexes is enough to inhibit Z-ring formation and that minimally MinC and MinD heterodimer formation is sufficient in a genetic setup that tests for complete cell division inhibition. As the authors point out, this does not explain why the molecules have the ability to form polymers and seem to act as dimers only, with unoccupied binding sites . Also, if recruitment of MinCD complex on to the membrane is enough for activation, then why is the MinD (switch-II region) mutant that could recruit MinC to the membrane is still inactive? And, finally, how does one MinC inhibit on average 40 FtsZ molecules? Clearly, more work is needed to elucidate the precise molecular mechanism of FtsZ inhibition by MinC.

Structure of FtsZ-rings *in vivo*

Z-ring Tethering to Membrane

The Z-ring must be anchored in order to transmit constriction force to the membrane or at least to stay synchronised with other events at the membrane or in the cell envelope. In *E. coli,* tethering is supported by either FtsA or ZipA, or the combined effects of both (Hale and de Boer 1999; Pichoff and Lutkenhaus 2005). Indeed, the *ftsA* gene of *E. coli* is essential. In contrast, *ftsA* can be disrupted in *B. subtilis* by insertional inactivation (Beall and Lutkenhaus 1992; Jensen et al. 2005). Here, it seems that tethering to the membrane is additionally provided by SepF (Duman et al. 2013) and EzrA (Singh et al. 2007).

FtsA

The *ftsA* gene is often found in an operon with *ftsZ* within the *dcw* (division and cell wall) cluster. The two proteins interact at the earliest stage of Z-ring formation and are essential for recruitment of the next set of proteins of the divisome. FtsA

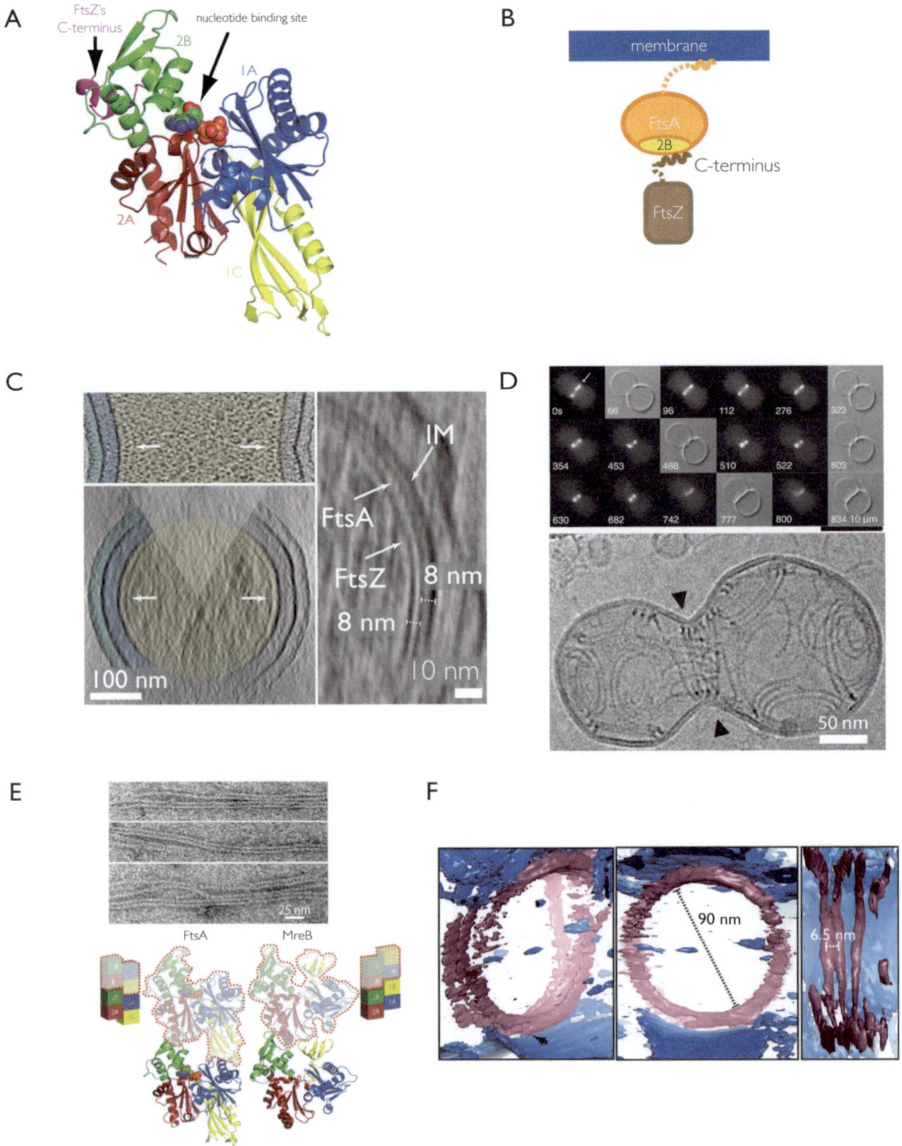

Fig. 7.3 Recent advances in understanding of the Z-ring architecture. (**a**) FtsA molecule with the C-terminal 16-residue FtsZ peptide (*purple*) bound on the surface of subdomain 2B (*green*). FtsA has been colour-coded according to the conserved actin family of proteins subdomain architecture (Modified from Szwedziak et al. 2012). (**b**) Architecture of the FtsA:FtsZ membrane-bound complex, the basic subunit of the Z-ring. (**c**) *Left: C. crescentus* NA1000/CB15N division site with filaments near the membrane (*top panel*, *black dots* highlighted by *arrow*). *Bottom panel* shows the same cell rotated 90° around the short axis of the cell. The Z-ring (*arrow*) is continuous and only invisible where there is no image because of the missing wedge (*shaded triangle*). *Right: E. coli* cells co-expressing FtsZ and FtsA (bicistronic, 1:1). FtsA filaments are in the middle between FstZ filaments and the inner membrane (IM), at a distance of 8 nm from both (Modified from Szwedziak et al. 2014). (**d**) A comparison of *in vitro* reconstitution experiments where FtsA and FtsZ

contains a conserved C-terminal motif, which is predicted to form an amphipathic helix and interact with the membrane (Pichoff and Lutkenhaus 2005).

Essential Role of FtsA in *E. coli*

If FtsA is prevented from localising at midcell, cell division is severely impaired and FtsZ rings are unable to contract (Addinall et al. 1996). Furthermore, a proper ratio of FtsZ to FtsA is required for cell division to occur (Dai and Lutkenhaus 1992; Dewar et al. 1992). An effect of the overproduction of either protein is diminished by overproduction of the other. These findings prompted an idea that FtsZ and FtsA may interact directly. Subsequently, FtsZ and FtsA were shown to colocalise in living *E. coli* cells by means of green fluorescent protein (Ma et al. 1996) and yeast two-hybrid analysis detected the interaction (Wang et al. 1997). Interestingly, FtsZ and FtsA from evolutionarily distant species such as *E. coli* and *B. subtilis* (Wang et al. 1997) or *Rhizobium meliloti* and *Agrobacterium tumefaciens* (Ma et al. 1997) were found to form complexes.

An *in vivo* genetic screen identified FtsA mutants on the surface of subdomain 2B impaired for FtsA-FtsZ interaction and these mutants retained the ability to bind to the membrane (Pichoff and Lutkenhaus 2007). For FtsZ, the last 70 residues were demonstrated to be involved in FtsA binding (Haney et al. 2001). Later, it was shown that a gene with a deletion of just 12 amino acids from the highly conserved C-terminal tail of *E. coli* FtsZ failed to complement chromosomal *ftsZ* mutants and failed to interact with FtsA (Ma and Margolin 1999). Finally, a crystal structure of FtsA bound with the C-terminal FtsZ peptide confirmed the data and provided structural details of this important interaction (Szwedziak et al. 2012) (Fig. 7.3a).

Fig. 7.3 (continued) were encapsulated inside liposomes. *Top panel* (Modified from Osawa and Erickson 2013) represents the FLM approach where the diffraction limit imposes constraints in determining liposomes sizes (please consult the scale bar) but can deliver time-lapse data. CryoEM approach (*bottom panel*, Modified from Szwedziak et al. 2014) is capable of visualising single FtsA and FtsZ filaments in compartments of sizes more similar to bacterial cells. It is clear to see that constriction sites are only formed where a ring made of the two proteins is present (*black arrowheads*) and not at other sites where filaments are located. (**e**) FtsA polymerises on a lipid monolayer forming long filaments, which often form doublets. X-ray protein crystallography provided structural basis of filament formation. A comparison of FtsA and MreB dimers. Monomeric subunits are rimmed in *red*. FtsA and MreB adopt a similar subdomain architecture; however, the 1B subdomain of MreB (*yellow*) is missing in FtsA and the inserted subdomain 1C in FtsA is located on the other side of the molecule. It therefore appears that FtsA is a subdomain variation of the actin fold that still enables the formation of canonical protofilaments (Modified from Szwedziak et al. 2012). (**f**) Close-up view of the FtsZ ring (*purple*) attached to the membrane (*blue*). The filaments overlap and interact laterally (*left panel*). View along the long axis shows that the ring is a perfect, closed circle (*middle panel*). Individual filaments are resolved (*right panel*) (Modified from Szwedziak et al. 2014)

Many Ways to Connect FtsZ to the Membrane

Interestingly, the same FtsZ interaction site within the last ~15 FtsZ residues is used by another protein, ZipA (Mosyak et al. 2000), which is a bitopic transmembrane protein that shows some functional overlap with FtsA, despite being very different in sequence. The exact functions of these proteins, FtsA and ZipA, and their relationship with FtsZ have remained enigmatic. A single residue alteration in *E. coli* FtsA, R286W (FtsA*), allows efficient division in the absence of ZipA (Geissler et al. 2003). Furthermore, although the C-terminal amphipathic helix of FtsA is essential for its function *in vivo* and plays a key role in targeting both FtsA and FtsZ to the membrane (Pichoff and Lutkenhaus 2005), FtsA's C-terminal membrane-targeting sequence can be substituted by MinD's membrane targeting sequence (MTS) and *vice versa* without obviously affecting functionality. Success in substitution also applies to the first transmembrane domain of MalF (Shiomi and Margolin 2008). This strongly suggests that the sequence of the membrane-targeting motif is not important, indicating that it serves only as a generic membrane anchor. Deletions as short as 5 residues from the C-terminus destroy the biological function of FtsA further emphasising the importance of membrane binding (Yim et al. 2000). Since FtsA is quite well conserved in bacterial species it probably constitutes the primary membrane anchor for the Z-ring, together with SepF (Fig. 7.3b). The outstanding question of why FtsZ is not anchored to the membrane directly will be discussed later.

Assembly of the Divisome

The interaction network between divisome proteins has been studied using various approaches, most notably bacterial two-hybrid assay (B2H) (Karimova et al. 2005) or Förster Resonance Energy Transfer (FRET) (Alexeeva et al. 2010), where binary protein-protein interactions may be tested. In conjunction with systematic mutagenesis and knockouts of divisome protein genes and subsequent localisation, these studies revealed the temporal hierarchy of protein recruitment and assembly regulation in *E. coli* (Vicente and Rico 2006). Assembly of the divisome is a linear two-step process: the early-assembling proteins, directly interacting with FtsZ, are first recruited to midcell and constitute the inner divisome. This includes FtsA, ZipA and ZapABCD and other auxiliary proteins that are not essential. Then, after a substantial delay, late division proteins are recruited (FtsK, FtsQ, FtsL, FtsB, FtsW, FtsI and finally FtsN) and they constitute the outer divisome. These proteins include components that are cytoplasmic, periplasmic and membrane-embedded (Goehring and Beckwith 2005). Recently, an unexpected link between early FtsA and late FtsN proteins has been discovered (Busiek et al. 2012). Despite being temporally resolved, they were shown to interact in the mature Z-ring. Formation of the FtsA:FtsN complex might be an answer to the elusive question of what triggers final constriction and peptidoglycan remodelling at the cell division site (Busiek and Margolin 2014), in a way that could be described as an assembly checkpoint of the divisome.

Discovery of the Z-ring Using Immuno-Gold and GFP Imaging

The term Z-ring was coined in a landmark publication (Bi and Lutkenhaus 1991) showing, by means of anti-FtsZ immuno-gold labelling electron microscopy of dividing *Escherichia coli* cells, that FtsZ localises to division sites and forms a cytoskeletal filament. Subsequently, Levin and Losick (Levin and Losick 1996) using immunofluorescence microscopy of fixed *Bacillus subtilis* cells visualised FtsZ as bands of fluorescence across the short axis of the cells and positioned between replicated nucleoids. Additionally, these bands were demonstrated to move close to one cell pole as the cells switch from vegetative growth to sporulation. Early attempts to visualise the Z-ring coincide with the onset of the green fluorescent protein (GFP) revolution in optical cellular imaging. Since protein filaments should be readily visible and would display strong fluorescent signal above the background level, FtsZ was the first bacterial protein to be expressed as a GFP fusion in live cells. The first three-dimensional images of FtsZ appearing as a closed ring at the midcell emerged in 1996 (Ma et al. 1996). Later, some cells were reported to contain incomplete rings and artificially high levels of FtsZ-GFP due to overexpression resulted in other structures such as spiral tubules, reflecting the ability of FtsZ filaments to coalesce into different arrangements.

Therefore, according to conventional FLM, using either immunofluorescence on fixed cells or FtsZ-GFP fusion proteins in live cells, the Z-ring appears to be a more-or-less complete circular entity with a clear lumen, which decreases in diameter to a diffraction-limited size over several minutes (Sun and Margolin 1998). FtsZ can also, under special conditions (e.g. asymmetric cell division in sporulating *B. subtilis*), form highly dynamic helical structures, oscillating between the midcell and the cell pole (Ben-Yehuda and Losick 2002). These pioneering studies revealed also the main caveat of this method, as FtsZ-GFP fusion proteins do generally not fully complement an *ftsZ* null mutant.

Z-ring Models

As detailed in Chap. 6, three models of FtsZ protofilament arrangement in a continuous Z-ring are consistent with estimates of the number of FtsZ molecules per cell: (A) a flat-ribbon model, where filaments overlap and interact laterally but radially the ring is just one protofilament thick; (B) a bundle model where the protofilaments are loosely arranged within the ring; (C) a random model, similar to the ribbon model but without strong lateral associations as there are large spaces between FtsZ protofilaments.

A Model of the Z-ring from PALM Imaging

Using the single molecule-based, super resolution fluorescent-light microscopy (SR-FLM) method PALM (photoactivated localisation microscopy), the Z-ring in *E. coli* has been visualised with a nominal (resolving) resolution of 35 nm (Fu et al. 2010). The ring was 110 nm wide, consistent with being a few protofilaments thick (Erickson et al. 1996; Huecas et al. 2008; Löwe and Amos 2000).

The FtsZ-ring May Appear to Be Discontinuous

STED

Somewhat different images of the Z-ring were produced with stimulated emission depletion (STED) microscopy analysis of *Bacillus subtilis* (Jennings et al. 2010). Here, at a nominal resolution of 80–90 nm the Z-ring appeared to be very irregular, erratic and discontinuous. The authors concluded that this could be caused by the FtsA protein, which has been proposed to snap FtsZ polymers into short protofilaments (Beuria et al. 2009).

3D-SIM

Another form of super resolution microscopy, known as 3D–structured illumination microscopy (3D–SIM) was used to examine the architecture of the Z-ring in two Gram positive bacteria of different shapes: rod-shaped *B. subtilis* and coccoid *Staphylococcus aureus* (Strauss et al. 2012). In both organisms, gaps or strong fluctuations of fluorescence intensity along the ring suggested that the ring is discontinuous and the distribution of FtsZ is at least heterogeneous around the ring. Time-lapse analyses confirmed that, once formed, the Z-ring is a dynamic structure. 3D–SIM in *E. coli* (Rowlett and Margolin 2014) and a recent high-throughput PALM study in *Caulobacter crescentus* (Holden et al. 2014) revealed similar patchy architectures. Additionally, FtsA and ZipA, the membrane anchors, were shown to mostly co-localise in similar patches around the cell circumference.

FRAP

According to FRAP (Fluorescence Recovery After Photobleaching) results, the Z-ring undergoes constant remodelling with a half-recovery in the order of 30 s (Stricker et al. 2002).

Effects of FtsZ-Bundling Proteins on the Ring Structure

The role of FtsZ–associated proteins (ZapA, ZapB, ZapC and ZapD) (Ebersbach et al. 2008; Hale et al. 2011) on the ring architecture might be important, although clearly, under normal circumstances these proteins are not essential. Some of the proteins have been proposed to promote FtsZ bundling *in vitro*. PALM analyses revealed (Buss et al. 2013) that their role *in vivo* might be to align FtsZ clusters containing multiple FtsZ protofilaments and participate in organising the ring structure as in the absence of ZapA and ZapB FtsZ bundles disperse and lead to abnormal septa, although this could also be caused by ZapB's interaction with the MatP system (Espéli et al. 2012).

Limitations of FLM for the Investigation of the FtsZ-ring Structure

Several problems may have arisen in previous studies of the Z-ring and divisome using GFP derivatives as fluorescent fusions to FtsZ and other proteins. Bulky tags such as GFP very often make the protein of interest non-functional (Margolin 2012) or trigger artefacts due to dimerisation (Swulius and Jensen 2012). These effects are particularly prevalent with filaments because of the high density of tags. For example, the result can be dominant negative effects, which are particularly obvious when protein filaments are poisoned with non-functional subunits at the ends. In some studies, the proteins were ectopically expressed and therefore their cellular level were far from native. Often, overexpression is used in order to overcome non-functionality, but this often leads to non-functional protein accumulating in non-native locations. In many studies, the native, unlabelled protein is still present but invisible. And, except when single molecule counting techniques are used, FLM is not able to demonstrate the absence of something and a general problem arises when a faint signal is assigned as background through scaling of the images. At best, fixation of cells will more-or-less negate the increase in resolution that can be achieved through the use of super resolution methods. And localisation super resolution methods can be limited by counting statistics of all the fluorophores in the sample.

CryoEM Studies of Z-ring Structure

To employ cryoEM techniques, unique and unstained structures, such as cells, are plunge-frozen in liquid ethane in order to preserve cellular structures in their native state at the exact moment of vitrification. When a technique called electron cryotomography (cryoET) is utilised, a series of transmission 2D images of the sample tilted in the electron beam is recorded and a 3D reconstruction is calculated. Recent technological advances in cryoEM hardware, specimen preparation and automated data collection have allowed studies of supramolecular assemblies in their natural environment in intact cells at resolutions approaching 3 nm. It should be stressed that there is currently no high-resolution labelling technology, analogous to GFP for FLM available for cryoET, limiting all imaging to natural contrast, which is very low for biological materials containing only light elements.

First CryoEM Study of Z-ring Structure

Due to its small size, *C. crescentus* cell specimens produced the first good tomograms (Li et al. 2007). This pioneering study showed individual, overlapping, scattered FtsZ filaments ~5 nm in diameter along the inner curvature of the cell. Most of the filaments were 40–120 nm long. The filaments were seen approximately perpendicular to the long axis of the cell and at a distance of 16 nm from the inner membrane, leaving some room for the membrane anchor and long tail of FtsZ. Interestingly, the lateral spacing between filaments turned out to be ~9.3 nm. According to the authors, no complete rings were visible unless a mutant was overexpressed and the number of filaments on one side of the cell was different from the number on the opposite side.

More CryoEM Studies of Z-ring Structure

The advent of direct electron detectors (Faruqi et al. 2015) made it possible to obtain a better-resolved *C. crescentus* (Fig. 7.3c**, left**) Z-ring images by cryoET and to tackle the problem in much thicker *E. coli* cells (Szwedziak et al. 2014), traditionally a serious obstacle for cryoET that struggles with thicknesses above ~600 nm when electrons at 300 kV are used. In both species the Z-ring was made of a single-layered, small band of filaments parallel to the membrane creating a most likely continuous ring. The problem of the lack of a specific cryoEM label was addressed by an innovative approach. Systematic perturbations of the system were introduced by modifying residues at the C-terminus of FtsZ and subsequent imaging by cryoET, for example: introducing a Q-rich linker between FtsZ's globular domain and the small C-terminal helix that binds FtsA placed the Z-ring at a distance of 21 nm from the inner membrane, as compared to 16 nm for the wild-type protein, demonstrating that FtsZ had been imaged.

Effects of the Missing Wedge of CryoET

The missing wedge in cryoET is a purely technical limitation arising from the inability to tilt the specimen in electron cryomicroscopes to angles much higher than 60–70°. After reconstruction, this causes features along the direction of the electron beam to be elongated and some features perpendicular to both tilt axis and electron beam to be distorted or completely invisible (Palmer and Löwe 2013). However, protein filaments were visible and uninterrupted except where the missing wedge affected the imaging, which can be safely gauged from the disappearance of the cell membrane and envelope. The existence of complete rings is supported by perfect matching of apparent gaps in constricting rings with the missing wedges in all of the analysed tomograms, in entirely random orientations; for this to occur just by coincidence would be remarkably implausible. Additionally, by providing both FtsA and FtsZ in excess simultaneously, many extra constriction events became visible and importantly, these septa were functional, as they produced minicells, consistent with a previous study (Begg et al. 1998). Altering the FtsA:FtsZ ratio to be close to 1:1 made FtsA filaments visible between the inner membrane and the layer corresponding to FtsZ, at a distance of 8 nm from each (Fig. 7.3c**, right**). It should also be noted that a more careful investigation of different stages of the cell cycle and the division process need to be performed in order to understand the ring's genesis from its seeding points, whatever and where ever they are, but clearly the structure cannot initially be a closed ring.

In vitro Z-ring Reconstitution Experiments

The conclusion that one could produce extra constriction sites and divisions by providing more FtsZ and FtsA prompted questions of whether the constriction force might be generated solely by targeting FtsZ to the membrane (the Z-centric hypothesis) and whether this could be done by a generic anchor or whether FtsZ and specifically FtsA are necessary and sufficient to carry out some constriction by themselves. Therefore, to uncouple these possibilities from the potential effect of cell wall synthesis (the scaffolding model), reconstitution experiments were initiated (Loose and Schwille 2009; Martos et al. 2012; Osawa et al. 2008; Rivas et al. 2014). Purified components, including the membrane bilayer, were used in well-defined experimental setups where experimental parameters can be controlled and measured precisely and additional components, as in cells, are absent.

FtsZ filament assembly has been studied *in vitro* by negative stain EM and atomic force microscopy (AFM). Purified protein was polymerised in the presence of different nucleotides and/or various accessory proteins and stained with uranyl acetate (Bramhill and Thompson 1994; Erickson et al. 1996; Popp et al. 2009; Small et al. 2007). A variety of structures, depending on the conditions applied, have been imaged. In addition to straight filaments, small, ring-shaped polymers represent a prominent species of FtsZ. They adopt a highly curved conformation with a 22°

bend between subunits and a ring diameter of 24 nm (Erickson et al. 1996). Alternatively, a moderately curved FtsZ filament conformation with a 2.5° bend has been reported and they were obtained in the presence of GTP or GMPCPP – a slowly hydrolysable GTP analogue (Mingorance et al. 2005).

Membrane Constriction by FtsZ-MTS

The earliest attempts to reconstitute the Z-ring *in vitro* made use of a hybrid FtsZ made by replacing FtsZ's tail with yellow fluorescent protein (YFP) and an MTS (membrane targeting sequence) from MinD, resulting in FtsZ-MTS (Osawa et al. 2008). *In vitro*, FtsZ-MTS assembled into multiple Z-rings when mixed with large multilamellar vesicles in the presence of GTP. These perfectly closed rings were oriented perpendicular to the long axis of the tube and triggered partial constrictions. However, complete liposome divisions were not observed. Interestingly, when bound to the outside of liposomes, the protein formed concave invaginations of the liposome surface (Osawa et al. 2009). Moving the MTS to the N-terminus triggered formation of convex bulges on the liposomes. The idea that intrinsic FtsZ protofilament curvature could generate the constriction force favours the bending model. Study of FtsZ's GTPase activity showed that nucleotide hydrolysis is essential for rapid exchange of FtsZ-MTS within the ring but without GTP turnover, Z-rings could still assemble and generate only an initial constriction (Osawa et al. 2009; Osawa and Erickson 2011). This suggests that the dynamics of the ring are indispensable for its function under these conditions. How the proteins ended up on the inside of the liposomes was not investigated.

Liposome Constriction by FtsZ/FtsA

More recently, FtsZ was used together with FtsA* as the membrane anchor (Osawa and Erickson 2013). FtsZ-YFP and FtsA were mixed with unilamellar liposomes (Fig. 7.3d**, top**). Here, at least some constrictions appeared to complete abscission (division), unlike FtsZ-MTS, which indicated that FtsA is critical for abscission and probably proper ring closure. *E. coli* FtsZ and FtsA, labelled with small organic dyes that impose less steric burden than GFP, were incorporated inside giant unilamellar vesicles (GUVs) (Jimenez et al. 2011). Confocal microscopy of such vesicles revealed the spatial distribution of filaments and liposome's deformation upon GTP addition. Although clear deformations were visible, no ring like structures were discernible, possibly due to the vesicles' sizes, which were one order of magnitude larger than regular bacterial cells. Similar results were obtained by encapsulating FtsZ and ZipA (Cabre et al. 2013), which induced vesicle shrinkage.

Rings of Both FtsZ and FtsA Filaments?

More detailed studies using electron microscopy and purified components showed that FtsA indeed binds to the membrane and forms actin-like filaments, on 2D lipid bilayers (Fig. 7.3e) (Szwedziak et al. 2012), despite possessing an unusual actin-domain architecture (van den Ent and Löwe 2000). This prompted an idea that FtsA filaments might have a more structural role in Z-ring function. As there is not enough FtsA in cells to form a complete A-ring, at least at the beginning of constriction, it has been envisaged that FtsA polymers might help with bending through the repeat mismatch (FtsZ ~4 nm, FtsA ~5 nm), although this clearly can only happen locally around the ring, initially, but might become more important at the end of constriction when all FtsA will be concentrated in a much smaller ring.

FtsZ and FtsA Reconstituted on Flat Membranes

Total internal reflection fluorescence microscopy (TIRFM) provided more insights into the dynamics of the FtsZ-FtsA system (Loose and Mitchison 2014). High temporal resolution imaging of FtsA-dependent recruitment of FtsZ polymers to supported membranes unveiled rotating rings, consisting of a polar arrangement of FtsA:FtsZ filaments, with the FtsZ filaments treadmilling. However, constricting rings were not observed, most likely because of the geometry of the membrane, with the membrane attachment 90° rotated when compared to cells. Interestingly, similar spiral filament arrangements on a flat support was obtained by investigating the FtsA:FtsZ filaments using negative stain EM (Szwedziak et al. 2014).

3D Structure of Reconstituted Rings in Liposomes

Unmodified FtsA and FtsZ from *Thermotoga maritima*, for their superior biochemical properties, were incorporated into lipid vesicles made from *E. coli* lipid extract and imaged using cryoET (Szwedziak et al. 2014) (Fig. 7.3d**, bottom**). Analysis of the three-dimensional arrangements of FtsZ and FtsA filaments reported that the Z-rings were composed of a small, single-layered band of filaments parallel to the membrane, creating a continuous ring through lateral filament contacts, with the filaments being minimally 6.5 nm apart. The ring was always a perfectly round circle suggesting that it was exerting force everywhere around the ring (Fig. 7.3f). Although, clearly, single snapshots like those obtained by cryoET make it difficult to deduce a timeline for the generation and progression of the observed rings, it was concluded that for the liposomes, rings of FtsA and FtsZ seemed to constrict the liposomes and that this could only happen if filaments slid against each other as the rings became smaller. It was noted that a complete ring structure would be self-regulating, as significant and symmetrical force generation would only be able to progress to much smaller diameters when a complete ring is formed.

Interestingly, when the same *Thermotoga* FtsA and FtsZ proteins were added to the outside of liposomes, they did not form ring-like structures but rather intrinsically bent co-polymers were bound to areas of negative curvature, having created negative membrane curvature from the outside. It was proposed that the intrinsic bending is caused by the mismatch of spacing between the proteins subunits in the two filaments as mentioned above, as FtsZ by itself was demonstrated to form straight polymers by cryoEM when not on membrane (Szwedziak et al. 2014).

Possible Drivers of Constriction

FtsZ has been demonstrated by negative stain EM to form 2D crystals with tight, continuous lateral interactions (Erickson et al. 1996; Löwe and Amos 2000) and this was done using FtsZ only, without the membrane anchor. The lateral spacing of 6.5–8 nm between filaments reported by Szwedziak and colleagues is slightly larger than the thickness of FtsZ filaments and presumably also FtsA filaments (Szwedziak et al. 2012). Previous *in vitro* work reported an interfilament distance of only 5 nm using FtsZ-MTS; however, this was after negative staining and no constrictions were observed, possibly due to lateral interactions being crystalline and hence too tight for any movement in the structure (Milam et al. 2012). In *C. crescentus* cells, the lateral spacing between filaments was found to be 9.3 nm (Li et al. 2007) and in the later study ~7.8 nm in *C. crescentus* and ~6.8 nm in *E. coli* (Szwedziak et al. 2014). AFM on a flat substrate, using only FtsZ, but observing spirals, provided an even larger distance of 12 nm (Mingorance et al. 2005). All of these measurements are averages with large variances. One may conclude that the filaments in the FtsZ ring interact transiently and direct contact is localised to only a few small regions within the ring at a time. This could facilitate the constriction process since the filaments have to be free to slide, if sliding is to happen. It was suggested previously that instead of forming many intermolecular solid bonds, which would lead to avidity and a barrier to sliding, an attractive force over a longer distance would allow the filaments to remain apart while interacting (Horger et al. 2008), although it is unclear what that force could be if the filaments are strongly negatively charged, as suggested from crystal structures.

Role of FtsA in Constriction

Loose and Mitchison have demonstrated that FtsZ's dynamic behaviour, including large-scale reorganisation of the filament network, is not its intrinsic property but rather emerges from the interaction with FtsA (Loose and Mitchison 2014). Alternatively, or simultaneously, it has been proposed that the repeat length mismatch of FtsA and FtsZ can be vital for the constriction process (Szwedziak et al. 2012, 2014). Given that the FtsZ:FtsA ratio needs to be around 5:1 at least at the start, there would not be enough FtsA to encircle the cell circumference at the very beginning of constriction. But at later stages less protein overall is needed to make a ring, so it is conceivable that there would be enough FtsA at the end when the septum is smaller. Also, the decreasing diameter of the ring, accompanied by increasing membrane curvature, would enable more and more FtsA subunits to polymerise, while still matching up 1:1 with FtsZ subunits, until a complete FtsA ring has been achieved. This may affect the activity of other divisomal components and could also provide additional energy incrementally. The mismatch between the two polymers may explain why FtsA exists at all, instead of FtsZ being directly attached to the membrane. Of course, the prediction of an FtsA ring at later stages of constriction is testable but currently speculation.

Role of Nucleotides During Constriction

Finally, what is the role of nucleotide hydrolysis during constriction of the inner divisome? Clearly, cell wall synthesis can generate force through the incorporation of lipid-II cell wall precursors, a process that releases energy. But both FtsZ and FtsA bind nucleotides and harbour hydrolase active sites for them. FtsA was found to crystallise with an ATP molecule bound in the active site (van den Ent and Löwe 2000) and biochemical studies with purified FtsA from *Streptococcus pneumoniae* confirmed that the protein preferentially binds ATP (Lara et al. 2005). However, no ATPase activity was detected. Work with *B. subtilis* FtsA, however, detected some trace hydrolytic activity but it is not clear if this is due to the protein itself or represents contamination (Feucht et al. 2001). It seems fair to say that there is currently no conclusive evidence that FtsA turns over ATP on its own. Perhaps the protein requires a yet unidentified binding partner for activation.

In contrast, FtsZ's GTPase activity was described already in the early 1990s during the Z-ring discovery era (de Boer et al. 1992) and has formed the basis of several theories of constriction force generation, particularly the bending model, as discussed in Chap. 6. Results obtained *in vitro* using a non-hydrolysable GTP analogue (Osawa and Erickson 2011) suggest that even without GTP hydrolysis Z-rings assemble on liposomes *in vitro* and generate initial constrictions but constriction stopped at later stages. A GTPase-deficient FtsZ mutant also appeared to function surprisingly well in cell division (Mukherjee et al. 2001). Recent *in vitro* work seems to suggest that liposome constriction through FtsZ and FtsA is also nucleotide hydrolysis independent since no nucleotides were added or present in some experiments (Szwedziak et al. 2014). The implication of this could be that nucleotide binding is required solely for filament growth and nucleotide hydrolysis for shrinkage as these are essential to maintain the highly dynamic state of the Z-ring. When constriction starts at large diameters, GTP-induced FtsZ polymerisation is needed to produce longer filaments that can reliably reach around the cell and also produce overlap. However, as the constriction progresses, the overlap increases and this could lead to a kinetic barrier either of sliding or any other process such as bending or pinching. Hydrolysis might be needed to keep filaments short enough for the system to remain dynamic. This process would be self-regulating as hydrolysis is an intrinsic property of the FtsZ polymer. Since polymer yield depends on the ring diameter (limited by membrane surface availability), fewer active sites for GTP hydrolysis would be present at a later stage so that overall dynamics remain constant.

Conclusions and Future Directions

One of the prerequisites for the existence of successful cellular forms of life is accurate and controlled division, after DNA replication and distribution of the replicated DNA to the two new daughter cells.

In order to obtain mechanistic insights into membrane dynamics during bacterial cell division a plethora of methods drawn from biochemistry and biophysics has been employed. FLM has produced data on how the Z-ring is remodelled, EM and cryoEM have elucidated the arrangement of filaments within the ring, X-ray crystallography has provided high-resolution structures of many components, making it possible to deduce insights from our chemical understanding of proteins.

However, as will have become obvious, many conflicting pieces of evidence currently make it impossible to produce a unifying model of FtsZ-based bacterial cell division.

Several key advances will be needed. Much better cryoET data of division sites should be able to resolve the most pressing questions about ring architecture. Carefully timing cells before imaging will also enable to produce a timeline of constriction and abscission. Better FLM labelling technology will be needed to produce strains that harbour fully functional fluorescent protein fusions before more conclusions are drawn from flawed experimental setups. Super-resolution techniques will provide many advances but must also make use of fully functional labelling and live cells without fixation. Bottom up, reconstitution approaches will play a vital role, but some caution is warranted as the divisome is complex and many proteins so far have proven to be biochemically problematic. And as so far, the involvement of membrane, peptidoglycan and outer membrane makes reconstitution a truly formidable and challenging task. More model organisms will be beneficial, especially those where different cell envelopes are involved. Cells without cell envelope such as L-forms and naturally ‘naked’ cells exist and might offer valuable insights in terms of cell division research. In short, there is much to come but in the end, a beautiful mechanism will emerge that will also offer possibilities and answers for evolutionary biology, antibiotic research and synthetic biology.

References

Adams DW, Errington J (2009) Bacterial cell division: assembly, maintenance and disassembly of the Z ring. Nat Rev Microbiol 7:642–653

Addinall SG, Bi E, Lutkenhaus J (1996) FtsZ ring formation in fts mutants. J Bacteriol 178:3877–3884

Adler HI et al (1967) Miniature *Escherichia coli* cells deficient in DNA. Proc Natl Acad Sci U S A 57:321–326

Alexeeva S et al (2010) Direct interactions of early and late assembling division proteins in *Escherichia coli* cells resolved by FRET. Mol Microbiol 77:384–398

Beall B, Lutkenhaus J (1992) Impaired cell division and sporulation of a Bacillus subtilis strain with the ftsA gene deleted. J Bacteriol 174:2398–2403

Begg K, Donachie WD (1998) Division planes alternate in spherical cells of *Escherichia coli*. J Bacteriol 180:2564–2567

Begg K et al (1998) Roles of FtsA and FtsZ in activation of division sites. J Bacteriol 180:881–884

Ben-Yehuda S, Losick R (2002) Asymmetric cell division in *B. subtilis* involves a spiral-like intermediate of the cytokinetic protein FtsZ. Cell 109:257–266

Bernhardt TG, de Boer PA (2005) SlmA, a nucleoid-associated, FtsZ binding protein required for blocking septal ring assembly over Chromosomes in *E. coli*. Mol Cell 18:555–564

Beuria TK et al (2009) Adenine nucleotide-dependent regulation of assembly of bacterial tubulin-like FtsZ by a hypermorph of bacterial actin-like FtsA. J Biol Chem 284:14079–14086

Bi EF, Lutkenhaus J (1991) FtsZ ring structure associated with division in *Escherichia coli*. Nature 354:161–164

Bork P, Sander C, Valencia A (1992) An ATPase domain common to prokaryotic cell cycle proteins, sugar kinases, actin, and hsp70 heat shock proteins. Proc Natl Acad Sci U S A 89:7290–7294

Bramhill D, Thompson CM (1994) GTP-dependent polymerization of *Escherichia coli* FtsZ protein to form tubules. Proc Natl Acad Sci U S A 91:5813–5817

Busiek K, Margolin W (2014) A role for FtsA in SPOR-independent localization of the essential *Escherichia coli* cell division protein FtsN. Mol Microbiol 92:1212–1226

Busiek KK et al (2012) The early divisome protein FtsA interacts directly through its 1c subdomain with the cytoplasmic domain of the late divisome protein FtsN. J Bacteriol 194:1989–2000

Buss J et al (2013) In vivo organization of the FtsZ-ring by ZapA and ZapB revealed by quantitative super-resolution microscopy. Mol Microbiol 89:1099–1120

Cabre EJ et al (2013) Bacterial division proteins FtsZ and ZipA induce vesicle shrinkage and cell membrane invagination. J Biol Chem 288:26625–26634

Cordell SC, Anderson RE, Lowe J (2001) Crystal structure of the bacterial cell division inhibitor MinC. EMBO J 20:2454–2461

Dai K, Lutkenhaus J (1992) The proper ratio of FtsZ to FtsA is required for cell division to occur in *Escherichia coli*. J Bacteriol 174:6145–6151

Dajkovic A et al (2008) MinC spatially controls bacterial cytokinesis by antagonizing the scaffolding function of FtsZ. Curr Biol 18:235–244

Davis BK et al (2002) Molecular evolution before the origin of species. Prog Biophys Mol Biol 79:77–133

de Boer PA, Crossley RE, Rothfield L (1988) Isolation and properties of minB, a complex genetic locus involved in correct placement of the division site in *Escherichia coli*. J Bacteriol 170:2106–2112

de Boer PA, Crossley RE, Rothfield L (1989) A division inhibitor and a topological specificity factor coded for by the minicell locus determine proper placement of the division septum in *E. coli*. Cell 56:641–649

de Boer PA et al (1991) The MinD protein is a membrane ATPase required for the correct placement of the *Escherichia coli* division site. EMBO J 10:4371–4380
de Boer PA, Crossley R, Rothfield L (1992) The essential bacterial cell-division protein FtsZ is a GTPase. Nature 359:254–256
Dewar SJ, Begg KJ, Donachie WD (1992) Inhibition of cell division initiation by an imbalance in the ratio of FtsA to FtsZ. J Bacteriol 174:6314–6316
Duman R et al (2013) Structural and genetic analyses reveal the protein SepF as a new membrane anchor for the Z ring. Proc Natl Acad Sci U S A 110:E4601–E4610
Ebersbach G et al (2008) Novel coiled-coil cell division factor ZapB stimulates Z ring assembly and cell division. Mol Microbiol 68:720–735
Erickson HP et al (1996) Bacterial cell division protein FtsZ assembles into protofilament sheets and minirings, structural homologs of tubulin polymers. Proc Natl Acad Sci U S A 93:519–523
Erickson HP, Anderson DE, Osawa M (2010) FtsZ in bacterial cytokinesis: cytoskeleton and force generator all in one. Microbiol Mol Biol Rev 74:504–528
Espéli O et al (2012) A MatP-divisome interaction coordinates chromosome segregation with cell division in : chromosome-divisome interplay. EMBO J 31:3198–3211
Faguy DM, Doolittle WF (1998) Cytoskeletal proteins: the evolution of cell division. Curr Biol 8:R338–R341
Faruqi AR, Henderson R, McMullan G (2015) Progress and Development of Direct Detectors for Electron Cryomicroscopy. Adv Ima Ele Physics 190:103–141
Feucht A et al (2001) Cytological and biochemical characterization of the FtsA cell division protein of Bacillus subtilis. Mol Microbiol 40:115–125
Fleurie A et al (2014) MapZ marks the division sites and positions FtsZ rings in Streptococcus pneumoniae. Nature 516:259–262
Fu G et al (2010) In vivo structure of the *E. coli* FtsZ-ring revealed by photoactivated localization microscopy (PALM). *PLoS One* **5**:e12682
Geissler B, Elraheb D, Margolin W (2003) A gain-of-function mutation in ftsA bypasses the requirement for the essential cell division gene zipA in *Escherichia coli*. Proc Natl Acad Sci U S A 100:4197–4202
Ghasriani H et al (2010) Appropriation of the MinD protein-interaction motif by the dimeric interface of the bacterial cell division regulator MinE. Proc Natl Acad Sci U S A 107:18416–18421
Ghosal D, Löwe J (2015) Collaborative protein filaments. EMBO J 34:2312–2320
Ghosal D et al (2014) MinCD cell division proteins form alternating copolymeric cytomotive filaments. Nat Commun 5:5341
Gierer A, Meinhardt H (1972) A theory of biological pattern formation. Kybernetik 12:30–39
Goehring NW, Beckwith J (2005) Diverse paths to midcell: assembly of the bacterial cell division machinery. Curr Biol 15:R514–R526
Guizetti J et al (2011) Cortical constriction during abscission involves helices of ESCRT-III-dependent filaments. Science 331:1616–1620
Hale CA, de Boer PA (1999) Recruitment of ZipA to the septal ring of *Escherichia coli* is dependent on FtsZ and independent of FtsA. J Bacteriol 181:167–176
Hale CA et al (2011) Identification of *Escherichia coli* ZapC (YcbW) as a component of the division apparatus that binds and bundles FtsZ polymers. J Bacteriol 193:1393–1404
Haney SA et al (2001) Genetic analysis of the *Escherichia coli* FtsZ.ZipA interaction in the yeast two-hybrid system. Characterization of FtsZ residues essential for the interactions with ZipA and with FtsA. J Biol Chem 276:11980–11987
Harry EJ, Monahan L, Thompson L (2006) Bacterial cell division: the mechanism and its precison. Int Rev Cytol 253:27–94
Hirota V, Ryter A, Jacob F (1968) Thermosensitive mutants of *E. coli* affected in the process of DNA synthesis and cellular division. Cold Spring Harb Symp Quant Biol 33:677–693

Holden SJ et al (2014) High throughput 3D super-resolution microscopy reveals Caulobacter crescentus in vivo Z-ring organization. *Proc Natl Acad Sci U S A* 111(12):4566–4571
Holečková N et al (2015) LocZ is a new cell division protein involved in proper septum placement in Streptococcus pneumoniae. MBio 6:e01700–e01714
Horger I et al (2008) FtsZ bacterial cytoskeletal polymers on curved surfaces: the importance of lateral interactions. Biophys J 94:L81–L83
Hu Z, Lutkenhaus J (1999) Topological regulation of cell division in *Escherichia coli* involves rapid pole to pole oscillation of the division inhibitor MinC under the control of MinD and MinE. Mol Microbiol 34:82–90
Hu Z, Lutkenhaus J (2000) Analysis of MinC reveals two independent domains involved in interaction with MinD and FtsZ. J Bacteriol 182:3965–3971
Hu Z, Saez C, Lutkenhaus J (2003) Recruitment of MinC, an inhibitor of Z-ring formation, to the membrane in *Escherichia coli*: role of MinD and MinE. J Bacteriol 185:196–203
Huecas S et al (2008) Energetics and geometry of FtsZ polymers: nucleated self-assembly of single protofilaments. Biophys J 94:1796–1806
Ivanov V, Mizuuchi K (2010) Multiple modes of interconverting dynamic pattern formation by bacterial cell division proteins. Proc Natl Acad Sci U S A 107:8071–8078
Jennings PC et al (2010) Super-resolution imaging of the bacterial cytokinetic protein FtsZ. Micron 42:336–341
Jensen SO, Thompson LS, Harry EJ (2005) Cell division in Bacillus subtilis: FtsZ and FtsA association is Z-ring independent, and FtsA is required for efficient midcell Z-Ring assembly. J Bacteriol 187:6536–6544
Jimenez M et al (2011) Reconstitution and organization of *Escherichia coli* proto-ring elements (FtsZ and FtsA) inside giant unilamellar vesicles obtained from bacterial inner membranes. J Biol Chem 286:11236–11241
Kang GB et al (2010) Crystal structure of Helicobacter pylori MinE, a cell division topological specificity factor. Mol Microbiol 76:1222–1231
Karimova G, Dautin N, Ladant D (2005) Interaction network among *Escherichia coli* membrane proteins involved in cell division as revealed by bacterial two-hybrid analysis. J Bacteriol 187:2233–2243
Katzmann DJ, Babst M, Emr SD (2001) Ubiquitin-dependent sorting into the multivesicular body pathway requires the function of a conserved endosomal protein sorting complex, ESCRT-I. Cell 106:145–155
Kiekebusch D et al (2012) Localized dimerization and nucleoid binding drive gradient formation by the bacterial cell division inhibitor MipZ. Mol Cell 46:245–259
Koonin EV (1993) A superfamily of ATPases with diverse functions containing either classical or deviant ATP-binding motif. J Mol Biol 229:1165–1174
Kruse K, Howard M, Margolin W (2007) An experimentalist's guide to computational modelling of the Min system. Mol Microbiol 63:1279–1284
Lan G et al (2009) Condensation of FtsZ filaments can drive bacterial cell division. Proc Natl Acad Sci U S A 106:121–126
Lara B et al (2005) Cell division in cocci: localization and properties of the Streptococcus pneumoniae FtsA protein. Mol Microbiol 55:699–711
Leger MM et al (2015) An ancestral bacterial division system is widespread in eukaryotic mitochondria. Proc Natl Acad Sci U S A 112:10239–10246
Levin PA, Losick R (1996) Transcription factor Spo0A switches the localization of the cell division protein FtsZ from a medial to a bipolar pattern in Bacillus subtilis. Genes Dev 10:478–488
Li Z et al (2007) The structure of FtsZ filaments in vivo suggests a force-generating role in cell division. EMBO J 26:4694–4708
Lindas AC et al (2008) A unique cell division machinery in the Archaea. Proc Natl Acad Sci U S A 105:18942–18946

Loose M, Schwille P (2009) Biomimetic membrane systems to study cellular organization. J Struct Biol 168. doi:10.1016/j.jsb.2009.03.016
Loose M, Mitchison TJ (2014) The bacterial cell division proteins FtsA and FtsZ self-organize into dynamic cytoskeletal patterns. Nat Cell Biol 16:38–46
Loose M et al (2008) Spatial regulators for bacterial cell division self-organize into surface waves in vitro. Science 320:789–792
Loose M et al (2011) Min protein patterns emerge from rapid rebinding and membrane interaction of MinE. Nat Struct Mol Biol 18:577–583
Löwe J, Amos LA (1998) Crystal structure of the bacterial cell-division protein FtsZ. Nature 391:203–206
Löwe J, Amos LA (2000) Helical tubes of FtsZ from Methanococcus jannaschii. Biol Chem 381:993–999
Lu C, Reedy M, Erickson HP (2000) Straight and curved conformations of FtsZ are regulated by GTP hydrolysis. J Bacteriol 182:164–170
Lutkenhaus J (2007) Assembly dynamics of the bacterial MinCDE system and spatial regulation of the Z ring. Annu Rev Biochem 76:539–562
Lutkenhaus J (2012) The ParA/MinD family puts things in their place. Trends Microbiol 20:411–418
Ma X, Margolin W (1999) Genetic and functional analyses of the conserved C-terminal core domain of *Escherichia coli* FtsZ. J Bacteriol 181:7531–7544
Ma X, Ehrhardt DW, Margolin W (1996) Colocalization of cell division proteins FtsZ and FtsA to cytoskeletal structures in living *Escherichia coli* cells by using green fluorescent protein. Proc Natl Acad Sci U S A 93:12998–13003
Ma X et al (1997) Interactions between heterologous FtsA and FtsZ proteins at the FtsZ ring. J Bacteriol 179:6788–6797
Margolin W (2005) FtsZ and the division of prokaryotic cells and organelles. Nat Rev Mol Cell Biol 6:862–871
Margolin W (2012) The price of tags in protein localization studies. J Bacteriol 194:6369–6371
Marston AL et al (1998) Polar localization of the MinD protein of Bacillus subtilis and its role in selection of the mid-cell division site. Genes Dev 12:3419–3430
Martos A et al (2012) Towards a bottom-up reconstitution of bacterial cell division. Trends Cell Biol 22:634–643
Mercier R, Kawai Y, Errington J (2013) Excess membrane synthesis drives a primitive mode of cell proliferation. Cell 152:997–1007
Michie KA, Löwe J (2006) Dynamic filaments of the bacterial cytoskeleton. Annu Rev Biochem 75:467–492
Milam SL, Osawa M, Erickson HP (2012) Negative-stain electron microscopy of inside-out FtsZ rings reconstituted on artificial membrane tubules show ribbons of protofilaments. Biophys J 103:59–68
Mingorance J et al (2005) Visualization of single *Escherichia coli* FtsZ filament dynamics with atomic force microscopy. J Biol Chem 280:20909–20914
Monahan LG, Harry EJ (2013) Identifying how bacterial cells find their middle: a new perspective. Mol Microbiol 87:231–234
Mosyak L et al (2000) The bacterial cell-division protein ZipA and its interaction with an FtsZ fragment revealed by X-ray crystallography. EMBO J 19:3179–3191
Mukherjee A, Lutkenhaus J (1994) Guanine nucleotide-dependent assembly of FtsZ into filaments. J Bacteriol 176:2754–2758
Mukherjee A, Saez C, Lutkenhaus J (2001) Assembly of an FtsZ mutant deficient in GTPase activity has implications for FtsZ assembly and the role of the Z ring in cell division. J Bacteriol 183:7190–7197
Oliva MA, Cordell SC, Löwe J (2004) Structural insights into FtsZ protofilament formation. Nat Struct Mol Biol 11:1243–1250

Osawa M, Erickson HP (2011) Inside-out Z rings–constriction with and without GTP hydrolysis. Mol Microbiol 81:571–579

Osawa M, Erickson HP (2013) Liposome division by a simple bacterial division machinery. Proc Natl Acad Sci U S A 110:11000–11004

Osawa M, Anderson DE, Erickson HP (2008) Reconstitution of contractile FtsZ rings in liposomes. Science 320:792–794

Osawa M, Anderson DE, Erickson HP (2009) Curved FtsZ protofilaments generate bending forces on liposome membranes. EMBO J 28:3476–3484

Palmer CM, Löwe J (2013) A cylindrical specimen holder for electron cryo-tomography. Ultramicroscopy 137:20–29

Park KT et al (2011) The Min oscillator uses MinD-dependent conformational changes in MinE to spatially regulate cytokinesis. Cell 146:396–407

Park KT, Du S, Lutkenhaus J (2015) MinC/MinD copolymers are not required for Min function. *Mol Microbiol* 98(5):895–909

Pichoff S, Lutkenhaus J (2005) Tethering the Z ring to the membrane through a conserved membrane targeting sequence in FtsA. Mol Microbiol 55:1722–1734

Pichoff S, Lutkenhaus J (2007) Identification of a region of FtsA required for interaction with FtsZ. Mol Microbiol 64:1129–1138

Popp D et al (2009) FtsZ condensates: an in vitro electron microscopy study. Biopolymers 91:340–350

Ramirez-Arcos S et al (2001) Expression of Neisseria gonorrhoeae cell division genes ftsZ, ftsE and minD is influenced by environmental conditions. Res Microbiol 152:781–791

Ramos D et al (2006) Conformation of the cell division regulator MinE: evidence for interactions between the topological specificity and anti-MinCD domains. Biochemistry 45:4593–4601

Raskin DM, de Boer PA (1999) MinDE-dependent pole-to-pole oscillation of division inhibitor MinC in *Escherichia coli*. J Bacteriol 181:6419–6424

Reeve JN et al (1973) Minicells of Bacillus subtilis. J Bacteriol 114:860–873

Rivas G, Vogel SK, Schwille P (2014) Reconstitution of cytoskeletal protein assemblies for large-scale membrane transformation. Curr Opin Chem Biol 22. doi:10.1016/j.cbpa.2014.07.018. Epub 2014 Aug 12

Rothfield L, Taghbalout A, Shih YL (2005) Spatial control of bacterial division-site placement. Nat Rev Microbiol 3:959–968

Rowlett VW, Margolin W (2014) 3D-SIM super-resolution of FtsZ and its membrane tethers in *Escherichia coli* cells. Biophys J 107:L17–L20

Samson RY et al (2008) A Role for the ESCRT System in Cell Division in Archaea. Science 322:1710–1713

Shen B, Lutkenhaus J (2009) The conserved C-terminal tail of FtsZ is required for the septal localization and division inhibitory activity of MinC(C)/MinD. Mol Microbiol 72:410–424

Shiomi D, Margolin W (2008) Compensation for the loss of the conserved membrane targeting sequence of FtsA provides new insights into its function. Mol Microbiol 67:558–569

Singh JK et al (2007) A membrane protein, EzrA, regulates assembly dynamics of FtsZ by interacting with the C-terminal tail of FtsZ. Biochemistry 46:11013–11022

Small E et al (2007) FtsZ polymer-bundling by the *Escherichia coli* ZapA orthologue, YgfE, involves a conformational change in bound GTP. J Mol Biol 369:210–221

Strauss MP et al (2012) 3D-SIM super resolution microscopy reveals a bead-like arrangement for FtsZ and the division machinery: implications for triggering cytokinesis. PLoS Biol 10:e1001389

Stricker J et al (2002) Rapid assembly dynamics of the *Escherichia coli* FtsZ-ring demonstrated by fluorescence recovery after photobleaching. Proc Natl Acad Sci U S A 99:3171–3175

Sun Q, Margolin W (1998) FtsZ dynamics during the division cycle of live *Escherichia coli* cells. J Bacteriol 180:2050–2056

Swulius MT, Jensen GJ (2012) The helical MreB cytoskeleton in *Escherichia coli* MC1000/pLE7 is an artifact of the N-Terminal yellow fluorescent protein tag. J Bacteriol 194:6382–6386

Szeto TH et al (2003) The MinD membrane targeting sequence is a transplantable lipid-binding helix. J Biol Chem 278:40050–40056
Szwedziak P et al (2012) FtsA forms actin-like protofilaments. EMBO J 31:2249–2260
Szwedziak P et al (2014) Architecture of the ring formed by the tubulin homologue FtsZ in bacterial cell division. *eLife* 3. doi:10.7554/eLife.04601
Thanbichler M, Shapiro L (2006) MipZ, a spatial regulator coordinating chromosome segregation with cell division in Caulobacter. Cell 126:147–162
Treuner-Lange A et al (2013) PomZ, a ParA-like protein, regulates Z-ring formation and cell division in Myxococcus xanthus. Mol Microbiol 87:235–253
Turing AM (1952) The chemical basis of morphogenesis. Philos Trans R Soc Lond Ser B Biol Sci 237:37–72
van den Ent F, Löwe J (2000) Crystal structure of the cell division protein FtsA from Thermotoga maritima. EMBO J 19:5300–5307
Veiga H, Jorge AM, Pinho MG (2011) Absence of nucleoid occlusion effector Noc impairs formation of orthogonal FtsZ rings during *Staphylococcus aureus* cell division. Mol Microbiol 80:1366–1380
Vicente M, Rico AI (2006) The order of the ring: assembly of *Escherichia coli* cell division components. Mol Microbiol 61:5–8
Wang X et al (1997) Analysis of the interaction of FtsZ with itself, GTP, and FtsA. J Bacteriol 179:5551–5559
Ward JEJ, Lutkenhaus J (1985) Overproduction of FtsZ induces minicell formation in *E. coli*. Cell 42:941–949
Willemse J et al (2011) Positive control of cell division: FtsZ is recruited by SsgB during sporulation of Streptomyces. Genes Dev 25:89–99
Woldringh C et al (1991) Toporegulation of bacterial division according to the nucleoid occlusion model. Res Microbiol 142:309–320
Wu LJ et al (2009) Noc protein binds to specific DNA sequences to coordinate cell division with chromosome segregation. EMBO J 28:1940–1952
Wu W et al (2011) Determination of the structure of the MinD-ATP complex reveals the orientation of MinD on the membrane and the relative location of the binding sites. Mol Microbiol 79:1515–1528
Wu LJ, Errington J (2012) Nucleoid occlusion and bacterial cell division. Nat Rev Mircobiol, 10:8–12
Yim L et al (2000) Role of the carboxy terminus of *Escherichia coli* FtsA in self-interaction and cell division. J Bacteriol 182:6366–6373

Chapter 8
Bacterial Actins

Thierry Izoré and Fusinita van den Ent

Abstract A diverse set of protein polymers, structurally related to actin filaments contributes to the organization of bacterial cells as cytomotive or cytoskeletal filaments. This chapter describes actin homologs encoded by bacterial chromosomes. MamK filaments, unique to magnetotactic bacteria, help establishing magnetic biological compasses by interacting with magnetosomes. Magnetosomes are intracellular membrane invaginations containing biomineralized crystals of iron oxide that are positioned by MamK along the long-axis of the cell. FtsA is widespread across bacteria and it is one of the earliest components of the divisome to arrive at midcell, where it anchors the cell division machinery to the membrane. FtsA binds directly to FtsZ filaments and to the membrane through its C-terminus. FtsA shows altered domain architecture when compared to the canonical actin fold. FtsA's subdomain 1C replaces subdomain 1B of other members of the actin family and is located on the opposite side of the molecule. Nevertheless, when FtsA assembles into protofilaments, the protofilament structure is preserved, as subdomain 1C replaces subdomain IB of the following subunit in a canonical actin filament. MreB has an essential role in shape-maintenance of most rod-shaped bacteria. Unusually, MreB filaments assemble from two protofilaments in a flat and antiparallel arrangement. This non-polar architecture implies that both MreB filament ends are structurally identical. MreB filaments bind directly to membranes where they interact with both cytosolic and membrane proteins, thereby forming a key component of the elongasome. MreB filaments in cells are short and dynamic, moving around the long axis of rod-shaped cells, sensing curvature of the membrane and being implicated in peptidoglycan synthesis.

Keywords FtsA • MreB • MamK • Actin superfamily • Elongasome • Membrane curvature • RodZ • Bacterial cell shape • Filament structure • Dynamic filament • Magnetosome

---

The original version of this chapter was revised. An erratum to this chapter can be found at DOI 10.1007/978-3-319-53047-5_16

T. Izoré • F. van den Ent (✉)
MRC Laboratory of Molecular Biology, Francis Crick Avenue, Cambridge CB2 0QH, UK
e-mail: thierry.izore@monash.edu; fent@mrc-lmb.cam.ac.uk

J. Löwe, L.A. Amos (eds.), *Prokaryotic Cytoskeletons*, Subcellular Biochemistry 84, DOI 10.1007/978-3-319-53047-5_8

Introduction

In the early 1990s the first cytoskeletal structure in bacteria was discovered in *Escherichia coli*: FtsZ, a tubulin-like protein that was shown to locate as a ring-like structure at the division site (Bi and Lutkenhaus 1991; Erickson 1995; Löwe and Amos 1998). FtsZ triggered a shift away from the widespread dogma that only eukaryotes contain a cytoskeleton and that the cells of bacteria are not spatially organized. Since then, homologs of other cytosekeletal elements, including actin (Bork et al. 1992; Jones et al. 2001; van den Ent et al. 2001b), most likely intermediate filaments (Bork et al. 1992; Jones et al. 2001; van den Ent et al. 2001b) and filaments not present in eukaryotes have been discovered (Kühn et al. 2010), revealing long-range internal organization of bacterial cells that thus far was entirely unknown. Actin-like proteins form the most diverse polymers and they engage in a wide array of biological functions in bacteria as well as archaea (for the latter see Chap. 10). Despite a shared sequence identity of below 20%, the three-dimensional structures of the monomers of bacterial actin-like proteins are reminiscent of the structure of eukaryotic G-actin (van den Ent et al. 2001a). The large differences in amino acid composition reflect the fact that the bacterial proteins fulfill diverse and separate functions and also their large evolutionary distances (Doolittle and York 2002). The resemblance of their monomeric structures extends to the structure of a single protofilament (strand), responsible for the conserved polymerisation-dependent ATPase activity of actin-like proteins (van den Ent et al. 2001b). In contrast, there are many different ways protofilaments/strands come together to make filaments, producing a surprising array of different filament architectures, presumably tailored to their different functions (Ozyamak et al. 2013a).

The following chapter will focus on bacterial homologues of actin involved in endogenous cellular processes, namely MamK, FtsA and MreB.

MamK Aligns Magnetosomes

Magnetosomes are intracellular membrane invaginations (Scheffel et al. 2006) associated with about 20 specific proteins (Grunberg et al. 2004). They contain biomineralized crystals of magnetite (Fe_3O_4) and/or greigite (Fe_3S_4) (Bazylinski et al. 1995). Each of these crystals is large enough (30–120 nm wide) to possess a permanent dipole moment. In the magnetotactic cells, magnetosome vesicles are attached to the cytoplasmic membrane and are arranged in a straight line where they produce a large combined dipole, reminiscent of a compass needle. This allows bacteria to swim along the Earth's magnetic field. Magnetotaxis is thought to facilitate the search for optimal oxygen levels in the aquatic environments. Magnetosome formation has mainly been studied in two magnetotactic bacteria: *Magnetospirillum magneticum* AMB-1 and *Magnetospirillum gryphiswaldense* MSR-1 (hereafter referred to as AMB-1 and MSR-1 respectively). Proteins involved in the assembly of magnetosomes are encoded by the genome of these bacteria in a large region named MAI for "Magnetosome Island". Within the MAI four different clusters, the *mam* (magnetosome membrane) and *mms* (magnetic particle membrane specific) clusters are of key importance for the formation and the functionality of magnetosomes (Murat et al. 2010). Electron cryo tomography of magnetotactic bacteria revealed that magnetosomes are aligned by interacting with the filamentous actin-like protein MamK (Komeili et al. 2006; Scheffel et al. 2006).

MamK Forms Cytoskeletal Filaments

MamK is an actin-like protein that shares a sequence identity of ~18% with eukaryotic actin and MreB, each, and has been reported to hydrolyze ATP and GTP at similar rates (Sonkaria et al. 2012). The protein is encoded within the *mamAB* cluster together with at least seven other conserved genes. These genes have been shown to encode proteins that localize at the magnetosomes membrane and are essential for magnetite formation as well as the organization into chains (Grunberg et al. 2004; Katzmann et al. 2011). The cytoskeletal filaments of MamK are ~ 250 nm long and run along 4–5 magnetosomes, as seen by cryo-electron microscopy (Komeili et al. 2006; Scheffel et al. 2006). The precise molecular function of MamK filaments is not understood and major differences exist between the two model organisms AMB-1 and MSR-1. In a *MamK* deletion strain, AMB-1 cells were no longer able to produce a stable magnetosome chain. Indeed, magnetosomes vesicles appeared scattered throughout the cell, either isolated or paired (Fig. 8.1) (Komeili et al. 2006). However, in MSR-1 cells, the loss of MamK filaments resulted in a reduced efficiency of magnetite crystal formation. Magnetosome chains would still form, but they were shorter in length and were mis-localized (Katzmann et al. 2010). Hence the idea emerged that instead of providing a backbone onto which vesicles align, MamK filaments might be implicated in sorting, concatenating (assembling small stretches of magnetosomes) and positioning a fully built magnetosome chain (Katzmann et al. 2010, 2011).

Along with MamK, MamJ is of critical importance (and its paralogue LimJ), as it is believed to tether MamK filaments to magnetosomes. In wild type strains, MamK and MamJ co-localize and form an extended cytoskeletal network that spans the entire length of the cell, as seen by cryo-electron tomography and light microscopy (Katzmann et al. 2010; Komeili et al. 2006; Pradel et al. 2006; Scheffel et al. 2006).

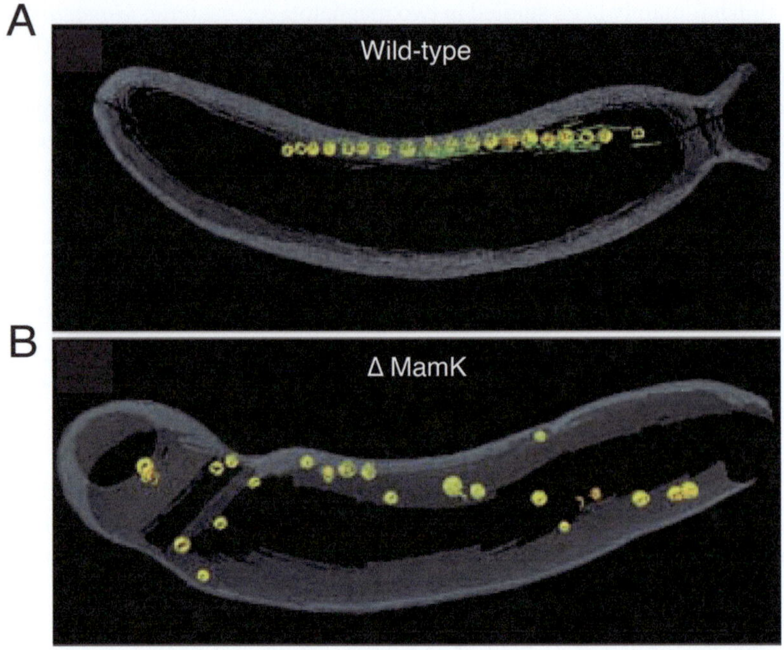

Fig. 8.1 Electron cryotomogram (cryoET) segmentation of (**a**) a wild-type *M. magneticum AMB-1* and (**b**) a *ΔmamK* mutant. Magnetosomes in *yellow*, MamK filaments in *green* and cell membrane in *grey* (From Komeili et al. 2006, Science, 311, 242–5)

MamK Filaments Are Dynamic in Cells

MamK filaments show relatively slow ATP hydrolysis-driven dynamic behavior in AMB-1 cells (Draper et al. 2011). Although, its ATPase activity is ten times lower than that of other actin-like proteins, it is essential for filament dynamics: a mutation of the conserved glutamate residue in the ATP catalytic site (E143A) results in filament formation, but no filament depolymerization (Ozyamak et al. 2013b). Possibly because of its slow ATPase activity, MamK filaments are highly stable in the cell. This is in contrast to the inherently unstable filaments of the plasmid-segregating protein ParM that are stabilized only by the interaction with adaptor complex ParR-*ParC*. Proteins affecting MamK dynamics are MamJ and LimJ (a MamJ paralogue in *M. magneticum* AMB-1). However, it is unclear if they interact directly or if other partners are needed (Draper et al. 2011). MamK dynamics are reminiscent of that of MreB, with an apparent lack of monomer movement within a filament, as seen by light microscopy. Subunit turnover without treadmilling could be explained by the mosaic nature of the filaments as seen in the cell, resulting from the overlap of many smaller filaments. Filament formation occurs at multiple sites within the cell from where the seeds extend up to 200–250 nm in length and are ~ 6 nm wide (Komeili et al. 2006), a diameter in agreement with other actin like proteins such as MreB or ParM (van den Ent et al. 2014; Salje et al. 2011). These small filaments then associate to form a larger, straight bundle (Pradel et al. 2006) that extend from one pole to the other, lining up with the concave side of the cell. The magnetosome chain is always maintained at mid-cell by a mechanism that is not well understood but involves MamK.

Cytokinesis and Maintenance of the Magnetic Field Orientation

Magnetotactic bacteria preserve magnetic orientation throughout their cell cycle. To do so, the cell undergoes a cell division process during which FtsZ creates an arc-shaped structure leading to an asymmetric constriction of the cell. Upon constriction, the cell rapidly bends and a snapping mechanism triggers division. The force involved in the membrane constriction has been proposed to be enough to break the magnetosome chain and disassemble the MamK filaments at the site of division (Katzmann et al. 2011; Lin and Pan 2011; Staniland et al. 2010). Immediately after cytokinesis, the magnetosome chain is localized at the new pole of the daughter cell, the old division site, and gets rapidly relocated to mid-cell, via a mechanism relying on MamK filaments (Katzmann et al. 2011). The role of MamK filaments is thus of importance both for aligning and also positioning the magnetosome chain at midcell for most of the cell cycle.

MamK: Monomer and Polymer Structure

The atomic structure of MamK has not been determined. However a number of 3D computational models of the monomer of MamK have been proposed, based on its amino-acid sequence and using actin or MreB as templates (Lin et al. 2014; Ozyamak et al. 2013b; Rioux et al. 2010; Sonkaria et al. 2012). Despite the low sequence similarity to actin and other actin-like proteins, MamK is certain to possess the canonical actin fold, with the typical ATP-binding pocket and conserved catalytic residues (Sonkaria et al. 2012).

Electron cryo microscopy revealed the structure of MamK filaments *in vitro* to 6.5 Å (Ozyamak et al. 2013b; Bergeron et al. 2016). Similarly to most actin-like proteins, MamK filaments assemble into parallel, two-stranded helical polymers. However, unlike actin or ParM, the subunits in the two protofilaments (strands) of the MamK filament are in register, that is non-staggered. This feature is shared with MreB that forms flat, un-staggered filaments (van den Ent et al. 2014; Salje et al. 2011) (see below). The helix formed by MamK is right-handed with a rotational angle and a subunit axial rise of 23.3° and 53 Å per subunit, respectively. MamK filaments have additional rotational C2 symmetry along the filament axis, when compared to filaments with staggered subunits, such as actin. The atomic homology models generated for MamK explain the cryoEM density maps well, suggesting that the longitudinal contacts seen in actin, crenactin, ParM and MreB are also conserved in MamK.

MamK-Like Protein Is a Homologue of MamK

Recently, a homologue of MamK, called MamK-like, has been discovered in the genome of *M. magneticum* (Rioux et al. 2010). MamK-like protein is translated from a gene located outside the MAI region and shares 54.5% sequence identity with MamK. The protein is able to interact with MamK and together they polymerize into mixed-filaments. It seems that in some conditions the function of MamK and that of MamK-like are redundant towards magnetosome chain assembly.

FtsA Is Involved in Bacterial Cell Division

Discovery of Filamenting Bacterial Phenotypes

Cell division is a complex process: cell morphogenesis, genome replication and segregation must all be performed. In many bacteria, the genes regulating cell division and cell wall synthesis are located in the *dcw* gene cluster (standing for division and cell wall). For a review see (Mingorance and Tamames 2004). This cluster comprises 16 and 13 genes in *E. coli* and *B. subtilis*, respectively (Tamames et al. 2001), including the *mur* genes involved in the synthesis of the lipid II precursor of the cell wall and other genes for which some mutations result in a temperature sensitive phenotype. The observed phenotype of these mutants is filamenting cells that are unable to achieve division while still elongating. These mutants were named *fts* for "filamentous temperature sensitive". The best-known mutants responsible for the *fts* phenotype contain modifications of the bacterial tubulin homologue FtsZ that polymerizes into a ring-like structure, the Z-ring (Bi and Lutkenhaus 1991). The Z-ring can be regarded as a bacterial analogue of the eukaryotic cytokinetic ring formed during cytokinesis. Z-ring formation has mostly been studied in the model organisms *Escherichia coli* and *Bacillus subtilis*.

Another well-studied group of *fts* mutants have mutations in the gene encoding FtsA, an actin-like protein that is involved in the positioning and regulation of the Z-ring and cell division machinery (divisome).

FtsA's Role in Cell Division

The observation that *E. coli* cells carrying a point mutation in the *ftsA* gene were not able to divide when incubated at the non-permissive temperature gave the first indication that FtsA was involved in cell division. Cells lacking functional FtsA were elongated and showed evidence that septum formation was initiated at several locations but was then rapidly aborted. It is interesting to note that when the thermosensitive *ftsA* mutant cells were grown as filamentous cells at the non-permissive temperature, and subsequently shifted to the permissive temperature, cell division resumed at a higher rate than usual until cells returned to a normal size (Tormo et al. 1980; Walker et al. 1975). In *E. coli*, FtsA is essential for cell division (Lutkenhaus and Donachie 1979) and needs to be efficiently expressed 10–15 min before the cell actually divides (Donachie et al. 1979). In *B. subtilis*, FtsA is not essential; cells lacking FtsA show a filamentous phenotype during vegetative growth, but they still divide at polar division sites and subsequently sporulate, albeit at a lower rate (Beall and Lutkenhaus 1992). This difference between the two organisms, FtsA being essential versus non essential, is explained by the presence of SepF in *B. subtilis*, a protein that plays a role somewhat similar to that of FtsA (Duman et al. 2013; Ishikawa et al. 2006), but that is structurally unrelated.

FtsA was first recognized as a potential actin-like protein through sequence-specific searches based on actin signature residues by Peer Bork and colleagues (Bork et al. 1992). Whether FtsA's ATPase activity is necessary or not for cell division is still a matter of debate; ATPase activities for *B. subtilis* and *P. aeruginosa* FtsA have been reported (Feucht et al. 2001; Paradis-Bleau et al. 2005), but no clear ATPase activity has been shown for FtsA from either *E. coli*, *T. maritima* or *S. pneumoniae* despite the presence of a seemingly complete ATPase active site (Lara et al. 2005; Martos et al. 2012; Szwedziak et al. 2012). A recent study showed that FtsA needs to be in a nucleotide-bound state to recruit FtsZ to the membrane (Loose and Mitchison 2014). Unlike actin, MamK or ParM, but similar to MreB (Salje et al. 2011), FtsA is bound to the cytoplasmic membrane (Pla et al. 1990). This feature is due to the presence of an amphipathic helix located at the C-terminus of the protein, encompassing the last 15 residues, that serves as a Membrane Targeting Sequence (MTS, not resolved in the crystal structure shown in Fig. 8.2a) and allows the protein to interact with the phospholipid bilayer. Indeed, an *E. coli* strain expressing a GFP-FtsA fusion without the MTS exhibits long fluorescent filaments visible in the cytoplasm that are no longer attached to the membrane. It has been shown that the interaction of FtsA with the membrane is of prime importance for its function since a single mutation in the MTS leads to misshaped cells and mislocalization of FtsA (Pichoff and Lutkenhaus 2005). Moreover, FtsA needs to self-interact to achieve its function *in vivo* and it is now known that this most likely relates to filament formation (Szwedziak et al. 2012). FtsA is an early division protein and is necessary for the proper function of the divisome. In cases where FtsA is absent or non-functional, the Z-ring still assembles, but the constriction process is aborted soon after (Addinall and Lutkenhaus 1996). The protein plays a triple role in the cell division process: it

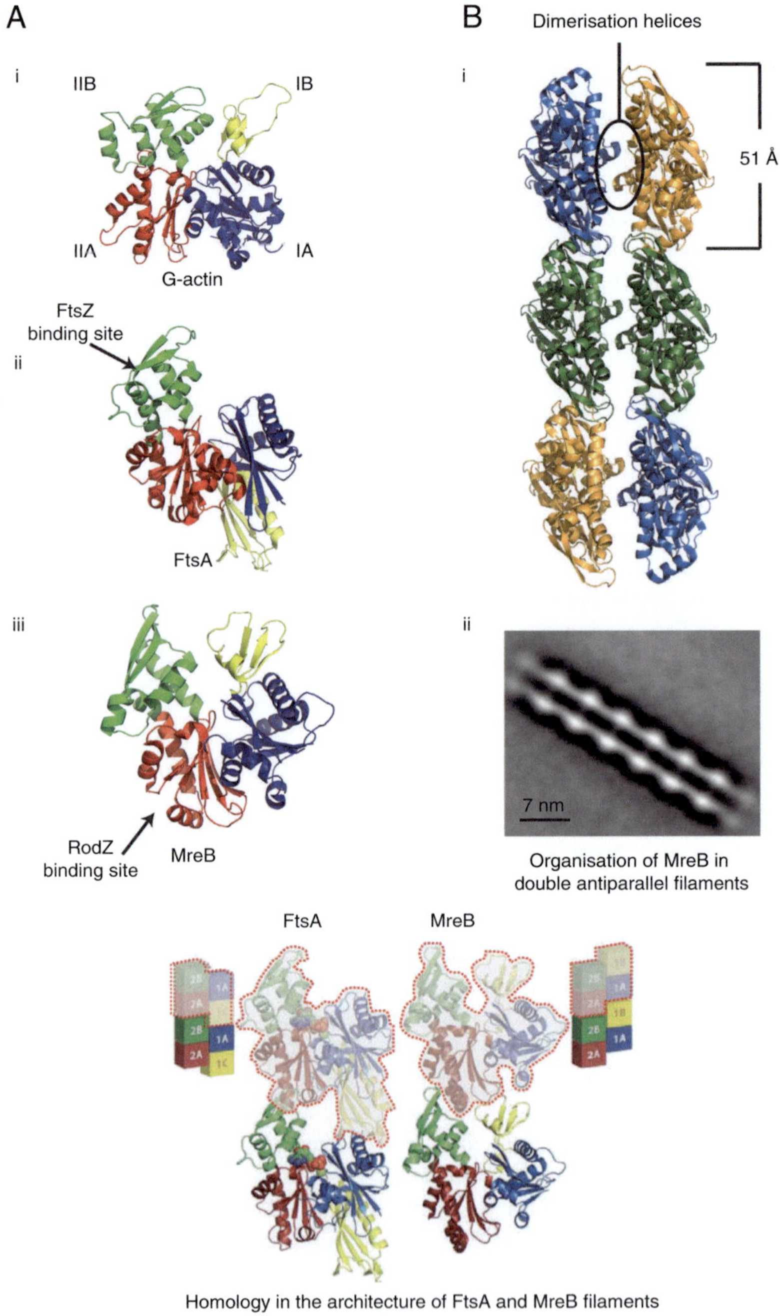

Fig. 8.2 (**a**) X-ray structures of actin (i), FtsA (ii) and MreB (iii). Each equivalent domain is color-coded according to the actin structure. (**b**) Crystal structure of an MreB antiparallel filament (i) and 2D class averages of MreB filaments obtained on lipid monolayers and imaged by negative staining (ii). (**c**) Comparison of filament architecturse between MreB and FtsA (From Szwedziak et al. 2012)

tethers the FtsZ-ring to the membrane, it is required to recruit other proteins to the Z-ring for efficient cytokinesis and finally it is able to interact with the peptidoglycan synthesis machinery (Tormo et al. 1986).

FtsA Interacts with FtsZ and Is Part of the Z-Ring

The first evidence that FtsA and FtsZ interact came from studies where the localization of FtsA was tracked in an FtsZ mutant strain (Addinall et al. 1996; Ma et al. 1996, 1997). In a genetic background where the FtsZ ring was not able to assemble, FtsA was not able to localize to the division site and similarly when FtsZ formed helical patterns in the cell, FtsA would adopt a similar arrangement. These studies revealed that FtsA interacts directly with FtsZ and that localization of FtsA is dependent on that of FtsZ. Indeed, FtsA and FtsZ are the first two proteins to localize to midcell and no time difference has been reported between the formation of the FtsZ-ring and the FtsA-ring (Den et al. 1999). The interaction mode of the two proteins is now well established: the C-terminal tail of FtsZ (in fact its last ~15 residues) is necessary for its function and for the interaction with FtsA. It has been demonstrated that a single point mutation in one of these C-terminal residues is enough to disrupt the interaction between the two proteins and produce misshaped cells (Ma and Margolin 1999). The C-terminal tail of FtsZ is also important for ZipA binding. ZipA also tethers FtsZ to the membrane and competes with FtsA for its binding site on FtsZ (Mosyak et al. 2000). The molecular details of the FtsA:FtsZ interaction were revealed by an X-ray structure of the complex between FtsA and the C-terminal 16 amino acids of FtsZ from *Thermotoga maritima* and also revealed by *in vivo* analysis of FtsA mutants in *E. coli* (Pichoff and Lutkenhaus 2007; Szwedziak et al. 2012). FtsZ utilizes its C-terminal helix to bind to FtsA subdomain 2B (Fig. 8.2a) via mostly hydrophilic interactions with a ~ 50 μM affinity. The length of the C-terminal helix in FtsZ dictates the distance between the FtsZ ring and its membrane anchor FtsA, as was shown by electron tomography of *E. coli* cells: FtsA and FtsZ form two concentric rings (upon overexpressing FtsA) with FtsZ filaments being located 16 nm away from the inner membrane. Cellular tomography also helped revealing that FtsZ formed a complete ring at the mid-cell but the actual structure produced by FtsA filaments when at physiological FtsA levels remains unclear. (Szwedziak et al. 2015) However, it is unlikely that FtsA forms a full ring in cells, especially not at early stages of constriction, due to a much lower concentration of FtsA compared to that of FtsZ. Indeed, early experiments have shown that a specific ratio between FtsA and FtsZ is required for the correct functioning of the division machinery. The number of FtsA molecules varies in cells and different groups have reported different figures ranging from 740 to 150 (Rueda et al. 2003; Wang and Gayda 1992) respectively). However, it seems that as long as an approximate ratio of one FtsA molecule to five FtsZ molecules is maintained, cell division proceeds (Dai and Lutkenhaus 1992).

FtsA and FtsZ form dynamic ring- or helix-like structures even in the absence of volume constraints, suggesting a natural tendency to co-assemble into curved structures. This has been suggested to have consequences for a possible mechanism of cell constriction. Currently, two dynamic mechanisms are being discussed, besides the effect of filaments bending the membrane: a sliding model where filaments of FtsA and FtsZ slide along each other leading to the closure and subsequent constriction of the FtsZA ring, or a treadmilling mechanism (Loose and Mitchison 2014; Szwedziak et al. 2015). A combination of these mechanisms seems also possible.

Structure of FtsA and Formation of Filaments

Proteins of the actin/HSP70 family adopt the canonical actin fold consisting of two domains with a common core forming an inter-domain groove that contains the nucleotide-binding site. Each domain is subdivided into two subdomains, namely 1A, 1B and 2A, 2B (see Fig. 8.1a). Among these, subdomains 1A and 2A are the most conserved, whereas subdomain 1B tends to have a more protein-specific fold.

As mentioned already, the primary sequence analysis of FtsA suggested some similarities to the actin/HSP70 family (Bork et al. 1992), but the sequence identity is very low. Hence, for some time it was a matter of debate whether FtsA was a real actin-like protein. With its crystal structure it became clear that *Thermotoga maritima* FtsA does adopt an actin-like fold (Szwedziak et al. 2012; van den Ent and Löwe 2000). Indeed, the structures of subdomains 1A, 2A, and 2B are very well preserved (Fig. 8.1a) as well as the residues lining the nucleotide-binding site. However, subdomain 1B is absent and a new subdomain, named 1C, occurs on the opposite side of the protein. No significant homology with any other known domains has been reported, but the divergence in the fold of subdomain 1C might be explained by a protein-specific adaptation for FtsA to fulfill its role in the division process. Indeed, a deletion of subdomain 1C blocks its function in cell division but not its localization to the Z-ring (Corbin et al. 2004). Protofilaments of all other actin-like proteins (ParM, MreB, crenactin, Ta0583) are mainly formed through the interaction of the two top subdomains (1B and 2B) of one subunit with the lower subdomain 2A of the next subunit. This cannot be the case for FtsA since subdomain 1C is located on the other side of the protein. Surprisingly, because subdomain 1B is missing, subdomain 1C of the next subunit takes its place in protofilament formation (Fig. 8.1c). This rearrangement of subdomains does not impair the protein's ability to form protofilaments when exposed to a lipid bilayer, as was shown by cryo tomography on reconstituted liposomes and by negative staining electron microscopy on a lipid monolayer (Szwedziak et al. 2015). In addition to filaments imaged by electron microscopy, single actin-like protofilaments of FtsA occurred in crystals, revealing the 48 nm spacing between the subunits.

It is currently unclear what kind of filaments FtsA forms in cells: is it a single protofilament, as has been proposed for crenactin (Braun et al. 2015; Izoré et al.

2014; Lindås et al. 2014), a double helical filament, like actin and ParM, or a flat ribbon like MreB (van den Ent et al. 2014)?

FtsA Recruits Other Proteins of the Divisome

It is believed that the proteins involved in cell division of *E. coli* localize to midcell in a more or less sequential way. In addition, interactions between early and late division proteins have been reported, suggesting that the divisome might be less of a transient complex of proteins than initially thought (Di 2003).

At the very early stage of the division process, FtsA interacts with ZipA and together they recruit FtsZ to the membrane. Later, many other proteins join the Z-ring, some of them interacting directly with FtsA. For example, full length FtsA, or indeed FtsA's subdomain 1C alone, fused to the polar protein DivIVA localizes to the cell poles and no longer to the Z-ring, dragging with it divisome proteins FtsI (also called PBP3) and FtsN (Corbin et al. 2004; Tormo et al. 1986), highlighting the role of subdomain 1C as an interaction site for other cell division proteins (Busiek et al. 2012; Busiek and Margolin 2014). Recent results led to a model where the late division protein FtsN interacts with FtsA once the Z-ring is fully assembled. Doing so, FtsN might activate FtsA and the FtsQLB complex that would in turn activate the synthesis of cell wall material by an unknown mechanism, eventually leading to the formation of the septum and the division of the cell (Weiss 2015).

MreB

MreB is widely distributed in bacteria and most non-spherical bacteria possess at least one MreB homologue. However, a few exceptions exist such as the Actinobacteria, Mycobacteria, and bacteria from the genuses *Rhizobium* and *Agrobacterium* that lack MreB. (Brown et al. 2012; Cameron et al. 2014; Kang et al. 2008; Letek et al. 2008). In MreB-containing rod-shaped bacteria, cell elongation occurs by lateral intercalation of newly synthesized peptidoglycan, whereas the aforementioned organisms grow by addition of new material at the cell poles, a mechanism independent of MreB (Daniel and Errington 2003).

MreB Is Involved in Cell Shape Maintenance

MreB, MreC and MreD are cell shape determinants and belong to the same *mre* operon (murein formation gene cluster E) (Doi et al. 1988; Figge et al. 2004; Levin et al. 1992; Varley and Stewart 1992; Wachi et al. 1987). Mutations in one of these genes result in gross morphological defects, and the round, bloated cells eventually

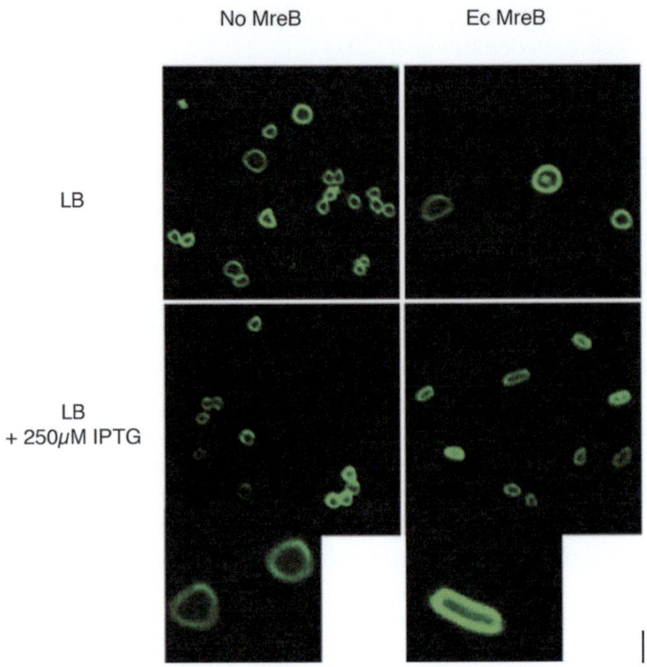

Fig. 8.3 *E. coli* cells grown in presence and absence of MreB. Absence of MreB causes a round and enlarged cell phenotype. Induction of MreB synthesis by addition of IPTG restores the wild-type cell morphology (From Salje et al. 2011)

lyse (Fig. 8.3). In the three model organisms *B. subtilis*, *E. coli* and *C. crescentus*, MreB, MreC and MreD are essential proteins. Under laboratory conditions, by overexpressing the cell division components FtsQAZ, compensating for the larger circumference of the round cells, these cells survive (Kruse et al. 2005).

MreB Is a Cytoskeletal Element

The first experimental evidence that MreB may indeed be a cytoskeletal element came from a study in *B. subtilis* in which the group of Jeff Errington showed that MreB and its paralog Mbl (MreB-like protein) seemingly formed an extended helical structure that wrapped around the long axis of the cell. Similar to actin, it had a role in morphogenesis and its phylogenetic distribution made sense with MreB's function being related to cell shape (Jones et al. 2001). At that time, it was proposed that the long helical MreB pattern would serve as guide for the spatially organized insertion of newly synthesized peptidoglycan strands. This model was/is attractive because this structure, spanning the entire cell, would provide long-range organization of cell-wall synthesis. Being very well accepted by the field, this model lasted for more than a decade until new imaging techniques and functional MreB-GFP fusions provided more insights into the architecture of these filaments in the cell. Super-resolution microscopy methods such as TIRF (Total Internal Reflection Fluorescence) and SIM (Structured Illumination Microscopy) allowed more detailed picture of MreB filaments in the cell. Instead of forming an elongated, helical structure, MreB was shown to localize in patches or dots of short filaments (Domínguez-Escobar et al. 2011; Garner et al. 2011). How these small patches of short filaments support long-range coherent maintenance of cell shape is more difficult to comprehend. It is now more or less accepted that the long cell-spanning helices seen by microscopy may have been an artifact (Bendezú et al. 2009; Swulius and Jensen 2012).

MreB Is an Actin-Like Protein

Along with FtsA, MreB was identified as a potential bacterial actin in 1992 by sequence alignments (Bork et al. 1992). However, its ability to form filaments by itself was only demonstrated experimentally in 2001 when the structure of *Thermotoga maritima* MreB was solved by X-ray crystallography (van den Ent et al. 2001b). It was then realized that despite a very low sequence identity with cytoplasmic actin (< 20%) MreB shared a similar fold with its eukaryotic counterpart (Fig. 8.1a), composed of four subdomains (1A, 1B and 2A, 2B) with a relatively high structural similarity. The MreB crystals contained a single protofilament (strand) that showed that the longitudinal repeat and the interacting surfaces between subunits in the filament are shared between MreB and actin (and all other actin-like proteins). Extensive contacts exist between subdomains IB/IIB of one subunit and subdomain IIA of the next subunit, leading to an interaction surface of 1700 $Å^2$ (van den Ent et al. 2001b). A more recent crystal structure of MreB from *C. crescentus* contains double, antiparallel protofilaments (van den Ent et al. 2001b). In contrast to actin or other helical actin-like proteins, MreB forms flat, double, non-staggered protofilaments arranged in an antiparallel fashion, as was shown by X-ray crystallography and electron microscopy (Fig. 8.2b) (van den Ent et al. 2014). The crystallography study of MreB in different nucleotide states (either as a monomer or as a polymer) revealed that the protein undergoes relatively small structural changes upon polymerization (van den Ent et al. 2014). It has been shown that from monomeric to polymeric (double protofilament) state, the inter domain cleft containing the active site narrows due to the movement of domain I towards domain II. This is accompanied by a rotation of domain I leading to flattening of one side of the protein, thus creating the newly formed inter-protofilament interface and the interaction of dimerization helices (Fig. 8.2b). These small changes are crucial for the active site of the protein, as they lead to the correct positioning of important residues involved in nucleotide binding and of the catalytically active water molecule for the hydrolysis of ATP. The observed mechanistic changes are reminiscent of those of the plasmid segregation ParM-family of proteins (Colavin et al. 2014; Gayathri et al. 2012, 2013; van den Ent et al. 2014) and also of actin.

Two related drugs are known to affect MreB's function by acting on the protein's active site: S-(3, 4-dichlorobenzyl) isothiourea (A22) and its derivative MP265 (Bean et al. 2009; Dye et al. 2011; Gitai et al. 2005; IWAI et al. 2002; Takacs et al. 2010; van den Ent et al. 2014). Co-crystal structures of *C. crescentus* MreB with A22 or MP265 revealed that these compounds act by blocking the exit channel of the gamma-phosphate of the cleaved nucleotide and by preventing the correct positioning of the catalytic water. Moreover, the presence of the compound in the catalytic pocket seems to block the movement of the dimerization helix, thus impairing the formation of functional MreB double protofilaments.

MreB Binds to Membranes and Is Dynamic in vivo

The ability of MreB to interact with the membrane bilayer is facilitated by two features. MreB in Gram positive bacteria has an exposed hydrophobic loop in subdomain IA and in Gram negative bacteria it has an additional N-terminal amphipathic helix (Maeda et al. 2012; Salje et al. 2011). In the presence of lipid membranes, MreB forms isolated filaments, either on a lipid-coated EM grid (Fig. 8.2b-ii) or on liposomes (Salje et al. 2011; van den Ent et al. 2014). Electron cryotomography and subtomogram averaging of a lipid-tube coated with MreB filaments showed that these filaments resemble the double filament structure solved by X-ray crystallography (Salje et al. 2011; van den Ent et al. 2014). MreB filaments sit very close to the bilayer, forming a structure only 5.5 nm away from the membrane, as was shown by tomography of MreB filaments bound to lipid tubes. Therefore, it is maybe not surprising that such small structures cannot be seen by tomography of whole cells (Swulius and Jensen 2012). For technical reasons, tomograms have extremely low signal to noise ratios so that MreB filaments are probably masked by the strong signal produced by the membrane bilayer and the proteins embedded within.

The antiparallel arrangement of MreB filaments combined with the fact that they are flat and membrane associated is most likely pivotal for MreB's function. The filament architecture produces non-polar filaments, meaning that both filament ends are structurally identical. Subsequently, the filaments cannot treadmill as was suggested before (Kim et al. 2006). In contrast, the finding of non-polar filaments support bidirectional movements of MreB filaments as was observed *in vivo* (Domínguez-Escobar et al. 2011; Garner et al. 2011; Reimold et al. 2013). The dynamic behavior of MreB filaments, circumferentially moving in weakly aligned circles around the long axis of cells, depends on the growth phase and varies between 22 and 50 nm/s for *B. subtilis* and up to 85 nm/s for *E. coli*. It has been shown that a mutation in *B. subtilis* MreB's ATP-binding site leads to static filaments (Domínguez-Escobar et al. 2011; Garner et al. 2011; Reimold et al. 2013). In contrast, earlier reports showed that despite an inactive catalytic site, MreB filaments are still capable of moving at a velocity similar to that of the wild-type protein by interacting with the cell wall synthesis machinery (Domínguez-Escobar et al. 2011; Garner et al. 2011; Teeffelen et al. 2011).

MreB Interacts With Both Cytosolic and Membrane Proteins

Yeast and bacteria two-hybrid screens suggest that MreB has an extended network of interaction partners. These putative interaction candidates include soluble cytoplasmic proteins involved in the synthesis of peptidoglycan precursors (MurD/E/F/G and DapI), membrane proteins involved in cell wall synthesis (MreC, MreD, RodZ and RodA) and also organism-specific proteins. For instance, MreBH from *B. subtilis*, interacts with LytE, an essential autolysin and *C. crescentus* MreB forms a complex with MbiA (Bendezú et al. 2009; Carballido-Lopez et al. 2006; Divakaruni et al. 2007; Favini-Stabile et al. 2013; Kruse et al. 2005; Mohammadi et al. 2007; Rueff et al. 2014; White et al. 2010; Yakhnina and Gitai 2012). However, one has to be somewhat cautious since other data shows that even with widely conserved proteins, the interaction network can be different from organism to organism, as suggested by a different localization of MreB and MreC in *C. crescentus* (Dye et al. 2005). These inconsistencies might also be due to the use of (large fluorescent) tags that might have perturbed MreB's function and the association with other proteins. Of all the putative complexes, until now only one interaction has been shown unambiguously in a co-crystal structure of MreB in complex with RodZ (van den Ent et al. 2010). RodZ is an important player in the regulation of the cell wall synthesis machinery (Bendezú et al. 2009; Gerdes 2009) and helps in tethering MreB to the membrane via the interaction between RodZ's N-terminal cytoplasmic domain and subdomain IIA of MreB. A recent report has established RodZ as being the sole link between MreB in the cytoplasm and the cell wall synthesis machinery located in the periplasm by interacting with both MreB and Pbp2/RodA (Morgenstein et al. 2015).

MreB Motion Is Linked to the Activity of the Cell Wall Synthesis Machinery

In the early days of MreB studies it was believed that MreB motion in the cell was driven by its own ATPase activity (and by consequence, filament polymerization/depolymerisation rate). However, there is currently evidence that in *B. subtilis* and *E. coli,* MreB movement is due to or limited by the activity of the cell wall synthesis machinery itself. Indeed, antibiotics targeting cell wall synthesis cause loss of MreB dynamics (Domínguez-Escobar et al. 2011; Garner et al. 2011; Teeffelen et al. 2011). Furthermore, treating the cell with MreB inhibitor A22 does not affect MreB motion, MreB and the peptidoglycan synthesis machinery continue to follow the same tracks in the cell. MreB and the peptidoglycan synthesis machinery follow a path mostly perpendicular to the long axis of the cell, circumferential motion, mainly due to the interaction with RodZ (Morgenstein et al. 2015). Even if nonessential for cell survival, the interaction between MreB and RodZ enables MreB motion and was shown to be important for rod shape determination in non-standard osmotic pressure growth. Due to the new observations, it was suggested that the long helical structures formed by MreB described in the early days were actually a series of moving patches that never cross each other but move both clockwise and anticlockwise (Morgenstein et al. 2015). The idea of a helical arrangement has, however been defended (Errington 2015) and there remains convincing evidence for it in some bacterial species (Nan et al. 2014).

MreB Senses Membrane Curvature

It is surprising and not well understood how very short filaments might regulate a uniform distribution of newly synthesized cell wall material. Recent reports suggest that localization of MreB (and subsequently the cell wall synthesis machinery) could be caused by MreB's preference of association with a negative cell membrane curvature (Billings et al. 2014; Renner et al. 2013; Ursell et al. 2014). Indeed, when cells are forced to grow into a particular V-shape, MreB consistently localizes at the tip of the V (Renner et al. 2013) and this indicates the possibility of interplay between MreB localization and the overall topology and shape of the membrane. A new model has emerged where MreB could sense a defect or feature in the membrane curvature and trigger the synthesis of new peptidoglycan at the site in order to straighten out the curvature. This way MreB might support the maintenance of cell shape without the need of a global mechanism for its regulation (Ursell et al. 2014). Similarly to its eukaryotic counterpart, it was suggested that MreB might organize membrane composition leading to patches of increased fluidity that might facilitate the formation of complexes involved in cell-wall synthesis (Strahl et al. 2014). Membrane composition seems to play a crucial role in MreB's function since the presence/absence of cell-wall precursors, lipid II, at the MreB binding locus has been shown to promote assembly/disassembly of the filaments in association with the membrane, respectively (Schirner et al. 2015).

Other Roles for MreB

As discussed above, the most established role of MreB is maintenance of the cell-shape as part of the elongasome, a generic term for those proteins involved in elongation of the bacteria. Surprisingly, MreB has also been reported to provide the guiding filaments for cell division in *Chlamydiae* (Ouellette et al. 2012) or cell competency in *B. subtilis* (Mirouze et al. 2015).

References

Addinall SG, Lutkenhaus J (1996) FtsA is localized to the septum in an FtsZ-dependent manner. J Bacteriol 178:7167–7172

Addinall SG, Bi E, Lutkenhaus J (1996) FtsZ ring formation in fts mutants. J Bacteriol 178:3877–3884

Bazylinski DA, Frankel RB, Heywood BR, Mann S, King JW, Donaghay PL, Hanson AK (1995) Controlled Biomineralization of Magnetite (Fe(inf3)O(inf4)) and Greigite (Fe(inf3)S(inf4)) in a Magnetotactic Bacterium. Appl Environ Microbiol 61:3232–3239

Beall B, Lutkenhaus J (1992) Impaired cell division and sporulation of a Bacillus subtilis strain with the ftsA gene deleted. J Bacteriol 174:2398–2403

Bean GJ, Flickinger ST, Westler WM, McCully ME, Sept D, Weibel DB, Amann KJ (2009) A22 disrupts the bacterial actin cytoskeleton by directly binding and inducing a low-affinity state in MreB. Biochemistry 48:4852–4857

Bendezú FO, Hale CA, Bernhardt TG, de Boer PAJ (2009) RodZ (YfgA) is required for proper assembly of the MreB actin cytoskeleton and cell shape in E. coli. EMBO J 28:193–204

Bergeron JR, Hutto R, Ozyamak E, Hom N, Hansen J, Draper O, Byrne ME, Keyhani S, Komeili A, Kollman JM (2016) Structure of the Magnetosome-associated actin-like MamK filament at sub-nanometer resolution. Protein Sci 25. doi:10.1002/pro.2979

Bi EF, Lutkenhaus J (1991) FtsZ ring structure associated with division in Escherichia coli. Nature 354:161–164

Billings G, Ouzounov N, Ursell T, Desmarais SM, Shaevitz J, Gitai Z, Huang KC (2014) De novo morphogenesis in L-forms via geometric control of cell growth: de novo morphogenesis in bacterial L-forms. Mol Microbiol 93:883–896

Bork P, Sander C, Valencia A (1992) An ATPase domain common to prokaryotic cell cycle proteins, sugar kinases, actin, and hsp70 heat shock proteins. Proc Natl Acad Sci U S A 89:7290–7294

Braun T, Orlova A, Valegård K, Lindås A-C, Schröder GF, Egelman EH (2015) Archaeal actin from a hyperthermophile forms a single-stranded filament. PNAS 112:9340–9345

Brown PJB, Pedro MAD, Kysela DT, Henst CVD, Kim J, Bolle XD, Fuqua C, Brun YV (2012) Polar growth in the Alphaproteobacterial order Rhizobiales. Proc Nat Acad Sci 109:1697–1701

Busiek KK, Margolin W (2014) A role for FtsA in SPOR-independent localization of the essential Escherichia coli cell division protein FtsN. Mol Microbiol 92:1212–1226

Busiek KK, Eraso JM, Wang Y, Margolin W (2012) The early divisome protein FtsA interacts directly through its 1c subdomain with the cytoplasmic domain of the late divisome protein FtsN. J Bacteriol 194:1989–2000

Cameron TA, Anderson-Furgeson J, Zupan JR, Zik JJ, Zambryski PC (2014) Peptidoglycan synthesis machinery in Agrobacterium tumefaciens during unipolar growth and cell division. mBio 5:e01219–e01214

Carballido-Lopez R, Formstone A, Li Y, Ehrlich SD, Noirot P, Errington J (2006) Actin homolog MreBH governs cell morphogenesis by localization of the cell wall hydrolase LytE. Dev Cell 11:399–409

Colavin A, Hsin J, Huang KC (2014) Effects of polymerization and nucleotide identity on the conformational dynamics of the bacterial actin homolog MreB. Proc Natl Acad Sci U S A 111:3585–3590

Corbin BD, Geissler B, Sadasivam M, Margolin W (2004) Z-ring-independent interaction between a subdomain of FtsA and late septation proteins as revealed by a polar recruitment assay. J Bacteriol 186:7736–7744

Dai K, Lutkenhaus J (1992) The proper ratio of FtsZ to FtsA is required for cell division to occur in Escherichia coli. J Bacteriol 174:6145–6151

Daniel RA, Errington J (2003) Control of cell morphogenesis in bacteria: two distinct ways to make a rod-shaped cell. Cell 113:767–776

Den BT, Buddelmeijer N, Aarsman MEG, Hameete CM, Nanninga N (1999) Timing of FtsZ Assembly in Escherichia coli. J Bacteriol 181:5167–5175
Di LG (2003) Use of a two-hybrid assay to study the assembly of a complex multicomponent protein machinery: bacterial septosome differentiation. Microbiology 149:3353–3359
Divakaruni AV, Baida C, White CL, Gober JW (2007) The cell shape proteins MreB and MreC control cell morphogenesis by positioning cell wall synthetic complexes. Mol Microbiol 66:174–188
Doi M, Wachi M, Ishino F, Tomioka S, Ito M, Sakagami Y, Suzuki A, Matsuhashi M (1988) Determinations of the DNA sequence of the mreB gene and of the gene products of the mre region that function in formation of the rod shape of Escherichia coli cells. J Bacteriol 170:4619–4624
Domínguez-Escobar J, Chastanet A, Crevenna AH, Fromion V, Wedlich-Söldner R, Carballido-Lopez R (2011) Processive movement of MreB-associated cell wall biosynthetic complexes in bacteria. Science 333:225–228
Donachie WD, Begg KJ, Lutkenhaus JF, Salmond GP, Martinez-Salas E, Vincente M (1979) Role of the ftsA gene product in control of Escherichia coli cell division. J Bacteriol 140:388–394
Doolittle RF, York AL (2002) Bacterial actins? An evolutionary perspective. BioEssays 24:293–296
Draper O, Byrne ME, Li Z, Keyhani S, Barrozo JC, Jensen G, Komeili A (2011) MamK, a bacterial actin, forms dynamic filaments in vivo that are regulated by the acidic proteins MamJ and LimJ: dynamic MamK filaments are regulated by MamJ and LimJ. Mol Microbiol 82:342–354
Duman R, Ishikawa S, Celik I, Strahl H, Ogasawara N, Troc P, Löwe J, Hamoen LW (2013) Structural and genetic analyses reveal the protein SepF as a new membrane anchor for the Z ring. PNAS 110:E4601–E4610
Dye NA, Pincus Z, Theriot JA, Shapiro L, Gitai Z (2005) Two independent spiral structures control cell shape in Caulobacter. PNAS 102:18608–18613
Dye NA, Pincus Z, Fisher IC, Shapiro L, Theriot JA (2011) Mutations in the nucleotide binding pocket of MreB can alter cell curvature and polar morphology in Caulobacter. Mol Microbiol 81:368–394
Erickson HP (1995) FtsZ, a prokaryotic homolog of tubulin. Cell 80:367–370
Errington J (2015) Bacterial morphogenesis and the enigmatic MreB helix. Nat Rev Micro 13:241–248
Favini-Stabile S, Contreras-Martel C, Thielens N, Dessen A (2013) MreB and MurG as scaffolds for the cytoplasmic steps of peptidoglycan biosynthesis. Environ Microbiol 15:3218–3228
Feucht A, Lucet I, Yudkin MD, Errington J (2001) Cytological and biochemical characterization of the FtsA cell division protein of Bacillus subtilis. Mol Microbiol 40:115–125
Figge RM, Divakaruni AV, Gober JW (2004) MreB, the cell shape-determining bacterial actin homologue, co-ordinates cell wall morphogenesis in Caulobacter crescentus. Mol Microbiol 51:1321–1332
Garner EC, Bernard R, Wang W, Zhuang X, Rudner DZ, Mitchison T (2011) Coupled, circumferential motions of the cell wall synthesis machinery and MreB filaments in B subtilis. Science 333:222–225
Gayathri P, Fujii T, Müller-Jensen J, van den Ent F, Namba K, Löwe J (2012) A bipolar spindle of antiparallel ParM filaments drives bacterial plasmid segregation. Science 338:1334–1337
Gayathri P, Fujii T, Namba K, Löwe J (2013) Structure of the ParM filament at 8.5 Å resolution. J Struct Biol 184:33–42
Gerdes K (2009) RodZ, a new player in bacterial cell morphogenesis. EMBO J 28:171–172
Gitai Z, Dye NA, Reisenauer A, Wachi M, Shapiro L (2005) MreB actin-mediated segregation of a specific region of a bacterial chromosome. Cell 120:329–341
Grunberg K, Muller E-C, Otto A, Reszka R, Linder D, Kube M, Reinhardt R, Schüler D (2004) Biochemical and proteomic analysis of the magnetosome membrane in magnetospirillum gryphiswaldense. Appl Environ Microbiol 70:1040–1050

Ishikawa S, Kawai Y, Hiramatsu K, Kuwano M, Ogasawara N (2006) A new FtsZ-interacting protein, YlmF, complements the activity of FtsA during progression of cell division in Bacillus subtilis. Mol Microbiol 60:1364–1380

Iwai N, Nagai K, Wachi M (2002) Novel S-Benzylisothiourea compound that induces spherical cells in escherichia coli probably by acting on a rod-shape-determining protein(s) other than penicillin-binding protein 2. Biosci Biotechnol Biochem 66:2658–2662

Izoré T, Duman R, Kureisaite-Ciziene D, Löwe J (2014) Crenactin from Pyrobaculum calidifontis is closely related to actin in structure and forms steep helical filaments. FEBBS Lett 588:776–782

Jones LJF, Carballido-Lopez R, Errington J (2001) Control of cell shape in bacteria: helical, actin-like filaments in Bacillus subtilis. Cell 104:913–922

Kang C-M, Nyayapathy S, Lee J-Y, Suh J-W, Husson RN (2008) Wag31, a homologue of the cell division protein DivIVA, regulates growth, morphology and polar cell wall synthesis in mycobacteria. Microbiology 154:725–735

Katzmann E, Scheffel A, Gruska M, Plitzko JM, Schüler D (2010) Loss of the actin-like protein MamK has pleiotropic effects on magnetosome formation and chain assembly in Magnetospirillum gryphiswaldense: MamK function in M. gryphiswaldense. Mol Microbiol 77:208–224

Katzmann E, Müller FD, Lang C, Messerer M, Winklhofer M, Plitzko JM, Schüler D (2011) Magnetosome chains are recruited to cellular division sites and split by asymmetric septation: segregation of magnetosome chains during cytokinesis. Mol Microbiol 82:1316–1329

Kim SY, Gitai Z, Kinkhabwala A, Shapiro L, Moerner WE (2006) Single molecules of the bacterial actin MreB undergo directed treadmilling motion in Caulobacter crescentus. Proc Natl Acad Sci U S A 103:10929–10934

Komeili A, Li Z, Newman DK, Jensen GJ (2006) Magnetosomes are cell membrane invaginations organized by the actin-like protein MamK. Science 311:242–245

Kruse T, Bork-Jensen J, Gerdes K (2005) The morphogenetic MreBCD proteins of Escherichia coli form an essential membrane-bound complex. Mol Microbiol 55:78–89

Kühn J, Briegel A, Mörschel E, Kahnt J, Leser K, Wick S, Jensen GJ, Thanbichler M (2010) Bactofilins, a ubiquitous class of cytoskeletal proteins mediating polar localization of a cell wall synthase in Caulobacter crescentus. EMBO J 29:327–339

Lara B, Rico AI, Petruzzelli S, Santona A, Dumas J, Biton J, Vicente M, Mingorance J, Massidda O (2005) Cell division in cocci: localization and properties of the Streptococcus pneumoniae FtsA protein. Mol Microbiol 55:699–711

Letek M, Ordóñez E, Vaquera J, Margolin W, Flärdh K, Mateos LM, Gil JA (2008) DivIVA is required for polar growth in the MreB-lacking rod-shaped actinomycete corynebacterium glutamicum. J Bacteriol 190:3283–3292

Levin PA, Margolis PS, Setlow P, Losick R, Sun D (1992) Identification of Bacillus subtilis genes for septum placement and shape determination. J Bacteriol 174:6717–6728

Lin W, Pan Y (2011) Snapping magnetosome chains by asymmetric cell division in magnetotactic bacteria: magnetotactic bacterial division. Mol Microbiol 82:1301–1304

Lin W, Deng A, Wang Z, Li Y, Wen T, Wu L-F, Wu M, Pan Y (2014) Genomic insights into the uncultured genus, 'Candidatus Magnetobacterium' in the phylum Nitrospirae. ISME J 8:2463–2477

Lindås A-C, Chruszcz M, Bernander R, Valegård K (2014) Structure of crenactin, an archaeal actin homologue active at 90°C. Acta Crystallogr D Biol Crystallogr 70:492–500

Loose M, Mitchison TJ (2014) The bacterial cell division proteins FtsA and FtsZ self-organize into dynamic cytoskeletal patterns. Nat Cell Biol 16:38–46

Löwe J, Amos LA (1998) Crystal structure of the bacterial cell-division protein FtsZ. Nature 391:203–206

Lutkenhaus JF, Donachie WD (1979) Identification of the ftsA gene product. J Bacteriol 137:1088–1094

Ma X, Margolin W (1999) Genetic and functional analyses of the conserved C-terminal core domain of escherichia coli FtsZ. J Bacteriol 181:7531–7544
Ma X, Ehrhardt DW, Margolin W (1996) Colocalization of cell division proteins FtsZ and FtsA to cytoskeletal structures in living Escherichia coli cells by using green fluorescent protein. PNAS 93:12998–13003
Ma X, Sun Q, Wang R, Singh G, Jonietz EL, Margolin W (1997) Interactions between heterologous FtsA and FtsZ proteins at the FtsZ ring. J Bacteriol 179:6788–6797
Maeda YT, Nakadai T, Shin J, Uryu K, Noireaux V, Libchaber A (2012) Assembly of MreB filaments on liposome membranes: a synthetic biology approach. ACS Synth Biol 1:53–59
Martos A, Monterroso B, Zorrilla S, Reija B, Alfonso C, Mingorance J, Rivas G, Jiménez M (2012) Isolation, characterization and lipid-binding properties of the recalcitrant FtsA division protein from Escherichia coli. PLoS One 7:e39829
Mingorance J, Tamames J (2004) The bacterial dcw gene cluster: an island in the genome? In: Vicente M, Tamames J, Valencia A, Mingorance J (eds) Molecules in time and space. Kluwer, Boston, pp 249–271
Mirouze N, Ferret C, Yao Z, Chastanet A, Carballido-Lopez R (2015) MreB-Dependent inhibition of cell elongation during the escape from competence in bacillus subtilis. PLoS Genet 11:e1005299
Mohammadi T, Karczmarek A, Crouvoisier M, Bouhss A, Mengin-Lecreulx D, den Tanneke B (2007) The essential peptidoglycan glycosyltransferase MurG forms a complex with proteins involved in lateral envelope growth as well as with proteins involved in cell division in Escherichia coli. Mol Microbiol 65:1106–1121
Morgenstein RM, Bratton BP, Nguyen JP, Ouzounov N, Shaevitz JW, Gitai Z (2015) RodZ links MreB to cell wall synthesis to mediate MreB rotation and robust morphogenesis. Proc Natl Acad Sci U S A 201509610
Mosyak L, Zhang Y, Glasfeld E, Haney S, Stahl M, Seehra J, Somers WS (2000) The bacterial cell-division protein ZipA and its interaction with an FtsZ fragment revealed by X-ray crystallography. EMBO J 19:3179–3191
Murat D, Quinlan A, Vali H, Komeili A (2010) Comprehensive genetic dissection of the magnetosome gene island reveals the step-wise assembly of a prokaryotic organelle. Proc Natl Acad Sci U S A 107:5593–5598
Nan B, McBride MJ, Chen J, Zusman DR, Oster G (2014) Bacteria that glide with helical tracks. Curr Biol 24:R169–R173
Ouellette SP, Karimova G, Subtil A, Ladant D (2012) Chlamydia co-opts the rod shape-determining proteins MreB and Pbp2 for cell division: chlamydial division is dependent on MreB and Pbp2. Mol Microbiol 85:164–178
Ozyamak E, Kollman JM, Komeili A (2013a) Bacterial actins and their diversity. Biochemistry 52:6928–6939
Ozyamak E, Kollman J, Agard DA, Komeili A (2013b) The bacterial actin MamK: in vitro assembly behavior and filament architecture. J Biol Chem 288:4265–4277
Paradis-Bleau C, Sanschagrin F, Levesque RC (2005) Peptide inhibitors of the essential cell division protein FtsA. Protein Eng Des Sel 18:85–91
Pichoff S, Lutkenhaus J (2005) Tethering the Z ring to the membrane through a conserved membrane targeting sequence in FtsA. Mol Microbiol 55:1722–1734
Pichoff S, Lutkenhaus J (2007) Identification of a region of FtsA required for interaction with FtsZ. Mol Microbiol 64:1129–1138
Pla J, Dopazo A, Vicente M (1990) The native form of FtsA, a septal protein of Escherichia coli, is located in the cytoplasmic membrane. J Bacteriol 172:5097–5102
Pradel N, Santini CL, Bernadac A, Fukumori Y, Wu LF (2006) Biogenesis of actin-like bacterial cytoskeletal filaments destined for positioning prokaryotic magnetic organelles. Proc Natl Acad Sci U S A 103:17485–17489

Reimold C, Defeu S, Joel H, Dempwolff F, Graumann PL (2013) Motion of variable-length MreB filaments at the bacterial cell membrane influences cell morphology. Mol Biol Cell 24:2340–2349

Renner LD, Eswaramoorthy P, Ramamurthi KS, Weibel DB (2013) Studying biomolecule localization by engineering bacterial cell wall curvature. PLoS ONE 8:e84143

Rioux J-B, Philippe N, Pereira S, Pignol D, Wu L-F, Ginet N (2010) A second Actin-Like MamK Protein in Magnetospirillum magneticum AMB-1 encoded outside the genomic magnetosome island. PLoS One 5:e9151

Rueda S, Vicente M, Mingorance J (2003) Concentration and assembly of the division ring proteins FtsZ, FtsA, and ZipA during the Escherichia coli cell cycle. J Bacteriol 185:3344–3351

Rueff A-S, Chastanet A, Domínguez-Escobar J, Yao Z, Yates J, Prejean M-V, Delumeau O, Noirot P, Wedlich-Söldner R, Filipe SR, Carballido-López R (2014) An early cytoplasmic step of peptidoglycan synthesis is associated to MreB in Bacillus subtilis. Mol Microbiol 91:348–362

Salje J, van Fusinita DE, de Piet B, Löwe J (2011) Direct membrane binding by bacterial actin MreB. Mol Cell 43:478–487

Scheffel A, Gruska M, Faivre D, Linaroudis A, Plitzko JM, Schüler D (2006) An acidic protein aligns magnetosomes along a filamentous structure in magnetotactic bacteria. Nature 440:110–114

Schirner K, Eun Y-J, Dion M, Luo Y, Helmann JD, Garner EC, Walker S (2015) Lipid-linked cell wall precursors regulate membrane association of bacterial actin MreB. Nat Chem Biol 11:38–45

Sonkaria S, Fuentes G, Verma C, Narang R, Khare V, Fischer A, Faivre D (2012) Insight into the assembly properties and functional organisation of the magnetotactic bacterial actin-like homolog, MamK. PLoS One 7:e34189

Staniland SS, Moisescu C, Benning LG (2010) Cell division in magnetotactic bacteria splits magnetosome chain in half. J Basic Microbiol 50:392–396

Strahl H, Bürmann F, Hamoen LW (2014) The actin homologue MreB organizes the bacterial cell membrane. Nat Commun 5

Swulius MT, Jensen GJ (2012) The Helical MreB Cytoskeleton in Escherichia coli MC1000/pLE7 is an Artifact of the N-Terminal Yellow Fluorescent Protein Tag. J Bacteriol 194:6382–6386

Szwedziak P, Wang Q, Freund SMV, Lowe J (2012) FtsA forms actin-like protofilaments. EMBO J 31:2249–2260

Szwedziak P, Wang Q, Bharat TAM, Tsim M, Löwe J (2015) Architecture of the ring formed by the tubulin homologue FtsZ in bacterial cell division. eLife Sci 3:e04601

Takacs CN, Poggio S, Charbon G, Pucheault M, Vollmer W, Jacobs-Wagner C (2010) MreB Drives De Novo Rod Morphogenesis in Caulobacter crescentus via Remodeling of the Cell Wall. J Bacteriol 192:1671–1684

Tamames J, González-Moreno M, Mingorance J, Valencia A, Vicente M (2001) Bringing gene order into bacterial shape. Trends Genet 17:124–126

Teeffelen SV, Wang S, Furchtgott L, Huang KC, Wingreen NS, Shaevitz JW, Gitai Z (2011) The bacterial actin MreB rotates, and rotation depends on cell-wall assembly. PNAS 108:15822–15827

Tormo A, Martínez-Salas E, Vicente M (1980) Involvement of the ftsA gene product in late stages of the Escherichia coli cell cycle. J Bacteriol 141:806–813

Tormo A, Ayala JA, Pedro MAD, Aldea M, Vicente M (1986) Interaction of FtsA and PBP3 proteins in the Escherichia coli septum. J Bacteriol 166:985–992

Ursell TS, Nguyen J, Monds RD, Colavin A, Billings G, Ouzounov N, Gitai Z, Shaevitz JW, Huang KC (2014) Rod-like bacterial shape is maintained by feedback between cell curvature and cytoskeletal localization. Proc Natl Acad Sci U S A 111:E1025–E1034

van den Ent F, Löwe J (2000) Crystal structure of the cell division protein FtsA from Thermotoga maritima. EMBO J 19:5300–5307

van den Ent F, Amos L, Löwe J (2001a) Bacterial ancestry of actin and tubulin. Curr Opin Microbiol 4:634–638
van den Ent F, Amos LA, Lowe J (2001b) Prokaryotic origin of the actin cytoskeleton. Nature 413:39–44
van den Ent F, Johnson CM, Persons L, de Piet B, Löwe J (2010) Bacterial actin MreB assembles in complex with cell shape protein RodZ. EMBO J 29:1081–1090
van den Ent F, Izore T, Bharat TAM, Johnson CM, Löwe J (2014) Bacterial actin MreB forms antiparallel double filaments. eLife Sci 3:e02634
Varley AW, Stewart GC (1992) The divIVB region of the Bacillus subtilis chromosome encodes homologs of Escherichia coli septum placement (minCD) and cell shape (mreBCD) determinants. J Bacteriol 174:6729–6742
Wachi M, Doi M, Tamaki S, Park W, Nakajima-Iijima S, Matsuhashi M (1987) Mutant isolation and molecular cloning of mre genes, which determine cell shape, sensitivity to mecillinam, and amount of penicillin-binding proteins in Escherichia coli. J Bacteriol 169:4935–4940
Walker JR, Kovarik A, Allen JS, Gustafson RA (1975) Regulation of bacterial cell division: temperature-sensitive mutants of Escherichia coli that are defective in septum formation. J Bacteriol 123:693–703
Wang H, Gayda RC (1992) Quantitative determination of FtsA at different growth rates in Escherichia coli using monoclonal antibodies. Mol Microbiol 6:2517–2524
Weiss DS (2015) Last but not least: new insights into how FtsN triggers constriction during Escherichia coli cell division. Mol Microbiol 95:903–909
White CL, Kitich A, Gober JW (2010) Positioning cell wall synthetic complexes by the bacterial morphogenetic proteins MreB and MreD. Mol Microbiol 76:616–633
Yakhnina AA, Gitai Z (2012) The small protein MbiA interacts with MreB and modulates cell shape in Caulobacter crescentus. Mol Microbiol 85:1090–1104

Chapter 9
Bacterial Nucleoid Occlusion: Multiple Mechanisms for Preventing Chromosome Bisection During Cell Division

Maria A. Schumacher

Abstract In most bacteria cell division is driven by the prokaryotic tubulin homolog, FtsZ, which forms the cytokinetic Z ring. Cell survival demands both the spatial and temporal accuracy of this process to ensure that equal progeny are produced with intact genomes. While mechanisms preventing septum formation at the cell poles have been known for decades, the means by which the bacterial nucleoid is spared from bisection during cell division, called nucleoid exclusion (NO), have only recently been deduced. The NO theory was originally posited decades ago based on the key observation that the cell division machinery appeared to be inhibited from forming near the bacterial nucleoid. However, what might drive the NO process was unclear. Within the last 10 years specific proteins have been identified as important mediators of NO. Arguably the best studied NO mechanisms are those employed by the *Escherichia coli* SlmA and *Bacillus subtilis* Noc proteins. Both proteins bind specific DNA sequences within selected chromosomal regions to act as timing devices. However, Noc and SlmA contain completely different structural folds and utilize distinct NO mechanisms. Recent studies have identified additional processes and factors that participate in preventing nucleoid septation during cell division. These combined data show multiple levels of redundancy as well as a striking diversity of mechanisms have evolved to protect cells against catastrophic bisection of the nucleoid. Here we discuss these recent findings with particular emphasis on what is known about the molecular underpinnings of specific NO machinery and processes.

Keywords NO factor • Noc • SlmA • Bacterial cytokinesis • Nucleoid occlusion • DNA spreading • *Staphylococcus aureus* • *Bacillus subtilis* • ParB family. FtsZ regulators • Nucleoid associated proteins (NAPs) • MatP • *matS* • SlmA binding sequence (SBS) • *Streptomyces* • MipZ • *Myxococcus xanthus* • PomZ • *Streptococcus pneumonia* • MapZ

The original version of this chapter was revised. An erratum to this chapter can be found at DOI 10.1007/978-3-319-53047-5_16

M.A. Schumacher (✉)
Department of Biochemistry, Duke University School of Medicine,
243 Nanaline H. Duke, Durham, NC 27710, USA
e-mail: maria.schumacher@duke.edu

J. Löwe, L.A. Amos (eds.), *Prokaryotic Cytoskeletons*, Subcellular Biochemistry 84, DOI 10.1007/978-3-319-53047-5_9

Introduction

Bacteria typically contain a single chromosome and divide by binary fission as mediated by the tubulin homolog, FtsZ. FtsZ is an ancient protein that, in addition to bacteria, drives cell division in multiple archaea, primitive mitochondria and chloroplasts (Bi and Lutkenhaus 1991; Margolin 2005; Adams and Errington 2009; Erickson et al. 2010). FtsZ proteins have a conserved structural organization that includes a short N terminal disordered region, a globular tubulin-like domain, a flexible linker ranging from ~30–300 residues that is also variable in sequence and a C-terminal tail (CTT) that consists of ~14 conserved residues followed by a short, variable region (Löwe and Amos 1998; Löwe and van den Ent 2001). GTP binding to the FtsZ globular domains nucleates interactions between subunits in which the plus end of one FtsZ protomer binds the minus end of another. These interactions lead to the generation of continuous linear protofilaments (pfs) (Bi and Lutkenhaus 1991; Lutkenhaus et al. 2012; Adams and Errington 2009) (Fig. 9.1a, b). FtsZ pfs composed of ~20–40 FtsZ protomers combine to form a septal ring-like structure called the Z ring, which encircles the cell and functions as a dynamic platform for the assembly of the divisome (Lutkenhaus et al. 2012; Egan and Vollmer 2013). Included in the divisome are peptidoglycan remodeling proteins that catalyze membrane cleavage to create two daughter cells (Egan and Volmer 2013). Although a conceptually simple process, accumulating evidence indicates that sophisticated and often redundant machinery are present in bacteria to ensure the proper temporal and spatial timing of Z ring formation (Adams and Errington 2009; Bailey et al. 2014; Monahan et al. 2014). Ultimately, this involves a tight synchronization of chromosome replication, segregation and septum formation. For the organism to survive, this process must ensure the equal parsing of chromosomes to daughter cells. Recent findings have revealed specific mechanisms that safeguard against chromosome fragmentation by the cell division machinery. This process called nucleoid occlusion (NO) is the subject of this chapter.

In order to deduce the mechanisms involved in NO, a detailed understanding of the cytokinetic machinery is needed, in particular, Z ring formation. Indeed, how FtsZ filaments are organized to form a stable yet dynamic Z ring has been a central question in the field. Recent work has started to shed light on this issue. To construct the Z ring, FtsZ, which is a soluble cytosolic protein, must be targeted to the membrane. Depending on the bacterium, this role is played by FtsA, ZipA or SepF (Pichoff and Lutkenhaus 2002; Romberg and Levin 2003; Duman et al. 2013; Huang et al. 2013). Studies in which FtsZ was artificially tethered to liposomes by fusing it to YFP and a FtsA membrane interacting amphipathic helix provided important evidence that FtsZ can form rings in the absence of other factors (Osawa et al. 2008; Erickson et al. 2010). Notably, these artificially formed Z rings are also capable of mediating constriction and even septation in simplified artificial membrane systems (Osawa and Erickson 2013). Studies in which FtsZ rings were reconstituted on liposomes suggested a model for Z ring formation involving tight and close lateral contacts between FtsZ pfs (Milan et al. 2012). However, recent *in vivo*

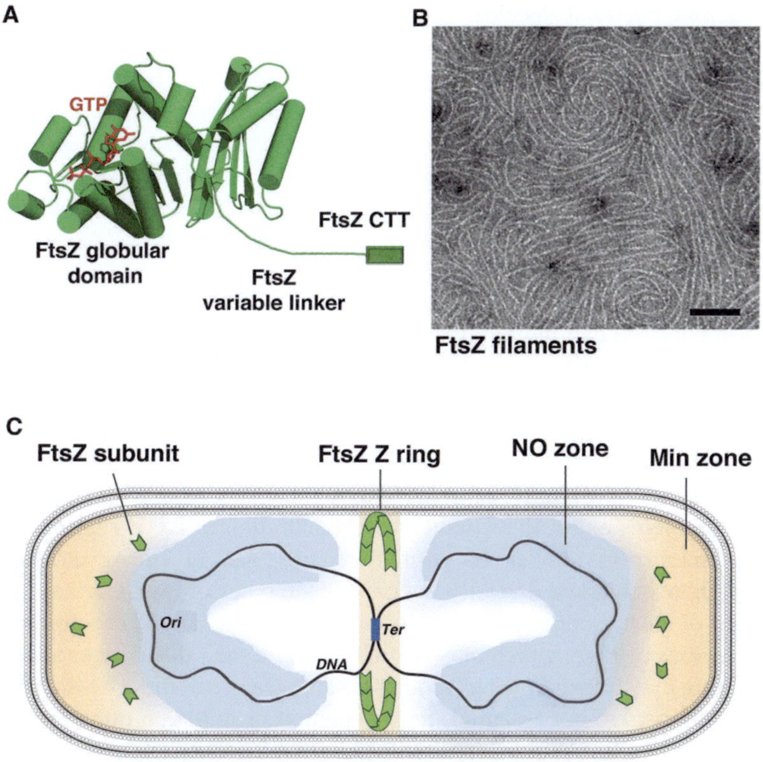

Fig. 9.1 FtsZ structure and regulation of its polymerization in cell division site placement. (**a**) Ribbon diagram of one subunit of FtsZ. Shown are the key structural elements (the short N-terminal disordered region is not displayed), including the globular domain, the extended linker domain and C-terminal tail, which is comprised of a conserved region and a short C-terminal variable region. GTP (*red*) binding to the globular domains drives interactions between FtsZ subunits, leading to protofilament formation. The C-tail binds effector proteins, which either sequester it to the membrane for Z ring formation or alter its polymerization activity. (**b**) The TEM of FtsZ is from *Caulobacter* (Current Opinion 2014 (https://www.ncbi.nlm.nih.gov/pubmed/24529242)). (**c**) Cell division site placement mechanisms. Shown is a cartoon of a *rod-shaped* bacterial cell indicating processes that direct Z ring formation to the cell center. In *E. coli*, the cell poles are protected from Z ring assembly by the well characterized Min inhibitory system. This system establishes a high local concentration of the FtsZ inhibitor MinC at the poles, thus preventing the generation of anucleate minicells. The replicated chromosomes are spared from Z ring bisection by nucleoid occlusion (NO). The zone of NO is indicated by *light blue* regions surrounding the chromosomal DNA

cryo-EM tomography and super-resolution imaging studies indicate that FtsZ polymers in the Z ring do not form a continuous structure around the cell but rather dispersed assemblages (Li et al. 2007; Fu et al. 2010; Strauss et al. 2012). Interestingly, although the combined imaging analyses suggest that bacteria all assemble loosely arranged Z rings, the precise configuration of its structure appears to vary across bacterial species. Specifically, photo-activated light microscopy (PALM) images of *Escherichia coli* Z rings revealed that they are composed of loose bundles of randomly overlapping FtsZ pfs (Fu et al. 2010). Three-dimensional structured illumination microscopy (3D–SIM) in *Bacilli*, by contrast, showed the Z ring as a discontinuous structure arranged as a bead-like pattern where densely populated zones of FtsZ polymers are interspersed with regions of few or no FtsZ polymers (Jennings et al. 2011; Strauss et al. 2012). In *Caulobacter crescentus*, studies indicate that short, nonoverlapping FtsZ pfs form a discontinuous Z ring at midcell (Li et al. 2007; Biteen et al. 2012).

While emerging data have begun to elucidate the Z ring structure, an enduring question in bacterial cell biology is how the position of Z ring formation is determined. Indeed, in most bacteria the intracellular levels of FtsZ remain essentially unchanged during the cell cycle and exceed the critical concentration required for Z ring formation (Lutkenhaus et al. 2012; Erickson et al. 2010). Thus, the spatial and temporal regulation of cytokinesis must be controlled via Z ring assembly. In the last few decades a bewildering array of FtsZ regulatory effectors have been identified. These proteins aid in the assembly and maturation of the Z ring or act in an inhibitory manner to prevent its formation (reviewed in Huang et al. 2013). Strikingly, there is marked diversity not only in the types of FtsZ effectors but also the number of such proteins found between bacterial species. These proteins bind different regions of FtsZ, which may partially explain their diversity as such multivalent binding may be synergistic. Another notable feature of these binding proteins is that while most are not essential, their depletion or mutation often impairs cell division under specific cellular conditions. Thus, the diversity might also reflect that these proteins are only required in a specific bacterial niche, which may not be accurately recapitulated in the laboratory. In fact, recent work has identified FtsZ regulators that function in response to particular environmental cues such as specific nutritional (OpgH, KidO and UgtP) and developmental states (MciZ and Maf) (Weart et al. 2003; Handler et al. 2008; Radhakrishnan et al. 2010; Briley et al. 2011; Hill et al. 2013; Bisson-Filho et al. 2015).

While the above proteins function as general regulators of FtsZ assembly, there are specialized protein based machineries that specifically prevent Z ring formation at the cell poles and over the nucleoid DNA (Fig. 9.1c). The former process is mediated in many bacteria by the *min* system. Cells that harbor impaired *min* systems form Z rings near the cell poles, leading to the generation of DNA-less "minicells" (Adler et al. 1967; Davie et al. 1984; de Boer et al. 1989; de Boer et al. 1990; Raskin and de Boer 1999a; Raskin and de Boer 1999b; Dajkovic et al. 2008). The Min system has been extensively studied in *E. coli* where it is comprised of MinC, which is a FtsZ inhibitor, MinD, a membrane-associated Walker box ATPase and MinE, a factor that interacts with and organizes the MinCD complex (Hu and Lutkenhaus 2001; Hu et al. 2002; Shih et al. 2003; Lutkenhaus 2007). MinE binds MinCD and oscillates from pole to pole (Raskin and de Boer 1999a; Raskin and de Boer 1999b). The net result of this oscillatory process is the formation of a zone of FtsZ inhibition at the cell poles. The *min* system also exists in *Bacillus subtilis*, however the role of MinE is played by MinJ/DivIVA, which are anchored at the cell poles and recruit the MinCD complex (Bramkamp et al. 2008). In contrast to cell pole inhibition of Z ring formation, mechanisms by which the nucleoid is spared from fragmentation by the Z ring, the process termed nucleoid occlusion (NO), have only begun to be elucidated in the last 10 years (Woldringh et al. 1990; Woldringh et al. 1991; Woldringh 2002). Indeed, it had been unknown if specific compounds or factors were involved in NO until 2004 when Noc and subsequently in 2005, when SlmA were identified as NO effector proteins in *B. subtilis* and *E. coli,* respectively (Wu and Errington 2004; Bernhardt and De Boer 2005). This chapter discusses what has been learned to date about NO, the recently identified NO factors SlmA and Noc and other proteins and mechanisms that contribute to NO.

The Bacterial Nucleoid and the Nucleoid Occlusion (NO) Model

More than three decades ago Woldringh and coworkers made the key observation that the cell division machinery appeared to be inhibited from forming over or near the bacterial nucleoid. Based on this finding they proposed the "nucleoid occlusion" (NO) hypothesis, which in simplest terms states that the nucleoid inhibits cell division wherever it occupies space (Woldringh et al. 1990; Woldringh et al. 1991). A later refinement of the model proposed that molecular crowding arising from transertion, which involves the concurrent translation and insertion of proteins into the membrane, was responsible for this inhibitory effect (Zaritsky and Woldringh 2003). Short or long range "inhibitors", of an unknown nature present near the nucleoid were also proposed (Woldringh et al. 1991). The nucleoid is the bacterial equivalent of the eukaryotic nucleus and is comprised of a highly condensed chromosome. However, unlike the eukaryotic nucleus, it is not surrounded by a nuclear envelope and hence is open to the cytosol and all its attendant proteins. But, it was unclear what molecules on or around the nucleoid might mediate the NO effect or if the DNA itself could somehow inhibit Z ring formation. Hence, understanding the molecular architecture and organization of the nucleoid is important in addressing the NO question.

A fully extended bacterial chromosome of 2–8 mega base pairs (Mb) is several thousand-fold larger than the cell it is contained within. Therefore, the DNA must be significantly condensed while also maintained in a state that permits dynamic processes such as replication, transcription and translation. Although early studies indicated that the bacterial nucleoid is a highly compacted structure, recent studies have shown that the organization of the bacterial chromosome is mediated at several levels and by multiple forces (Wang et al. 2013; Gruber 2014). Macromolecular crowding limits the boundaries of the nucleoid while local folding of the chromosome into superhelices, called plectonemes, is mediated by negative DNA supercoiling, caused primarily by the activity of DNA gyrase (Drlica 1992; Gubaev and Klostermeier 2014). Supercoiling alone does not lead to the observed dramatic compaction of the chromosome and it has been estimated that half of the chromosome is constrained by abundant nucleoid associated proteins (NAPs) (Hardy and Cozzarelli 2005). These proteins bind specifically and nonspecifically to DNA and introduce kinks and bends in the DNA. Well known *E. coli* NAPs include HU, IHF, Fis and H-NS (Azan et al. 2000). NAPs primarily mediate short range DNA condensation. A further level of organization was revealed that consists of higher order compacted regions (800 kb–1 Mb) called Macrodomains (MDs), which are insulated from other DNA regions. While these domains have been identified and characterized primarily in *E. coli*, data suggests similar higher order organizations are likely present in other bacterial chromosomes (Le et al. 2013). In *E. coli,* the presence of MDs was demonstrated by elegant fluorescence in situ hybridization (FISH) assays as well as experiments based on recombination, which assessed the frequency of random collisions between different sites on the chromosome (Niki et al., 2000; Valens et al. 2004). These studies identified four primary MDs in *E. coli*, the Ori (origin containing), Ter (terminus containing) and two additional insulated domains, the left and right MDs. Two unstructured regions are located on either side of the Ori MD (Niki et al. 2000; Valens et al. 2004).

The precise molecular mechanism(s) that give rise to specific organization of MDs is still unclear. However, proteins that bind DNA sites contained within one or a few MDs, so-called MD specific or selective proteins, have been postulated to play some role in this organization. To date, only one MD specific factor has been identified, which is the Macrodomain Ter binding protein, MatP (Mercier et al. 2008). MatP was discovered through bioinformatic analyses that identified a DNA sequence (*matS)* found only in the Ter MD (Mercier et al. 2008). The Ter MD contains ~26 *matS* sites. This sequence was subsequently used to isolate the MatP protein. MatP was found to localize with the Ter MD throughout the cell cycle. Recent structures of MatP-*matS* complexes provided key insight into how it functions as an MD organizer (Dupaigne et al., 2012). These structures revealed that MatP contains an unusual tripartite fold with a novel four-helix bundle DNA-binding domain, ribbon-helix-helix dimerization region and a long C-terminal coiled coil domain. The latter domain mediates extended tetrameric linkages between MatP dimers that allow it to bridge distant *matS* sites (Dupaigne et al. 2012). *In vivo* studies on MatP tetramer deficient mutants showed that these proteins were unable to mediate proper Ter MD condensation, thus supporting that the DNA bridging function of the protein has a crucial role in Ter compaction and organization (Dupaigne et al. 2012). Studies to identify additional MD specific DNA binding proteins are ongoing. Such factors might be predicted to function in multiple nucleoid processes, including NO. In fact, as will be described in a subsequent section, recent data revealed that MatP itself plays a role in NO by interacting with and affecting the localization of the cell division machinery. MatP's property as a protein that binds specifically to the Ter MD, which is the last DNA region to segregate, is key to this function (Espéli et al. 2012). While MatP remains the only known MD specific DNA binding protein, the recently discovered NO factors, *B. subtilis* Noc and *E. coli* SlmA, were both shown to bind the chromosome in a MD selective manner. As described below, this MD DNA-binding selectivity was demonstrated to be crucial for their NO functions.

B. subtilis Noc, a Nucleoid Occlusion Factor of the ParB Family

The first identified NO factor was the *B. subtilis* protein originally called YyaA (Sievers et al. 2002). Interestingly, YyaA was initially studied for a possible role in DNA segregation as it displays strong sequence similarity to ParB proteins (Sievers et al. 2002). ParB proteins are DNA binding factors that collaborate with ParA proteins to mediate DNA segregation (Hayes and Barillà 2006; Schumacher 2012). Specifically, ParB proteins bind centromere DNA to form partition complexes, which then recruit and activate their partner ParA proteins to physically separate replicated DNA substrates (Hayes and Barillà 2006). Par proteins are required for the segregation of low copy number plasmids. However, only a few chromosomally encoded ParB-ParA proteins have been shown to be essential for nucleoid DNA segregation (Ptacin et al. 2010). Indeed, the *B. subtilis soj-spo0J* operon encodes chromosomal ParA-ParB-like factors, however their precise roles in segregation are somewhat unclear (Lee and Grossman 2006). Hence, ongoing studies have attempted to identify additional players in *B. subtilis* DNA segregation. One possible candidate gene was *yyaA*, which is located near the origin of replication (*oriC*), just upstream of the operon encoding the chromosomal ParB-ParA proteins, which are called Spo0J-Soj. The 283-residue YyaA protein harbors 35% sequence identity to the *B. subtilis* ParB homolog, Spo0J, which suggested that it arose from a partial gene duplication event and might be involved in DNA segregation. But, studies on YyaA showed that it played no role in chromosome segregation (Sievers et al. 2002). However, a dramatic effect on sporulation was observed upon its overexpression, which suggested that it may be involved in the regulation of septum formation.

The function of YyaA was subsequently revealed in genetic cross experiments in which certain combinations of null mutations significantly affected cell viability (Wu and Errington 2004). These effects were narrowed down to the loss of both the *min* system (*divIVA, minC, minD* or *ezrA*) and the *yyaA* gene. In particular, when both MinD and YyaA were depleted, cell division was arrested and the bacteria formed elongated filaments. This was distinct from the minicell phenotype observed upon *min* depletion alone and indicated that the combined loss of both *minD* and *yyaA* caused severe cell division defects. Closer examination of the double knockout strains revealed that FtsZ became dispersed in the cell and hence was unable to reach a sufficiently high concentration to form a productive Z ring (Wu and Errington 2004). Moreover, under conditions when DNA replication was perturbed, the absence of *yyaA* led to the guillotining of chromosomal DNA. Importantly, it was shown that the effect of Noc on cell division was direct and not due to transcription regulation; microarray studies revealed that although the Noc protein appeared to localize over much of the chromosome, this binding had little affect on the expression of any known gene, including those involved in cell division. In addition, the *noc* gene is constitutively expressed, in a cell cycle independent manner (Wu and Errington 2004). The nonspecific binding of YyaA to the nucleoid was a property deemed important for an NO factor. Thus, the combined findings of these initial experiments indicated that YyaA was a bone fide NO factor and it was therefore renamed, Noc (Wu and Errington, 2004).

Although initial studies indicated that Noc bound nonspecifically to the nucleoid, they revealed a gap in Noc localization near the Ter regions of the segregating chromosomes (Wu and Errington 2004). A subsequent interrogation of Noc nucleoid binding by ChAP-on-Chip uncovered a consensus Noc-binding DNA sequence (NBS), 5′-ATTTCCCGGGAAAT-3′ (Wu et al. 2009). Strikingly, NBS were found to be scattered over the nucleoid at ~70 sites, with the notable exception of the Ter region of the DNA (Fig. 9.2a). As the Ter region is the last to segregate, the absence of the Noc, which is an inhibitor of Z ring formation, would allow the ring to form concomitant with segregation. The finding explained the observation that Noc was absent from the midcell near segregating DNA and suggested that Noc not only serves as a spatial regulator of cell division by protecting its bound DNA against fragmentation but also a timing device that coordinates chromosome segregation and cell division (Wu et al. 2009). However, a central question that remained unanswered from these studies was how Noc inhibits FtsZ from forming Z rings over the DNA. Indeed, as will be discussed, a NO factor in *E. coli* called SlmA had at the time been discovered and was shown to bind directly to FtsZ and affect its polymerization properties (Bernhardt and de Boer, 2005). By contrast, despite exhaustive analyses, no evidence for an interaction between Noc and FtsZ or any component of the cell division machinery was demonstrated (Wu et al. 2009).

Noc Binds Both Chromosomal NBS Sites and the Membrane

Although Noc showed no interactions with cell division proteins that might help explain its NO function, in addition to binding to Non-Ter regions of the chromosome, localization studies revealed that it appeared to form dynamic foci near or at the periphery of the bacterial membrane (Adams et al. 2015). The membrane potential has been shown to effect the localization of peripheral membrane proteins. Hence, treatment with carbonyl cyanide *m*-chlorophenyl hydrazine (CCCP), which dissipates the membrane potential, was employed to ascertain any affect on Noc localization (Adams et al., 2015). These studies revealed that CCCP treatment led to a dramatic alteration in Noc localization away from the membrane. Such localization effects were also observed in a F_1F_0 ATP synthase deletion strain indicating that the resultant Noc delocalization was caused by the loss of the membrane potential and not ATP depletion. To try and identify possible membrane interacting motifs, Adams and coworkers analyzed the Noc sequence and found that residues 2–9 in the N-terminal region of the protein were predicted to fold into an amphipathic helix (Cornell and Taneva 2006; Sapay et al. 2006; Adams et al. 2015). It is notable that this would form a very short helix, with only ~2 turns. Nonetheless, an N-terminal Noc mutant lacking the first 10 residues (NocΔ10) did not localize to the membrane and was nonfunctional.

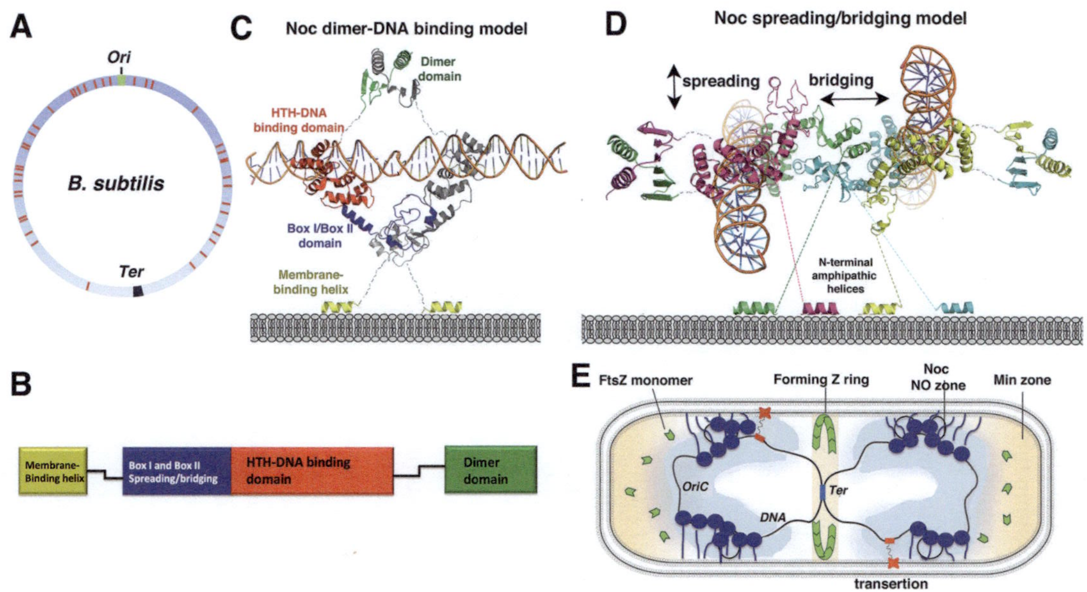

Fig. 9.2 Noc mediated NO in *B. subtilis*. (**a**) Schematic of the *B. subtilis* chromosome (modified from Adams et al. 2014) indicating binding sites for Noc. Notably these interacting regions are localized throughout most of the chromosome with the exception of the Ter region, thus allowing synchronous DNA segregation and cell division. (**b**) Schematic showing domain organization of the *B. subtilis* Noc protein. The protein includes an N-terminal putative amphipathic helix, linker, arginine rich box I and box II region, HTH DNA binding domain and C-terminal dimer domain. (**c**) Molecular model of Noc (colored coded according to domain organization in (**b**) generated by combining structural elements of the P1 ParB and *H. pylori* Spo0J proteins. (**d**) Molecular model for box I/box II bridging in Noc based on the interactions observed in the *H. pylori* Spo0J-DNA structure (Chen et al. 2015) within the arginine rich region. Note, the box I/box II bridging contacts are in proximity to the linker region that extends the N-terminal, putative amphipathic helices towards the membrane suggesting a compact protein-DNA-membrane complex. (**e**) Schematic model for Noc mediated NO in a dividing cell. A segregating cell is shown that is just starting division (the Z ring has begun forming at the cell center). Noc dimers are represented as *blue circles* with N-terminal regions projecting and binding the membrane (via the N-terminal amphipathic helices). Noc molecules both bridge and spread along the DNA to create a compact structure near the membrane that would exclude FtsZ molecules from forming polymeric structures in their vicinities

Examination of sequence alignments of multiple Noc proteins show that the N-terminal region, which encompasses the putative amphipathic helix, is highly conserved, while the ~20 residue region linking this stretch to the remainder of the protein is not. The region of strong conservation extends from residues 1–14, which suggests the possibility that the amphipathic helix might be longer than the eight residue helix predicted by the AmphipaSeeK program (Sapay et al. 2006; Adams et al. 2015). Interestingly, however, multiple sequence alignments show that some Noc homologs contain a proline within this region. Prolines are known to be helix breakers and the location of this residue, near the center of the putative helix, indicates a possibility that this region may not be helical or perhaps forms a distorted helix. However, experiments showing that the Noc N-terminal region could be swapped with a known amphipathic helix from the hepatitis C virus protein NS4B and maintain membrane targeting supported the helical hypothesis (Adams et al. 2015). But, when GFP was fused with the Noc N-terminal helix only weak membrane localization was observed. The combined data were subsequently explained by the finding that multiple Noc proteins must be bound and "spread" along the DNA to establish a strong interaction with the membrane and the weakly binding Noc N-terminal helix. DNA spreading is a characteristic feature of ParB proteins and occurs when the protein binds it cognate site and nucleates interactions with surrounding non-specific DNA (Lynch and Wang 1995; Rodionov et al. 1999; Kusiak et al. 2011). Spreading was initially observed as a gene silencing effect that occurred when ParB proteins bound their *parS* centromere site and spread to surrounding genes (Lynch and Wang 1995; Rodionov et al., 1999). In these studies when ParB levels were artificially elevated, gene silencing extended several kb from the *parS* site. Spreading can also be inferred, as was the case for Noc, by chromatin immunoprecipitation (ChIP) assays (Wu et al. 2009), which reveal peaks of protein binding at the cognate site as well as surrounding DNA. Spreading leads to the formation of large and presumably more stable protein-nucleic acid complexes. In the case of DNA segregation this is beneficial as the presence of multiple ParB proteins provides a high local concentration of binding sites for its partner protein ParA. For Noc, it would similarly generate a localized accumulation of membrane binding anchor regions, as there are an estimated 4500 Noc molecules/cell (Wu and Errington 2004). Consistent with this, when residues in regions of Noc known to be involved in spreading by ParB homologs were mutated, membrane localization was disrupted (Chen et al. 2015; Adams et al. 2015). Another unusual DNA binding property exhibited by ParB proteins is DNA bridging. Data indicate that the same N-terminal regions that mediate spreading are also involved in bridging (Chen et al. 2015). Given that Noc can presumably spread and perhaps even bridge DNA sites, it might be expected that it could play a role in chromosome organization and/or compaction. Interestingly, however, studies indicated that Noc depletion apparently does not alter nucleoid organization. However, it is clearly important for its NO function (Sievers et al. 2002; Adams et al. 2015).

A Molecular Model for Noc-Mediated NO

To date there is no structural information available for a Noc protein. However, as noted the sequences of Noc proteins clearly place them in the ParB/Spo0J family. Multiple structures of ParB/Spo0J proteins in the absence and presence of DNA have been solved that can be used to provide insight into Noc structure (Schumacher and Funnell 2005; Schumacher et al. 2010; Chen et al. 2015). These ParB structures encompass all but the N-terminal region, which contains the putative amphipathic helix in Noc. Specifically, the structure of P1 ParB containing the DNA-binding and C-terminal dimer domains in complex with DNA was solved and recently a structure of the *Helicobacter pylori* Spo0J protein that included the N-terminal arginine-rich domain, implicated in DNA spreading, connected to the HTH region was obtained bound to DNA (Schumacher and Funnell 2005; Schumacher et al. 2010; Chen et al. 2015). By combining these structures, a model for Noc can be constructed as shown in Fig. 9.2b, c. According to this model, the N-terminal helices would be connected via a sufficiently long linker to allow them to contact the membrane without clash from the remainder of the protein. The long, flexible linker connects to an arginine-rich region (also called the ParB-box I and box II regions), which mediates DNA spreading and bridging. This region is followed by the central HTH, DNA binding domain. Finally, flexibly attached to the HTH domain is the C-terminal dimer domain (Fig. 9.2b, c). Notably, data have demonstrated that the C-terminal region also mediates dimerization in Noc (Adams et al. 2015). The resultant Noc model provides insight into how it is organized on the nucleoid near the membrane and how the amphipathic helix may be positioned relative to the rest of the molecule. Interestingly, it suggests that binding of multiple Noc to the nucleoid would lead to a significant accretion of tethered bridged and spread molecules near the membrane (Chen et al. 2015). This bridging model is based on the Spo0J arginine rich interactions. However, while the box I-II containing region clearly mediates long range bridging interactions, the P1 and F ParB-DNA structures revealed short range bridging between residues in the HTH domains. While not mentioned in the Chen et al. work, similar HTH-mediated bridging contacts are observed in the *H. pylori* Spo0J-DNA complex (Chen et al. 2015). Such HTH mediated bridging interactions provide close apposition of ParB-DNA molecules and would enable a higher degree of DNA compaction. Hence, it is possible that both the box I-II and HTH regions may contribute to DNA organization by ParB and Noc proteins. This Noc structural model suggests an NO mechanism in which the close localization of multiple, bulky Noc protein-nucleic acid aggregates cause considerable crowding effects, thus precluding Z ring formation at the membrane in their vicinity, while allowing Z ring assembly near the midcell at the Ter regions. Notably, this Noc-mediated crowding model (Fig. 9.2d) is highly analogous to the transertion-based NO hypothesis proposed by Woldringh and coworkers decades ago. According to that model, large complexes formed during coupled transcription-translation with membrane protein insertion (transertion) were predicted to physically occlude the cell division machinery from properly assembling near the nucleoid (Woldringh

2002). Noc mediated crowding at the membrane may be further enhanced by ongoing transertion events within the vicinity of Noc assemblages (Adams et al. 2015) (Fig. 9.2e).

Noc Function in Other Bacteria: Noc Is Essential for NO in S. aureus

Noc proteins appear to be widespread in the Firmicutes, with 310, 18 and 7 putative members found in *Bacillaceae, Listeria* and *Planococcaceae*, respectively. Although Noc plays an important role in protecting the *B. subtilis* nucleoid from bisection it is not essential, except under certain conditions such as perturbation of DNA replication (Wu and Errington 2004). Recent studies, however, revealed that Noc is required for proper NO in *Staphylococcus aureus*, underscoring the importance of studying proteins and processes in multiple model systems (Veiga et al. 2011). Unlike its rod shaped brethren, *S. aureus* is spherical and divides by the generation of septa over three consecutive division cycles. This complex cell division mechanism suggests that accurate cell division site placement would be especially crucial in these bacteria. *S. aureus* lack a Min system but encode a Noc protein. Thus, following the discovery of the *B. subtilis* Noc as an NO factor, studies on the homologous 279 amino acid protein in *Staphylococcus aureus* (48% identity to the *B. subtilis* Noc) were performed to assess its NO role in this organism. Strikingly, these studies revealed that deletion of *S. aureus noc* resulted in Z ring formation over the nucleoid and irreparable DNA damage (Veiga et al. 2011). Specifically, *noc-* cells showed a significant increase in cell size and, in ~15% of the cells, division septa were formed over the nucleoid. Importantly, detailed analyses showed that the DNA in these cells was, in fact, bisected by the division machinery and not just trapped by it. Thus, unlike the case in *B. subtilis,* the Noc protein plays an essential NO function in *S. aureus,* safeguarding against nucleoid bisection even during normal growth conditions.

The binding properties and sequence of *S. aureus* Noc suggests it may employ a NO mechanism similar to the *B. subtilis* protein. *S. aureus* Noc shares high sequence identity within its predicted DNA binding domain with the *B. subtilis* protein and data indicated that the *S. aureus* Noc binds the same NBS sequence as the *B. subtilis* protein (Adams and Errington 2009). Moreover, as in in *B. subtilis*, these DNA sites are located within Non-Ter regions on the *S. aureus* chromosome (Adams and Errington 2009). Consistent with this, studies using a functional *S. aureus* Noc-YFP fusion showed that it localized with the nucleoid in a manner similar to the *B. subtilis* protein whereby it was not present at the midcell region between segregating chromosomes (Adams and Errington 2009). Staphylococcal Noc proteins are among the Noc homologs that harbor a proline within their putative N-terminal amphipathic helices. Hence, whether this region of *S. aureus* Noc forms an amphipathic helix and localizes to the membrane remains to be tested.

Nucleoid Occlusion in *E. coli*: Identification of SlmA as an NO Factor

Studies on Noc clearly indicate a role in NO for this protein in Firmicutes. However, Noc is not found in Gram-negative bacteria. To isolate an NO factor in *E. coli*, Bernhardt and de Boer utilized a screen similar to that used to identify Noc, in which mutations that were synthetically lethal with a defective Min system were characterized. The screen revealed a clear phenotype with defective septation when mutations in a previously uncharacterized gene, *ttk*, which stands for twenty-three-kDa protein, were obtained (Bernhardt and de Boer 2005). Based on the fact that it caused a synthetic lethality with a defective Min system, the protein was renamed SlmA. Strikingly, SlmA is structurally distinct from Noc; SlmA encodes a TetR-like protein, not a ParB-like homolog. The TetR transcription regulatory family is one of the largest in bacteria with more than 200,000 members identified in 115 genera of Gram-positive, α-, β- and γ-proteobacteria, cyanobacteria and archaea (Ramos et al. 2005; Cuthbertson and Nodwell 2013). Thus, similar to how Noc represents a repurposing of a fold commonly used for DNA segregation, SlmA appears to repurpose a transcription regulator fold for NO. TetR proteins share a common structure, typically consisting of 9 α-helices, with helices 1–3 forming the N-terminal HTH DNA-binding domain. Following the DNA binding region is an all-helical domain that mediates dimerization as well as ligand binding. A characteristic feature of TetR protein is that they sense metabolites, small peptides or drugs, which affect their DNA binding function. The small molecules bind to the C-terminal dimerization domain to cause structural changes, typically resulting in "inducing" (removing) the protein from its cognate DNA site. Notably, SlmA was the first TetR member identified that does not function in transcription regulation and it does not appear to bind any small molecule metabolite or effector.

The studies on SlmA revealed a striking similarity in NO properties compared to Noc. In particular, *slmA-minD-* cells were defective in preventing Z ring formation near the nucleoids when grown in minimal media and SlmA was shown to localize to the nucleoid (Bernhardt and de Boer 2005). Thus, while Noc and SlmA harbor completely different folds, the data suggested they might function analogously to affect NO. However, SlmA does not localize to the membrane. In addition, a final experiment by Bernhardt and de Boer revealed a key distinction between SlmA and Noc. Namely unlike Noc, SlmA was shown to bind directly to FtsZ (Bernhardt and de Boer 2005). In these initial studies the addition of SlmA to FtsZ in the presence of GTP led to the formation of large ribbon-like structures compared to the small protofilaments formed by FtsZ-GTP alone. While the relevance of the SlmA-FtsZ interaction was not clear at this time, it provided a direct link between an NO factor, SlmA, and the protein at the heart of the cell division machinery, FtsZ.

Characterization of SlmA as a Specific DNA-Binding Protein

The initial studies on SlmA indicated a chromosomal localization of the protein and also revealed the important finding that it interacts directly with FtsZ. To delve deeper into the molecular function(s) of SlmA, studies interrogating its structure, DNA binding and FtsZ interacting properties were performed (Tonthat et al. 2011; Cho et al. 2011). Tonthat and coworkers initiated experiments aimed at ascertaining if SlmA showed sequence specific DNA binding using a method called restriction endonuclease protection, selection and amplification (REPSA). These studies revealed that SlmA recognized a DNA site with the palindromic sequence, 5′-GTGAGTACTCAC-3′. Fluorescence polarization binding studies confirmed that SlmA bound the palindromic site with high affinity (50–100 nM) and specificity. Hence, the sequence was named the SlmA binding sequence (SBS) (Tonthat et al. 2011). Perusal of the *E. coli* chromosome revealed SBS sites within all MDs with the exception of the Ter region and ChIP analyses confirmed this pattern (Tonthat et al. 2011; Cho et al. 2011) (Fig. 9.3a). Notably, SlmA homologs are widespread and appear to be present in γ- and β-proteobacteria with the exceptions of *Francisella, Legionella, Pseudomonas, Stenotrophomonas, Xanthomonas* and *Neisseria*. The DNA binding domains of these proteins are all highly conserved, which suggests that they likely bind the same SBS consensus. In fact, when SBS sites were mapped onto the chromosomes of these organisms the same pattern emerged; SBS are found scattered throughout these chromosomes with the exception of the Ter region.

The studies showing that SlmA binds the nucleoid in a MD-selective manner indicated that, like Noc, it also acts as a timing device in coordinating DNA segregation with cell division. Furthermore, this suggested that, like Noc, SlmA must function as an inhibitor of Z ring assembly through the nucleoid, allowing the Z ring to form near the segregating Ter regions. The initial study on SlmA that showed an interaction with FtsZ suggested a direct link to this inhibition (Bernhardt and de Boer 2005). However, these studies implied that the SlmA-FtsZ interaction stimulated the formation of large FtsZ polymers or ribbons, rather than inhibiting FtsZ polymer formation (Bernhardt and de Boer, 2005). But subsequent experiments revealed the key finding that when SlmA is bound to SBS DNA, it negatively affected FtsZ polymer formation (Cho et al. 2011; Du and Lutkenhaus 2014). Different studies produced somewhat conflicting results in that they showed that SlmA-SBS complexes antagonized FtsZ pfs formation or prevented the formation of higher order FtsZ assemblages (Cho et al. 2011; Tonthat et al. 2011). The apparent contrasting results appear to be explained, at least in part, by the length of the SBS used in various studies; experiments with longer SBS sites showed clear disassembly of single pfs while this effect was less obvious in studies using shorter SBS (Du and Lutkenhaus 2014). Nonetheless, these findings showed that SlmA antagonized FtsZ polymerization, revealing it as the only known FtsZ regulator that requires DNA binding to mediate its effect on FtsZ polymerization.

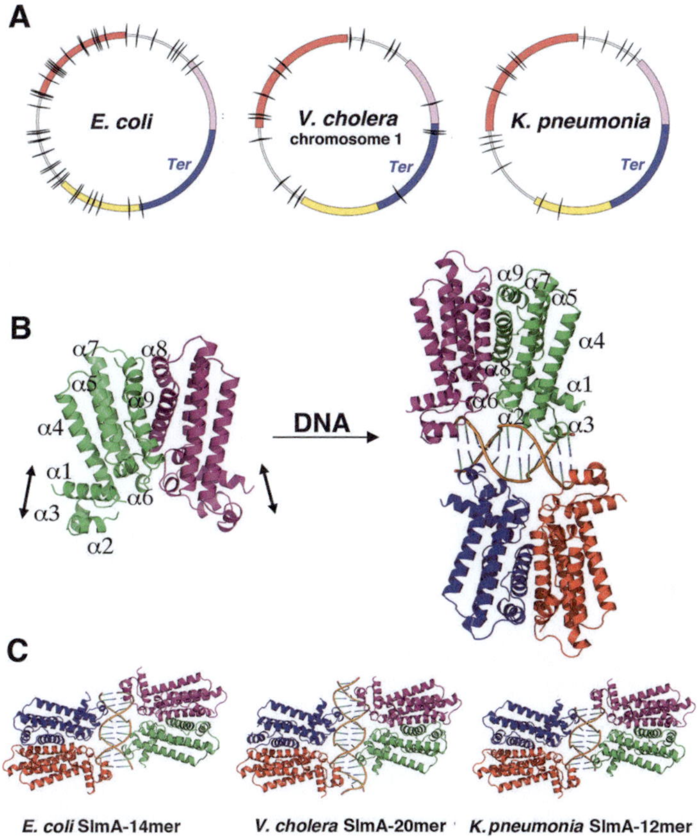

Fig. 9.3 SlmA structure and effects of specific DNA binding. (**a**) Schematic of the *E. coli* MG1655, *V. cholera* chromosome 1 and *Klebsiella pneumonia* chromosomes with the Ori, Ter, *left* and *right* macrodomains (MD) colored *red, blue, yellow and pink*, respectively (the Ter MD is also labeled). Mapped onto the chromosomes as tic marks are the predicted locations of the consensus SlmA binding sites (SBS). (**b**) Consequences of specific DNA binding by SlmA. The apo *E. coli* SlmA structure is shown with one subunit of the dimer colored *magenta* and the other *green*. Secondary structural elements are labeled for one subunit. Specific DNA binding leads to the formation a SlmA dimer-of-dimer complex (Tonthat et al. 2014). DNA binding also leads to the stabilization of the DNA binding domains in one conformation. The DNA binding domains are highly flexible (denoted by the *double headed arrows*) in the absence of specific DNA as observed by multiple apo SlmA structures (Tonthat et al. 2014). (**c**) Structures of SlmA-SBS complexes. Shown are SlmA-DNA structures from *E. coli*, *K. pneumonia* and *V. cholera* (Tonthat et al. 2014). All complexes reveal the same SlmA dimer-of-dimer binding mode onto the specific SBS DNA site

SlmA Must Be Bound to the SBS/Nucleoid to Inhibit Z Ring Formation

In addition to mapping the nucleoid binding sites of SlmA, Tonthat and coworkers also aimed to gain insight into the molecular mechanism of NO by SlmA and why

SBS DNA is necessary for the NO function of SlmA. Hence, they solved its structure by X-ray crystallography. As SlmA does not function in transcription regulation, it was unclear if it contained a canonical TetR fold. In particular, while the N-terminal domain of SlmA was highly homologous to TetR proteins, its C-terminal domain shows no sequence similarity to TetR proteins and was, in fact, predicted to contain a coiled coil. However, the SlmA structure revealed that it indeed contained a TetR fold within both N and C-terminal domains (Schumacher et al. 2001; Bellinzoni et al. 2009; Ramos et al. 2009; Cuthbertson and Nodwell 2013) (Fig. 9.3b). Thus, the structure was consistent with SlmA's role as a sequence specific DNA binding protein yet it provided no insight into how it might interact with FtsZ. Multiple experiments were carried out to try and address the question of how SlmA-SBS inhibits FtsZ assembly. One study suggested that SlmA enhanced the GTPase activity of SlmA. Stimulation of FtsZ GTPase activity would be predicted to favor its depolymerization. These studies also indicated that SBS binding converted a SlmA monomer to a dimer (Cho et al. 2011). These combined findings would provide an explanation for both the requirement that SlmA be bound to the SBS and how the SlmA-SBS complex would effect FtsZ polymerization. However, later studies carried out in multiple labs revealed no significant alteration in the GTP activity of FtsZ in the presence of SlmA or SlmA-SBS (Tonthat et al. 2013; Du and Lutkenhaus 2014; Cabré et al. 2015). Moreover, subsequent studies also showed that like all other TetR proteins, SlmA is a dimer even in its apo state (Tonthat et al. 2011; Cabré et al. 2015).

To obtain molecular insight into the SBS requirement for SlmA's NO function, structures were solved of SlmA and SlmA-SBS complexes from multiple Gram-negative bacteria including *E. coli*, *Vibrio cholera* and *Klebsiella pneumonia* (Fig. 9.3c). Notably, consistent with bioinformatic analyses showing that *V. cholera* and *K. pneumonia* chromosomes harbour similarly dispersed SBS sites (Fig. 9.3a), biochemical studies confirmed that their SlmA proteins bound SBS sequences with high affinity and specificity (Tonthat et al. 2013). In addition to the different SlmA proteins used, these structures were obtained under very different crystallization conditions. The results were very surprising as, in all cases, the SlmA proteins bound the SBS as dimer-of-dimers (Fig. 9.3c). The dimer-of-dimer binding arrangement was subsequently supported by multiple biochemical assays (Tonthat et al. 2013). The combined structural studies showed that DNA binding has two important outcomes. First, it stabilizes the DNA binding domain structure. Indeed, in apo SlmA structures the DNA binding domains are highly flexible and in some of the structures, largely disordered. Secondly, DNA binding leads to the formation of dimer-of-dimers, with a highly conserved binding orientation (Fig. 9.3b).

Interestingly, ChIP analyses revealed that SBS sites often cluster close together on the nucleoid and SlmA enriched regions extend adjacent to the SBS (Cho et al. 2011). The finding that SlmA binds DNA as a dimer-of-dimers helps explain these data. Given that SlmA binds as a dimer-of-dimer, the majority of the SlmA, estimated at 300–400 molecules/cell, would be bound to SBS or SBS-like sites. The ChIP data suggested the possibility that the remaining SlmA molecules might be nucleated onto nearby DNA leading to spreading, similar to Noc, which could further enhance its inhibitory action on FtsZ polymerization along the nucleoid. Subsequent biochemical experiments supported this hypothesis (Tonthat et al. 2014). Altogether, the biochemical and structural data indicate that SlmA binds to SBS sites as a dimer-of-dimers and spreads, but does not bridge, to adjacent DNA. Spreading of SlmA dimers on the DNA suggested possibilities for how it may act as an FtsZ antagonist. However, understanding the SlmA-FtsZ interaction was required to gain molecular insight into this process.

Attempts to identify the full complement of interacting regions between SlmA and FtsZ proved difficult. Initial studies using small angle X-ray scattering (SAXS) revealed a SlmA dimer sandwiched between two FtsZ molecules (Tonthat et al. 2011; Tonthat et al. 2013). But, due to the low resolution of this structure, it is not possible to discern specific residues and surfaces involved in the interaction between the FtsZ core and SlmA. Moreover, subsequent studies showed that the FtsZ C-tail appears to mediate key SlmA contacts with FtsZ (Du and Lutkenhaus 2014; Du et al. 2015). Interestingly, this binding may have initially been overlooked, as it involves a multivalent type of contact. Specifically, multiple FtsZ CTTs must be presented on a polymer to provide a high enough local concentration to allow binding to SlmA (Du et al. 2015). Hence, the SlmA-SBS-FtsZ interaction is highly complex and multivalent. SlmA residues that bind FtsZ were identified genetically by the Bernhardt lab (Cho and Bernhardt 2013). Their studies revealed that FtsZ interacting residues in SlmA cluster together in a region between α4 and α5 on the SlmA structure. However, it is currently unknown whether these SlmA regions bind the FtsZ globular domain or FtsZ CTT or both.

SlmA-Mediated NO: A Molecular Mechanism

Although several questions remain, the combined studies thus far suggest a SlmA-mediated NO mechanism based on sequestration and depolymerization of FtsZ. According to this model, extended SlmA molecules spread along the DNA and bind FtsZ C-tails thus, capturing FtsZ pfs (Fig. 9.4a, b). Binding of the FtsZ CTT would aid in the inhibition of FtsZ Z ring formation by competing with FtsA for binding to this region, an interaction required to localize FtsZ to the membrane for Z ring formation. Bioinformatic data indicate that many of the SlmA binding sites are near genes that encode transmembrane proteins, suggesting that SlmA sites may be brought close to the membrane by transertion events, which would further aid in FtsZ binding (Tonthat et al. 2013). However, data indicating that SlmA-DNA complexes actively disrupt FtsZ pfs, suggests a more complex mechanism of NO than FtsZ displacement from the membrane (Cho et al. 2011; Du and Lutkenhaus, 2014). While studies have indicated that the FtsZ CTT may play some role in polymer assembly in *B. subtilis*, this does not appear to be the case in *E. coli* (Buske and Levin 2012; Buske and Levin 2013). Thus, the dissolution of FtsZ pfs by SlmA-SBS may require the second binding interaction involving the FtsZ globular domain (Fig. 9.4a, b). Such a complex multivalent interaction whereby multiple SlmA

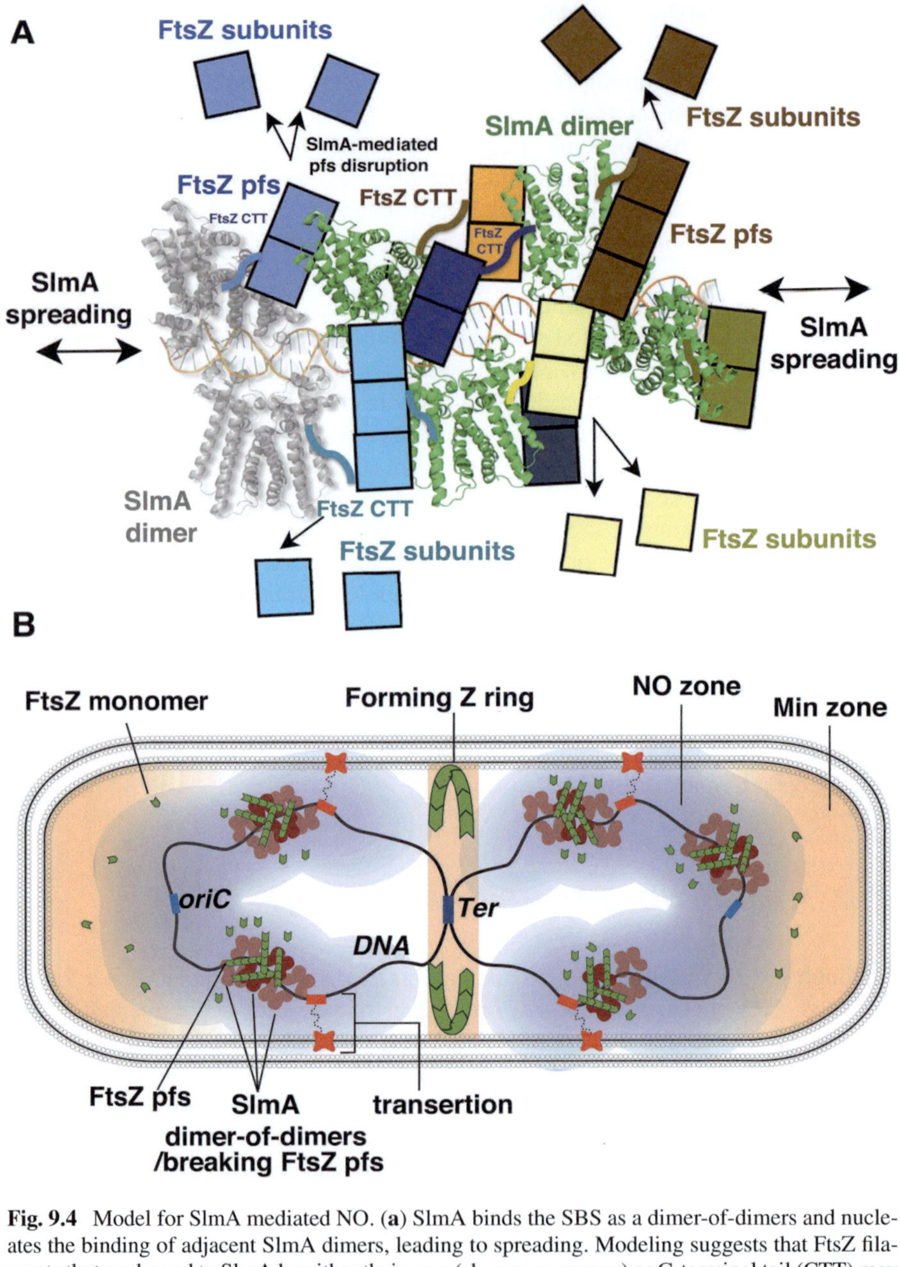

Fig. 9.4 Model for SlmA mediated NO. (**a**) SlmA binds the SBS as a dimer-of-dimers and nucleates the binding of adjacent SlmA dimers, leading to spreading. Modeling suggests that FtsZ filaments that are bound to SlmA by either their core (shown as *squares*) or C-terminal tail (CTT) may introduce strain that ultimately fractures the protofilaments. The resultant bound FtsZ subunits would dissociate as the interactions from individual core domains or CTTs alone do not bind SlmA with high affinity. (**b**) Schematic model of SlmA mediated NO. Shown is a segregating cell just starting division in which the Z ring has begun formation at the cell center concomitant with segregation of the two chromosomes (*black lines*). SlmA dimer-of-dimers bind and spread on non-Ter regions. SlmA-DNA complexes bind FtsZ protofilaments via contacts with FtsZ CTT and core domain. These interactions dissociate FtsZ protofilaments in the vicinity of the DNA. SBS sites are predicted to be near genes encoding transmembrane proteins, which could transiently localize the inhibitory SlmA-DNA complexes near the membrane to further sequester FtsZ pfs and inhibit their growth. Thus, SlmA creates an effective inhibitory zone against FtsZ pfs around the nucleoid

molecules grasp the FtsZ pfs could disrupt the FtsZ polymer (Fig. 9.4a, b). Furthermore, recent data indicate that the dynamic properties of the nucleoid serve mechanical functions such as movement and force (Lim et al. 2014). It is possible that such dynamics might aid in fracturing the FtsZ pfs grasped between multiple SlmA dimers bound along the DNA. Indeed, an important feature of this model is that the individual contacts to SlmA, which involve the FtsZ CTT and globular region, are individually very weak and hence likely stabilized only in the context of an assembled FtsZ pfs or polymer. As a result, disruption of FtsZ pfs would lead to FtsZ monomers that are tenuously bound to SlmA-DNA and hence readily dissociated from these nucleoid regions (Fig. 9.4a, b). Disruption of FtsZ pfs would therefore prevent Z ring formation near the SlmA clusters on the nucleoid.

Noc- and SlmA-Independent Mechanisms of NO

Noc and SlmA play important roles in NO under specific conditions, however, they are not essential proteins (Sievers et al. 2002; Bernhardt and de Boer 2005). Indeed, subsequent studies revealing that accurate division site place occurred in *min-slmA- E. coli* cells and *min-noc- B. subtilis* cells clearly indicated that other NO mechanisms must exist in these bacteria (Rodrigues and Harry 2012; Bailey et al. 2014). Recent work by Bailey and coworkers uncovered an NO role for MatP in *E. coli*. These studies revealed that MatP interacts with the cell division machinery at mid-cell prior to septation. The interaction partner of MatP was revealed to be ZapB, which is a coiled coil protein that binds an FtsZ effector called ZapA. ZapA-ZapB forms a complex with FtsZ during Z ring formation in *E. coli* (Espéli et al. 2012; Galli and Gerdes 2012; Buss et al. 2013; Buss et al. 2015). Studies from several laboratories revealed that the ZapB-ZapA complex stabilizes the Z ring by anchoring clusters of FtsZ at the membrane. MatP apparently binds directly to ZapB when it is assembled with ZapA and FtsZ at the septation site (Espéli et al. 2012; Buss et al. 2013; Buss et al. 2015). The assembly of this multi-protein system not only appears to stabilize Z ring formation, but also affects its constriction dynamics (Buss et al. 2015). Importantly in terms of its role in NO, this interaction network anchors the Ter region of the nucleoid to the Z ring during DNA segregation. Thus, similar to Noc and SlmA, the MatP-Ter interaction with ZapB-ZapA appears to serve as a timing device to coordinate DNA segregation with cytokinesis.

While NO mechanisms other than those mediated by SlmA and Noc are clearly at play during normal cell growth, recent studies indicate that specific cell cycle regulated mechanisms for NO also exist. For example, experiments that examined cell division during *B. subtilis* sporulation uncovered a sporulation-specific protein called RefZ (Regulator of FtsZ) that effected Z ring assembly in a cell cycle dependent manner (Wagner-Herman et al. 2012). RefZ production was shown to be required for the asymmetric localization of the Z ring during *B. subtilis* sporulation (Wagner-Herman et al. 2012). Consistent with an important role in division site placement, overexpression of RefZ during vegetative growth significantly altered the assembly of FtsZ, ultimately leading to cell filamentation and lysis, while sporulating cells that lacked RefZ were delayed in polar FtsZ assembly. Interestingly, RefZ is predicted to be a TetR-like protein and ChIP studies indicated that it binds specific DNA sites that are primarily located in the Ori region of the chromosome with some sites also located in the Ter region (Wagner-Herman et al. 2012). Studies demonstrated that these specific binding sites are essential for the proper localization and function of RefZ. The combined data indicate that RefZ likely plays a role in promoting the switching of Z ring formation from a medial to a polar location during *B. subtilis* sporulation. But more studies are needed to determine the precise mechanism by which RefZ controls Z ring assembly during sporulation. Based on the current data, RefZ could act as an activator of Z ring assembly near the cell poles or alternatively, it could function as an inhibitor of ring formation at midcell.

The Control of Septal Placement in Sporulating *Streptomyces*

Gram-positive bacteria of the *Streptomyces* species are soil-dwelling organisms with a complex life cycle that is strikingly similar to filamentous fungi. This life cycle includes both unicellular and reproductive/multicellular stages. During the unicellular phase, these organisms form a vegetative mycelium that consists of syncytial cells separated by septa spaced at ~5–10 mm (Wildermuth and Hopwood 1970). Their transition to the reproductive stage involves the formation of long spore chains and a complex and dramatic cell division event where up to a hundred Z rings are generated, precisely equidistant from each other, in the extended chains (Chater 2001; Flärdh and Buttner 2009). Notably, this developmental switch is also accompanied by the production of secondary metabolites, which include antibiotics; *Streptomyces* produce >50% of all known antibiotics and a plethora of anticancer compounds (Hopwood 2007).

Although the cell division machinery in these organisms is generally similar to that of typical unicellular bacteria, there are some important distinctions (McCormick and Losick 1996). For example, while *Streptomyces* contain an FtsZ protein, canonical cell division regulatory proteins such as FtsA, ZipA, MinC and Noc are missing in these organisms. And interestingly, although two MinDs and a DivIVA homolog are present in these organisms, they do not appear to play a role in Z ring localization. In fact, the *Streptomyces* DivIVA protein carries out a completely different function than in other bacteria in that it controls cell polarity or tip growth (Flärdh 2003; Flärdh and Buttner 2009; Flärdh 2010). Given these distinctions, it was postulated that cell division control in *Streptomyces* likely involves unique proteins. This hypothesis was confirmed when studies showed that null mutants of the *ssgA* and *ssgB* genes, which are found only in *Streptomyces*, are blocked at a step preceding cell division during sporulation (Keijser et al. 2003; Noens et al. 2005; Traag and van Wezel 2008).

Unlike the canonical NO and Min regulatory systems in *E. coli* and *B. subtilis*, SsgB and SsgA regulate division site placement in a positive manner. In particular,

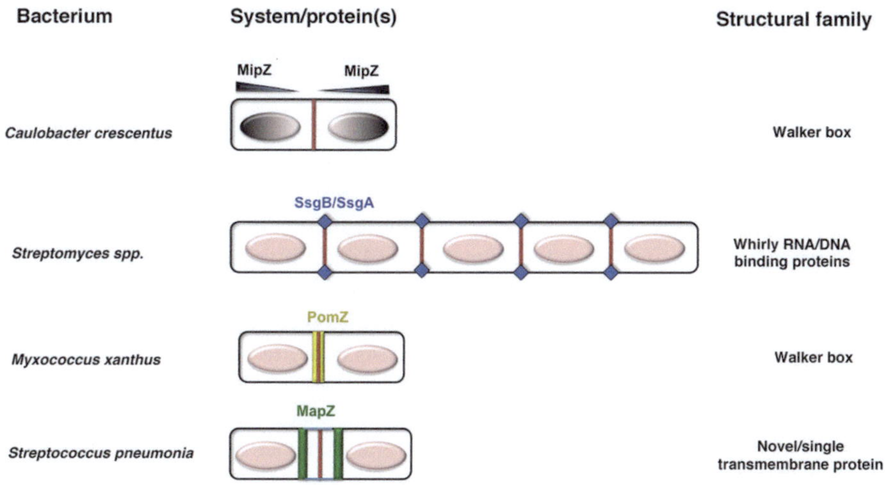

Fig. 9.5 Multiple mechanisms by which selected bacteria mediate division site placement to prevent nucleoid bisection by the Z ring. Shown are the systems in which NO mechanisms have been established. The nucleoids are shown as *pink spheres*. The structural families that the proteins belong to are indicated. Notable is the diversity of proteins and the mechanisms by which they prevent Z rings from forming through nucleoid DNA

SsgA orchestrates division by mediating the correct localization of SsgB, which then binds and recruits FtsZ to future cell division sites. SsgB also appears to tether the Z ring to the cytoplasmic membrane (Fig. 9.5). Interestingly, the structure of SsgB revealed it to be a member of the whirly ssDNA/guide RNA-binding protein family (Desveaux et al. 2002; Schumacher et al. 2006; Xu et al. 2009). These proteins have been found in the mitochondria of parasitic protozoa and plants and interact with nucleic acids. But studies performed thus far have revealed no evidence for nucleic acid binding by SsgB. Furthermore, SsgB lacks the concave face used for nucleic acid binding by other whirly proteins (Desveaux et al. 2002; Schumacher et al. 2006; Xu et al. 2009). Instead, a conserved hydrophobic surface of the protein has been implicated in protein-protein interactions (Xu et al. 2009). Thus, while it is clear that the SsgB/SsgA proteins mediate cell division site placement in *Streptomyces*, the specific molecular mechanism remains to be determined. Moreover, a fascinating question that has not yet been resolved is how the precise spacing of so many Z rings is mediated by this protein machinery during sporulation-specific cell division in *Streptomyces*.

MipZ, a Negative Regulator of Cell Division Site Placement in *Caulobacter crescentus* with a ParA-Like Walker Fold

C. crescentus is an oligotrophic, Gram-negative bacterium that has served as an important model system in the study of the bacterial cell cycle, cell differentiation and asymmetric cell division (Kelly et al. 1998; Sackett et al. 1998; Tsokos and Laub 2012). *Caulobacter* has a complex, biphasic life cycle involving a flagellated swarmer cell and sessile stalk stage. *Caulobacter* lacks obvious homologs of MinCD or NO factors. Thus, to try and gain insight into cell division in *Caulobacter*, Thanbichler and Shapiro carried out a bioinformatics search for cell cycle regulated genes that are conserved in α proteobacteria. This query led to the identification of an open reading frame that encoded a 278 residue protein, which was subsequently called MipZ (Midcell positioning of FtsZ). The deduced amino acid sequence of MipZ placed it in the Walker-box superfamily and sequence comparisons showed it to be distantly related to partition ParA proteins and MinD, which also contain Walker-box motifs. Strikingly, attempts to delete the *mipZ* gene failed indicating that it was essential (Thanbichler and Shapiro 2006). Studies showed that MipZ overexpression prevented FtsZ assembly throughout the cell, causing a complete block in cell division (Thanbichler and Shapiro 2006). By contrast, depletion of MipZ resulted in Z ring formation at random locations throughout the cell. Thus, these findings revealed that MipZ plays a central role in cell division and placement of the Z ring in *Caulobacter*.

Subsequent studies on MipZ showed that, like its plasmid ParA counterparts, it binds nonspecifically to the nucleoid DNA and that this interaction is regulated by contacts with ParB (Kiekebusch et al. 2012). Moreover, MipZ was also shown to bind FtsZ. Structural studies confirmed that MipZ contains a Walker-box fold that is similar to that of Soj and other ParA homologs (Kiekebusch et al. 2012). Also, similar to other ParA proteins, MipZ forms ATP-dependent dimers. Studies showed that MipZ dimers interact with and antagonize FtsZ pfs by stimulating the GTPase activity of FtsZ (Kiekebusch et al. 2012). Unexpectedly, distinct from ParA dimers, which are destabilized by interactions with ParB, MipZ dimer formation is stimulated by ParB proteins, which are bound to the origin at the *Caulobacter* cell poles (Kiekebusch et al. 2012). Because of the intrinsic ATPase activity of MipZ, dimers eventually dissociate into monomers, which no longer bind DNA. It was shown that these monomers are then recaptured and re-dimerized by ParB thus, regaining DNA binding capability. Because of this continual recycling, a gradient of the MipZ dimer on the nucleoid is created, which is highest near ParB, at the cell poles, and decreases as the distance from ParB increases (Fig. 9.5). This localization engenders MipZ with a Min system like role, where it prevents Z ring formation near the poles. At the same time, MipZ also functions as an NO factor as its nonspecific binding to DNA allows it to inhibit Z ring formation over the nucleoid. Thus, in *Caulobacter*, polar cell division inhibition and NO are performed by one protein, MipZ. This amalgamation of division site selection functions into one protein may be advantageous to *Caulobacter* cells, which unlike *B. subtilis* and *E. coli* have tightly coupled DNA segregation and cell division events.

PomZ, a ParA-Like Positive Regulator of Division Site Placement in *M. xanthus*

The δ-proteobacterium *Myxococcus xanthus* is an unusual bacterium that divides by binary fission in the presence of nutrients but ceases growth during nutrient deprivation, when it forms multicellular fruiting bodies and eventually differentiates into spores (Konovalova et al. 2010). Similar to *C. crescentus*, *M. xanthus* lacks clear homologs of the NO and Min systems. Insight into the cell division by *M. xanthus* was obtained by work carried out by Treuner-Lange and coworkers (Treuner-Lange et al. 2013). Their studies revealed that deletion of *agmE,* a gene previously implicated in motility, resulted in severe defects in cell division, as exemplified by the formation of filamentous and nucleoid free cells. These division defects were shown to be caused by the inappropriate spatial regulation of Z ring formation. Hence, the authors renamed AgmE to PomZ (Positioning at midcell of FtsZ) (Treuner-Lange et al. 2013). Although the structure of the 319-residue PomZ protein is not yet known, sequence homology analyses place it in the ParA-like family of proteins. In particular, BLAST searches reveal that is shares the strongest sequence-similarity with ParA proteins in *Myxococcal* species (>80%) followed by *Bacilli* ParA proteins (~35%).

The fact that PomZ harbors a ParA-like fold similar to MipZ suggested that it may function in an analogous manner. Consistent with this, PomZ was found to binds ATP. Strikingly, however, despite the fact that PomZ binds ATP and harbors a ParA-like fold, studies clearly showed that it functions in a manner contrary to MipZ. Specifically, unlike MipZ, which is primarily found at the cell poles, PomZ localizes to the incipient division site where the Z ring forms. Moreover, it arrives at the cell division site before FtsZ. Subsequent experiments showed that PomZ actually recruits FtsZ to the septal site and functions to stabilize Z ring formation, rather than disrupt it, as does MipZ (Treuner-Lange et al. 2013) (Fig. 9.5). PomZ does appears to localize over the nucleoid similar to MipZ, but it performs a positive role in recruitment and stabilization of the Z ring in contrast to the antagonistic function on Z ring formation carried out by MipZ. It is currently unclear how PomZ recruits FtsZ to the midcell as although a PomZ-FtsZ interaction has been detected it appears to be indirect (Treuner-Lange et al. 2013). However, in that regard a very unexpected finding from the study was that *M. xanthus* FtsZ behaves very differently from other characterized FtsZ proteins. In particular, the *M. xanthus* FtsZ is unable to polymerize in vitro in the presence of GTP and/or polymer stimulating conditions, such as high concentrations of calcium. The authors therefore suggest that the FtsZ filaments might be too small to be detected in these experiments. Nonetheless, given this finding, it would seem that if the *M. xanthus* FtsZ is also unable to form filaments productively in vivo at native concentrations, a positive rather than negative regulatory system would be required to ensure proper temporal and spatial FtsZ assembly. Perhaps PomZ fulfills that role. Clearly, future studies will be needed to address the specific molecular functions of this unique FtsZ regulator.

Interestingly, ParA-like proteins similar to PomZ and MipZ are found in many bacteria and hence, may play similar roles in division site selection. For example, the *Corynebacterium glutamicum* PldP protein, which is a putative ParA-like protein, shows functionalities similar to MipZ. It localizes over the nucleoid and a null mutant exhibits a division defect (Donovan et al. 2010). Thus, PomZ, MipZ and PldP may be added to a growing list of ParA-like proteins that have been shown to be general determinants of macromolecular localization in bacteria. Well-characterized functions of these proteins include movement and localization of chromosomes and plasmids (Hayes and Bariilà 2006; Schumacher 2012), polar protein complexes (Ringgaard et al. 2011) and carboxysomes (Savage et al. 2010). Thus, an ancient bacterial ParA-like ATPase could have functioned as a general spatial regulator of many systems and eventually evolved to perform specialized macromolecular localization functions. This apparently includes cell division site placement in several bacterial species.

MapZ, a Division Site Anchor/Assembly Protein in *Streptococcus pneumonia*

The human pathogen, *S. pneumonia,* is an oval Gram-positive bacterium that divides along its long axis, similar to rod-shaped bacteria. Yet the biology of *S. pneumonia* appears unique in many respects. First, it maintains it shape by interchanging peripheral peptidoglycan (PG) synthesis and septal PG synthesis, which take place during cell elongation and cell division, respectively. Also, *S. pneumonia* lacks all the standard cell division regulatory machinery, including NO and Min. Furthermore, a specific Ser/Thr kinase, StkP, has been shown to be essential for cell division and morphogenesis of this organism. StkP is essential for septal formation and localization of PG synthesis. To gain insight into the function of StkP in *S. pneumonia* cell division, Fleurie et al. examined the roles of StkP substrates in cytokinesis. Strikingly, a StkP substrate was identified that was essential for the correct placement and assembly of the Z ring, hence this protein was named MapZ for midcell anchored protein Z (Fleurie et al. 2014; Bramkamp 2015). Simultaneously, Holečková et al. identified the same protein and also found it was essential for correct division site placement in *S. pneumonia* (Holečková et al. 2015).

The equatorial plane of *S. pneumonia* is marked by an unusual PG ring, often called a piecrust, based on its appearance in electron microscopic images. During cellular elongation, the PG piecrust structure splits and becomes separated by PG synthesis. It has been postulated that these equatorial rings could act as positional landmarks for cell division (Zapun et al. 2008). MapZ contains a novel protein sequence with a predicted single transmembrane segment connecting an extracellular N-terminal domain and cytosolic C-terminal domain. MapZ was shown to bind PG through its extracellular domain and then co-migrate with equatorial rings while the cytoplasmic domain of MapZ binds FtsZ (Fleurie et al. 2014) (Fig. 9.5). The tubulin containing core of FtsZ appears to interact with MapZ; the FtsZ C-tail was not required for MapZ binding. Thus, a simple mechanism was suggested in which MapZ is stably tethered to the PG ring structure by its N-terminal domain and coordinates division site selection with elongation.

A further important finding from studies on MapZ was that it precedes FtsZ and FtsA, to the nascent division site (Fleurie et al. 2014). In fact, FtsZ and FtsA were delocalized in cells lacking MapZ, supporting the notion that MapZ spatially regulates the localization and positioning of the Z ring, likely through its interaction with FtsZ. As noted MatZ was originally studied because it is a substrate for StkP. Two threonine residues in MapZ are phosphorylated by StkP. While mutation of these threonines did not affect MapZ positioning to the site of septation, it did affect FtsZ stability and constriction. Interestingly, this finding suggests that MapZ might also play a role in later steps of cell division. MapZ is conserved amongst *Streptococci, Lactococci*, and *Enterococci*. Moreover, none of these bacteria harbor NO or Min systems, suggesting that they all likely use a cell division placement mechanism involving MapZ similar to *S. pneumonia*.

No NO: Nucleoid Equidistribution by Random Cell Division in *Mycobacterium*

Mycobacterium are rod-shaped Gram-positive bacteria that are able to adapt to variety of lifestyles in multiple environments ranging from water to human hosts (Wayne 1994; Carter et al. 2003; Chacon et al. 2004; Monack et al. 2004; Vaerewijck et al. 2005). To gain insight into the regulatory mechanisms controlling *Mycobacterium* diversity, Singh and coworkers followed the changes in morphology and DNA during the life cycles of several model *Mycobacteria* (Singh et al. 2013). Because *Mycobacterium* spp. lack Min and NO systems, the authors were particularly interested in the mechanisms controlling cell division. What they found was remarkable; Z ring assembly in exponentially growing cells was essentially random. Many division sites were off center to the extent that noticeably unequal daughter cells were produced. In addition, multiple dividing cells assembled Z rings over their nucleoids. Yet, the cells continued to divide and produce viable offspring with intact nucleoids. This seeming disastrous cell division strategy appeared to be compensated by unequal polar growth, which maintained the size of the resulting daughter cells. In addition, post-division DNA transport through the septum by a FtsK homolog, MSMEG2690 rescued nucleoids from bisection by the cell division machinery (Singh et al. 2013). FtsK proteins perform similar roles in other bacteria, typically being called upon during cases of emergency, such as resolving chromosome dimers (Männik and Bailey 2015). Thus, remarkably, the data from Singh and coworkers suggest that *Mycobacterium* have evolved mechanisms to ensure accurate and faithful genomic transmission that take place after cell division involving unequal polar growth and DNA translocation.

Concluding Remarks

The faithful inheritance of genetic material from one generation to the next necessitates that chromosomes are not only reliably replicated and segregated but that the cell division machinery accurately forms between replicated DNA elements without their fragmentation or bisection. While this process, called nucleoid occlusion, was formulated decades ago, it has only been recently that proteins and mechanisms have been identified that mediate this process. In particular, in the last 10 years specific NO factors have been identified in the model organisms, *B. subtilis* and *E. coli*, and insight into the mechanisms by which these factors mediate NO has been obtained. In the case of SlmA, such information has been gleaned at the molecular level. Strikingly, the two identified NO factors from *B. subtilis* and *E. coli,* Noc and SlmA belong to entirely different protein structural families and studies indicate that they employ very different mechanisms of NO. However, as surprising as this initially appeared, recent data identifying NO factors and mechanisms in other bacteria have revealed a shocking diversity in not only the types of proteins employed in NO but their mechanisms of action as well. Some of the diversity may be explained by the redundancy of NO mechanisms within a given bacterial species. Perhaps redundancy in NO should not be surprising as this process protects against chromosome fragmentation, which would lead to immediate cell death (Rodrigues and Harry 2012). Hence, employing a single NO protein would present a system that is highly vulnerable to disabling mutations, with resulting catastrophic results. As we have thus far characterized the cell division machinery from a fraction of bacteria, it seems likely that many additional and completely unpredicted mechanisms of bacterial NO await discovery.

References

Adams DW, Errington J (2009) Bacterial cell division: assembly, maintenance and disassembly of the Z ring. Nat Rev Microbiol 7:642–653

Adams DW, Wu LJ, Errington J (2014) Cell cycle regulation by the bacterial nucleoid. Curr Opin Microbiol 22:94–101

Adams DW, Wu LJ, Errington J (2015) Nucleoid occlusion protein Noc recruits DNA to the bacterial cell membrane. EMBO J 34:491–501

Adler HI, Fisher WD, Cohen A, Hardigree AH (1967) Miniature *Escherichia coli* cells deficient in DNA. Proc Natl Acad Sci U S A 57:321–326

Azam TA, Hiraga S, Ishihama A (2000) Two types of localization of the DNA-binding proteins within the *Escherichia coli* nucleoid. Genes Cells 5:613–626

Bailey MW, Bisicchia P, Warren BT, Sherratt DJ, Männik J (2014) Evidence for divisome localization mechanisms independent of the *min* system and SlmA in *Escherichia coli*. PLoS Genet 10:e1004504

Bellinzoni M, Buroni S, Schaeffer F, Riccardi G, DE Rossi E, Alzari PM (2009) Structural plasticity and distinct drug-binding modes of the LfrR, a mycobacterial efflux pump regulator. J Bacteriol 191:7531–7537

Bernhardt TG, de Boer PA (2005) SlmA, a nucleoid-associated, FtsZ binding protein required for blocking septal ring assembly over chromosomes in *E. coli*. Mol Cell 18:555–564

Bi E, Lutkenhaus J (1991) FtsZ ring structure associated with division in *Escherichia coli*. Nature 354:161–164

Bisson-Filho AW, Discola KF, Castellen P, Blasios V, Martins A, Sforça ML, Garcia W, Zeri AC, Erickson HP, Dessen A, Gueiros-Filho FJ (2015) FtsZ filament capping by MciZ, a developmental regulator of bacteria division. Proc Natl Acad Sci U S A 112:E2130–E2138

Biteen JS, Goley ED, Shapiro L, Moerner WE (2012) Three-dimensional super-resolution imaging of the midplane protein FtsZ in live *Caulobacter crescentus* cells using astimatism. ChemPhysChem 13:1007–1012

Bramkamp M (2015) Following the equator: division site selection in *Streptococcus pneumonia*. Trends Microbiol 23:121–122

Bramkamp M, Emmins R, Weston L, Donovan C, Daniel RA, Errington J (2008) A novel component of the division-site selection system of *Bacillus subtilis* and a new mode of action for the division inhibitor MinCD. Mol Microbiol 70:1556–1569

Briley K Jr, Prepiak P, Dias MJ, Hahn J, Dubnau D (2011) Maf acts downstream of ComGA to arrest cell division in competent cells of *B. subtilis*. Mol Microbiol 81:23–39

Buske PJ, Levin PA (2012) The extreme C-terminus of the bacterial cytoskeletal protein FtsZ plays a fundamental role in assembly independent of modulatory proteins. J Biol Chem 287:10945–10957

Buske PJ, Levin PA (2013) A flexible C-terminal linker is required for proper FtsZ assembly in vitro and cytokinetic ring formation in vivo. Mol Microbiol 89:249–263

Buss J, Coltharp C, Huang T, Pohlmeyer C, Wang SC, Hatem C, Xiao J (2013) In vivo organization of the FtsZ-ring by ZapA and ZapB revealed by quantitative super-resolution microscopy. Mol Microbiol 89:1099–1120

Buss J, Coltharp C, Shtengel G, Yang X, Hess H, Xiao J (2015) A multilayered protein network stabilized the *Escherichia coli* FtsZ-ring and modulates constriction dynamics. PLoS Genet 11:e1005128

Cabré EJ, Monterroso B, Alfonso C, Sánchez-Gorostiaga A, Reija B, Jiménez M, Vicente M, Zorrilla S, Rivas G (2015) The nucleoid occlusion SlmA protein accelerates the disassembly of the FtsZ protein polymers without affecting their GTPase activity. PLoS One 10:e0126434

Carter G, Wu M, Drummond DC, Bermudez LE (2003) Characterization of biofilm formation by clinical isolates of *Mycobacterium avium*. J Med Microbiol 52:747–752

Chacon O, Bermudez LE, Barletta RR (2004) Johne's disease, inflammatory bowel disease, and *Mycobacterium paratuberculosis*. Annu Rev Microbiol 58:329–363

Chater KF (2001) Regulation of sporulation in *Streptomyces coelicolor* A3(2): a checkpoint multiplex? Curr Opin Microbiol 4:667–673

Chen B-W, Lin M-H, Chu C-H, Hsu C-E, Sun Y-J (2015) Insights into ParB spreading from the complex structure of Spo0J and *parS*. Proc Natl Acad Sci U S A 112:6613 Epub

Cho H, Bernhardt TG (2013) Identification of the SlmA active site responsible for blocking bacterial cytokinetic ring assembly over the chromosome. PLoS Genet 9:e1003304

Cho H, McManus HR, Dove SL, Bernhardt TG (2011) Nucleoid occlusion factor SlmA is a DNA-activated FtsZ polymerization antagonist. Proc Natl Acad Sci U S A 108:3773–3778

Cornell RB, Taneva SG (2006) Amphipathic helices as mediators of the membrane interaction of amphitropic proteins, and as modulators of bilayer physical properties. Curr Protein Pept Sci 7:539–552

Cuthbertson L, Nodwell JR (2013) The TetR family of regulators. Microbiol Mol Biol Rev 77:440–475
Dajkovic A, Lan G, Sun SX, Wirtz D, Lutkenhaus J (2008) MinC spatially controls bacterial cytokinesis by antagonizing the scaffolding function of FtsZ. Curr Biol 18:235–244
Davie E, Snydor K, Rothfield LI (1984) Genetic basis of minicell formation in *Escherichia coli* K-12. J Bacteriol 158:1202–1203
de Boer PAJ, Crossley RE, Rothfield LI (1989) A division inhibitor and a topological specificity factor coded for by the minicell locus determine proper placement of the division septum. Cell 56:641–649
de Boer PAJ, Crossley RE, Rothfield LI (1990) Central role for the *Escherichia coli minC* gene product in two different cell division-inhibition systems. Proc Natl Acad Sci U S A 87:1129–1133
Desveaux D, Allard J, Brisson N, Sygusch J (2002) A new family of plant transcription factors displays a novel ssDNA-binding surface. Nat Struct Biol 9:512–517
Donovan C, Schwaiger A, Krämer R, Bramkamp M (2010) Subcellular localization and characterization of the ParAB system from *Cornebacterium glutamicum*. J Bacteriol 192:3441–3451
Drlica K (1992) Control of bacterial DNA supercoiling. Mol Microbiol 6:425–433
Du S, Lutkenhaus J (2014) SlmA antagonism of FtsZ assembly employs a two pronged mechanism like MinCD. PLoS Genet 10:e1004460
Du S, Park KT, Lutkenhaus J (2015) Oligomerization of FtsZ converts the FtsZ tail motif (conserved carboxy-terminal peptide) into a multivalent ligand with high avidity for partners ZipA and SlmA. Mol Microbiol 95:173–188
Duman R, Ishikawa S, Celik I, Strahl H, Ogasawara N, Troc P, Löwe J, Hamoen LW (2013) Structural and genetic analyses reveal the protein SepF as a new membrane anchor for the Z ring. Proc Natl Acad Sci U S A 110:E4601–E4610
Dupaigne P, Tonthat NK, Espéli O, Whitfill T, Boccard F, Schumacher MA (2012) Molecular basis for a protein-mediated DNA-bridging mechanism that functions in condensation of the *E. coli* chromosome. Mol Cell 48:560–571
Egan AJF, Vollmer W (2013) The physiology of bacterial cell division. Ann N Y Acad Sci 1277:8–28
Erickson HP, Anderson DE, Osawa M (2010) FtsZ in bacterial cytokinesis: cytoskeleton and force generator all in one. Microbiol Mol Biol Rev 74:504–528
Espéli O, Borne R, Dupaigne P, Thiel A, Gigant E, Mercier R, Boccard F (2012) A MatP-divisome interaction coordinates chromosome segregation with cell division in *E. coli*. EMBO J 31:3198–3211
Flärdh K (2003) Growth polarity and cell division in *Streptomyces*. Curr Opin Microbiol 6:564–571
Flärdh K (2010) Cell polarity and the control of apical growth in *Streptomyces*. Curr Opin Microbiol 13:758–765
Flärdh K, Buttner MJ (2009) *Streptomyces* morphogenetics: dissecting differentiation in a filamentous bacterium. Nat Rev Microbiol 7:36–49
Fleurie A, Lesterlin C, Manuse S, Zhao C, Cluzel C, Lavergne JP, Franz-Wachtel M, Macek B, Combet C, Kuru E, VanNieuwenhze MS, Brun YV, Sherratt D, Grangeasse C (2014) MapZ marks the division sites and positions FtsZ rings in *Streptococcus pneumonia*. Nature 516:259–262
Fu G, Huang T, Buss J, Coltharp C, Hensel Z, Xiao J (2010) In vivo structure of the *E. coli* FtsZ-ring revealed by photoactivated localization microscopy (PALM). PLoS One 5:e12682
Galli E, Gerdes K (2012) FtsZ-ZapA-ZapB interactome of *Escherichia coli*. J Bacteriol 194:292–302
Gruber S (2014) Multilayer chromosome organization through DNA bending, bridging and extrusion. Curr Opin Microbiol 22:102–110
Gubaev A, Klostermeier D (2014) The mechanism of negative DNA supercoiling: a cascade of DNA-induced conformational changes prepares gyrase for strand passage. DNA Repair 16:23–34

Handler AA, Lim JE, Losick R (2008) Peptide inhibitor of cytokinesis during sporulation in *Bacillus subtilis*. Mol Microbiol 68:588–599
Hardy CD, Cozzarelli NR (2005) A genetic selection for supercoiling mutants of *Escherichia coli* reveals proteins implicated in chromosome structure. Mol Microbiol 57:1636–1652
Hayes F, Barillà D (2006) Assembling the bacterial segrosome. Trends Biochem Sci 31:247–250
Hill NS, Buske PJ, Shi Y, Levin PA (2013) A moonlighting enzyme links *Escherichia coli* cell size with central metabolism. PLoS Genet 9:e1003663
Holečková N, Doubravová L, Massidda O, Molle V, Buriánková K, Benada O, Kofroňová O, Ulrych A, Branny P (2015) LocZ is a new cell division protein involved in proper septum placement in *Streptococcus pneumonia*. MBio 6:1–13
Hopwood DA (2007) Therapeutic treasures from the deep. Nat Chem Biol 3:457–458
Hu Z, Lutkenhaus J (2001) Topological regulation of cell division in *E. coli* spatiotemporal oscillation of MinD requires stimulation of its ATPase by MinE and phospholipid. Mol Cell 7:1337–1343
Hu Z, Gogol EP, Lutkenhaus J (2002) Dynamic assembly of MinD on phospholipid vesicles regulated by ATP and MinE. Proc Natl Acad Sci U S A 99:6761–6766
Huang KH, Durand-Heredia J, Janakiraman A (2013) FtsZ ring stability: of bundles, tubules, crosslinks and curves. J Bacteriol 195:1859–1868
Jennings PC, Cox GC, Monahan LG, Harry EJ (2011) Super-resolution imaging of the bacterial cytokinetic protein FtsZ. Micron 42:336–341
Keijser BJ, Noens EE, Kraal B, Koerten HK, van Wezel GP (2003) The *Streptomyces coelicolor ssgB* gene is required for early stages of sporulation. FEMS Microbiol Lett 225:59–67
Kelly AJ, Sackett MJ, Din N, Quardokus E, Brun YV (1998) Cell cycle-dependent transcriptional and proteolytic regulation of FtsZ in Caulobacter. Genes Dev 12:880–893
Kiekebusch D, Michie KA, Essen LO, Löwe J, Thanbichler M (2012) Localized dimerization and nucleoid binding drive gradient formation by the bacterial cell division inhibitor MipZ. Mol Cell 46:245–259
Konovalova A, Petters T, Søgaard-Andersen L (2010) Extracellular biology of *Myxococcus xanthus*. FEMS Microbiol Rev 34:89–106
Kusiak M, Gapczynska A, Plochocka D, Thomas CM, Jagura-Burdzy G (2011) Binding and spreading of ParB on DNA determine its biological function in *Pseudomonas aeruginosa*. J Bacteriol 193:3342–3355
Le TB, Imakaev MV, Mimy LA, Laub MT (2013) High-resolution mapping of the spatial organization of a bacterial chromosome. Science 342:731–734
Lee PS, Grossman AD (2006) The chromosome partitioning proteins Soj (ParA) and Spo0J (ParB) contribute to accurate chromosome partitioning, separation of replicated sister origins and regulation of replication initiation in *Bacillus subtilis*. Mol Microbiol 60:853–869
Li Z, Trimble MJ, Brun YV, Jensen GJ (2007) The structure of FtsZ filaments in vivo suggests a force-generating role in cell division. EMBO J 26:4694–4708
Lim HC, Surovtsev IV, Beltran BG, Huang F, Bewersdorf J, Jacobs-Wagner C (2014) Evidence for a DNA-relay mechanism in ParABC-mediated chromosome segregation. Elife 3:e02758
Löwe J, Amos LA (1998) Crystal structure of the bacterial cell-division protein FtsZ. Nature 391:203–206
Löwe J, van den Ent F (2001) Conserved sequence motif at the C-terminus of the bacterial cell division protein FtsZ. Biochemie 83:117–120
Lutkenhaus J (2007) Assembly dynamics of the bacterial MinCDE system and spatial regulation of the Z ring. Annu Rev Biochem 76:539–562
Lutkenhaus J, Pichoff S, Du S (2012) Bacterial cytokinesis: from Z ring to discovery. Cytoskeleton (Hoboken) 69:778–790
Lynch AS, Wang JC (1995) SopB protein-mediated silencing of genes linked to the *sopC* locus of *Escherichia coli* F plasmid. Proc Natl Acad Sci U S A 92:1896–1900
Männik J, Bailey MW (2015) Spatial coordination between chromosomes and cell division proteins in *Escherichia coli*. Front Microbiol 14:6.306

Margolin W (2005) FtsZ and the division of prokaryotic cells and organelles. Nat Rev Mol Cell Biol 6:862–871
McCormick JR, Losick R (1996) Cell division gene ftsQ is required for efficient sporulation but not growth and viability in *Streptomyces coelicolor* A3(2). J Bacteriol 178:5295–5301
Meier EL, Goley ED (2014) Form and function of the bacterial cytokinetic ring. Curr Opin Cell Biol 26:19–27. doi:10.1016/j.ceb.2013.08.006
Mercier R, Petit MA, Schbath S, Robin S, El Karoui M, Boccard F, Espéli O (2008) The MatP/matS site-specific system organizes the terminus region of the *E. coli* chromosome into a macrodomain. Cell 135:475–485
Milan SL, Osawa M, Erickson HP (2012) Negative-stain electron microscopy of inside-out FtsZ rings reconstituted in artifical membrane tubules show ribbons of protofilaments. Biophys J 103:59–68
Monack DM, Mueller A, Falkow S (2004) Persistent bacterial infections: the interface of the pathogen and the host immune system. Nat Rev Microbiol 2:747–765
Monahan JG, Liew AT, Bottomley AL, Harry EJ (2014) Division site positioning in bacteria: one size does not fit all. Front Microbiol 5:19
Niki H, Yamaichi Y, Hiraga S (2000) Dynamic organization of chromosomal DNA in *Escherichia coli*. Genes Dev 14:212–223
Noens EE, Mersinias V, Traag BA, Smith CP, Koerten HK, van Wezel GP (2005) SsgA-like proteins determine the fate of peptidoglycan during sporulation of *Streptomyces coelicolor*. Mol Microbiol 58:929–944
Osawa M, Erickson HP (2013) Liposome division by a simple bacterial division machinery. Proc Natl Acad Sci U S A 110:11000–11004
Osawa M, Anderson DE, Erickson HP (2008) Reconstitution of contractile FtsZ rings in liposomes. Science 320:792–794
Pichoff S, Lutkenhaus J (2002) Unique and overlapping roles for ZipA and FtsA in septal ring assembly in *Escherichia coli*. EMBO J 21:685–693
Ptacin JL, Lee SF, Garner EC, Toro E, Eckart M, Comolli LR, Moemer WE, Shapiro L (2010) A spindle-like apparatus guides bacterial chromosome segregation. Nat Cell Biol 12:791–798
Radharkrishnan SK, Pritchard S, Violllier PH (2010) Coupling prokaryotic cell fate and division control with a bifunctional and oscillating oxidoreductase homolog. Dev Cell 18:90–101
Ramos A, Letek M, Campelo AB, Vaquera J, Mateos LM, Gil JA (2005) Altered morphology produced by *ftsZ* expression in *Corynebacterium glutamicum* ATCC 13869. Microbiology 151:2563–2572
Ramos JL, Martinez-Bueno M, Molina-Henares AJ, Teran W, Watanabe K, Zhang X, Gallegos MT, Brennan RG, Tobes R (2009) The TetR family of transcriptional repressors. Microbiol Mol Biol Rev 69:326–356
Raskin DM, de Boer PA (1999a) Rapid pole-to-pole oscillation of a protein required for directing division to the middle of *Escherichia coli*. Proc Natl Acad Sci U S A 96:4971–4976
Raskin DM, de Boer PA (1999b) MinDE-dependent pole-to-pole oscillation of division inhbiitor MinC in *Escherichia coli*. J Bacteriol 181:6419–6424
Ringgaard S, Schirner K, Davis BM, Waldor MK (2011) A family of ParA-like ATPases promoters cell poles maturation by facilitating polar localization of chemotaxis proteins. Genes Dev 25:1544–1555
Rodionov O, Lobocka M, Yarmolinsky M (1999) Silencing of genes flanking the P1 plasmid centromere. Science 283:546–549
Rodrigues CD, Harry EJ (2012) The Min system and nucleoid occlusion are not required for identifying the division site in *Bacillus subtilis* but ensure its efficient utilization. PLoS Genet 8:e1002561
Romberg L, Levin PA (2003) Assembly dynamics of the bacterial cell division protein FtsZ: poised at the edge of stability. Annu Rev Microbiol 57:125–154
Sackett MJ, Kelly AJ, Brun YV (1998) Ordered expression of *ftsQA* and *ftsZ* during the *Caulobacter crescentus* cell cycle. Mol Microbiol 28:421–434

Sapay N, Guermerur Y, Deleage G (2006) Prediction of amphipathic in-plane membrane anchors in monotopic proteins using a SVM classifier. BMC Bioinf 7:255

Savage DDF, Afonso B, Chen AH, Silver PA (2010) Spatially ordered dynamics of the bacterial carbon fixation machinery. Science 327:1258–1261

Schumacher MA (2012) Bacterial plasmid partition machinery: A minimalist approach to survival. Curr Opin Struct Biol 22:72–79

Schumacher MA, Funnell BE (2005) Structures of ParB bound to DNA reveal mechanism of partition complex formation. Nature 438:516–519

Schumacher MA, Miller MC, Grkovic S, Brown MH, Skurray RA, Brennan RG (2001) Structural mechanisms of QacR induction and multidrug recognition. Science 294:2158–2163

Schumacher MA, Karamooz E, Zíková A, Trantírek L, Lukeš J (2006) Crystal structures of *T. brucei* MRP1/MRP2 guide-RNA binding complex reveal RNA matchmaking mechanism. Cell 126:701–711

Schumacher MA, Piro KM, Xu W (2010) Insight into F plasmid DNA segregation revealed by structures of SopB and SopB-DNA complexes. Nucleic Acids Res 38:4514–4526

Shih Y-L, Le T, Rothfield L (2003) Division site selection in *Escherichia coli* involves dynamic redistribution of Min proteins within coiled structures that extend between the two cell poles. Proc Natl Acad Sci U S A 100:7865–7870

Sievers J, Raether B, Perego M, Errington J (2002) Characterization of the *parB-like yyaA* gene of *Bacillus subtilis*. J Bacteriol 184:1102–1111

Singh B, Nitharwal RG, Ramesh M, Pettersson BM, Kirsebom LA, Dasgupta S (2013) Asymmetric growth and division in *Mycobacterium* spp: compensatory mechanisms for non-medial septa. Mol Microbiol 88:64–76

Strauss MP, Liew AT, Turnbull L, Whitchurch CB, Monahan LG, Harry EJ (2012) 3D-SIM super resolution microscopy reveals a bead-like arrangement for FtsZ and the division machinery: implications for triggering cytokinesis. PLoS Biol 10:e1001389

Thanbichler M, Shapiro L (2006) MipZ, a spatial regulator coordinating chromosome segregation with cell division in *Caulobacter*. Cell 126:147–162

Tonthat NK, Arold ST, Pickering BF, Van Dyke MW, Liang S, Lu Y, Beuria TK, Margolin W, Schumacher MA (2011) Molecular mechanism by which the nucleoid occlusion factor, SlmA, keeps cytokinesis in check. EMBO J 30:154–164

Tonthat NK, Milam SL, Chinnam N, Whitfill T, Margolin W, Schumacher MA (2013) SlmA forms a higher-order structure on DNA that inhibits cytokinetic Z-ring formation over the nucleoid. Proc Natl Acad Sci U S A 110:10586–10591

Tonthat NK, Milam SL, Chinnam N, Whitfill T, Margolin W, Schumacher MA (2014) SlmA forms a higher-order structure on DNA that inhibits cytokinetic Z-ring formation over the nucleoid. Proc Natl Acad Sci U S A 110:10586–10591

Traag BA, van Wezel GP (2008) The SsgA-like proteins in actinomycetes: small proteins up to a big task. Antonie Van Leeuwenhoek 94:85–97

Treuner-Lange A, Aguiluz K, van der Does C, Gómez-Santos N, Harms A, Schumacher D, Lenz P, Hoppert M, Kahnt J, Muñoz-Dorado J, Søgaard-Andersen L (2013) PomZ, a ParA-like protein, regulates Z-ring formation and cell division in *Myxococcus xanthus*. Mol Microbiol 87:235–253

Tsokos CG, Laub MY (2012) Polarity and cell fate asymmetry in *Caulobacter crescentus*. Curr Opin Microbiol 15:744–740

Vaerewijck I, Huys G, Palomino JC, Swings J, Portaels F (2005) Mycobacteria in drinking water distribution systems: ecology and significance or human health. FEMS Microbiol Rev 29:911–934

Valens M, Penaud S, Rossignol M, Cornet F, Boccard F (2004) Macrodomain organization of the *Escherichia coli* chromosome. EMBO J 23:4330–4341

Veiga H, Jorge AM, Pinho MG (2011) Absence of nucleoid occlusion effector Noc impairs formation of orthogonal FtsZ rings during *Staphylococcus aureus* cell division. Mol Microbiol 80:1366–1380

Wagner-Herman JK, Bernard R, Bunne R, Bisson-Filho AW, Kumar K, Nguyen T, Mulcahy L, Koullias J, Gueiros-Filho FJ, Rudner DZ (2012) RefZ facilitates the switch from medial to polar division during spore formation in *Bacillus subtilis*. J Bacteriol 194:4608–4618

Wang X, Loopis PM, Rudner DZ (2013) Organization and segregation of bacterial chromosomes. Nat Rev Genet 14:191–203

Wayne LG (1994) Dormancy of *Mycobacterium tuberculosis* and latency of disease. Eur J Clin Microbiol Infect Dis 13:908–914

Weart RB, Lee AH, Chien AC, Haeusser DP, Hill NS, Levin PA (2003) A metabolic sensor governing cell size in bacteria. Cell 130:335–347

Wildermuth H, Hopwood DA (1970) Septation during sporulation in *Streptomyces coelicolor*. J Gen Microbiol 60:51–59

Woldringh CL (2002) The role of co-transcriptional translation and protein translocation (transertion) in bacterial chromosome segregation. Mol Microbiol 45:17–29

Woldringh CL, Mulder E, Valkenburg JA, Wientjes FB, Zaritsky A, Nanninga N (1990) Role of the nucleoid in the toporegulation of division. Res Microbiol 141:39–49

Woldringh CL, Mulder E, Huls PG, Vischer N (1991) Toporegulation of bacterial division according to the nucleoid occlusion model. Res Microbiol 142:309–320

Wu LJ, Errington J (2004) Coordination of cell division and chromosome segregation by a nucleoid occlusion protein in *Bacillus subtilis*. Cell 117:915–925

Wu L, Ishikawa S, Kawai Y, Oshima T, Ogasawara N, Errington J (2009) Noc protein binds to specific DNA sequences to coordinate cell division with chromosome segregation. EMBO J 28:1940–1952

Xu Q et al (2009) Structural and functional characterizations of SsgB, a conserved activator of developmental cell diviison in morphologically complex actinomycetes. J Biol Chem 284:25268–25279

Zapun A, Vernet T, Pinho MG (2008) The different shapes of cocci. FEMS Microbiol Rev 32:345–360

Zaritksy A, Woldringh CL (2003) Localizing cell division in spherical *Escherichia coli* by nucleoid occlusion. FEMS Microbiol Lett 226:209–214

Chapter 10
Structure and Dynamics of Actin-Like Cytomotive Filaments in Plasmid Segregation

Pananghat Gayathri and Shrikant Harne

Abstract One of the well-known functions of the bacterial cytoskeleton is plasmid segregation. Type II plasmid segregation systems, among the best characterized with respect to the mechanism of action, possess an actin-like cytomotive filament as the motor component. This chapter describes the essential components of the plasmid segregation machinery and their mechanism of action, concentrating on the actin-like protein family of the bacterial cytoskeleton. The structures of the actin-like filaments depend on their nucleotide state and these in turn contribute to the dynamics of the filaments. The components that link the filaments to the plasmid DNA also regulate filament dynamics. The modulation of the dynamics facilitates the cytomotive filament to function as a mitotic spindle with a minimal number of components.

Keywords ParM • Actin-like • Alps • Plasmid segregation type II • Centromere • ParMRC system • R1 plasmid • Cytomotive filament • Treadmilling • ParR • PALM • ParRC helix • AlfA • AlfB • Alp12A • Alp7A • pSK41 • Bipolar spindle

Plasmid Segregation

Unlike high-copy number plasmids, low-copy number plasmids need to ensure that they are inherited to both daughter cells before or during bacterial cell division. To achieve this, they encode dedicated machinery that keeps the replicated plasmids apart within the bacterial cell. This greatly enhances the probability of equal partitioning to the daughter cells and facilitates the maintenance of the plasmids, extra-chromosomal genetic material, across generations of bacteria.

P. Gayathri (✉) • S. Harne
Indian Institute of Science Education and Research (IISER),
Dr. Homi Bhabha Road, Pune 411008, India
e-mail: gayathri@iiserpune.ac.in

J. Löwe, L.A. Amos (eds.), *Prokaryotic Cytoskeletons*, Subcellular Biochemistry 84, DOI 10.1007/978-3-319-53047-5_10

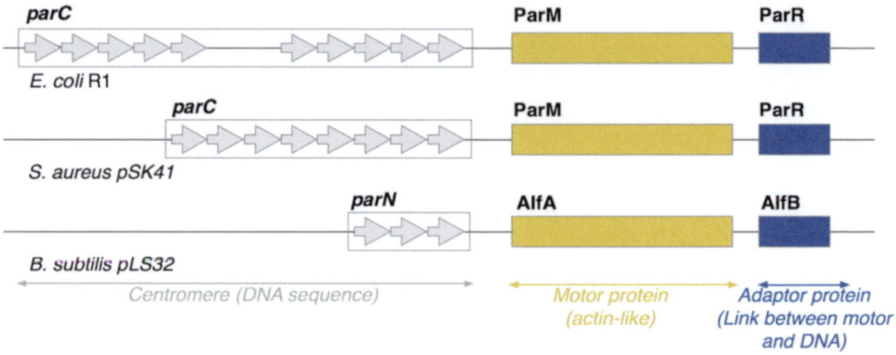

Fig. 10.1 Organization of *par* loci encoding actin-like protein homologs. In *E. coli*, *parC*, the centromere, is formed by two sets of five 11-base pair (bp) repeats separated by a 39 bp region. In *S. aureus* pSK41 and *B. subtilis* AlfA, the centromere is a continuous region formed by eight and three short repeats (iteron) respectively. The centromere region (each *arrow* represents an iteron) is followed by genes that encode a motor protein of the actin fold (*yellow*) and an adapter protein (*blue*). In general, the promoter region of the motor protein is located within the centromere sequence. Adapter proteins can thus repress the expression of motor proteins by binding to the centromere region. The lengths of the DNA segments shown for the centromere sequences and the protein-encoding regions are not drawn to scale

Components of a Plasmid Segregation System

Plasmid segregation systems were discovered in the early 1980s, well before the existence of filamentous proteins was established, and found to confer stability to plasmids containing them (Gerdes and Molin 1986). The first plasmid segregation locus to be identified was the ParMR*C* plasmid stability locus, first named *stb* (for stability). Studies carried out to understand the plasmid stability locus of *E. coli* R1 plasmid identified three components – (i) ParM – an actin-like protein that assembles into filaments in the presence of ATP as nucleotide (ii) *parC* – a centromeric DNA segment on the plasmid where ParM is attached, and (iii) ParR – an adaptor protein that couples ParM to *parC,* and also acts as a transcriptional repressor of ParM (Gerdes and Molin 1986; Dam and Gerdes 1994).

Later, characterization of a variety of plasmid and bacterial chromosome segregation loci led to the conclusion that the minimal number of necessary and sufficient components for a plasmid or DNA segregation system (Fig. 10.1) include – (i) a motor component that can actively push or pull the DNA, by utilizing energy derived from nucleotide binding/hydrolysis, (ii) a point of attachment on the plasmid DNA, a locus that functions as a centromere, and (iii) an adaptor protein that serves as a link between the DNA and the motor (Gerdes et al. 2010).

Motor

In some of the plasmid or DNA segregation systems, the motor component functions through assembly of the monomeric protein into a filamentous structure (Møller-Jensen et al. 2002). Typically, the adaptor protein mediates the binding of the centromeric DNA to the end of the filament (Jensen and Gerdes 1997). The elongation/shortening of the filament facilitates the movement of the DNA, thereby resulting in a pushing or a pulling mechanism (Møller-Jensen et al. 2002, 2003). Since it is the filament itself that acts as a linear motor, as opposed to the molecular motors of eukaryotic cytoskeletal systems such as myosins, kinesins or dyneins that walk on the cytoskeletal tracks, the bacterial cytoskeletal filaments involved in DNA segregation fall under the category of cytomotive filaments (Löwe and Amos 2009). The dynamic assembly and disassembly of the filaments is driven by changes in the nucleotide state. Interactions with the adaptor and/or the adaptor/centromere complex modulate these dynamics (Møller-Jensen et al. 2003; Garner et al. 2004).

The segregation systems have been classified into different types (I – III) based on the fold of the motor protein. The motor proteins of Type I, II and III comprise ATPases of the WACA family (deviant Walker A motif containing Cytoskeletal ATPases), ATPases of the actin family (Fig. 10.2), and GTPases of the tubulin family, respectively (Gerdes et al. 2010). Type I is the most common system that also functions in bacterial chromosome segregation but is the least understood. Type II systems will be described in detail in the current chapter on actin-like cytomotive filaments in plasmid segregation, while Type III will be discussed in the next chapter (Chap. 11).

Centromere

The locus on the plasmid that allows for the attachment of the filament motor via an adaptor protein can be regarded a centromere-like region of the plasmid DNA (Dam and Gerdes 1994). The centromeres of plasmid segregation systems comprise repetitive segments of DNA, termed iterons. In different systems there are 3–12 iterons in their centromeric sequences, each of 6–12 bp length (Figure 10.1). Each iteron binds a functional unit (either monomer or dimer) of the adaptor protein (Dam and Gerdes 1994; Møller-Jensen et al. 2007; Schumacher et al. 2007). The arrangement of the iterons allows for cooperative binding of the adaptor on the plasmid DNA (Møller-Jensen et al. 2003). The resulting protein-DNA complex is sometimes called the 'segrosome'. The structure of the segrosome (Fig. 10.3) is typically organized with the DNA on the outer surface of an extended super-helical structure of the adaptor proteins (Møller-Jensen et al. 2007; Schumacher et al. 2007).

A.

```
                Phosphate I       Connect I          Phosphate II      Connect II
Actin_Human     DETTALVCDNGSGL    MYVAIQAVLSLYASGR   -TTGIVLDSGDGVTH   ANNVMSGGTTMY
MreB_Ec         MFSNDLSIDLGTAN    VFLIEEPMAAAIGAGL   ATGSMVVDIGGGTTE   RGMVLTGGGALL
ParM_Ec_R1      ---MLVFIDDGSTN    VKVMPESIPAGYEVLQ   LDSLLIIDLGGTTLD   THVMVIGGGAEL
ParM_Sa_pSK41   -NVYVMALDFGNGF    VKIVAQPMGTLLDLNM   EGKYSVLDFGSGTTI   DRIIVTGGGANI
Alp12A_Bs_pE88  TNEYVMTLDAGKYE    ITIKAEGSGVLFLEQ-   NKNVAVIDFGGLNMG   DSLIFIGGTTQK
Alp7A_Bs_pLS32  -NISRMNVDFGNSM    AACRIESEVARWAIKK   NYDVVFCDLGGGTDD   YKYLYFGGVGEV
AlfA_Bs_pLS20   --TLTTVIDIGNFS    IVMAAEGLGALNFSDS   -LNCVIVDAGSKTLN   YPIVCTGGKAEE
                        D G            E/Q                 D G              GG
```

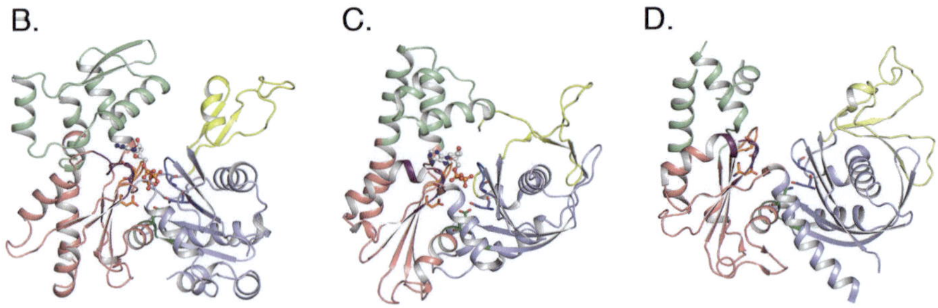

Fig. 10.2 Motor proteins in Type II plasmid segregation systems are actin-like. (**a**) Sections of the structure-based sequence alignment obtained using PROMALS3D (Pei et al. 2008) of human and bacterial actins (UNIPROT ID's- Human actin: P68133, *E. coli* MreB: P0A9X4, *E. coli* R1 ParM: P11904, *S. aureus* pSK41 ParM: O87364, *B. subtilis* pE88 Alp12A: Q89A01, *B. subtilis* pLS32 Alp7A: C7F6X5, *B. subtilis* pLS20 AlfA: S5DTP8) shows that the signature motifs of the ASKHA fold (Bork et al. 1992) are conserved in the bacterial actins involved in plasmid segregation. Monomeric structures of (**b**) actin (PDB: 1YAG), (**c**) R1 ParM (PDB: 1MWM) and (**d**) pSK41 ParM (PDB: 3JS6). The four domains are color coded as IA/1 (*blue*), IB/2 (*yellow*), IIA/3 (*red*), IIB/4 (*green*). The signature motifs are colored according to the labels in (a), and the conserved residues highlighted in stick representation. Bound nucleotides, if any, are shown in ball and stick representation

Adaptor Protein

Adaptor proteins, as the name implies, mediate the attachment of the filament motors to the centromere-like sequences on the DNA. They have a surface that interacts with the DNA, and another binding site for the motor protein (Salje and Löwe 2008). Cooperative binding of the adaptor protein to the repetitive sequences of the centromere results in the formation of the segrosome complex (Fig. 10.3). The adaptor protein, on its own and in the centromere-bound form, is capable of altering the dynamics of assembly of the cytomotive filament (see later sections) (Møller-Jensen et al. 2003; Garner et al. 2004; Gayathri et al. 2012; Garner 2007). This is a significant feature of plasmid segregation systems, which helps with dispensing of alternative nucleation factors for filament assembly, checkpoints for spindle assembly, and disassembly factors. In many if not all cases, the centromeric sequence overlaps with the promoter-binding region for the motor. Hence, the adaptor protein also acts as a transcriptional repressor for the motor and adaptor proteins (Dam and Gerdes 1994). The repression activity regulates the number of monomers

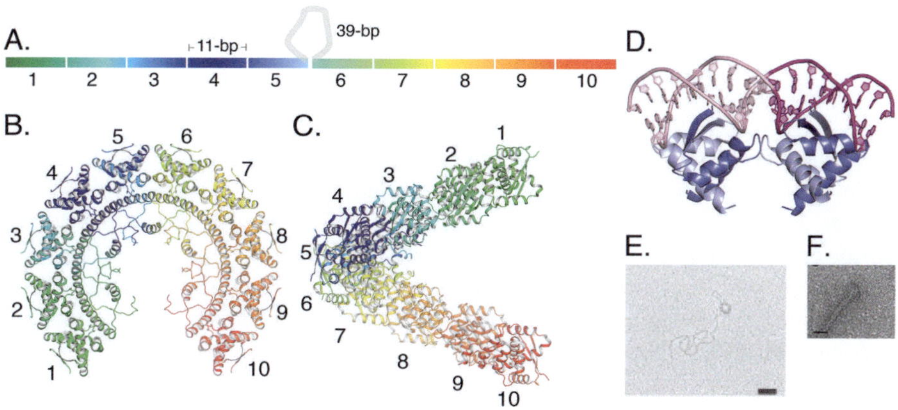

Fig. 10.3 Architecture of the segrosome complex in plasmid segregation. (**a**) Organization of the centromeric sequence into 10 iterons. The 39-bp insertion between the iterons is shown as a grey loop. (**B–C**) pB171 ParR (PDB ID: 2JD3) forms a super-helical structure resembling a lockwasher. A dimer of ParR binds to each iteron of *parC* (color-coded and numbered correspondingly) (**b**) View of ParR helix along the screw axis and (**c**) side view. (**d**) ParR binds DNA as a dimer to one iteron of the centromeric sequence, as shown by the crystal structure of the N-terminal region of dimeric ParR (PDB ID: 2Q2K) of pSK41 plasmid (*light* and *dark blue*) in complex with a 20-bp DNA sequence of two iterons of *parC* (magenta and pink). (**e**) Electron micrographs of R1 ParR assembled on *parC*-containing plasmid (Adapted from Møller-Jensen et al. (2007)). (**f**) Negatively stained EM image of R1 ParR bound to the tip of a ParM filament (Adapted from Salje and Löwe (2008))

of the motor protein available in the cell, and thus controls filament dynamics. This leads to an efficient mechanism of disassembly and/or control over excessive filament formation and bundling.

Actin-Like Cytomotive Filaments in Plasmid Segregation

A large number of actin-like proteins have been identified in bacterial genomes (Derman et al. 2012). Some of these proteins have been shown to be involved in plasmid partitioning while the functions of others remain elusive. The best-characterized plasmid segregation system involving an actin-like cytomotive filament is the ParMR*C* system (Salje et al. 2010). Other actin-like plasmid segregation systems are AlfA, Alp7A, Alp12, pB171 and pSK41. The following sections discuss in detail how the structure and dynamics of the cytomotive filaments lead to plasmid partitioning as, currently, the function of the ParMRC system is probably the best-understood in bacterial systems involving cytomotive filaments.

ParMRC System of E. coli *R1 Plasmid*

A number of studies utilizing cytological investigations (Gerdes and Molin 1986; Dam and Gerdes 1994), in vitro biochemical assays (Møller-Jensen et al. 2002; Jensen and Gerdes 1997; Møller-Jensen et al. 2003, 2007; Schumacher et al. 2007; Salje and Löwe 2008), reconstitution experiments in vitro (Garner et al. 2004; Garner 2007), live cell fluorescence microscopy (Campbell and Mullins 2007), X-ray crystallography (Gayathri et al. 2012; van den Ent et al. 2002), electron microscopy and tomography experiments (Salje et al. 2009; Orlova et al. 2007; Popp et al. 2008; Bharat et al. 2015) have led to deep mechanistic insights into the action of the ParMR*C* system of plasmid segregation.

Structural Features of ParMR*C*

The structural organization of the three components of the plasmid segregation machinery has been elucidated for ParMR*C*.

Architecture of the ParR*C* Complex

The *parC* locus of the ParMR*C* system consists of 10 iterons of 11-base pair sequences of DNA, five each on either side of a 39-base pair region (Fig. 10.3a) (Dam and Gerdes 1994). The structure of ParR demonstrates that it forms a lockwasher-like super-helical structure at high concentrations and in the presence of DNA containing the *parC* sequence (Fig. 10.3b–d) (Møller-Jensen et al. 2007). The positively charged surface on the outside of the lockwasher provides a complementary electrostatic surface for the negatively charged DNA backbone, as well as a repeating structural match to the iterons; each dimer of ParR binds to an iteron comprising 11-base pairs of *parC* (Møller-Jensen et al. 2007).

Architecture of ParM Filament

In the presence of ATP, ParM forms helical filaments with a left-handed twist (Gayathri et al. 2012; Orlova et al. 2007; Popp et al. 2008). These can also be described as a two-start helix with two protofilaments/strands (Fig. 10.4a). The arrangement of the monomers within the protofilament of ParM is similar to that in actin, though the helical parameters and the twist are different. The rotation between adjacent subunits in the protofilament is 27° for actin (right-handed), and −30° (left-handed) for ParM (Gayathri et al. 2012; Fujii et al. 2010). Table 10.1 contains a summary of the helical parameters of all the actin-like filaments in plasmid segregation for which a reliable reconstruction is available. The polar structure (filament

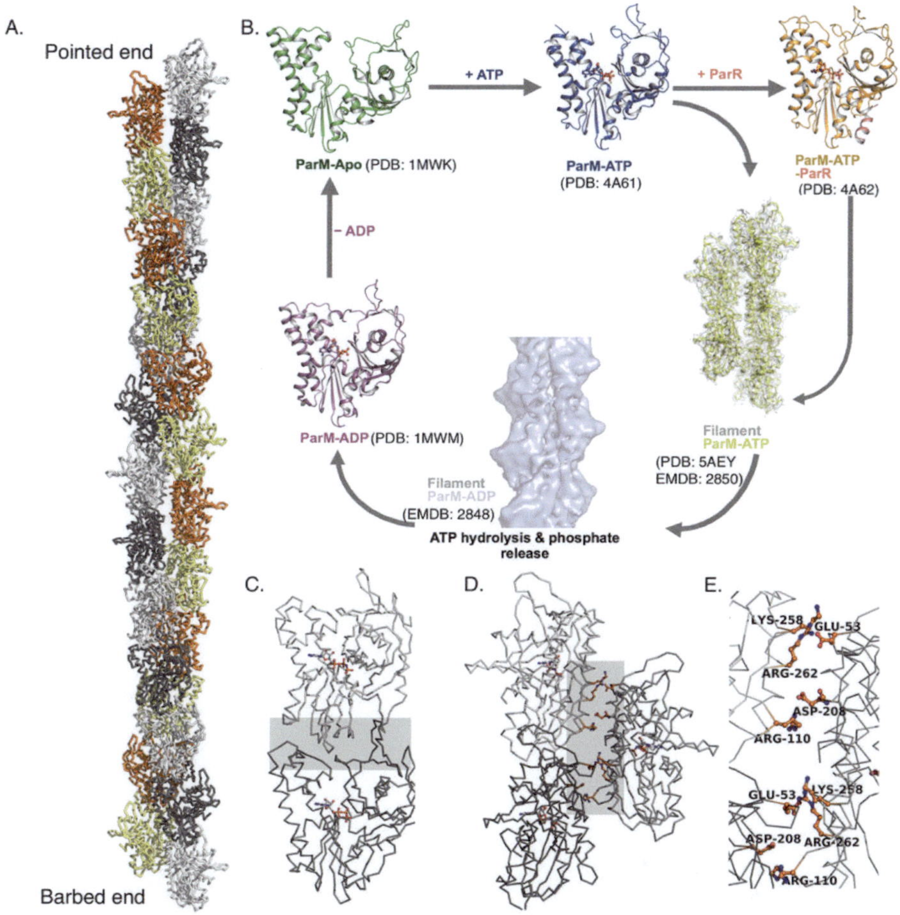

Fig. 10.4 Conformational cycle of ParM. (**a**) The ParM filament is composed of two protofilaments (PDB ID: 4A6J). Alternate monomers are colored as grey and black in one protofilament/strand and yellow and orange in the other. Monomers in one protofilament are offset by approximately half their length (~ 24 Å) with respect to the other protofilament – the filament is staggered. The filament is polar, with one end termed as the barbed end and the other as the pointed end. The standard view of the monomers, shown in Fig. 10.2, and in (**b**), is such that the bottom of the monomer faces the barbed end of the filament. (**b**) The nucleotide-dependent conformational cycle of ParM: ParM in its apo-form binds ATP resulting in a ~25° domain closure between domains 1 and 2. The conformational change in ParM-ATP can either allow its assembly to form filaments or it binds with ParR to forms a ParM-ATP-ParR complex. The ParR-bound conformation of ParM and the monomer conformation within the ParM-ATP filament are equivalent – a more compact conformation where ATP hydrolysis is stimulated. Upon release of phosphate, the filaments dissociate to form ParM-ADP monomers, which can release ADP to go back to the apo-form. The ATP-bound conformations have been captured using AMPPNP, a non-hydrolysable analogue of ATP. (**c**) Intra-protofilament contacts between R1 ParM monomers with the polymerization interface highlighted in *grey*. (**d**) Inter-protofilament contacts between monomers from adjacent protofilaments/strands. (**e**) Enlarged view of the contacts in (**d**) shows that the inter-protofilament contacts include a few salt bridges (PDB ID for C-E: 5AEY)

Table 10.1 Helical parameters and Dynamics of actin-like filaments

Actin-Like Protein	Structure (helical parameters)						Dynamics		
	PDB	EMDB	Resolution (Å)	Handedness	Phi (°)[a]	Azimuthal increment (Å)	Instability ratio (ccADP/ccATP)[b]	Behavior	Nucleus
Actin	3MFP	5168	6.6	Right	26.8	27.6	1.6 (Pollard 1986)	Treadmilling	3 (Pollard 1986)
R1 ParM	5AEY	2850 (AMPPNP)	4.3	Left	30	23.4	>160	Dynamic instability	3 (Garner et al. 2004)
	–	2848 (ADP)	11	Left	30	24.7			
pB171 ParM (Rivera et al. 2011)	–	–	19	Left	27.8	24.2	>140	Dynamic instability	2
AlfA (Polka et al. 2009)	–	–	15	Left	46	25	4.2	Treadmilling	3–4
Alp12A (Popp et al. 2012)	4APW	2068	19.7	Right	13.24	42.8	No polymerization with ADP	Dynamic instability	4

[a]Phi denotes the angle of rotation between adjacent subunits within a protofilament
[b]ccADP/ccATP denotes ratio of the critical concentrations of ADP to ATP

with dissimilar ends) of the ParM filament is described by the terms barbed end and pointed end, borrowing from the actin filament terminology (Fig. 10.4a).

Conformational Cycle of ParM

The different conformations of monomeric ParM in the nucleotide-free and in the bound states with the hydrolyzed (ADP) and non-hydrolyzed (AMPPNP, a non-hydrolysable analog of ATP) nucleotide states have been captured using X-ray crystallography (Gayathri et al. 2012; van den Ent et al. 2002). The conformations of the filament form of ParM in the different nucleotide states have been observed recently by high-resolution electron microscopic reconstructions (Bharat et al. 2015). These give a seemingly complete picture of the conformational cycle that ParM undergoes during nucleotide hydrolysis-dependent dynamics (Fig. 10.4b).

Upon nucleotide binding, there is ~25° domain closure between domains 1 and 2 (van den Ent et al. 2002). Comparison between ATP and ADP monomeric conformations does not reveal any prominent conformational changes. However, the assembly of the ATP-bound monomers into the filament form results in a further domain closure resulting in the formation of the most compact structure of ParM (Bharat et al. 2015). The conformation of ParM in the ParR peptide-bound state matches well the conformation of ParM monomers within the filament (Gayathri et al. 2012). This structure also shows that it binds to a pocket at the polymerization interface (Fig. 10.4b, c). Hence, ParR can be accommodated only at the barbed end of the ParM filament.

ParM Filament Dynamics

ParM filaments are preferentially formed in the presence of ATP. The critical concentration for filament formation is very high in the presence of ADP, and even higher without any nucleotide (Garner et al. 2004). Studies have shown that ParM can also utilize GTP for polymerization (Popp et al. 2008). Filament growth does not require external nucleation factors, and filaments are capable of nucleating on their own, above the critical concentration. Though the filaments are structurally asymmetric, they are kinetically symmetric, and grow, alone, equally at the barbed end and the pointed end (Garner et al. 2004). However, in the presence of ParR*C* complex, the critical concentration of ParM monomers for filament formation is lower, and the growth at the barbed end, which is the ParR*C*-bound end, is faster than at the pointed end (Fig. 10.5a) (Gayathri et al. 2012).

ParM filaments exhibit dynamic instability (Garner et al. 2004). They grow in the presence of ATP-bound monomers. Hydrolysis of ATP destabilizes the filaments, resulting in disassembly. Binding of ParR*C* prevents the disassembly of the filaments from the barbed end (Gayathri et al. 2012). ParM monomers are added by insertional polymerization at the ParR*C*-bound end (Møller-Jensen et al. 2003; Garner 2007). The pointed end is inhibited from disassembly by lateral stabilization

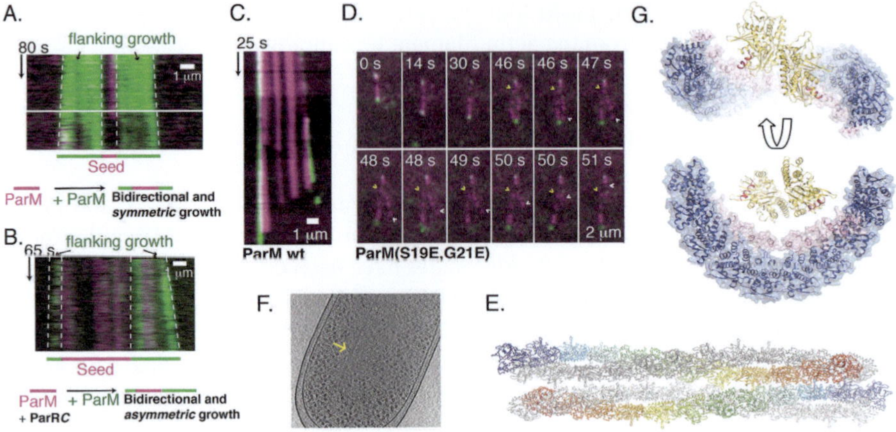

Fig. 10.5 Effect of ParR*C* on ParM filament assembly and bipolar spindle formation. Kymographs of ParM filaments by dual-label experiments (*purple* ParM seed and *green* growing filament) in the absence (**a**) and presence (**b**) of ParR*C* using TIRF microscopy. ParM filaments, in absence of ParR*C*, exhibit symmetric growth at both ends. The bidirectional growth becomes asymmetric upon addition of ParR*C*, which rapidly adds ParM monomers at one end, the barbed end. C & D) ParM filaments are stabilized by the binding of ParR*C* at the barbed end (**c**) and by pairing at the pointed end (**d**). Loss of ParR*C* (*green*) at the barbed end (**c**) leads to disassembly of the ParM (*purple*) filament. The white arrow highlights the initiation of disassembly, observed by the slope. The dissociation of a spindle (**d**), due to repulsive charges in a mutant of ParM (ParM-S19R, G21R), shows that a ParM spindle (*purple*) is composed of more than one filament, that they are oriented anti-parallel, and that un-pairing triggers disassembly from the pointed end. White and yellow arrowheads highlight the ends of each filament within the disassembling spindle. (**e**) Atomic model comprising one repeat of the antiparallel ParM filament pair. (**f**) A view of the bipolar spindle captured in a electron cryotomogram of an *E. coli* cell. The filament doublet shows a clear lack of superhelicity, consistent with the model in (**e**). (**g**) The overall dimensions of the ParR*C* complex and the helical pitch of ParM filaments are similar. Thus ParR*C* can facilitate nucleation and accelerate ParM growth at the barbed end. (**A–G** adapted from Gayathri et al. (2012) and Bharat et al. (2015))

upon pairing with another ParR*C*-bound filament in an antiparallel orientation (Fig. 10.5c–e) (Gayathri et al. 2012). Thus, an antiparallel arrangement of two filaments forms a stable bipolar spindle of ParMR*C* (Fig. 10.5d–e). An interesting feature of ParMR*C* spindle dynamics is that ParR*C* interacts only with the ATP-bound conformation of ParM (Møller-Jensen et al. 2003). This ascertains that ParR*C* remains bound only when there are sufficient free ParM monomers within the cell, where they will normally be ATP-bound. Decrease in the concentration of ATP-bound ParM monomers leads to the loss of ParR*C* from the tip, and subsequent disassembly of the bipolar spindle (Møller-Jensen et al. 2002; Gayathri et al. 2012).

Information on filament dynamics has been obtained by in vitro reconstitution experiments using fluorescently labeled proteins and DNA containing *parC* (Garner et al. 2004; Gayathri et al. 2012; Garner 2007). Light scattering experiments provide information on the kinetics of filament nucleation, growth and disassembly (Table 10.1) (Garner et al. 2004; Popp et al. 2008; Galkin et al. 2009). Plasmid segregation dynamics have also been observed using live cell imaging within *E. coli* using fluorescently labeled components (Campbell and Mullins 2007). A detailed account of the various methods utilized to study actin-like filaments is provided in reference (Petek and Mullins 2014).

Correlation Between ParM Structure and Dynamics

Observation of the complete conformational cycle provides a clear understanding of how structural features of the actin-like filament contribute to dynamics of assembly (Fig. 10.4b). ParM monomers assemble into the filament form in the presence of ATP. The conformational transition from the monomeric to the filament form stimulates ATP hydrolysis within the filament (Møller-Jensen et al. 2003; Gayathri et al. 2013). The monomers are held in the filament conformation through longitudinal lattice contacts involving adjacent monomers even post ATP hydrolysis, as is observed from the ParM filament reconstruction in presence of ATP and vanadate (Bharat et al. 2015). However, ParM filaments formed in the presence of ADP have a different conformation of monomers within the filament. Comparison of the ParM-ADP and ATP filament conformations shows that the inter-protofilament and intra-protofilament contacts are not optimal for the stability of the ADP-bound filament. Figure 10.4c–e shows the contacts that hold the filament in the AMPPNP-bound state of ParM (Bharat et al. 2015). The sub-optimal contacts lead to the disassembly of the filaments post ATP hydrolysis and loss of phosphate, unless the lattice contacts with ATP-bound ParM monomers is maintained at the tips, where ATP hydrolysis may not have occurred yet.

Dynamic instability exhibited by the ParM filaments implies that they remain stable as long as the ends of the filament are held in the ATP-bound filament conformation. If the rate of addition of monomers slows down, presumably due to reduction in the concentration of ATP-bound ParM monomers, the conversion of ATP to ADP conformation occurs in the ParM monomers towards the ends of the filament, leading to rapid disassembly. The observation of disassembly in the ParM spindles show that the process is directional (Fig. 10.5c, d) and proceeds from the destabilized end of the filament (barbed end where the destabilization is by the loss of ParR*C* (Fig. 10.5c), pointed end if the destabilization is by decoupling of a paired filament) (Fig. 10.5d) (Gayathri et al. 2012).

The structural arrangement of the ParR*C* helix might explain its processivity when polymerizing ParM filaments and also the insertional polymerization mechanism of ParR*C*-driven ParM filament elongation (Fig. 10.5g). The binding site at the polymerization interface always ensures that the ParR*C* helix is at the tip of the filament, thus providing a rationale for insertional polymerization (Gayathri et al. 2012). ParR*C* facilitates the conversion to the filament conformation and prevents disassembly by speeding up the rate of addition of ATP-bound monomers. The multimerization involved due to the repetitive segments of the ParR binding sites at the iterons of *parC* provides multiple ParM-binding sites for the segrosome complex. This probably contributes towards the processivity of movement of ParR*C* at the barbed end of ParM filaments. The amount of ATP-bound ParM monomers in the cell may also affect processivity. Finally, the architecture of the ParR*C* complex suggests that nucleation and acceleration of growth at the barbed end may occur due to similarity in the helical pitch of the ParM filament and that of the ParR*C* super-helix (Fig. 10.5g).

Antiparallel bipolar spindle assembly is facilitated by ParM's filament architecture. The groove of one filament fits into the groove of another in an antiparallel manner (Fig. 10.5e). Hence, surface residues on the filament contribute to the efficient functioning of the bipolar spindle, such that the interaction is not too strong to hinder the dynamics by antiparallel pairing, but not too weak to prevent them from forming a stable spindle, as has been demonstrated by mutational studies (Gayathri et al. 2012). These bipolar spindles have been observed in vitro (Fig. 10.5c, d) and also in vivo using electron tomography of *E. coli* cells (Fig. 10.5f) (Bharat et al. 2015). The electron tomography observations in vivo point out that the plasmid segregation is most likely 'asynchronous' – each bipolar spindle segregates a different pair of plasmid sisters (Bharat et al. 2015). The process of segregation is also not coordinated with the cell cycle of the bacterium. These facts point out that the mechanism of plasmid segregation in ParMR*C* strives to separate pairs of plasmid copies, and thus ensures that they are apart during cell division.

AlfA of Bacillus Subtilis pLS32

A partitioning system, active during vegetative growth and sporulation in *B. subtilis* was discovered on pBET31, the mini-plasmid of pLS32, and AlfA and AlfB were identified to be the actin-like protein and the adaptor protein components respectively (Becker et al. 2006).

Organization of the AlfA Segregation System

The *alf* operon consists of AlfA, forming an *a*ctin-*l*ike *f*ilament (*Alf*), AlfB, a DNA-binding protein that acts as the adapter coupling the filament to the plasmid, and *parN*, a DNA locus containing three binding sites for AlfB (Becker et al. 2006; Polka et al. 2009; Tanaka 2010). AlfA was found to be actin-like based on the identification of nucleotide binding and hydrolysis motifs (Fig. 10.2a) (Becker et al. 2006). The interaction between AlfB and *parN* has been characterized and the first two iterons are essential for the binding (Tanaka 2010). Since the *parN* sequence is located within the promoter region of *alfA*, binding of AlfB to *parN* represses its expression. It has also been shown using yeast two hybrid experiments that AlfB binds to AlfA. Thus, the three components AlfA, AlfB and *parN* constitute a plasmid segregation system for pLS32 of *Bacillus subtilis* (Tanaka 2010).

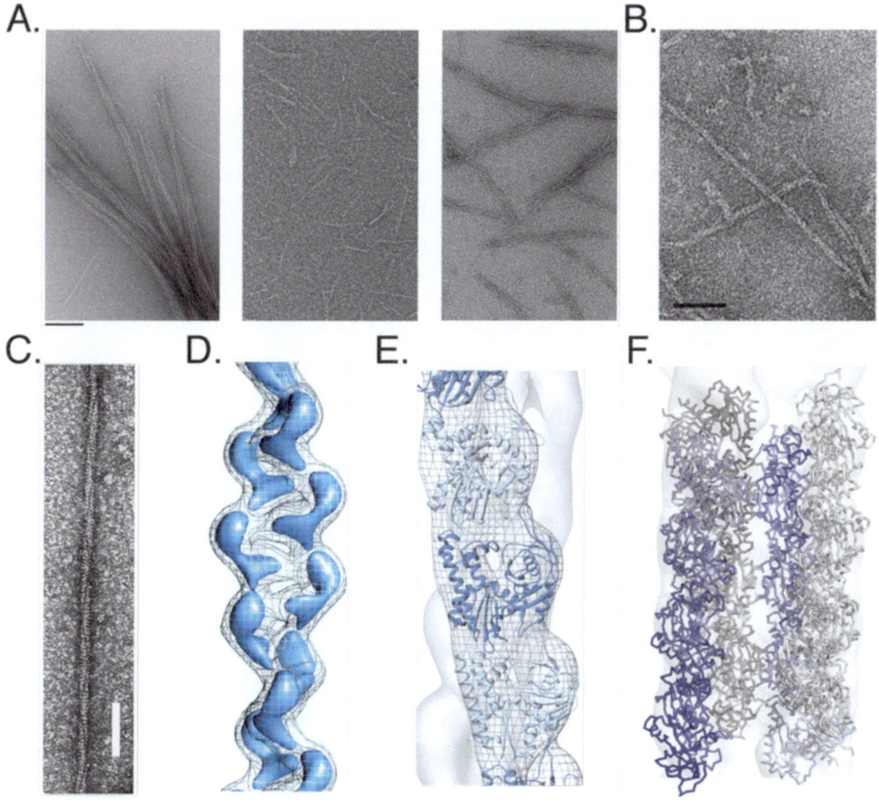

Fig. 10.6 Electron microscopy of actin-like proteins from Type II plasmid segregation systems. (**a**) AlfA alone forms bundles (*left*), the presence of AlfB debundles AlfA and decreases its length (*centre*), and bundles are restored upon addition of AlfB-parN complex (*right*) (Adapted from Becker et al. (2006)). (**b**) pB171 ParM forms filaments in presence of ATP (Adapted from Rivera et al. (2011)). (**c**) Filaments of Alp12A (Adapted from Popp et al. (2012)). (**d–f**) EM reconstruction of AlfA filaments (**d**, adapted from Polka et al. (2009)); pB171 ParM filament (e, adapted from Rivera et al. (2011)); Alp12A filaments (f, PDB: 4APW; EMDB: 2068). Alp12A exists as a four-stranded antiparallel filament. Each filament is shown as two protofilaments colored in *blue* and *grey* respectively. The shade of blue of the monomers varies from dark towards light from the barbed end to the pointed end, highlighting the antiparallel arrangement

Structure of AlfA

Though the monomeric structure of AlfA has not been determined yet, it is expected to be of actin fold. Contrary to the ParM filaments that exhibit dynamic instability, AlfA was observed to form stable filaments within the cell and also in vitro (Polka et al. 2009). The polymerization occurred in the presence of ADP or ATP, and also GDP or GTP, although filaments formed with ATP were more stable compared to those with GTP. Electron microscopy images showed the presence of bundles of AlfA filaments (Fig. 10.6a). The bundling of AlfA was dependent on the salt concentration. The bundles were disrupted in the presence of higher concentration of KCl (Polka et al. 2009).

Three-dimensional reconstruction of electron microscopy images of negatively-stained AlfA showed that the filaments are left-handed, similar to ParM (Fig. 10.6). The helical parameters vary between AlfA and ParM. AlfA exhibits a higher twist of 46° between adjacent monomers in the filament compared to 30° in ParM (Table 10.1). A monomer of ParM in the apo-state (with an open nucleotide cleft) was observed to be the best fit into the EM reconstruction map (Polka et al. 2009). However, a high-resolution EM reconstruction from electron cryomicroscopy images will be required in order to provide more detailed information.

AlfA Filament Dynamics

AlfA forms polymers in the presence of ATP, ADP, GTP or GDP (Polka et al. 2009). The kinetics of polymerization show that the critical concentration of ADP-bound AlfA monomers is only about 4 fold higher than that of ATP-bound monomers. This implies that AlfA most likely does not function by a dynamic instability driven mechanism as observed for ParM. The ability to form filaments with ADP or ATP results in the formation of stable bundles of AlfA filaments (Polka et al. 2009). Initial observations of stable filaments of AlfA that do not show dynamics were intriguing since a mechanism for plasmid segregation involving a stable filament is difficult to reconcile. For plasmid segregation to occur by a filament bound to a DNA there should be available a pool of monomers, maintained above the critical concentration for elongation. The pool is typically maintained by recycling monomers released by the breakdown of filaments that are not bound to DNA, as in the case of dynamic instability of unbound ParM filaments.

Later studies showed that AlfB reduces the bundle formation of AlfA polymers and destabilizes them (Fig. 10.6a). This effect was attributed to the sequestration of AlfA monomers by AlfB (Polka et al. 2014). Additionally, AlfB bound to *parN* was found to nucleate AlfA filaments. The filaments formed in the presence of the AlfB-*parN* complex were longer, and had a tendency to bundle (Fig. 10.6a). The AlfB-*parN* complex bound only to one end of the AlfA filaments (Polka et al. 2014). In vitro experiments showed that DNA moves processively along the growing tips of AlfA filaments, while disassembly occurs at the other end, giving the appearance of a comet-like structure (Polka et al. 2014). Thus, the combined effect of filament stabilization by the AlfB-*parN* complex at one end, and the destabilizing effect of free AlfB at the other end results in treadmilling of the AlfA filament. This treadmilling is unidirectional, and pushes the DNA in the direction of growth of the filament. Bundling of two filaments in antiparallel orientations results in movement of DNA in opposite directions. Reference-free averaging of images of AlfA filament pairs suggests that the filaments do show a preference for bundling in an antiparallel orientation, similar to that of ParM filaments (Polka et al. 2014).

Further studies of the structural details of how AlfB interacts with AlfA will be needed to understand the mechanism of action of AlfB as a destabilizing agent on its own, and as a nucleator and a promoter of filament elongation in the presence of *parN* DNA.

Other Actin-Like Filaments Involved in Plasmid Segregation

Many actin-like proteins (Alps) have been identified in bacteria (Derman et al. 2012). Some of these are found in extra-chromosomal elements such as plasmids, phage and conjugative DNA elements. The functions of most of these proteins are unknown. However, representative members of some of the actin-like protein families that have been characterized demonstrate that they form filaments in vitro and in vivo and have been implicated in plasmid segregation (Derman et al. 2012; Rivera et al. 2011; Popp et al. 2012). Other than ParM and AlfA, for which mechanistic details of plasmid segregation are emerging, the rest of the actin-like cytomotive filaments have only been partially characterized. A summary of the information on these systems is given in the following sections (Fig. 10.6 and Table 10.1).

Alp12A

Alp12A forms part of the plasmid segregation system of *Clostridium tetani* pE88 plasmid (Popp et al. 2012). In addition to Alp12A, some of the other genes harbored in pE88 are the tetanus toxin and the transcriptional repressor TetR. The DNA sequence equivalent to *parC* has not been identified yet.

Structure of Alp12A

Alp12A forms filaments in the presence of ATP or GTP. Interestingly, electron microscopy of Alp12A filaments formed in presence of ATP shows that they comprise four actin-like protofilaments or strands. Out of the four protofilaments, two of the protofilaments are arranged antiparallel in orientation to the other two. The arrangements of pairs of protofilaments that are oriented parallel are not similar to the inter-protofilament arrangement observed in other known actin-like filaments (Fig. 10.6c). A prominent cleft is found in between the two antiparallel pairs of filaments, which was observed as a dark line in the negatively stained images. The handedness has been estimated to be right-handed. It is interesting to note that there is an antiparallel arrangement of protofilaments, which will probably contribute towards moving DNA in opposite directions during plasmid segregation (Popp et al. 2012).

Alp12A Filament Dynamics

The kinetics of polymerization for Alp12A have been studied using light scattering experiments and kinetic analysis of the data. Accordingly, Alp12A exhibits dynamic instability. It utilizes either ATP or GTP for polymerization. However, similar to other actin-like proteins discussed so far, ATP-dependent polymerization is more efficient. Fitting of experimental light scattering data suggests that the nucleus for polymerization involves an initial dimer formation, followed by tetramerization (Table 10.1). This model for nucleation is consistent with the four-stranded antiparallel filament model (Popp et al. 2012).

Alp7A

Alp7A is an actin-like protein from the *Bacillus subtilis* plasmid pLS20. The dynamics of Alp7A and the associated Alp7R and *alp7C* have been observed in vivo (Derman et al. 2012). GFP labeling of Alp7A expressed in cells showed that these filaments grow and shrink, thus exhibiting dynamic instability. Mutation of the aspartate most likely involved in ATP hydrolysis resulted in the presence of long and stable filaments within the cell, and disrupted the dynamics, emphasizing the role of ATP hydrolysis in the process. Similar to ParR, Alp7R acts as a repressor for Alp7A expression and binds to a repetitive sequence of DNA, *alp7C*, positioned at the promoter region of Alp7A.

pSK41

pSK41 is a plasmid that confers multiple drug resistance to *Staphylococcus aureus*. ParM and ParR proteins and *parC* iterons have been identified for the pSK41 plasmid segregation system. Crystal structures of the pSK41 ParM (Popp et al. 2010), and a DNA-bound segrosome complex of the ParR (Schumacher et al. 2007) have been determined.

Structure and Dynamics of pSK41 ParM

The structure of pSK41 ParM shows that it is indeed an actin-like protein (Fig. 10.2d). The protein in its apo-state superposes best with the nucleotide-bound conformation of R1 ParM. The filaments of pSK41 have been observed by negative stain EM, although currently no near-atomic structure or 3D-reconstruction of the filaments is available. A mutational analysis has identified the probable residues involved at the interface, and this study highlights that the orientation of monomers in the pSK41 ParM will be similar to that of the monomers in other actin-like ParM protofilaments (Popp et al. 2010).

Light scattering studies to understand the dynamics have shown that pSK41 ParM polymerizes in the presence of ATP or GTP as nucleotide, and filaments were not observed with ADP or GDP. Hence, they exhibit dynamic instability similar to R1 ParM (Popp et al. 2010).

Structure of pSK41 ParR

Crystal structure of pSK41 ParR shows that the protein adopts a ribbon-helix-helix fold, and a dimer of ParR interacts with a 20-mer tandem repeat of 10 bp of parC DNA. It forms a super helical structure with the DNA wound around the outer surface of the helix (Fig. 10.3b–d). The electrostatic surface potential on the inside surface of the super helical structure is complementary to that of the pSK41 ParM monomer (Schumacher et al. 2007).

pB171 ParM

ParM from *E. coli* virulence plasmid pB171 shares a comparatively high degree of conservation with ParM from R1 plasmid, Though 41% identity between them is much less than the degree of conservation between eukaryotic actins, the structure and dynamics of pB171 ParM seem to be quite similar to that of R1 ParM. Fig. 10.6 and Table 10.1 summarize relevant features of pB171 ParM in comparison to the other members of the actin-like filaments in plasmid segregation (Rivera et al. 2011).

General Features of Actin-Like Plasmid Machineries

The sets of actin-like plasmid segregation machinery characterized till date permit us to compare with the eukaryotic actin cytoskeleton and associated proteins and highlight some of the features of the bacterial cytoskeleton required for the function of DNA segregation. The major significance emerging from studying these systems lies in their simplicity, in how the cytoskeletal systems are empowered to carry out DNA segregation faithfully with minimal numbers of components. Of course, this is not accidental in the case of extra-chromosomal genetic elements such as plasmids, mobile genetic elements, and phages because of their extremely limited genome size.

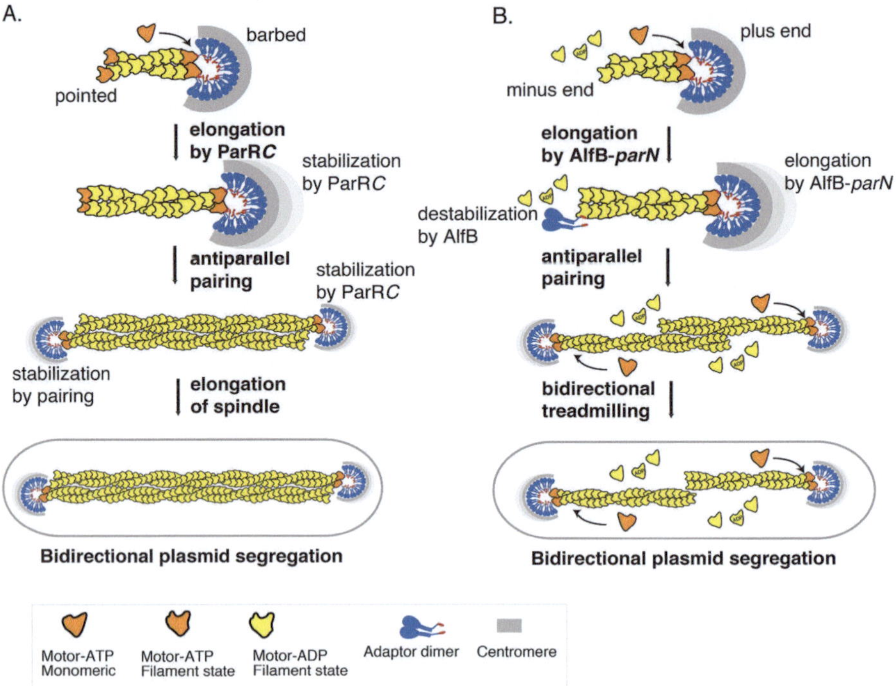

Fig. 10.7 Schematic representation of the mechanism of plasmid segregation in (**a**) R1 ParM, and (**b**), AlfA (Adapted from Gayathri et al. (2012) and Polka et al. (2014))

Possible Ways to Build a Bipolar Spindle

The actin-like cytomotive filaments functioning as DNA segregation machinery highlight that there are multiple ways in which a bipolar spindle can be constructed, using a basic building block or monomer of the same protein fold and the same protofilament architecture. Dynamics driven by nucleotide-hydrolysis play a major role in the formation of the bipolar spindle. The two major patterns of filament dynamics, as also observed in the eukaryotic cytoskeletal systems, are dynamic instability and treadmilling. How the plasmid segregation system manipulates either of these dynamics to achieve segregation is very interesting (Fig. 10.7).

A filament that exhibits dynamic instability requires both ends to be stabilized, else it will not survive. In the ParMR*C* system, the two ends of the filaments are rescued from dynamic instability by different means – (i) when ParR-bound DNA binds to one of the ends (ie, barbed end) (ii) when paired with another filament coming together in an antiparallel direction (at the pointed end). This arrangement ensures two requirements for DNA segregation – (i) capture of the DNA is achieved (ii) antiparallel pairing of two DNA-attached filaments facilitates coordinated movement of a pair towards the extremes of the cell, thus acting as a 'checkpoint' for DNA replication. The antiparallel orientation and elongation in the DNA-bound end ensures separation of the attached pair.

In treadmilling, elongation occurs at one end, while the other end shortens. If DNA is attached to the growing end, as in AlfA-driven plasmid segregation, the system functions via a pushing mechanism and pairs separate when two such filaments come together in an antiparallel orientation. However, the antiparallel arrangement does not contribute to increase the stability of the paired end. In the case of AlfA, the dynamics at the growing end (plus end) and the shortening end (minus end) are both modulated by AlfB. *parN*-bound AlfB at the growing end elongates the filament, while free AlfB destabilizes the minus end and ensures recycling of monomers for the growing end to elongate.

Both of the cases of actin-based plasmid segregation mechanisms mentioned above work by a pushing mechanism at filament plus ends. An interesting case of plasmid attaching to the minus end of a treadmilling filament, implicating a pulling mechanism, has been observed in TubZR*C*, a tubulin-like cytomotive plasmid segregation system (Fink and Löwe 2015) (discussed in the next chapter). Many of the actin-like filaments appear to prefer an antiparallel arrangement of protofilaments, which is a desired or an intuitive configuration, for a pushing mechanism for the separation of two plasmids. Characterization of the interaction with the adaptor protein and the segrosome complex for more actin-like plasmid segregation systems will throw light on the different combinations of possible modulations of the filament dynamics, required for building a plasmid segregation system.

Design of a Minimal Bipolar Spindle

One of the endearing features of plasmid segregation systems is the minimal number of components required to carry out their functions. This is possible by exploiting the principles of filament dynamics for dispensing with many of the components such as nucleators, and disassembly factors needed in eukaryotic cytoskeletal systems. The insertional polymerization mechanism ensures that DNA is always bound to one end only and not to the sides of the filament, thus accomplishing segregation. Additionally, the amounts of proteins involved in plasmid segregation, and thus filament dynamics, is maintained by strategies such as transcriptional repression by the adaptor protein.

In the ParMR*C* system, the number of ParM molecules available is regulated by the transcriptional repression by ParR. Hence, when ParR is bound to DNA and initiates the formation of a spindle, production of more ParM monomers is shut down. The existing ParM monomers get incorporated into the stabilized spindle and this further reduces the amount of free monomers present in the cell. Processivity depends on the amount of ATP-bound ParM. When the level of ATP-bound ParM goes low, the spindle disassembles automatically due to the deficiency of ATP-bound ParM. This is facilitated by the design of an adaptor protein that binds only to the ATP-bound ParM, and thus dispenses with the requirement of external disassembly factors.

Thus, the control of filament dynamics serves multiple functions: -(i) It facilitates the capture of DNA by nucleating filament formation, a role played by the segrosome complex. (ii) It dispenses with the requirement of additional factors for disassembly once the plasmids are moved apart within the cell. (iii) The requirement of two antiparallel filaments provides a measure for the presence of an even number of plasmids, as is the case after DNA replication.

Comparison with the Actin Cytoskeleton

The actin-like proteins in plasmid segregation and indeed all actin-like proteins of the bacterial cytoskeleton show a high degree of sequence divergence amongst themselves and with respect to the actin sequence. The sequence identities between these and actins are typically less than 20%. The sequence variation allows for the varying architectures of protofilaments/strands and the resulting filaments. Importantly, the structure of the intra-protofilament/strand interface is conserved throughout the family, including MreB (van den Ent et al. 2001), FtsA (Szwedziak et al. 2012), MamK (Ozyamak et al. 2013), ParM (Bharat et al. 2015), AlfA(Polka et al. 2009), Alp12A (Popp et al. 2012). Though a similar arrangement of the monomers is maintained within the protofilaments, the residues at the interface show variation, preventing the possibility of copolymerization among these filaments and also reflecting large evolutionary distances. The organization of the protofilaments gives rise to the different types of filament architectures. Each plasmid might have evolved and optimized its own plasmid partitioning system, since variability ensures that the machinery segregates only its own kind. The actin cytoskeleton in eukaryotes is highly conserved probably due to the actin pool required for a large number of functions within the cell, and the cell has to maintain a required pool of actin monomers for all these functions (Gunning et al. 2015). The number of protein interactions with actin is also very high, which results in a high degree of conservation. In contrast, plasmid segregation systems are autonomous, not interacting functionally with any cellular components and this means that the evolutionary pressure on the components involved is very low, as long as function is conserved.

An interesting feature that highlights a conserved feature of the actin fold in actin-like plasmid segregation systems is the region of interaction between ParM and ParR. ParR interacts with ParM through an amphipathic helix on the C-terminal end of ParR with a pocket in between the sub-domains 1A and 2A in ParM (Gayathri et al. 2012). The corresponding region in actin, between sub-domains 1 and 3, also called hydrophobic cleft, binds most of the modulators of actin polymerization such as nucleators (eg. formin, spire) and severing factors, for example cofilin and twinfilin, interact (Dominguez and Holmes 2011). For ParM, the same region in the fold has been exploited to regulate the filament dynamics in a bacterial actin. Determining the binding site on the actin fold will be required to confirm a similar mode of action for the other adaptor proteins such as AlfB, which acts both as a severing factor and a nucleator. The existence of nucleating factors and formin-like proteins in the bacterial cytoskeletal systems was for the first time established for plasmid segregation systems (Gayathri et al. 2012).

To conclude, the actin-like plasmid segregation machinery has evolved to perform the function of DNA segregation with a minimal number of components. Further mechanistic insights being gained from the large variety of actin-like plasmid segregation machineries will uncover the means by which a spindle can be built, and also provide invaluable information on the evolution of cytoskeletal systems from bacteria, archaea and eukaryotes. Recent work reporting an archaeal segregation system is a timely example for progress in these directions (Schumacher et al. 2015).

Acknowledgements The work in the lab is supported by INSPIRE Faculty Research Grant, Department of Science and Technology (DST), India, and Innovative Young Biotechnologist Award, Department of Biotechnology, India. PG and SH acknowledge INSPIRE (DST) and IISER-Pune fellowships.

References

Becker E, Herrera NC, Gunderson FQ et al (2006) DNA segregation by the bacterial actin AlfA during Bacillus subtilis growth and development. EMBO J 25:5919–5931. doi:10.1038/sj.emboj.7601443

Bharat TAM, Murshudov GN, Sachse C, Löwe J (2015) Structures of actin-like ParM filaments show architecture of plasmid-segregating spindles. Nature 523:106–110. doi:10.1038/nature14356

Bork P, Sander C, Valencia A (1992) An ATPase domain common to prokaryotic cell cycle proteins, sugar kinases, actin, and hsp70 heat shock proteins. Proc Natl Acad Sci U S A 89:7290–7294. doi:10.1073/pnas.89.16.7290

Campbell CS, Mullins RD (2007) In vivo visualization of type II plasmid segregation: bacterial actin filaments pushing plasmids. J Cell Biol 179:1059–1066. doi:10.1083/jcb.200708206

Dam M, Gerdes K (1994) Partitioning of plasmid R1 Ten direct repeats flanking the parA promoter constitute a centromere-like partition site parC, that expresses incompatibility. J Mol Biol 236:1289–1298. doi:10.1016/0022-2836(94)90058-2

Derman AI, Nonejuie P, Michel BC et al (2012) Alp7R regulates expression of the actin-like protein Alp7A in Bacillus subtilis. J Bacteriol 194:2715–2724. doi:10.1128/JB.06550-11

Dominguez R, Holmes KC (2011) Actin structure and function. Annu Rev Biophys 40:169–186. doi:10.1146/annurev-biophys-042910-155359

Fink G, Löwe J (2015) Reconstitution of a prokaryotic minus end-tracking system using TubRC centromeric complexes and tubulin-like protein TubZ filaments. Proc Natl Acad Sci U S A 2–7:E1845–E1870. doi:10.1073/pnas.1423746112

Fujii T, Iwane AH, Yanagida T, Namba K (2010) Direct visualization of secondary structures of F-actin by electron cryomicroscopy. Nature 467:724–728. doi:10.1038/nature09372

Galkin VE, Orlova A, Rivera C et al (2009) Structural polymorphism of the ParM filament and dynamic instability. Structure 17:1253–1264. doi:10.1016/j.str.2009.07.008

Garner EC (2007) Reconstitution of DNA Segregation. Science 315:1270–1274. doi:10.1126/science.1138527

Garner EC, Campbell CS, Mullins RD (2004) Dynamic instability in a DNA-segregating prokaryotic actin homolog. Science 306:1021–1025. doi:10.1126/science.1101313

Gayathri P, Fujii T, Møller-Jensen J et al (2012) A bipolar spindle of antiparallel ParM filaments drives bacterial plasmid segregation. Science 338:1334–1337. doi:10.1126/science.1229091

Gayathri P, Fujii T, Namba K, Löwe J (2013) Structure of the ParM filament at 8.5Å resolution. J Struct Biol 184:33–42. doi:10.1016/j.jsb.2013.02.010

Gerdes K, Molin S (1986) Partitioning of plasmid R1. J Mol Biol 190:269–279. doi:10.1016/0022-2836(86)90001-X

Gerdes K, Howard M, Szardenings F (2010) Pushing and pulling in prokaryotic DNA segregation. Cell 141:927–942. doi:10.1016/j.cell.2010.05.033

Gunning PW, Ghoshdastider U, Whitaker S et al (2015) The evolution of compositionally and functionally distinct actin filaments. J Cell Sci 128:2009–2019. doi:10.1242/jcs.165563

Jensen RB, Gerdes K (1997) Partitioning of plasmid R1. The ParM protein exhibits ATPase activity and interacts with the centromere-like ParR-parC complex. J Mol Biol 269:505–513. doi:10.1006/jmbi.1997.1061

Löwe J, Amos LA (2009) Evolution of cytomotive filaments: te cytoskeleton from prokaryotes to eukaryotes. Int J Biochem Cell Biol 41:323–329. doi:10.1016/j.biocel.2008.08.010

Møller-Jensen J, Jensen RB, Löwe J, Gerdes K (2002) Prokaryotic DNA segregation by an actin-like filament. EMBO J 21:3119–3127. doi:10.1093/emboj/cdf320

Møller-Jensen J, Borch J, Dam M et al (2003) Bacterial mitosis: ParM of plasmid R1 moves plasmid DNA by an actin-like insertional polymerization mechanism. Mol Cell 12:1477–1487. doi:10.1016/S1097-2765(03)00451-9

Møller-Jensen J, Ringgaard S, Mercogliano CP et al (2007) Structural analysis of the ParR/parC plasmid partition complex. EMBO J 26:4413–4422. doi:10.1038/sj.emboj.7601864

Orlova A, Garner EC, Galkin VE et al (2007) The structure of bacterial ParM filaments. Nat Struct Mol Biol 14:921–926

Ozyamak E, Kollman J, Agard DA, Komeili A (2013) The bacterial actin MamK: in vitro assembly behavior and filament architecture. J Biol Chem 288:4265–4277. doi:10.1074/jbc.M112.417030

Pei J, Tang M, Grishin NV (2008) PROMALS3D web server for accurate multiple protein sequence and structure alignments. Nucleic Acids Res 36:W30–W34. doi:10.1093/nar/gkn322

Petek NA, Mullins RD (2014) Bacterial actin-like proteins: purification and characterization of self-assembly properties. Methods Enzymol 540:19–34. doi:10.1016/B978-0-12-397924-7.00002-9

Polka JK, Kollman JM, Agard DA, Mullins RD (2009) The structure and assembly dynamics of plasmid actin AlfA imply a novel mechanism of DNA segregation. J Bacteriol 191:6219–6230. doi:10.1128/JB.00676-09

Polka JK, Kollman JM, Mullins RD (2014) Accessory factors promote AlfA-dependent plasmid segregation by regulating filament nucleation, disassembly, and bundling. Proc Natl Acad Sci 111:2176–2181. doi:10.1073/pnas.1304127111

Pollard TD (1986) Rate constants for the reactions of ATP-and ADP-actin with the ends of actin filaments. J Cell Biol 103:2747. doi:10.1083/jcb.103.6.2747

Popp D, Narita A, Oda T et al (2008) Molecular structure of the ParM polymer and the mechanism leading to its nucleotide-driven dynamic instability. EMBO J 27:570–579. doi:10.1038/sj.emboj.7601978

Popp D, Xu W, Narita A et al (2010) Structure and filament dynamics of the pSK41 actin-like ParM protein: implications for plasmid dna segregation. J Biol Chem 285:10130–10140. doi:10.1074/jbc.M109.071613

Popp D, Narita A, Lee LJ et al (2012) Novel actin-like filament structure from Clostridium tetani. J Biol Chem 287:21121–21129. doi:10.1074/jbc.M112.341016

Rivera CR, Kollman JM, Polka JK et al (2011) Architecture and assembly of a divergent member of the ParM family of bacterial actin-like proteins. J Biol Chem 286:14282–14290. doi:10.1074/jbc.M110.203828

Salje J, Löwe J (2008) Bacterial actin: architecture of the ParMRC plasmid DNA partitioning complex. EMBO J 27:2230–2238. doi:10.1038/emboj.2008.152

Salje J, Zuber B, Löwe J (2009) Electron cryomicroscopy of *E. coli* reveals filament bundles involved in plasmid DNA segregation. Science 323:509–512. doi:10.1126/science.1164346

Salje J, Gayathri P, Löwe J (2010) The ParMRC system: molecular mechanisms of plasmid segregation by actin-like filaments. Nat Rev Microbiol 8:683–692. doi:10.1038/nrmicro2425

Schumacher MA, Glover TC, Brzoska AJ et al (2007) Segrosome structure revealed by a complex of ParR with centromere DNA. Nature 450:1268–1271. doi:10.1038/nature06392

Schumacher MA, Tonthat NK, Lee J et al (2015) Structures of archaeal DNA segregation machinery reveal bacterial and eukaryotic linkages. Science 349:1120–1124. doi:10.1126/science.aaa9046

Szwedziak P, Wang Q, Freund SM, Löwe J (2012) FtsA forms actin-like protofilaments. EMBO J 31:2249–2260. doi:10.1038/emboj.2012.76

Tanaka T (2010) Functional analysis of the stability determinant AlfB of pBET131, a miniplasmid derivative of Bacillus subtilis (natto) plasmid pLS32. J Bacteriol 192:1221–1230. doi:10.1128/JB.01312-09

van den Ent F, Amos LA, Löwe J (2001) Prokaryotic origin of the actin cytoskeleton. Nature 413:39–44. doi:10.1038/35092500

van den Ent F, Møller-Jensen J, Amos LA et al (2002) F-actin-like filaments formed by plasmid segregation protein ParM. EMBO J 21:6935–6943. doi:10.1093/emboj/cdf672

Chapter 11
Tubulin-Like Proteins in Prokaryotic DNA Positioning

Gero Fink and Christopher H.S. Aylett

Abstract A family of tubulin-related proteins (TubZs) has been identified in prokaryotes as being important for the inheritance of virulence plasmids of several pathogenic Bacilli and also being implicated in the lysogenic life cycle of several bacteriophages. Cell biological studies and reconstitution experiments revealed that TubZs function as prokaryotic cytomotive filaments, providing one-dimensional motive forces. Plasmid-borne TubZ filaments most likely transport plasmid centromeric complexes by depolymerisation, pulling on the plasmid DNA, *in vitro*. In contrast, phage-borne TubZ (PhuZ) pushes bacteriophage particles (virions) to mid cell by filament growth. Structural studies by both crystallography and electron cryo-microscopy of multiple proteins, both from the plasmid partitioning sub-group and the bacteriophage virion centring group of TubZ homologues, allow a detailed consideration of the structural phylogeny of the group as a whole, while complete structures of both crystallographic protofilaments at high resolution and fully polymerised filaments at intermediate resolution by cryo-EM have revealed details of the polymerisation behaviour of both TubZ sub-groups.

Keywords TubZ • Plasmid segregation Type III • Tubulin superfamily • Filament motility • Dynamic instability • Treadmilling • tubZRC operon • TubR • PALM • Bacteriophage TubZ homologue • RepX • PhuZ • *Bacillus thuringiensis* • *Clostridium botulinum* bacteriophage c-st • Bacteriophage virion centring

G. Fink (✉)
Division of Structural Studies, Medical Research Council Laboratory of Molecular Biology, CB2 0QH Cambridge, UK
e-mail: gfink@mrc-lmb.cam.ac.uk

C.H.S. Aylett
Department of Biology, Institute for Molecular Biology and Biophysics, Eidgenössische Technische Hochschule (ETH) Zürich, Zürich, Switzerland
e-mail: christopher.aylett@mol.biol.ethz.ch

J. Löwe, L.A. Amos (eds.), *Prokaryotic Cytoskeletons*, Subcellular Biochemistry 84, DOI 10.1007/978-3-319-53047-5_11

Introduction

Protein filaments belonging to the recently-discovered bacterial cytoskeleton are building blocks of molecular machines that segregate bacterial DNA, in chromosomes, plasmids and even some viruses, to daughter cells. These transport mechanisms, which are crucial for viability and fitness of cells as they ensure that daughter cells inherit genetic material, probably arose long ago. Plasmids present at high-copy numbers are normally segregated in a random manner during cell division. In contrast, low-copy number plasmids carry dedicated segregation or partitioning systems for plasmid maintenance, encoded by loci that are sometimes called par (for partition). These systems consist of a cytomotive, filament-forming nucleotide triphosphatase (NTPase), an adaptor protein and a centromeric region on the plasmid DNA. Based on the nature of the NTPase, plasmid-borne segregation systems may be categorized: type I systems utilize Walker A-like adenosine triphosphatases (ATPase) of the ParA type; type II systems employ actin-like filaments that actively push plasmids to the cell poles prior to cell division; type III systems are a new family of tubulin-related guanosine triphosphatases (GTPases) named TubZ that has been identified on numerous plasmids carrying virulence factors of certain pathogens. TubZ proteins have been discovered on large virulence plasmids of Bacilli, in particular of the *B. cereus* group (Okinaka et al. 1999; Berry et al. 2002), and are also found in bacteriophages from *Clostridium* and *Pseudomonas* species (Sakaguchi et al. 2005; Kraemer et al. 2012). In contrast to chromosome encoded tubulin homologues of the FtsZ family, required during bacterial cell division, plasmid- and phage-borne tubulin-like proteins have different functions.

Among the first TubZ family members discovered was a factor required for replication of plasmid pXO1 of *B. anthracis* and therefore the protein was named RepX (Tinsley and Khan 2006) and TubZ was first used for the protein from Bacillus thuringiensis plasmid pBtoxis (Tang et al. 2006). Later TubZ proteins were shown to form intracellular filaments that are required for the segregation of plasmids (Larsen et al. 2007). However, RepX possesses properties very similar to those of the other TubZs, since it forms polymers (in a guanosine triphosphate dependent manner – Anand et al. 2008; Akhtar et al. 2009), which are also implicated in plasmid replication, (Tang et al. 2007; Ge et al. 2014a). This plasmid-borne tubulin related protein is therefore also referred to as TubZ.

Type III plasmid segregation systems can be subdivided into two orthologous groups, of plasmid- or phage-borne TubZs. Bacteriophage encoded tubulin-like TubZ proteins named PhuZ are implicated to foster the lytic cycle of the bacteriophage by concentrating phage DNA and virions to the middle of cells (Kraemer et al. 2012). The following sections give summaries of TubZ biology and Type III segregation systems. First we describe structural studies and in vitro reconstitution experiments that have led to the current understanding of how plasmid-borne and phage-encoded TubZs form filaments and are regulated. Later the translocation mechanisms of plasmids or phage DNA by filaments are described.

Phylogeny of the TubZ Group of Tubulins

Genes encoding TubZs, and operons encoding TubZ systems, have been discovered in a wide variety of organisms and contexts, many of which are separated by a considerable phylogenetic distance (Fig. 11.1). However, on consideration of the disposition of the first few proteins identified, an extremely clear pattern emerged. TubZs are predominantly encoded upon plasmids or on genomes of bacteriophages within eubacteria. Given the sequence diversity of TubZ-like proteins, assignment has therefore frequently been based upon the presence of a tubulin-like gene within a plasmid or bacteriophage. Successful identification of TubZs has in almost all cases been corroborated either by the identification of other proteins of the known TubZ systems, including TubR, TubX (winged helix DNA-binding regulators) or TubY (helix-turn-helix regulator), or by structural evidence supporting the formation of TubZ-type helical filaments.

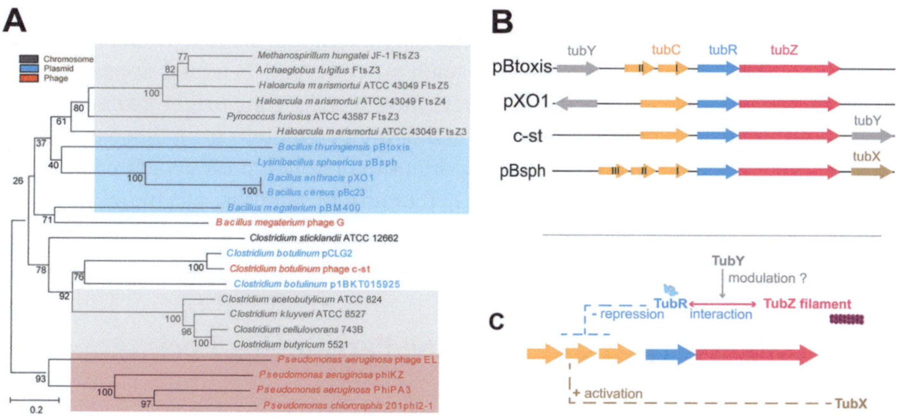

Fig. 11.1 Distribution and genomic organisation of TubZ systems within plasmids and bacteriophages (**a**) A phylogenetic tree of TubZ systems (from Oliva et al. 2012). Plasmid encoded systems are shown in *blue*, phage-borne ones in *red* and related chromosomal genes in *grey*. (**b**) Schematics of the TubZ encoding partitioning loci. *Arrows* indicate the coding region for GTPases TubZ (*magenta*) and centromere-binding proteins TubR (blue); *tubC* centromere-like blocks of iterons are represented as *short arrows* (*orange*). Note TubZ and *tubC* have synonyms in other systems but for simplicity are here named after the pBtoxis nomenclature. (**c**) Diagram of TubZ operon regulation combining the available information for the different operons from pBtoxis, bacteriophage c-st and pBsph. DNA-binding properties of TubR and TubX are indicated by *dashed blue* and *brown lines* respectively. Regulation of promoter activity is indicated by (+) enhancing or (–) repressing. Formation of a partitioning complex of TubR, *tubC* and TubZ is shown by *magenta double arrow line*. Modulation of the centromeric complex or filament by TubY represented by *grey arrow*

Origins and Dispersion

The wide disposition of the TubZ family of proteins and their presence overwhelmingly upon plasmids or bacteriophages (Fig. 11.1a) implies a mechanism for their dispersion, but also suggests that the question of determining their origin may well remain unanswered in perpetuity. The lateral separation of TubZ genes, and the lack of their presence in species phylogenetically intermediate, imply lateral transfer of the original TubZ-like tubulin into many separate eubacterial species. Their presence also in bacteriophage genomes suggests that a pseudolysogenic bacteriophage (i.e. one that is retained as a plasmid within the host bacterium) might have provided the transfer vector. Plasmids bearing TubZ genes or operons may then be interpreted as representing the, now symbiotic, remains of long-term pseudolysogens, whereas the chromosomal copies present within *Clostridium* species might constitute the equivalent residue of a truly lysogenic bacteriophage.

Given such a spread via lateral gene transfer, it is difficult to imagine a line of investigation that might resolve the question of the origin of the TubZ family within a particular trunk of the tree of life. We can only attempt to exclude branches with a low probability of having given rise to such an emigrant. It would appear highly unlikely, for instance, that the TubZs might have a eukaryotic origin given the substantial differences in both their protein and filament structure from those exhibited by eukaryotic tubulins, however it is important to note that proteins not overly dissimilar in sequence and structure, the CetZ family of tubulins, are found in abundance in many archaeal genomes.

Organization of the TubZ Encoding Partitioning Loci

Virulence plasmids of *Bacillus thuringiensis* pBtoxis, *Bacillus anthracis* pXO1, *Bacillus sphaericus* pBsph and *Bacillus cereus* pBc23 carry operons encoding proteins of the TubZ family (Fig. 11.1). The best investigated examples of the so-called *tubZRC* operons are from pBtoxis (Tang et al. 2006, 2007; Larsen et al. 2007) and pBsph (Ge et al. 2014a) plasmids. Bacteriophage c-st from *Clostridium botulinum* (Sakaguchi et al. 2005), also existing as a circular bacteriophage carries another well-studied TubZRC operon (Oliva et al. 2012).

The *tubZ* gene encodes for the tubulin-like GTPase TubZ, furthermore the *tubZRC* operon contains the DNA binding protein TubR, and *tubC* DNA repeats, which are located directly upstream of the *tubR* and *tubZ* genes (Fig. 11.1b). Both coding sequences and the cis-acting DNA sequence *tubC* are required for plasmid maintenance. This has been directly demonstrated by deletion analysis and construction of minireplicons of plasmids pBtoxis and pBsph. The requirement of the bacteriophage encoded TubZRC for segregating circular phages during host cell division remains to be shown.

tubC contains multiple degenerate DNA repeats of 12 base pair length (iterons), which are organized into blocks, it is required for replication initiation and functions as a centromere. For pBtoxis the seven centromeric iterons are spilt into two blocks of three and four, separated by 45 nucleotides (Aylett and Lowe 2012) whereas *tubC* of pBsph was found to consist of 11 iterons organized into three blocks (Ge et al. 2014a).

Regulation of the tubZRC *Operon*

Plasmid pBsph contains an imperfect and inverted iteron in the first block of the *tubC* locus. This iteron is suggested to function as operator of the pBsph *tubZRC* operon since parts of the operon promoter are also located within *tubC*. Interestingly, TubR generally binds cooperatively to *tubC* and acts as a transcriptional repressor for the *tubZRC* operons. Thus, fine-tuning of cellular TubZ and TubR levels is important for plasmid maintenance. Deregulation through overexpression of TubR or TubZ in trans reduced the frequency of daughter cells containing plasmids (Larsen et al. 2007).

In addition to the operon, two new genes located in proximity to the *tubZ* and *tubR* genes have recently been identified. They regulate the TubZ system either transcriptionally or directly on the protein level.

Regulator protein TubX has been found as transcriptional activator in pBsph (Ge et al. 2014b), also binding to five 8 bp operator repeats and displacing TubR from the third block of *tubC* iterons. TubX stimulates transcription of *tubZ* and *tubR*, leading to increased expression and thus stability of the minireplicons without affecting their copy number. Whether displacement of TubR by TubX leads to dissociation of a functional centromeric complex is unknown. TubX has so far only been described for pBsph. It remains to be seen whether functional homologues are present in other TubZ systems on plasmids or bacteriophages.

Another non-operon regulatory protein is TubY, first found in bacteriophage c-st (Oliva et al. 2012). It contains a helix-turn-helix motif and a carboxy-terminal domain similar to the MerR family of transcriptional regulators, plus a short stretch of positively charged hydrophobic residues suggested to act as amphipathic helix. Biochemical and electron microscopy experiments found that TubY modulates filaments directly rather than influencing expression. Interestingly, homologues of TubY proteins have been identified up and downstream of TubZRC operons in other species (Oliva et al. 2012). Whether this TubY plays a role in TubZ systems to partition plasmids or circular bacteriophages is unknown.

Further investigation is required to elucidate the specific roles of TubY and TubX proteins in modulating segregation and maintenance of other TubZ bearing plasmids.

Structural Insights into TubZ Biology

Structural studies by both crystallography and electron cryo-microscopy of multiple proteins, both from the plasmid partitioning sub-group and the bacteriophage virion centring sub-group of TubZ homologues (also referred to as PhuZs) allow the detailed consideration of the structural phylogeny of the group as a whole, while complete structures of both crystallographic protofilaments at high resolution and fully polymerised filaments at intermediate resolution by cryo-EM have revealed details of the polymerisation behaviour of both TubZ sub-groups.

Due to their interesting evolutionary history and key roles in the function of important virulence systems in known pathogens, the TubZ group of tubulin/FtsZ homologues have been subject to considerable structural investigation; fifteen structures are now available in the protein data bank, spanning five different proteins from two plasmids and three bacteriophages, allowing a detailed consideration of the structural phylogeny of the TubZ group without the great risk of errors entailed by extrapolation from only a few examples of an orthologous protein. Furthermore, the application of judicious mutations and biochemistry has allowed the majority of the important states of the most well-studied of these proteins to be captured, from apo-monomeric forms lacking the cognate guanosine nucleotide cofactor, through to complete filaments visualised at intermediate resolution by electron cryo-microscopy and interpreted through the modification of higher resolution protofilament crystal structures.

Two broad classes of TubZ genes, canonical TubZ orthologues involved in plasmid maintenance and a group of more distantly related TubZ homologues involved in bacteriophage virion centring (PhuZs), have been proposed based upon phylogeny, structural studies and biological activity. Given the many close functional parallels and structural similarities between these proteins, it remains unclear whether they represent different sub-groups within the TubZ homologues, or whether the studies carried out to date have simply examined in detail two relatively distant points within a spectrum of genes that bridge a broad swathe of roles and structural configurations. We therefore treat the two groups as sub-classes, providing a separate prototypical structure for each; *Bacillus thuringiensis subspecies israelensis* TubZ, from pBtoxis, is cast in this role for the plasmid partitioning TubZs, having been the first discovered, first resolved and most extensively investigated of the TubZ group, whereas *Pseudomonas chlororaphis* bacteriophage 210Φ2-1 TubZ (also denoted PhuZ) provides its counterpart within the bacteriophage virion centring TubZ homologues. Structures representing crystallographic protofilaments and cryo-EM reconstructions of full filaments are available for both groups of proteins, the crystal structures to high resolution, allowing clear interpretation and the production of an atomic structure, and the cryo-EM reconstructions to intermediate resolution, allowing the prior crystallographic protofilaments to be refitted unambiguously (Table 11.1).

While TubZ homologues share the absolutely conserved amino-terminal GTPase and carboxy-terminal activation domains that are the hallmark of tubulin/FtsZ-like

Table 11.1 Resolved structures of TubZ homologues

Name	Organism	PDB ID	Resolution	Notes	Study
Plasmid partitioning					
TubZ	*Bacillus thuringiensis*	3M8K	2.3 Å	Apo-monomer	Ni et al. (2010)
TubZ	*Bacillus thuringiensis*	3M89	2.0 Å	GTPγS monomer	Ni et al. (2010)
TubZ	*Bacillus thuringiensis*	2XKA	3.0 Å	GTPγS protofilament	Aylett et al. (2010)
TubZ	*Bacillus thuringiensis*	2XKB	3.0 Å	GDP protofilaments	Aylett et al. (2010)
TubZ	*Bacillus thuringiensis*	3J4S	6.8 Å	4-stranded filaments	Montabana and Agard (2014)
TubZ	*Bacillus thuringiensis*	3J4T	10.8 Å	2-stranded filaments	Montabana and Agard (2014)
RepX/TubZ	*Bacillus cereus*	4EI7	1.9 Å	GDP dimer	Hoshino and Hayashi (2012)
RepX/TubZ	*Bacillus cereus*	4EI8	2.1 Å	Apo-monomer	Hoshino and Hayashi (2012)
RepX/TubZ	*Bacillus cereus*	4EI9	3.3 Å	GTP dimer	Hoshino and Hayashi (2012)
TubZ	Bacteriophage C-ST	3V3T	2.3 Å	Apo monomer	Oliva et al. (2012)
Bacteriophage virion centring					
PhuZ	Bacteriophage 210Φ2-1	3RB8	2.6 Å	GDP protofilament	Kraemer et al. (2012)
PhuZ	Bacteriophage 210Φ2-1	3R4V	1.7 Å	GDP protofilament	Kraemer et al. (2012)
PhuZ	Bacteriophage 210Φ2-1	3J5V	7.1 Å	3-stranded filaments	Zehr et al. (2014)
TubZ	Bacteriophage ΦKZ	3ZBP	2.0 Å	GDP monomer	Aylett et al. (2013)
TubZ	Bacteriophage ΦKZ	3ZBQ	1.7 Å	GDP protofilament	Aylett et al. (2013)

proteins, and which, when combined, provide the active site for GTP hydrolysis allowing eventual filament destabilisation, they are generally characterised by much greater structural heterogeneity than is typical in their more staid, chromosomal counterparts. Given such variability, and the phylogenetic distance between members of the family mentioned earlier in the chapter, it might be possible to suppose that the group were simply a catchall for any non-FtsZ bacterial tubulin. Several absolutely conserved features that have so far defined TubZs of both bacteriophage and plasmid origin belie this interpretation however; all have so far shared an elongated helical carboxy-terminal extension essential for efficient polymerisation, which is intimately involved in the formation of the subunit-subunit interface, and polymerise to form substantially twisted protofilaments that therefore lead to helical filaments on their incorporation into a lattice. Other, subtler, commonalities will become apparent as we review the available structures below.

Plasmid Partitioning TubZ Orthologues

Structural efforts began with, and have focused upon, the prototypical TubZ from *Bacillus thuringiensis* plasmid pBtoxis. Thus the plasmid partitioning TubZs are the much better described of the two groups, ten structures having been described that encompass all the available nucleotide and polymerisation states of these proteins. However, the structures of *Bacillus cereus* pXO1 TubZ and bacteriophage c-st TubZ from *Clostridium botulinum* are also known (Fig. 11.2).

Bacillus thuringiensis TubZ

The first crystallographic study of the prototypical TubZ group member (Ni et al. 2010) resolved two structures in a monomeric state without subunit-subunit interactions within the lattice, however two crystal structures of protofilaments of the same protein, formed with the poorly hydrolysable GTP analogue GTPγS and with GDP respectively, swiftly followed (Aylett et al. 2010). *Bacillus thuringiensis* TubZ was carboxy-terminally truncated in both studies in order to remove the aforementioned characteristic "tail", which affected crystallisation.

Bacillus thuringiensis TubZ exhibits both amino- and carboxy-terminal extensions to the core fold; both are helical and have been denoted H0 (to retain the established secondary structure nomenclature (Lowe et al. 2001)) and H11 respectively. The α-helical carboxy-terminus projects from the activation domain, making contact with the GTPase domain and then bracketing the side of the subunit-subunit interface. This "tail" has significant freedom of movement, having occupied a significantly different conformation in every crystal structure resolved to date. Given that the TubZ "tail" has been implicated in binding to multimeric complexes formed by its cognate binding partner, the TubR*C* centromeric complex, this freedom of movement has been proposed to represent a means by which the radius of recruitment and avidity of binding of a larger complex might be increased.

The role of α-helix H0 at the amino-terminus, is also extremely interesting; it lies within the hydrophobic core of the protein, intercalated between the GTPase and activation domains. These two domains are rotated outwards around this point to incorporate its intrusion, effecting a shift in their relative orientations. Loop T7, which provides the key water-activating carboxylic acid residues in all tubulin/FtsZ-like proteins, bridges the core helix H7 and the subunit-interface surface of the

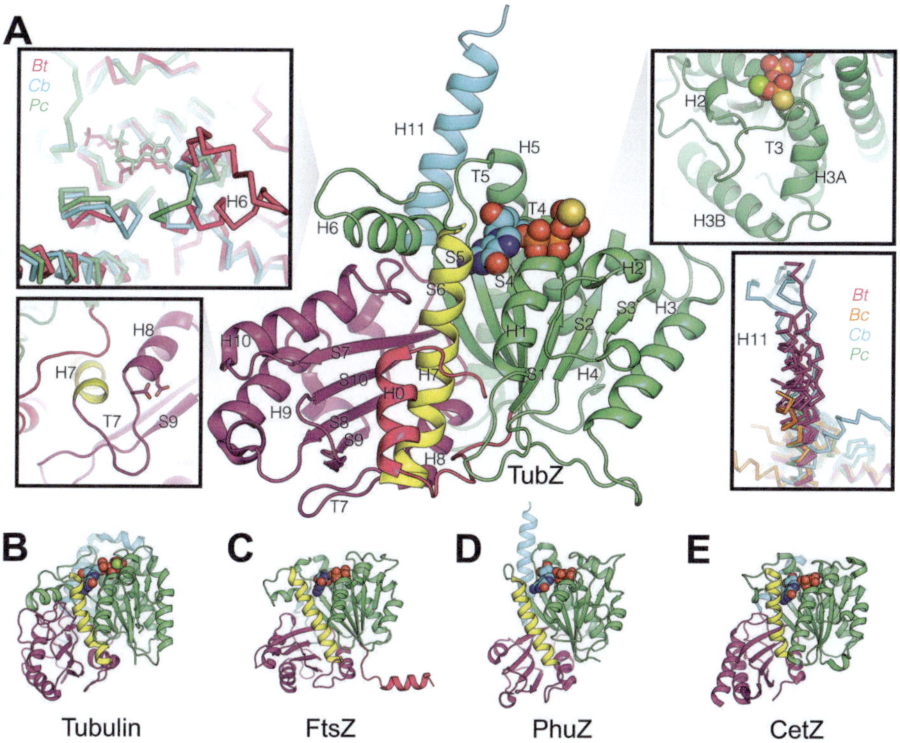

Fig. 11.2 Crystal structures of TubZ proteins (**a**) The crystal structure of the prototypical TubZ (*Bacillus thuringiensis* TubZ; PDB ID 3M89) is shown with the secondary structural elements indicated according to the tubulin/FtsZ nomenclature (Lowe et al. 2001). The inset panels indicate; variation in α-helix H6 between homologues, activation loop T7, helix H2 and loop T3, and changes in the position of helix H11 from a viewpoint rotated 180° respectively. The prototypical structures of other proteins are shown in cartoon representation in panels (**b**) *Bos taurus* α-tubulin, (**c**) *Methanocaldococcus jannaschii* FtsZ, (**d**) *Pseudomonas chlororaphis* bacteriophage 210Φ2-1 TubZ (**e**) *Haloferax volcanii* CetZ1. The structures are shown from the same viewpoint, with GTPase domains coloured green, activation domains coloured purple, the principal core helix H7 linking these two elements coloured *yellow* and the remaining amino- and carboxy-terminal extensions to the canonical fold coloured in *reds* and *blues* respectively

activation domain, and is concomitantly extended to account for the increased distance: both changes have significant affects on polymerisation as discussed in detail in section "Structure and architecture of TubZ filaments".

Bacillus cereus TubZ/RepX

A carboxy-terminal truncation of TubZ from *Bacillus cereus* (residues 1–389), was resolved crystallographically in apo-monomeric, and in both nucleotide states as a pseudo-protofilamentous dimer, in a study by Hoshino and Hayashi (Hoshino and

Hayashi 2012). Despite very weak sequence conservation (23%) between this TubZ and that from *Bacillus thuringiensis*, the r.m.s.d between Cα atoms on superimposition of the two structures is only 2.3 Å. Both H0 and H11 are also conserved in composition and position between the two structures; however there are several smaller differences in the conformations of H9 and H10, and disorder in loops T2 and T3, unaffected by the presence of soaked GTPγS, although it is most plausible that this is due to the inability of these regions to undergo conformational changes within the crystal lattice.

An area of considerable divergence between the two structures, on the other hand, was in nucleotide binding. Whereas the guanine base was sandwiched between aliphatic lysine residue side chains (K33 and K237) in *B. thuringiensis,* in *B cereus* TubZ this pocket is missing, resulting in a much more accessible and exchangeable nucleotide. One notable observation upon comparison of the apo- and GDP bound states of the molecule is the reorganisation of H5 on nucleotide binding. This has a large effect on H11, resulting in it moving it up and away from the inter-subunit interface. Given the previous observations of the weakened guanosine nucleotide-binding pocket, this movement would prevent the transient incorporation of apo-monomer into growing protofilaments of this particular TubZ; however, the concentrations of GTP and GDP in the cell make it unlikely that this mechanism would be of great importance as the active site would typically remain occupied.

Clostridium botulinum Bacteriophage c-st TubZ

TubZ from bacteriophage c-st represents the only structure to date of a plasmid partitioning TubZ homologue found within a bacteriophage genome; in this case it is involved in partitioning the lysogenic phage, which persists as a circular plasmid within its host cell. The structure of the full-length, 358 residue, protein bearing a single point mutation (T100A) within characteristic tubulin loop T4 of the GTPase domain, has been resolved (Oliva et al. 2012). Given that T4 is involved in nucleotide phosphate recognition, and that analogous mutations within other tubulin/FtsZ-like proteins inhibit nucleotide binding, the lack of a guanine nucleotide within the crystal lattice is not an unexpected consequence. Notably, there is no evidence of the unravelling of H5 as visualised in *Bacillus cereus* TubZ, however flexibility in T2, H2 and T3, which bind the nucleotide phosphate moieties, persists.

Comparison of bacteriophage c-st TubZ with that from *Bacillus thuringiensis* reveals, however, that it is the most divergent of the three orthologues to be resolved. Although the superimposition of the GTPase and activation domains reveals that they remain very well conserved (2.2 Å Cα r.m.s.d) despite sequence divergence (16% identity), whereas the presence and position of H0 is conserved between both the *Bacillus cereus* and *Bacillus thuringiensis* TubZs, it is entirely absent in *Clostridium botulinum* bacteriophage c-st TubZ. The relative rotation of the GTPase and activation domains persists between the structures however, leaving a shallow groove in the surface of the protein, and implying that the role of H0 is replaced by bulkier side-chains and repacking of the inter-domain interface. Another very important difference in this example of a TubZ is the truncation of H6, which lies at the subunit-subunit interface, to a single helical turn, a modification that might be expected to substantially decrease the available platform for the adjacent subunit. This notable modification to the core-fold persists into the bacteriophage virion-centring class of TubZs.

Bacteriophage Virion Centring TubZ Orthologues

The bacteriophage virion centring TubZ homologues remain comparatively poorly described in comparison to the plasmid partitioning TubZ homologues, although significant structural effort has also been expended despite less focus on their biology to date. Five structures have so far been described, including high-resolution crystal structures of a monomeric state, two different protofilaments from different organisms, and a cryo-EM reconstruction of the complete filament of the prototypical member of the family at intermediate resolution.

There exists some ambiguity in the nomenclature surrounding this subclass of proteins; the initial description of the first virion centring TubZ from *Pseudomonas chlororaphis* bacteriophage 210Φ2-1 denoted the protein using a new nomenclature "PhuZ", which was supported by several structural differences from the plasmid partitioning TubZs and the very different biological context within which these proteins act, being involved in the movement of assembled bacteriophage virions within the cell. However, the opinion of other groups within the field that there was insufficient evidence to support a splitting of the TubZs prevented this nomenclature being adopted for *Pseudomonas aeruginosa* bacteriophage ΦKZ, despite initial attempts to conform in order to prevent confusion. Within this chapter we resolve this ambiguity by referring to the proteins forming this class as a sub-group of the TubZ homologues, but retaining them within the wider TubZ group for further consideration. This interpretation is supported structurally by the presence of conserved structural elements in both TubZ sub-groups (Fig. 11.2).

Pseudomonas chlororaphis Bacteriophage 210Φ2-1 TubZ/PhuZ

The first structure of the prototypical TubZ from *Pseudomonas chlororaphis* bacteriophage 210Φ2-1 (PhuZ) was initially resolved using crystallography by Kraemer and colleagues in 2012 (Kraemer et al. 2012). The full-length protein crystallised in a GDP-occupied state in which crystallographic symmetry generated a weakly linked protofilament. The structure exhibits a number of parallels with TubZ from *Clostridium botulinum* bacteriophage c-st in particular; H0 is also entirely absent, whereas H11 is retained in both cases and occupies a very similar orientation to that in all TubZ proteins. Whereas a single turn of conserved secondary structural element α-helix H6 had been retained in the bacteriophage c-st TubZ homologue, it is completely absent in bacteriophage 210Φ2-1 TubZ. A further compaction of the core fold occurs through the reduction of H10 to a single helical turn. The most notable feature of the structure is the resolution of the extended C-terminus, ordered against the adjacent subunit (see section "Structure and architecture of TubZ filaments"); a feature that remains as of yet unresolved to high resolution in any of the plasmid partitioning TubZ proteins, probably due to the much longer H11 extensions within these proteins, and the fact that they have typically been truncated in the higher resolution crystal structures to facilitate lattice formation.

Pseudomonas aeruginosa Bacteriophage ΦKZ TubZ

The closely related TubZ from *Pseudomonas aeruginosa* bacteriophage ΦKZ has also been resolved, through the application of crystallography, in two states, one monomeric and the other polymeric, both containing GDP, by Aylett and colleagues (Aylett et al. 2013). The conformations of the protein and the features observed are very similar between the bacteriophage 210Φ2-1 and bacteriophage ΦKZ TubZ orthologues, encouraging direct extrapolation of the features of one homologue onto the other, an approach that has been of considerable use in advancing our understanding of the polymerisation-depolymerisation cycle.

Structural Comparisons to Other Groups of Tubulin/FtsZ Homologues

Given the extreme divergence within the TubZ group of tubulin/FtsZ homologues, it has proven difficult to place them evolutionarily further or nearer towards any particular group. Certain features are held in common between TubZs and both FtsZs and canonical tubulins as well as the newly discovered archaeal CetZs. A comparison to prototypical FtsZ and tubulin structures (we use the crystal structure of the FtsZ protofilament from *Staphylococcus aureus* (Tan et al. 2012) (PDB ID 4DXD) as an exemplar for the eubacterial FtsZs, *Bos taurus* tubulin (Lowe et al. 2001) (PDB ID 1JFF) as the prototype for the eukaryotic tubulins and *Haloferax volcanii* CetZ1 as the prototypical archaeal CetZ (Duggin et al. 2015)) reveals a series of intriguing parallels with each, but also divergence from any, of these groups. The CetZs, in particular, exhibit many of the structural features of TubZs, however too few structures of this novel group are available to definitively establish commonalities (Fig. 11.2).

The FtsZ group of tubulin/FtsZ homologues are involved in cell division in the eubacteria and many archaea. They typically exhibit the canonical amino-terminal GTPase and carboxy-terminal activation domain structure, without any significant modifications to the secondary structure or conformation of these two domains. The two-domain fold commonly possesses both amino- and carboxy-terminal extensions; helix H0 is an amino-terminal extension found in many FtsZ structures, which is mobile relative to the core fold, whereas the typical carboxy-terminal extension of FtsZ is a β-hairpin, comprising strands S11 and S12, beyond which the protein is typically extended coil without any known secondary structural elements. FtsZs assemble into straight, essentially untwisted, protofilaments, which appear to condense into double- or multi-stranded filaments to form the eubacterial "Z-ring" (Chaps. 5 and 7; Szwedziak et al. 2014, Tan et al. 2012).

Parallels with the Cell Division FtsZs

Although they form a distinct group, TubZs typically exhibit a higher degree of structural homology to FtsZs than to the eukaryotic tubulins; the r.m.s.d on Cα superimposition of *Bacillus thuringiensis* TubZ and FtsZ is 3.7 Å, whereas the same figure for α-tubulin is 5.5 Å. While the amino-terminal extension, H0, is not universally conserved between TubZs, this in fact exactly parallels the situation in FtsZs, although the role of the extension appears very different between the groups. A further structural commonality with the FtsZs is the absence in TubZs of any large surface extensions, such as the tubulin M and N-loops, involved in lateral interactions between protofilaments. This parallel relates to the greater simplicity of the filaments formed.

While the TubZ GTPase active site is typically divergent from that in both FtsZs and tubulins, the key residues involved in γ-phosphate hydrolysis are more closely related to those present in FtsZs in both the plasmid partitioning and virion centring TubZ sub-groups. Whereas the canonical eukaryotic tubulins rely upon a conserved glutamate residue, all TubZs resolved to date appear to utilise an aspartate to activate the attacking water molecule, as do the FtsZs, although two further glutamates are present in the immediate vicinity of the conserved residues in bacteriophage c-st TubZ, meaning that this cannot be stated unequivocally.

Parallels with the Eukaryotic Tubulins

The role of eukaryotic tubulins in chromosome segregation is analogous to the task performed by the plasmid partitioning TubZ proteins. The polymerising tubulin unit includes two different monomers, α- and β-tubulin, which are pre-assembled around a molecule of GTP bound within a catalytically incompetent subunit-subunit interface. In addition to the canonical amino-terminal GTPase and carboxy-terminal activation domain structure, all tubulins incorporate a carboxy-terminal extension comprising a helical hairpin (H11 and H12), which provides the binding platform for processive motor proteins, and several notable surface loops (H1-S2, H2-S3, M- and N-loops). These stabilise the lateral interactions between the thirteen straight protofilaments incorporated into a cylindrical microtubule (Lowe et al. 2001).

The characteristic α-helical carboxy-terminal extension of tubulins is more similar to H11 of TubZs than the β-hairpin of the FtsZs, and furthermore this region has been proposed to occupy a similar role in the TubZs and tubulins, providing a binding platform for the cognate cofactor TubR in the case of the plasmid partitioning TubZs. It is also worth noting that helix H11 occupies similar space in both cases, although H12 is absent in the TubZs.

The active site residues involved in forming the interface with loop T7 of the adjacent subunit are also partially conserved between these proteins. Within the conserved loop T1, whereas FtsZs retain a glycine residue that cannot contact the adjacent loop T7, both the eukaryotic tubulins and the plasmid partitioning TubZs exhibit a glutamine that makes contacts with the backbone of the adjacent loop T7. The bacteriophage virion-centring TubZs, on the other hand, adopt a glycine in this position in the same manner as the eubacterial FtsZs.

The recognition of the guanosine base, and the binding the guanosine nucleotide within the GTPase domain, are carried out similarly in TubZs and tubulins. Whereas in the FtsZs, recognition of the guanine base is accomplished by an asparagine and an aspartate residue, which complement the N2/N3 and N1/O6 positions respectively, in tubulins the second aspartate is replaced by an asparagine. In both subgroups of the TubZs, this arrangement is conserved, despite the considerable variation in other aspects of base recognition.

Parallels with the Archaeal CetZs

The CetZs represent the newest and currently least well understood of the tubulin/FtsZ-like groups of proteins. Multiple copies of CetZ genes are found within the genomes of many archaea, however to date only a single role has been ascribed to one them, governing cell shape and motility in *Haloferax volcanii*. Little is known about their dynamics, or the type of filaments that they might form, however three crystal structures are available, including a crystallographic protofilament retaining GTPγS (Chap. 12; Duggin et al. 2015). Although it is too early to draw definitive conclusions about fundamental similarities to or dissimilarities from this group, they already exhibit several parallels. Most notably, the CetZs exhibit the carboxy-terminal extension, H11, in a remarkably similar conformation and position to that observed within TubZ and its homologues. Features that are less informative at present include helix H6, an element conserved in both eukaryotic tubulins and eubacterial FtsZs, though relatively extended in tubulin (see Fig. 11.2); H6s seen in all plasmid-encoded TubZs resemble those in FtsZs and in all CetZ structures resolved to date but H6 is drastically reduced in all three bacteriophage-encoded TubZ homologues. Filament twist is another uncertain feature, the only observation of a CetZ in a polymeric form to date being of untwisted crystallographic protofilaments. We note that both bacteriophage virion-centring TubZs also exhibit untwisted protofilaments within crystal structures, while cryo-EM studies reveal that these proteins form helical filaments when released from the confines of crystallographic symmetry.

Structure and Architecture of TubZ Filaments

TubZ Protofilaments

Careful protein engineering, the application of poorly hydrolysable nucleotide analogues and fortuitous crystal lattice formation have allowed protofilaments of both sub-groups of the TubZ homologues to be resolved to high resolution within the confines of crystal lattices. This has proven extremely important for our understanding of the mechanisms of GTP mediated protofilament formation, of concomitant hydrolysis and phosphate release, and of eventual depolymerisation. The observation of subunit interactions in crystals has revealed details of the destabilisation of the active site, which is formed by the subunit-subunit interface in tubulin/FtsZ-like proteins through juxtaposition of the GTPase and activation domains of adjacent subunits. The intermediate resolution structures so far available of intact filaments by cryo-EM are of insufficient resolution to visualise directly, but confirm these features (Fig. 11.3).

Plasmid Partitioning TubZ Protofilaments

No significant reconfiguration of the monomeric subunit beyond the recruitment of GTP and concomitant ordering of the subunit-interface of the GTPase domain is required for TubZ polymerisation. *Bacillus thuringiensis* TubZ polymerises in the presence of GTP or its analogues to form helical protofilaments with a 44 Å repeat (Aylett et al. 2010). This occurs despite the conservation of the subunit-subunit interface from the straight protofilaments of FtsZ and tubulin necessary within the tubulin/FtsZ protein family to couple polymerisation to nucleotide hydrolysis; in *Bacillus thuringiensis* TubZ protofilaments, aspartate residues 266 and 269 are poised within the active site to activate a water molecule, catalysing the release of the γ-phosphate. The twisted protofilaments characteristic of the TubZs arise from the rotation of activation domain relative to the GTPase domain, applying an excess rotation within the subunit rather than across the subunit-subunit interface, which would require reconfiguration of the active site. A slight tilt across the subunit interface redirects the vector of polymerisation to add superhelicity to the protofilament, ensuring that it polymerises along a relatively linear path rather than describing wide loops. The reduced subunit interface entailed by this difference is offset by substantial contacts formed between H11 and the adjacent subunit.

Within the *Bacillus thuringiensis* TubZ protofilament guanine nucleotide exchange would appear to be impossible without significant rearrangement of the protein, whereas release of the hydrolysed γ-phosphate and associated Mg^{2+} ion is quite possible through either breathing of T3 or through a channel towards the activation domain of the adjacent subunit. This asymmetry of exchange will swiftly lead to a predominantly GDP-occupied state of the protofilament. Whereas initial crystallographic studies suggested that no significant changes might be expected on

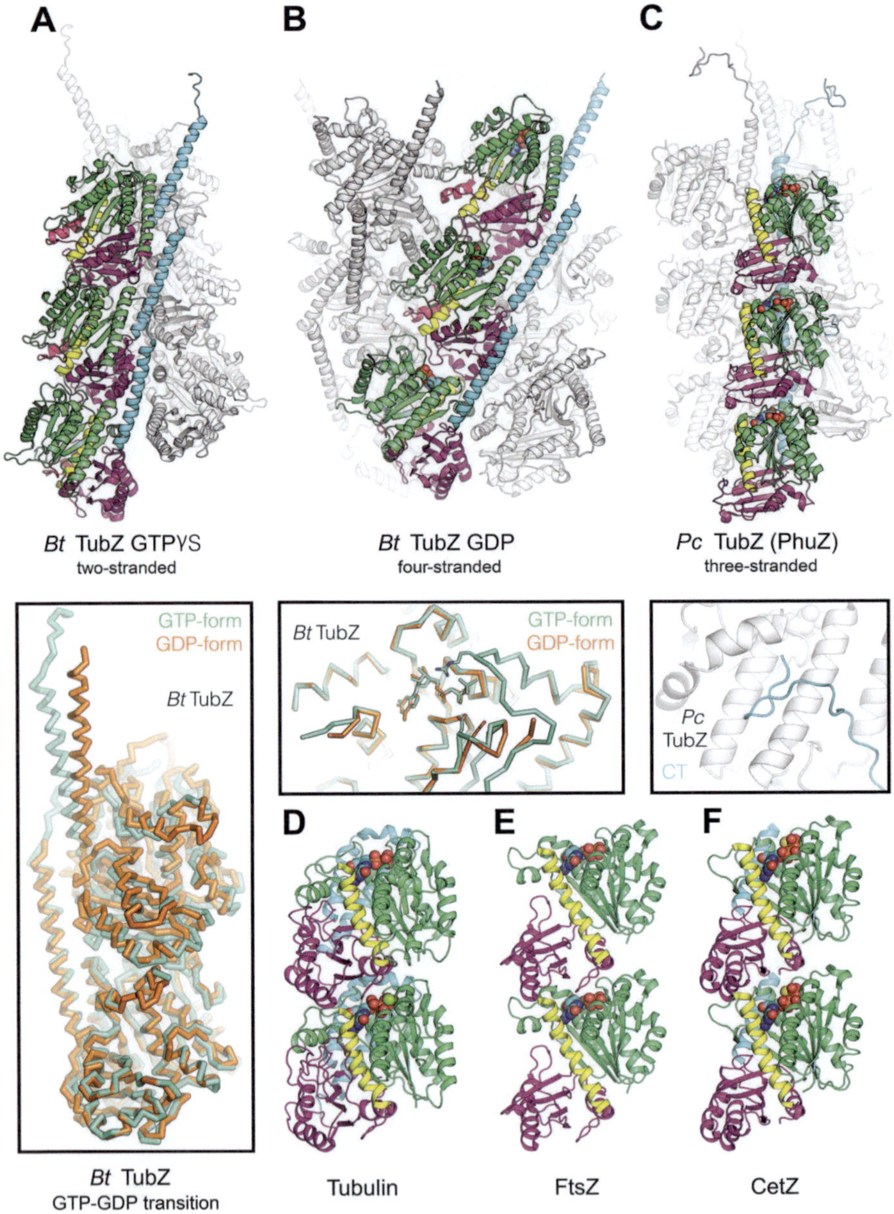

Fig. 11.3 Filaments types of TubZ proteins (**a**) The cryo-EM structure of the two-stranded GTPγS-form *Bacillus thuringiensis* TubZ filament, the inset panel shows the increase in the relative rotation of the adjacent subunit after phosphate and Mg^{2+} release. (**b**) The cryo-EM structure of the four-stranded *Bacillus thuringiensis* TubZ filament; the inset panels show the relaxation of the active site on phosphate and Mg^{2+} release. (**c**) The cryo-EM structure of the three-stranded *Pseudomonas chlororaphis* bacteriophage 210Φ2-1 TubZ filament; the inset panel showing the interaction site of the carboxy-terminus of a crystallographically related molecule in the protofilament crystal structure. Protofilaments of tubulin (**d**), FtsZ (**e**) and *Haloferax volcanii* CetZ (**f**) are shown for the purposes of comparison. The structures are shown according to an identical colour scheme and representation as those in Fig. 11.2; however, in order to facilitate interpretation, in **a–c** only one protofilament within each structure has been shown in colour, while the others are shown in shades of grey

nucleotide exchange (Ni et al.), these relied on soaking Mg-GTPγS into pre-formed crystals stabilised through lattice interactions in an apo-state. In fact, release of phosphate and Mg^{2+} is strongly coupled to destabilisation of the TubZ subunit-subunit interface. Arginine residue 87, which complements the charge of the γ-phosphate at the tip of loop T3, and aspartate residue 64 within helix H2, are both release upon the transition to GDP, resulting in relaxation of loop T3 and H3 beyond it to a poorly ordered state and of helix H2 away from loop T7 of the adjacent subunit, disrupting the subunit-subunit interface.

In addition to potentiating the release of subunits from the protofilament through disruption of a large part of the subunit-subunit interaction surface, the disruption of the active site results in a difference in the degree of rotation tolerated between the Mg-GTP and GDP states of *Bacillus thuringiensis* TubZ protofilaments. This rotation manifested itself within the crystal structure through the completion of a unit cell within twelve rather than fourteen subunits, however it has even more pronounced effects upon the morphology of the filament lattice itself as discussed in section "Filament structure" and in Montanaba and Agard (2014).

Bacteriophage Virion Centring TubZ Protofilaments

Whereas the twisted protofilaments necessary to form a helical filament have been resolved crystallographically in the case of the plasmid partitioning TubZ subgroup, this has not, to date, proved possible in the case of the bacteriophage virion centring TubZs. In the monomeric GDP state, bacteriophage virion centring TubZ subunits do not appear to be competent to polymerise, two secondary structural elements that are inimical to polymerisation form on either side of the subunit; both T3 and S9-H10 adopt α-helical conformations, while H11 projects over the top of the subunit, obscuring the entry of an incoming protein. Unfortunately a monomeric GTP-equivalent state is not yet available for comparison, so it remains impossible to tell whether these conformations are transient and relax on subunit formation, or whether the presence of γ-phosphate disfavours them, a likely hypothesis in particular in the case of the T3 helix as the tip of T3 directly contacts the nucleotide γ-phosphate.

Two states of bacteriophage virion centring TubZ protofilaments are known; one in which the active site is properly formed from ΦKZ TubZ, with a 43.5 Å subunit repeat, and another from 210Φ2-1 TubZ with a 47 Å subunit repeat, in which the subunit-subunit interface is solvent accessible. In both cases H11 and the carboxy-terminal extension to the canonical tubulin/FtsZ-like fold are resolved completely, revealing that the acidic region at the extremity of the tail becomes completely ordered upon the adjacent subunit, forming a short β-hairpin. This "knuckle" binds across conserved α-helices H3, H4 and H5 of the GTPase core fold, burying 1′226 $Å^2$ of exposed surface on each subunit, and implying a further source of stabilisation of the subunit-subunit interaction in addition to the contacts made by H11 buttressing the base of the adjacent subunit, and possibly a sensor for nucleotide state within the active site, given the relaxation of H3 on hydrolysis in the plasmid partitioning TubZ homologues. Unfortunately both states of the protofilaments were resolved with GDP bound, despite the inclusion of non-hydrolysable analogues, preventing one from determining whether or not any differences exist between sub-groups in the GTPase cycle and the means by which it is coupled to depolymerisation.

Although the formation of a canonical active site, akin to that in the ΦKZ TubZ structure, within the filament is almost certain, it remains quite possible that the relaxed state of the subunit-subunit interface from 210Φ2-1 TubZ represents an arrival/departure complex which has been stabilised by this secondary interaction. This possibility implies a regulatory role during polymerisation and depolymerisation for the TubZ carboxy-terminal extensions, most likely in both sub-groups, in addition to the previously established role in cargo recruitment.

Filament Structure

Whereas the protofilaments of both sub-groups of the TubZ proteins exhibit a series of commonalities, the structures of the different filaments formed on polymerisation contrast with one another. Major differences in morphology occur, even for a single protein, such as for plasmid partitioning TubZ filaments at different stages in the polymerisation pathway. Intermediate resolution structures of these filaments by cryo-EM have been interpreted with the aid of the crystallographic protofilament structures, providing pseudo-atomic models for all three filament morphologies established to date (Fig. 11.3).

Plasmid Partitioning TubZ Filaments

Initial negatively stained studies of TubZ filaments identified two- and four-stranded, right-handed helical filaments, but were of too low a resolution to generate a reliable model of the filament (Aylett et al. 2010). Cryo-EM studies by Montabana and colleagues resolved both types of TubZ filament at intermediate resolution and allowed their interpretation through the use of the prior protofilament crystal structures to provide molecular models of each filamentous state (Montabana and Agard 2014). Two-stranded *Bacillus thuringiensis* filaments form helical filaments ~100 Å across with a 44 Å subunit repeat and a half subunit offset between the parallel protofilaments. The four-stranded filaments, on the other hand, form a structure ~120 Å across with a ~ 20 Å central lumen and a shorter, 43.5 Å, subunit repeat, associated with increased twist, but without any offset between the parallel protofilaments.

Bacillus thuringiensis TubZ was observed to form predominately two-stranded filaments in the artificially stabilised GTPγS polymerised state, however four-stranded filaments came to predominate when the protein was polymerised in GTP. The observation of weak T3 density, and H2 movement in the four-stranded filament, both of which were visible in the GDP state of crystallographic protofilaments, support an interpretation in which phosphate release is coupled to this transition, while the greater twist (~32° versus ~24°) entailed by the transition to four-stranded filaments is also in agreement with this hypothesis. The transition has been proposed to occur through the growth of single protofilaments along the sides of a two-stranded filament, filling in the extra strands as the initial two-stranded filament hydrolyses GTP and releases phosphate and Mg^{2+}.

An extremely important observation made for the two-stranded structure was excess density corresponding to carboxy-terminal-tail interactions at the base of H3 of the adjacent subunit. This implies the conservation of "knuckle" contacts throughout the TubZ group of proteins. It appears probable that exposure of the knuckle at only one end of the filament may impose directionality upon binding of the cognate TubR*C* centromeric complex to the *Bacillus thuringiensis* TubZ filament.

Bacteriophage Virion Centring TubZ Filaments

The structure of filaments of *Pseudomonas chlororaphis* bacteriophage 210Φ2-1 TubZ has also been resolved to intermediate resolution, through the use of cryo-EM, by Zehr and colleagues (Zehr et al. 2014). In contrast to the situation for the *Bacillus thuringiensis* TubZs, the bacteriophage TubZ forms three-stranded, right-handed helical filaments with a subunit repeat of 43.2 Å. Within the filaments, a canonical "tight", GTPase competent interaction is observed at the subunit-subunit interface, comparable to that in bacteriophage ΦKZ TubZ protofilaments. Given that polymerisation was accomplished in the presence of a non-hydrolysable analogue of GTP, it remains impossible to identify any conformational changes that may take place upon hydrolysis in this subgroup of the TubZ proteins. The protofilaments making up the filament are parallel, but each is successively offset by one-third of a subunit (leading to a one-start left handed topology for the helical symmetry of the filament itself, disregarding the protofilaments).

The topology of bacteriophage 210Φ2-1 TubZ filaments is extremely interesting and was quite unexpected. The carboxy-terminus, including H11 and the knuckle that makes contact with the adjacent GTPase domain, is located on the interior of the filaments, a topology that has not been observed for any other tubulin/FtsZ-like protein, and one that is inimical to the means by which tubulin/FtsZ-like proteins are typically thought to undergo depolymerisation, subunits "peeling" away from the GDP-occupied side of the active site on relaxation of this part of the subunit-subunit interface following phosphate release. The lateral interactions observed within the centre of the filament heavily involve both the activation domain and H11, however it remains unclear how hydrolysis might affect this lattice, and new studies will be necessary to address this important open question.

Towards a Fuller Structural Understanding of TubZ Polymers

The TubZ field has begun to enter a state of relative maturity with the intermediate-resolution structures of the filaments of both well-studied sub-groups. TubZ protofilaments in both sub-groups appear to behave in very similar fashions, however filament diversity may well be a unifying theme of future studies. Given recent advances in the resolution attainable by cryo-EM, it is not unreasonable to expect great strides in the resolution of new protein filaments in the coming years, and many more surprises may yet lie just over the horizon.

Increasing Similarity of Protofilament Interactions

The observation of twisted protofilaments for both subgroups of the TubZs, in addition to release and increased freedom within the active site on nucleotide hydrolysis, implies a unified polymerisation and depolymerisation mechanism that is likely to be conserved throughout the TubZ group. An important observation within this context was that of the carboxy-terminal "knuckle" interactions within the two-stranded filaments of *Bacillus thuringiensis* TubZ. It appears highly likely that communication between subunits might generally be achieved by such a mechanism, while the relaxation of T3 on γ-phosphate release, if conserved into the bacteriophage virion centring TubZs, could provide a means by which nucleotide state might be communicated to the knuckle, T3 undergoing a significant relaxation on plasmid-partitioning TubZ phosphate and Mg^{2+} release.

Increasing Diversity in Filament Interactions

The observation of such different filament topologies for related proteins implies the possibility of substantial diversity within the eubacterial TubZs. Given that filament structures have thus far been established for only two TubZ group members, it does not appear improbable that known TubZs might even exhibit different filament lattices, particularly when one considers that the phage group-members, such as *Clostridium botulinum* bacteriophage c-st TubZ appear structurally intermediate between those found within *Bacillus* species and those in *Pseudomonas*. The future investigation of further TubZ filaments will be necessary not only to establish any lines of division within the group, but may also perhaps yield new and interesting dividends in its own right.

Nucleation and Controlled Polymerisation/Depolymerisation

The mechanisms by which TubZ filaments are nucleated, and their polymerisation controlled by their cognate cargo complexes, have only just begun to be investigated. The two- to four-stranded transition engendered by nucleotide hydrolysis in *Bacillus thuringiensis* TubZ remains a subject of great structural interest, while depolymerisation of these filaments has yet to be approached systematically at all. While the extension of the discovery of the role of the carboxy-terminus in the bacteriophage virion centring TubZs into the plasmid centring TubZs provides immediate indications of a means by which control might be exerted given the overlapping binding sites proposed for this region, how this might function at a structural level remains a subject of great interest for future studies.

Cytomotive TubZ Filaments

Generally, filaments performing work through nucleotide hydrolysis are able to transport cargo and hence are cytomotive. Structural investigation suggests that TubZs form dynamic filaments due to the distinct conformational states seen when bound to different nucleotides. Thus TubZ filaments are principally able to conduct mechanical work driven by the nucleotide hydrolysis dependent polymerisation and depolymerisation. Transduction of chemical energy by hydrolysis of guanosine triphosphate (GTP) stored in the filament lattice leads to a conformational change that destabilises subunit interactions within the polymer thus fostering disassembly and subsequently drives filament turnover. As a result, filaments can be dynamically unstable, that is stochastically alternating between cycles of growth and shrinkage, the hallmark of eukaryotic microtubules. Alternatively, dynamic filaments can treadmill, growing exclusively from one end only, whilst subunits are lost from the opposing end, as first described for filamentous actin. Growing filaments would be able to push, or shrinking filaments pull, through the cytosol, any DNA coupled to the filament ends.

Plasmid-Borne TubZ Filaments of Type III Segregation Systems

For the *tubZRC* operons interdependent roles in replication and plasmid segregation are described. It is until now not resolved whether TubZ filaments contribute to replication or segregation, or possibly both. Significant evidence provided by different studies shows that the GTP dependent polymerization activity and disassembly of TubZ is essential for plasmid stability. This has been independently and extensively demonstrated for plasmids pBtoxis and pBsph. Both plasmids were lost in populations in which GTP hydrolysis mutants of TubZ were expressed. In line with this argument, mutations predicted to interfere with the assembly of the filament also caused severe loss of plasmids (Larsen et al. 2007; Ge et al. 2014a).

Cell Biology of TubZ Filaments

Indeed TubZ proteins have been shown to form dynamic filaments in cells. In *Bacillus thuringiensis* hosting pBtoxis plasmids that express TubZ-GFP (green fluorescent protein) fusion proteins uniform green filaments were observed (Larsen et al. 2007). Other TubZ proteins fused to GFP such as RepX of pXO1 in *B. anthracis* or TubZ from pBsph in *B. sphaericus* also confirmed cellular filaments (Akhtar et al. 2009; Srinivasan et al. 2011; Ge et al. 2014a). Formation of cellular TubZ filaments depends on TubZ protein levels, which is expected for polymer forming proteins. They only formed once the cellular TubZ protein levels passed the critical concentration threshold. In cells bearing the natural TubZ-encoding plasmids or cells expressing recombinant TubZ, one or more filaments are seen to span through the entire cell from pole to pole.

Larsen and colleagues studying pBtoxis carrying *B. thuringiensis* cells performed the first and only description of cellular GFP-labelled TubZ filament dynamics. Filaments are mobile and translocate within cells, bending around the cell poles (Fig. 11.4). This motion is dependent on filament disassembly because GTP hydrolysis deficient TubZ mutants form stable and static cellular filaments. TubZ filaments lengthen from a distinct end whereas they retract and shorten from the opposing one thus exhibiting a clear polar behaviour. The addition and subtraction

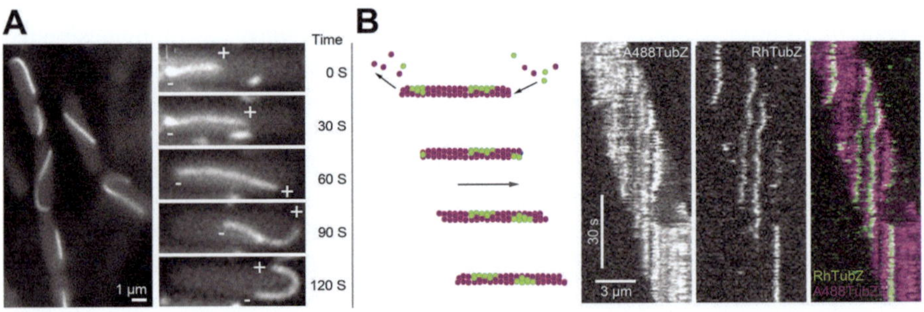

Fig. 11.4 Dynamics of TubZ filaments *in vivo* and *in vitro* (**a**) Still image of TubZ-GFP fusion forming filaments in *B. thuringiensis* cells. Filaments span through the cell from pole to pole (taken form Larsen et al. 2007). Right image sequence of a cellular TubZ filament growing from *left* to the *right* (+). The filament displays a clear behavioural polarity. At about 60s the filament retracts from the opposing end (−) while continuing growing. TubZ move through the cytosol via treadmilling and bend at the cell ends. (**b**) Schematics of treadmilling TubZ filament. Dual colour filament labelled sparsely with green monomers. While addition of GTP bound TubZ monomers occurs at the *right end* GDP bound subunits are lost at the *left*. Thus the *green* monomer islands within the filament are stationary, while the filament appears to translocate from *left* to the *right*. Next to it are dual colour *fluorescent* kymographs of a speckled TubZ filament (showing the spatial position of a filament within a region of a time-lapse acquisition over time) treadmilling across a glass slide *in vitro* in the presence of GTP, imaged by TIRF (total internal reflection fluorescence) microscopy (*Y-axis* represents time and the *X-axis* distance). Images taken from Fink and Löwe 2015 (Note that TubZ filaments do not display dynamic instability unlike eukaryotic microtubules or PhuZ (described later), although it is a tubulin. Also worth mentioning is that unlike bacterial-actin ParM TubZ filaments alone elongate in a polar fashion)

of subunits is observed almost exclusively at the filament ends. This is a clear indication that multi-stranded TubZ filaments or parallel bundles occur in cells, because filaments built out of only single protofilaments would undergo fragmentation into smaller pieces. The observation of dark speckles, introduced in cellular filament structures using fluorescence recovery after photobleaching (FRAP), revealed that treadmilling causes the intracellular translocation. Treadmilling as cause for rapid turnover of TubZ subunits was further visualized by fusing GFP to TubZ expressed in *Escherichia coli,* which hints that no other plasmid-borne factors are required for this filament behaviour. Because immobile non-dynamic filaments lead to plasmid loss, the described treadmilling and resulting motion of filaments trough the cytosol seems to be inevitably crucial for plasmid segregation and stability (Larsen et al. 2007).

The dynamics of other filament-forming TubZs such as RepX remains to be characterised in a cellular context in plasmid-bearing host cells. Whether they will display treadmilling or dynamic instability is unclear, although given homology and *in vitro* results of pBtoxis TubZ or phage-encoded PhuZ (described in detail later) they are likely to show some form of treadmilling, at least outside the cellular environment. In all cases where filament assembly has been confirmed in cells, their dynamic behaviour is critical for the stability of the plasmid. Also, for phage encoded PhuZ, the dynamic behaviour of filaments is crucial for its cellular role in positioning virions to the cell centre, a process described in more detail below.

In vitro *Properties of TubZ Filaments*

The assembly and dynamic behaviour of TubZ filaments has been described in best detail in the cytoplasm of *B. thuringiensis* or *E. coli* cells as outlined above. Independent of the cell biological description, *in vitro* reconstitution of filament assembly assures that the intrinsic filament characteristics are not fundamentally influenced by cytosolic factors. Additionally, reconstitution experiments which allow observation of individual filaments using single molecule total internal reflection microscopy (TIRF) also allow for analysis of the function of filament-binding proteins and their effect on filaments dynamics (Fig. 11.4). Most of the information of TubZ filament assembly and dynamics comes from studies of the TubZ system from plasmid pBtoxis *in vitro* but TubZs of *B. cereus, B. anthracis* and circular bacteriophage c-st are also fairly well biochemically investigated (Anand et al. 2008; Chen and Erickson 2008; Hoshino and Hayashi 2012; Oliva et al. 2012; Fink and Lowe 2015).

Purified TubZ proteins assemble into filaments upon supplementing with GTP and magnesium salts. Compared to the other bacterial tubulin homologue FtsZ, hydrolysis from GTP to GDP + Pi occurs faster and in a cooperative manner. Bulk biochemical assays such as light scattering or polymer-pelleting assays indicated that monomeric TubZs assemble into higher order filamentous structures upon GTP binding that disappear when GTP has been consumed. This has further been confirmed using negative stain transmission electron microscopy. As for all multi-stranded polymers, TubZ filaments assemble only above a critical concentration of available monomers (0.8–2 μM). The specific value varies for different TubZs and depends on the methodology used for determination. Doping filaments with small amounts of non-hydrolysable GTP nucleotide analogues leads to a dramatic increase in polymers suggesting that a cap consisting of several GTP bound TubZ subunits at the growing end is sufficient to prevent the filament from disassembling. This cap has been interpreted as analogous to the elongation of microtubules (Chen and Erickson 2008) that grow robustly from their ends until the GTP cap vanishes introducing the switch to catastrophic depolymerisation, related to dynamic instability.

Microscopic Dynamics of *de novo* Assembled TubZ Filaments of pBtoxis

TubZ filaments treadmill in cells, which has lead to the idea that they either pull or push plasmids through the cytosol (Barilla 2010). In agreement with the cellular treadmilling of TubZ filaments, filaments *in vitro* displayed this exact treadmilling action when isolated and characterised by means of fluorescent single molecule microscopy as recently demonstrated by Fink and Löwe (Fink and Lowe 2015) (Fig. 11.4). Growth never occurred from both ends simultaneously nor were previously shrinking ends seen to switch to growth or vice versa. Surprisingly, although belonging to the tubulin family of proteins, these filaments were never observed to exhibit dynamic instability, which is to switch stochastically between growth and shrinkage. The microscopy studies clearly demonstrated the polar nature of the filaments: with disassembly from one end only, referred to as the minus end. The previously mentioned stabilisation through a "GTP-TubZ" cap introduced by non-hydrolysable nucleotide analogues preventing TubZ subunit loss occurs at the minus end. It also confirms the idea formed by cellular observations that filaments consist of more than one protofilament, since the GDP lattice along the filament length does not lead to lateral breakage and consequent disassembly. TubZ filament growth at the plus end depends, as expected for filaments, on the accessible monomer concentration whereas shrinkage from the minus end is independent and therefore a zero order reaction. As TubZ filaments assemble and increase length when growth rates exceed the disassembly rate, they are highly prone to form bundles. Such bundles represent higher order structures and can be hundreds of micrometres long, forming a network. Due to the individual filament dynamics they separate again, dependent on the filament orientation within bundles. The structural nature of the growing plus end as compared to the shrinking minus end has so far not been identified and can only be speculated about. Due to recent structural data described in detail in section "Structure and architecture of TubZ filaments", it appears possible that one of the ends of four-stranded TubZ filaments has the carboxy-terminal extension of TubZ subunits exposed (Montabana and Agard 2014). This suggests the existence of structural features additional to the nucleotide state of TubZ within the lattice (GTP vs. GDP) specifying growing and shrinking ends.

The TubZ Carboxyterminal Extension After Helix 11

TubZ proteins require their carboxy-terminal extension for efficient polymerisation. Indications for these interesting observations came from labs noticing that carboxy-terminal tagged TubZ polymerises poorly compared to wildtype TubZ (Hoshino and Hayashi 2012; Montabana and Agard 2014). Carboxy-terminal TubZ mutants from plasmids pBtoxis or pXO1 revealed that deletion of only few residues impairs activity and completely abolishes it if all residues are cleaved after helix 11. Tail-dependent polymerisation is probably a common characteristic of TubZ proteins. Several possibilities can explain the observed inability to polymerize and subsequent deficient GTPase activity. It seems that the tail is either competent to stabilise the filament lattice preventing it from breaking up, reduces the rate of subunit loss from the minus end or increases the growth rate. Alternatively the tails of multiple TubZs may be required for the formation of a stable, nucleation-competent filament, seed.

The Centromeric TubRC Complex

In order to be separated and transported into daughter cells genetic material must be attached to its segregation machinery. Specialized structures composed of specific DNA sequence repeats that are recognized by distinct DNA binding proteins form this linkage, and constitute the centromeric complex. The function of the centromeric complex is to link the plasmid or circular bacteriophage to the cytomotive filaments. In TubZ systems, besides its role as transcriptional repressor, the DNA binding protein TubR (found upstream of the *tubZ* gene in the *tubZRC* operon), with centromeric *tubC* iterons, constitute the nucleoprotein adaptor (referred as "TubRC") that physically links the plasmid to TubZ filaments (Fig. 11.1).

The DNA Binding Adaptor Protein TubR

TubR from several systems (such as pBtoxis, pBsph and bacteriophage c-st), was shown to specifically bind *tubC* DNA by electrophoretic mobility gel shift assays and other biophysical techniques (Tang et al. 2007; Ni et al. 2010; Aylett and Lowe 2012; Oliva et al. 2012; Fink and Lowe 2015). The consensus DNA binding sites have been mapped in detail for a full *tubC* region in plasmids pBtoxis and pBsph using microarray hybridization analysis or DNA footprinting (Aylett and Lowe 2012; Ge et al. 2014a). For other plasmid and bacteriophage TubZ systems only the *tubC* region is known without knowledge of the exact number of iterons or the specific nucleotide sequences recognized by TubR. TubR of pBtoxis contains a winged helix-turn-helix (HTH) motif typical of DNA binding proteins. It was shown via using X-ray crystallography that TubR functions as a dimer, each pair binding to one *tubC* iteron (Ni et al. 2010; Aylett and Lowe 2012). TubR uses a non-canonical DNA binding mechanism to bind DNA compared to other HTH DNA binding proteins. Instead of forming a DNA-protein interaction the HTH motif is required for TubR dimerization. Interestingly, the amino-terminal ends of the paired recognition helices within one dimer together contribute DNA binding activity inserting into the same major groove. Together they produce the major contact with DNA. Beta hairpins within the “wings” of the dimer contact the adjacent minor grooves thus introducing a cleft between the paired recognition helices. The two clefts have positively charged amino acids exposed on the surface, further facilitating DNA binding activity. The structure of another homologue (*Bacillus megaterium)* is known, revealing the same protein architecture for recognition of the centromeric DNA (Aylett and Lowe 2012). Its analysis revealed a very similar binding mode and surface as for TubR from *Bacillus thuringiensis*. In contrast to both known structures, the TubR from the circular bacteriophage c-st seems to be monomeric, as suggested by analytical ultracentrifugation, and the mode of DNA binding is unresolved. Another apparent difference is revealed by titration experiments for c-st TubR suggesting that each of the four identified *tubC* (synonym *tubS*) sites in c-st recruit only one TubR molecule (Oliva et al. 2012).

Structure of the Centromeric Adaptor Complex

Aylett and Löwe took electron micrographs of *tubC* with all seven iterons in complex with adaptor TubRs that revealed a flexible DNA protein filament (Fig. 11.5), without a particular preference in shape or conformation (Aylett and Lowe 2012). Interestingly, this flexible nucleoprotein filament is able to interact with unpolymerised TubZ subunits, as has been shown for several TubZ systems using EMSA (electrophoretic mobility shift assays) and anisotropy measurements (Tang et al. 2007; Ni et al. 2010; Fink and Lowe 2015).

Similarly, micrographs showing TubR from plasmid pBM400 of *B. megaterium* in complex with the corresponding intragenic *tubZRC* operon DNA also revealed relatively flexible protein DNA filaments, although these favoured a more curved, open ring-like structure. In line with an open ring-like, helical protein DNA filament *B. megaterium* TubR crystalized in the absence of DNA in open helical filaments. Interestingly, these helical filaments expose positively charged residues at the exterior of the helix. In contrast, negatively charged residues locate to the inner face of the helix. This conformation is also seen when complexed with *tubC* DNA. Electron micrographs and crystallization of helical TubR filaments suggest that *tubC* DNA wraps around this open helix. The dimensions of a single open helix would allow winding of a single *tubC* DNA locus (with a contour length of 50 nm) around the helix outer circumference (Aylett and Lowe 2012).

The structural nature of TubR, when bound to *tubC* acting as repressor of the *tubZRC* operon, or as functional fully assembled centromeric complex is so far

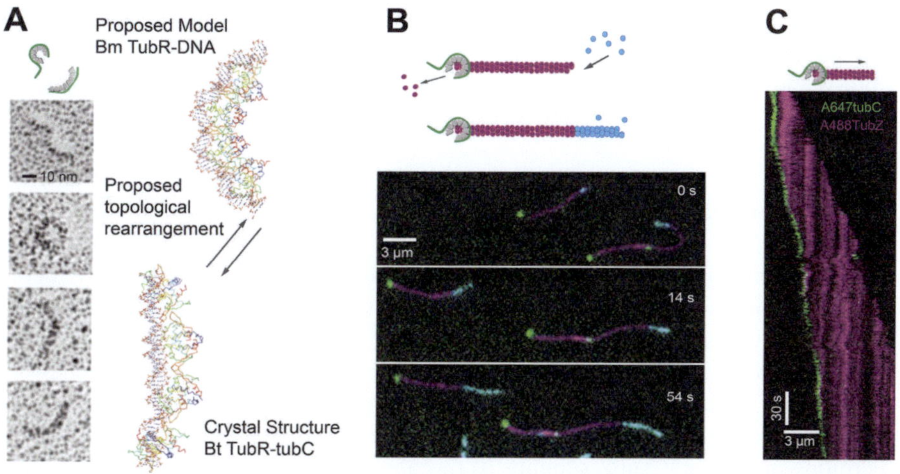

Fig. 11.5 Structure and TubZ filament binding of the centromeric complex composed of *tubC* and TubR (referred as "TubRC") TubRC. (**a**) TubRC complexes, schematics showing TubR dimers (*grey*) bound to tubC (*green*) forming a flexible DNA protein filament. *Below*: Electron microscopy images of rotary shadowed TubR from *B. thuringiensis* bound to all seven *tubC* iterons (modified from Aylett 2012) illustrating the flexibility. *Right*: The proposed conformational change occurring upon TubZ monomers or filament interaction, as deduced from structures form *B. megaterium* and *B. thuringiensis* TubRC complexes. Shown are crystal structures form TubRC complexes of both species. Protein is presented as ribbon, *blue* at N terminus and *red* at C terminus DNA sticks. (**b**) Schematics presenting minus end binding of TubRC (*green*) and polar growth of TubZ filaments from seeds (*magenta*) after the addition of differently labelled monomer (*cyan*). TubRC does not induce polymerisation at- nor does it track the growing plus end of TubZ filaments. *Below*: TIRF (total internal reflection fluorescence) microscopy micrographs showing single, labelled, *tubC* molecules (*green*) occupied by unlabelled TubR, binding to the minus ends of (*magenta*) labelled TubZ filament seeds from which TubZ extensions (*cyan*) grow. Growth occurs at the opposite end to TubRC binding. (**c**) Dual colour kymograph displaying minus end tracking of labelled TubRC complexes on treadmilling TubZ filaments. TubRC processively tracks depolymerising minus ends

enigmatic. The extended TubR filaments winding around DNA obtained for pBtoxis TubR when crystalized with two *tubC* iterons could represent the repressor state whilst the adaptor helix is the functional TubZ filament binding conformation. This second structure is reminiscent of the active centromeric complex forming an adaptor helix at growing filament ends described for the centromeric adaptor complex of actin-like ParMRC segregation system.

This comparison suggests the possibility that upon binding to the filament, or individual TubZ subunits, the centromeric complex of pBtoxis undergoes a conformational change suitable to capture TubZ filaments, whereas TubR from pBM400 may already favour an adaptor helix as conformation (Aylett and Lowe 2012) (Fig. 11.5).

Centromeric complexes of TubZ systems have so far been characterised for plasmids and for only one bacteriophage, c-st. Other Pseudomonas bacteriophages have putative homologues of TubR potentially fulfilling similar roles to known plasmid encoded TubR proteins. These are yet to be resolved.

Besides linking the plasmid to the cytomotive TubZ filament, TubR seems to also play a role in the initiation of plasmid replication (Tang et al. 2007). But this possibility and the exact contribution of TubR in replication is a matter of current research.

TubRC Interaction with TubZ Filaments

For tubulin-like proteins the carboxy-terminal tail presents a platform able to mediate interactions with proteins that modulate filament behaviour directly or anchorage to other cytosolic structures as described for FtsZs (Vaughan et al. 2004). This situation is also apparent for TubZs where the extended carboxy-terminus is bifunctional. The centromeric complex of plasmid borne TubZ systems from *B.cereus* or *B. thuringiensis* as well as bacterionphage c-st initiates filament assembly by recruiting monomeric TubZ subunits as revealed by co-sedimentation, light scattering and single molecule experiments conducted by several groups (Aylett and Lowe 2012; Hoshino and Hayashi 2012; Oliva et al. 2012; Fink and Lowe 2015). The TubZ carboxy-terminal tail is crucial for the formation of the tripartite complex consisting of *tubC* DNA, TubR and TubZ subunits. The interaction is mainly facilitated by the carboxy-terminal extension of TubZ monomers with the last about 20 amino acids having the greatest role in TubR binding activity. Interestingly, these residues are positively charged (basic) and could potentially bind to the partially negatively charged (acidic) TubR surface exposed opposite to its *tubC*-binding interface (Ni et al. 2010; Hoshino and Hayashi 2012). It is not clear whether the tails are also required for the TubRC interaction with assembled filaments, because filament formation is effected. However, this seems likely, given that for FtsZ protein filaments proteins the carboxy-terminal tail plays an important role in targeting interacting partners.

In vitro reconstitution experiments addressed the dynamic nature of the interaction of TubRC complexes with TubZ filaments using *B. thuringiensis* pBtoxis TubZRC system (Fink and Lowe 2015). These experiments found that the TubRC centromeric complexes do not target growing plus ends of treadmilling filaments, promoting TubZ subunit incorporation at growing ends. Instead the complex processively tracks the shrinking minus end, while still allowing the TubZ filament to treadmill reducing depolymerisation rates *in vitro*. The conducted experiments lead to the picture that TubRC centromeric complexes promote the formation of TubZ filaments by recruiting single TubZ monomers forming a stable growth competent seed (Fig. 11.5). Clearly, TubRC centromeric complexes are not filament polymerases incorporating subunits at the growing ends (as the centromeric counterparts from actin-like segregation systems) nor do they increase nucleotide turnover or asymmetric polar growth of filaments. What makes TubRC bind preferentially to shrinking filament ends? An obvious possibility is that TubR hangs on to the carboxy-terminal tails and that these are fully exposed at only one end of the TubZ filament, a prediction that can be made from filament structures resolved by electron microscopy (Montabana and Agard 2014). This explains the distinct end binding behaviour but also the transient lateral interactions of TubRC observed by Fink and Löwe (Fink and Lowe 2015). Biochemical reconstitution experiments leading to a structural picture of exactly how this complex is able to track-shrinking ends is matter of great interest in the field, revealing basic biophysical principles.

TubZRC in Comparison to ParMRC Segregation System

It has been possible to constitute TubZ mediated DNA transport outside cells. The formation of a centromeric complex enriches the adapter protein locally on *tubC* DNA to facilitate filament minus end binding through higher avidity. Partial *tubC* sequences are proven to be less effective. TubRC formation controls the local formation of filaments. Once TubZ filaments assemble in the presence of TubRC they extend from the centromeric complex. Because TubZ filaments are still capable of treadmilling they act as one-dimensional motors dragging the minus end bound TubRC centromeric complex through solution. For cargo movement to occur TubZ filaments must experience resistance to motion. Outside the cellular environment this resistance can be achieved through medium viscosity, non-specific surface binding or through inter-filament interaction. The ability to recognize and track shortening minus ends enables movement of the centromeric complex by treadmilling TubZ filaments. Depolymerisation of TubZ filaments from shrinking minus ends creates the force leading to pulling of TubRC centromeric complexes.

The behaviour of TubRC is perhaps unexpected given its apparent architectural similarity to the actin-like partitioning systems (Aylett and Lowe 2012). The centromeric complex of the ParMRC actin-based segregation system accelerates growth of filaments by binding at the growing (barbed) end, catalysing insertion of polymer subunits. Furthermore ParMRC pushes DNA molecules apart through the formation of a bipolar spindle. This bipolar spindle forms by a self-selecting process, requiring the dynamic instable nature of the ParM filaments (Garner et al. 2004, 2007; Gayathri et al. 2012). (Monopolar spindles are unstable whereas antiparallel arrangements are stable.) The thoroughly investigated TubZRC system of pBtoxis displays neither characteristic. Monopolar filament attachments are stable because the lack of dynamic instability in TubZ filaments means no self-selection occurs for bipolar spindles. On the contrary, TubZRC cannot separate DNA molecules in this way, as it would cluster DNA when forming bipolar filament structures (Fink and Lowe 2015).

Does TubZRC Segregate Plasmids?

It has been speculated that once TubZ filaments are anchored to the plasmid the dynamic properties change, switching from treadmilling to dynamic instability, in analogy to microtubule systems that exhibit dynamic instability. When the minus ends of microtubules are tightly anchored to centrosomes or microtubule organizing centres, only plus ends exchange subunits. Whereas, free microtubules have also been shown to treadmill in cells as well. Since *TubZRC* behaves so differently to known segregation systems it is likely that treadmilling TubZ filaments in the presence of the centromeric complex constitute a third mechanism in prokaryotic plasmid segregation.

A pulling mechanism by treadmilling TubZ filaments, rather than one pushing by formation of bipolar structures, is supported by observations that in *Bacillus* cell filaments treadmill and bend around cell poles. However, it is still unresolved how cells employ this end-tracking activity and whether it is only a part of a much more complex mechanism.

Complex mechanisms might involve the capturing and attachment of the growing plus end to a cellular structure, which then anchors the filament during minus end disassembly driving the translocation of DNA. Additional filament organizing centres, similar to spindle poles that anchor filaments as suggested for the PhuZ-based phage centring system, or other regulators may be required. A possible anchor could be the bacterial nucleoid and an additional Tub protein encoded by a gene, downstream or upstream of *tubZRC* operon, called *tubY* (Oliva et al. 2012). TubY might have DNA-binding activity and could therefore act as an anchor adaptor.

TubZ systems seem relatively unstable when creating synthetic minireplicons containing *tubZRC* loci. This particularity indicates the possibility that additional factors for proper plasmid segregation are required. The segregation of plasmids by TubZRC has so far not been monitored in living cells. Further cellular investigations of this special and important pathogenicity factor inheritance system are warranted.

Cytomotive Bacteriophage-Borne TubZ Filaments

Some bacteriophages contain TubZ homologues within their genomes. The c-st encoded TubZ system seems to be required for the segregation of the circular bacteriophages during the pseudolysogenic cycle in *Clostridium botulinum*. This TubZ system has already been described and compared to the related plasmid-borne systems in the above sections due to its likely similar role in segregating circular DNA molecules. However, it remains to be shown that this system is in fact important for the bacteriophage life cycle, thus far only a single structural and biochemical study by Oliva and co-workers has shed light on this system.

More generally the phage-borne tubulin-like proteins comprise a subgroup to the plasmid-borne TubZs due to their structural divergence and are named thereafter PhuZ. To date there is only one PhuZ phage-borne system (phage 201Φ2-1) well characterised within host cells.

Cell Biology of PhuZ

This system functions in a different cell biological context to the plasmid segregating TubZ systems. Researchers from the Pogliano and Agard groups performed cell biological groundwork on this system, showing that cellular filaments (visualized by expressing PhuZ-GFP fusion proteins) nucleate throughout the cytosol and organize into filament bundles (Kraemer et al. 2012; Erb et al. 2014). In bacteriophage infected cells PhuZ filaments seem to grow from the ends (as observed for TubZ), but do not reveal treadmilling behaviour in cells. PhuZ filaments seem to be focused or anchored with one end at cell poles (Fig. 11.6). Thus, free filament ends point towards the cell centre from opposing poles. Compared to the filament ends focused at cell poles, those ends pointing towards mid-cell are dynamic. However, limitations in light microscopy and the strong filament bundling prevented the observation and characterization of individual filament ends at cell poles. Non-focused ends do seem to undergo some kind of dynamic instability, when emanating from the cell poles (Erb et al. 2014). This prompted the suggestion that PhuZ filaments organize into a spindle-like structure, where cell pole focused and anchored filaments comprise a spindle half (Kraemer et al. 2012; Erb et al. 2014). This is in analogy to eukaryotic spindles, where microtubules focused and stabilised at each spindle pole emanate to the spindle equatorial complex of overlapping microtubules and microtubule-attached chromosomes.

Interestingly, DNA of the bacteriophage localised between the growing ends of PhuZ half-spindles during the lysogenic cycle. This centring is abrogated when expressing hydrolysis deficient mutants that alter the dynamics of the PhuZ filament. Centring activity through pushing by growing ends that attach to bacteriophage particles contributes to the efficient lysogenic cycle of the bacteriophage and is thus important for the phage's life cycle (Kraemer et al. 2012).

PhuZ Filament Dynamics

The notion of growing filaments pushing bacteriophage particles via dynamically unstable PhuZ filament plus ends was supported by *in vitro* analysis of purified PhuZ using TIRF microscopy (Erb et al. 2014). In contrast to the anticipated activity in cells, PhuZ filaments exhibited treadmilling. A better picture of PhuZ dynamics and the ability of plus ends to switch from growth to shrinkage was obtained by stabilising filament seeds (using non-hydrolysable GTP analogues) from which PhuZ was allowed to elongate. Filament extensions formed around the critical concentration (2 μM) in the presence of GTP were not stable. Periods of growth were interrupted by catastrophic disassembly of the extension followed by repeated elongation from the seed (Fig. 11.6). In principle PhuZ, filaments could constantly push macromolecules and thus centre bacteriophage particles in cells when minus ends

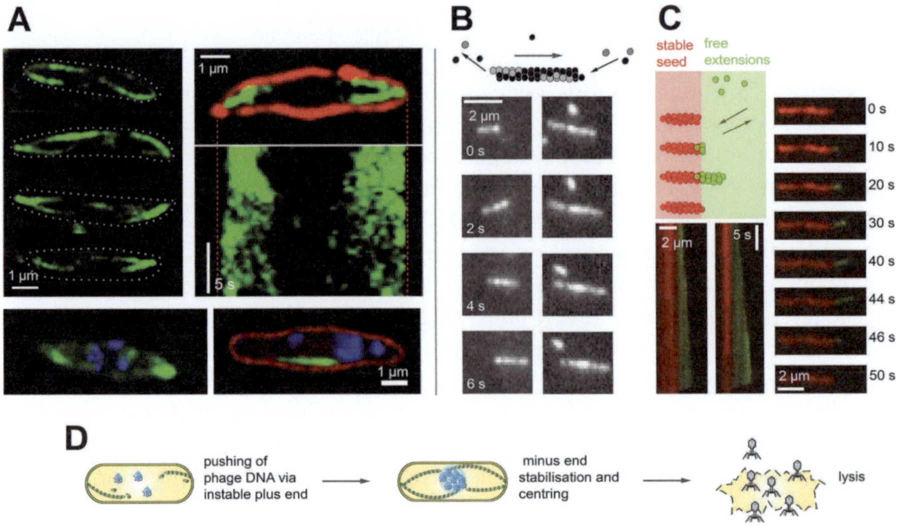

Fig. 11.6 Dynamic PhuZ (phage TubZ) filaments of phage 210Φ2-1 *in vivo* and *in vitro* (**a**) Fluorescence images show PhuZ filaments (*green*) in *Pseudomonas chlororaphis* cells (red FM4-64 stain or dotted line) and phage nucleoids by DAPI (*blue*). *Left panel:* Filaments extend from opposing poles to the cell centre with their free dynamic ends pointing to the middle whereas ends at the cell poles seem to be focused and stabilised. *Right panel*: Still image of PhuZ spindle, below a kymograph illustrating the dyanmic PhuZ ends at midcell of the cell above. *Bottom panel* pictures: *Left,* a wildtype cell which has, in between opposing free PhuZ filament ends, centered phage nucleoid particles. *Right:* When the dynamics of filaments is disturbed by expressing a hydrolysis deficient mutant of PhuZ phage particles become displaced and are not centred anymore. (**b**) Purified PhuZ assembles *in vitro* into filaments that crawl over the glass surface via treadmiling, reminiscent of filaments formed from purified TubZ. Depicted are image sequences of treadmilling filament that are visualised by TIRF microscopy. (**c**) Schematic representation of dynamic PhuZ filaments extensions (*green*) grown from a stabilised PhuZ filament seed (*red*). Filaments elongate only from one end when free subunits are added at concentrations close to the critical concentration. These extensions are not stabilised and stochastically depolymerise completely from the plus end. Dual colour kymographs and florescence TIRF images of PhuZ displaying catastrophic events of depolymerisation from the initially growing filament plus end. Depolymerisation occurs throughout the entire *green* elongation and stops at the stabilised *red* seed. (**d**) A model envisioning how phage nucleoids are centred by PhuZ filaments: PhuZ filaments are stabilised at minus ends pointing to opposing poles, so they grow with their plus ends towards the cell centre. Pushing by growing filaments plus ends positions attached phage particles at midcell. This centring seems to be beneficial for the lytic cycle of phage 201Φ2-1 (**a–d** modified form Erb et al. 2014 and Kraemer et al. 2012)

are stabilised and anchored. As for TubZ, PhuZ requires a carboxy-terminal extension to form polymers efficiently, presumably by making additional contacts with the adjacent subunits of a three-stranded filament, as described in detail above in section "Structure and architecture of TubZ filaments".

PhuZ filaments resemble other treadmilling TubZ filaments when both ends are free but they convert to dynamic instability when one end is stabilised. This contrasts with plasmid-borne TubZ filaments, which do not exhibit dynamic instability, even when stabilised *in vitro* by the centromeric complex. Thus, the different filament structures described below may be responsible for mechanisms that specifically allow or prohibit dynamic instability in each system.

What links the phage to the growing ends of filaments, and how filaments are anchored to the cell poles remains mysterious and demands further investigation.

References

Akhtar P, Anand SP, Watkins SC, Khan SA (2009) The tubulin-like RepX protein encoded by the pXO1 plasmid forms polymers in vivo in Bacillus anthracis. J Bacteriol 191:2493–2500

Anand SP, Akhtar P, Tinsley E, Watkins SC, Khan SA (2008) GTP-dependent polymerization of the tubulin-like RepX replication protein encoded by the pXO1 plasmid of Bacillus anthracis. Mol Microbiol 67:881–890

Aylett CH, Lowe J (2012) Superstructure of the centromeric complex of TubZRC plasmid partitioning systems. Proc Natl Acad Sci U S A 109:16522–16527

Aylett CH, Wang Q, Michie KA, Amos LA, Lowe J (2010) Filament structure of bacterial tubulin homologue TubZ. Proc Natl Acad Sci U S A 107:19766–19771

Aylett CH, Izore T, Amos LA, Lowe J (2013) Structure of the tubulin/FtsZ-like protein TubZ from Pseudomonas bacteriophage PhiKZ. J Mol Biol 425:2164–2173

Barilla D (2010) One-way ticket to the cell pole: plasmid transport by the prokaryotic tubulin homolog TubZ. Proc Natl Acad Sci U S A 107:12061–12062

Berry C, O'neil S, Ben-Dov E, Jones AF, Murphy L, Quail MA, Holden MT, Harris D, Zaritsky A, Parkhill J (2002) Complete sequence and organization of pBtoxis, the toxin-coding plasmid of Bacillus thuringiensis subsp. israelensis. Appl Environ Microbiol 68:5082–5095

Chen Y, Erickson HP (2008) In vitro assembly studies of FtsZ/tubulin-like proteins (TubZ) from Bacillus plasmids: evidence for a capping mechanism. J Biol Chem 283:8102–8109

Duggin IG, Aylett CH, Walsh JC, Michie KA, Wang Q, Turnbull L, Dawson EM, Harry EJ, Whitchurch CB, Amos LA, Lowe J (2015) CetZ tubulin-like proteins control archaeal cell shape. Nature 519:362–365

Erb ML, Kraemer JA, Coker JK, Chaikeeratisak V, Nonejuie P, Agard DA, Pogliano J (2014) A bacteriophage tubulin harnesses dynamic instability to center DNA in infected cells. Elife:3

Fink G, Lowe J (2015) Reconstitution of a prokaryotic minus end-tracking system using TubRC centromeric complexes and tubulin-like protein TubZ filaments. Proc Natl Acad Sci U S A 112:E1845–E1850

Garner EC, Campbell CS, Mullins RD (2004) Dynamic instability in a DNA-segregating prokaryotic actin homolog. Science 306:1021–1025

Garner EC, Campbell CS, Weibel DB, Mullins RD (2007) Reconstitution of DNA segregation driven by assembly of a prokaryotic actin homolog. Science 315:1270–1274

Gayathri P, Fujii T, Moller-Jensen J, van Den Ent F, Namba K, Lowe J (2012) A bipolar spindle of antiparallel ParM filaments drives bacterial plasmid segregation. Science 338:1334–1337

Ge Y, Hu X, Zhao N, Shi T, Cai Q, Yuan Z (2014a) A new tubRZ operon involved in the maintenance of the Bacillus sphaericus mosquitocidal plasmid pBsph. Microbiology 160:1112–1124

Ge Y, Zhao N, Hu X, Shi T, Cai Q, Yuan Z (2014b) A novel transcriptional activator, tubX, is required for the stability of Bacillus sphaericus mosquitocidal plasmid pBsph. J Bacteriol 196:4304–4314

Hoshino S, Hayashi I (2012) Filament formation of the FtsZ/tubulin-like protein TubZ from the Bacillus cereus pXO1 plasmid. J Biol Chem 287:32103–32112

Kraemer JA, Erb ML, Waddling CA, Montabana EA, Zehr EA, Wang H, Nguyen K, Pham DS, Agard DA, Pogliano J (2012) A phage tubulin assembles dynamic filaments by an atypical mechanism to center viral DNA within the host cell. Cell 149:1488–1499

Larsen RA, Cusumano C, Fujioka A, Lim-Fong G, Patterson P, Pogliano J (2007) Treadmilling of a prokaryotic tubulin-like protein, TubZ, required for plasmid stability in Bacillus thuringiensis. Genes Dev 21:1340–1352

Lowe J, Li H, Downing KH, Nogales E (2001) Refined structure of alpha beta-tubulin at 3.5 A resolution. J Mol Biol 313:1045–1057

Montabana EA, Agard DA (2014) Bacterial tubulin TubZ-Bt transitions between a two-stranded intermediate and a four-stranded filament upon GTP hydrolysis. Proc Natl Acad Sci U S A 111:3407–3412

Ni L, Xu W, Kumaraswami M, Schumacher MA (2010) Plasmid protein TubR uses a distinct mode of HTH-DNA binding and recruits the prokaryotic tubulin homolog TubZ to effect DNA partition. Proc Natl Acad Sci U S A 107:11763–11768

Okinaka RT, Cloud K, Hampton O, Hoffmaster AR, Hill KK, Keim P, Koehler TM, Lamke G, Kumano S, Mahillon J, Manter D, Martinez Y, Ricke D, Svensson R, Jackson PJ (1999) Sequence and organization of pXO1, the large Bacillus anthracis plasmid harboring the anthrax toxin genes. J Bacteriol 181:6509–6515

Oliva MA, Martin-Galiano AJ, Sakaguchi Y, Andreu JM (2012) Tubulin homolog TubZ in a phage-encoded partition system. Proc Natl Acad Sci U S A 109:7711–7716

Sakaguchi Y, Hayashi T, Kurokawa K, Nakayama K, Oshima K, Fujinaga Y, Ohnishi M, Ohtsubo E, Hattori M, Oguma K (2005) The genome sequence of Clostridium botulinum type C neurotoxin-converting phage and the molecular mechanisms of unstable lysogeny. Proc Natl Acad Sci U S A 102:17472–17477

Srinivasan R, Mishra M, Leong FY, Chiam KH, Balasubramanian M (2011) Bacillus anthracis tubulin-related protein Ba-TubZ assembles force-generating polymers. Cytoskeleton (Hoboken) 68:501–511

Szwedziak P, Wang Q, Bharat TA, Tsim M, Lowe J (2014) Architecture of the ring formed by the tubulin homologue FtsZ in bacterial cell division. Elife 3:e04601

Tan CM, Therien AG, Lu J, Lee SH, Caron A, Gill CJ, Lebeau-Jacob C, Benton-Perdomo L, Monteiro JM, Pereira PM, Elsen NL, Wu J, Deschamps K, Petcu M, Wong S, Daigneault E, Kramer S, Liang L, Maxwell E, Claveau D, Vaillancourt J, Skorey K, Tam J, Wang H, Meredith TC, Sillaots S, Wang-Jarantow L, Ramtohul Y, Langlois E, Landry F, Reid JC, Parthasarathy G, Sharma S, Baryshnikova A, Lumb KJ, Pinho MG, Soisson SM, Roemer T (2012) Restoring methicillin-resistant *Staphylococcus aureus* susceptibility to beta-lactam antibiotics. Sci Transl Med 4:126ra35

Tang M, Bideshi DK, Park HW, Federici BA (2006) Minireplicon from pBtoxis of Bacillus thuringiensis subsp. israelensis. Appl Environ Microbiol 72:6948–6954

Tang M, Bideshi DK, Park HW, Federici BA (2007) Iteron-binding ORF157 and FtsZ-like ORF156 proteins encoded by pBtoxis play a role in its replication in Bacillus thuringiensis subsp. israelensis. J Bacteriol 189:8053–8058

Tinsley E, Khan SA (2006) A novel FtsZ-like protein is involved in replication of the anthrax toxin-encoding pXO1 plasmid in Bacillus anthracis. J Bacteriol 188:2829–2835

Vaughan S, Wickstead B, Gull K, Addinall SG (2004) Molecular evolution of FtsZ protein sequences encoded within the genomes of archaea, bacteria, and eukaryota. J Mol Evol 58:19–29

Zehr EA, Kraemer JA, Erb ML, Coker JK, Montabana EA, Pogliano J, Agard DA (2014) The structure and assembly mechanism of a novel three-stranded tubulin filament that centers phage DNA. Structure 22:539–548

Chapter 12
The Structure, Function and Roles of the Archaeal ESCRT Apparatus

Rachel Y. Samson, Megan J. Dobro, Grant J. Jensen, and Stephen D. Bell

Abstract Although morphologically resembling bacteria, archaea constitute a distinct domain of life with a closer affiliation to eukaryotes than to bacteria. This similarity is seen in the machineries for a number of essential cellular processes, including DNA replication and gene transcription. Perhaps surprisingly, given their prokaryotic morphology, some archaea also possess a core cell division apparatus that is related to that involved in the final stages of membrane abscission in vertebrate cells, the ESCRT machinery.

Keywords ESCRT-III • Vps4 • cdvA • cdvB • Cytokinetic ring • Archaeal cell constriction • *Sulfolobus acidocaldarius* • AAA+ protein • MIT domain • Electron cryotomography • Cryo-ET • STIV

R.Y. Samson
Department of Molecular and Cellular Biochemistry, Indiana University,
Simon Hall MSB, 212 S Hawthorne Drive, Bloomington, IN 47405, USA
e-mail: rsamson@umail.iu.edu

M.J. Dobro
School of Natural Science, Hampshire College, Amherst, MA 01002, USA

G.J. Jensen
Division of Biology, California Institute of Technology, Pasadena, CA 91125, USA

Howard Hughes Medical Institute, California Institute of Technology,
Pasadena, CA 91125, USA
e-mail: jensen@caltech.edu

S.D. Bell (✉)
Department of Molecular and Cellular Biochemistry, Indiana University,
Simon Hall MSB, 212 S Hawthorne Drive, Bloomington, IN 47405, USA

Department of Biology, Indiana University,
Simon Hall MSB, 212 S Hawthorne Drive, Bloomington, IN 47405, USA
e-mail: stedbell@indiana.edu; sdbell123@gmail.com

J. Löwe, L.A. Amos (eds.), *Prokaryotic Cytoskeletons*, Subcellular Biochemistry 84, DOI 10.1007/978-3-319-53047-5_12

The Diversity of Archaea

Since their discovery in the late 1970's by Carl Woese, it has become apparent that archaea are abundant components of the biosphere (Woese and Fox 1977). Archaea have been isolated from a huge diversity of ecological niches, ranging from Antarctic surface waters to hydrothermal vent systems, where the organisms grow at temperatures in excess of 100 °C. Although the extremophilic archaea are probably the best known, there are many archaeal species adapted to mesophilic environments. By way of example, ammonia-oxidizing archaea are abundant in soil samples and the human microbiome is populated by a number of archaeal species, most notably the gut methanogen *Methanobrevibacter smithii* (Stahl and de la Torre 2012; Bang and Schmitz 2015). Although no archaeal pathogens have yet been described, there are examples of archaeal-bacterial consortia, archaea existing in symbiosis with marine sponges and there have been some links proposed between archaeal abundance and human health (Bang and Schmitz 2015; Schink 1997; Preston et al. 1996).

With the ever-increasing taxonomic sampling of archaea, phylogenetic analyses have revealed that the archaeal domain of life can be split into two main groupings or "super-phyla" – the Euryarchaea and the "TACK" superphylum (Guy and Ettema 2011). TACK comprises the Thaumarchaea, Aigarchaea, Crenarchaea and Korarchaea and has been proposed to be the closest grouping of archaea to the last common ancestor between Archaea and Eukaryotes (Guy and Ettema 2011). Indeed, recent work from Embley and colleagues has suggested that the divergence of Archaea and Eukaryotes occurred following the emergence of the Euryarchaeal lineage (Williams et al. 2013). This proposal has profound implications for the evolution of life on Earth, not least of which is the revelation that there are two, not three, domains of life (Fig. 12.1).

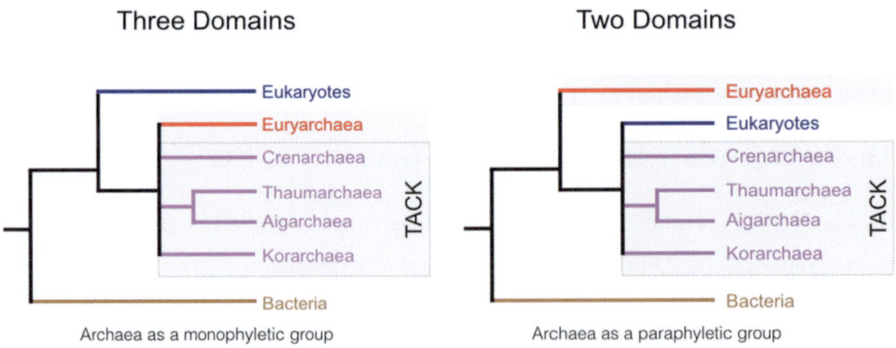

Fig. 12.1 Alternative phylogenetic trees for the evolution of cellular organisms. The *left* hand image shows the classical tree proposed by Carl Woese revealing three domains of life. The *right*-hand phylogenetic tree illustrates the recent proposal by Embley and colleagues that eukaryotes arose from within the archaeal lineage, following the divergence of the Euryarchaea. This topology is compatible with earlier proposals from the Lake laboratory (Rivera and Lake 2004; Williams et al. 2013; Woese and Fox 1977)

Interestingly, examination of the limited data available on archaeal chromosome copy number and cell-cycle organization provides hints that these processes within TACK organisms bear closer resemblance to eukaryotes than do their counterparts in euryarchaea (Samson and Bell 2011). More specifically, TACK organisms have defined gap phases between DNA synthesis and cell division (Samson and Bell 2011, 2014; Bernander 1998). Indeed, the eukaryotic cell cycle nomenclature of G1, S, G2 and M phase is commonly applied to archaeal cell cycle parameters. However, it is important to emphasize that, at the mechanistic level, archaeal M-phase is completely unrelated to eukaryotic mitosis. In contrast to the defined cell cycle phases seen in TACK species, the limited number of euryarchaea that have been studied reveal less obvious partitioning of the cell cycle. In fact, it appears that some species may have overlapping DNA replication and cell division phases. In agreement with this latter proposition, the members of the Euryarchaea generally have high chromosome copy numbers. For example, *Haloferax volcanii* has about 15 copies of its chromosome during exponential growth (Zerulla and Soppa 2014; Breuert et al. 2006). The polyploidy of euryarchaea contrasts with the TACK superphylum organisms that have been characterized, all of which show a simple one to two chromosome copy number oscillation during their cell cycle (Lundgren et al. 2008). Furthermore, studies of the crenarchaeon *Sulfolobus solfataricus* provided evidence for cohesion of sister chromatids following the completion of DNA replication (Robinson et al. 2007). Thus, at both phylogenetic and organizational levels, members of the TACK superphylum show remarkable similarities to eukaryotes.

Cell Division Machineries

Initial studies published in 1996 of the euryarchaea *Halobacterium salinarum* and *Pyrococcus woesei* by the Jackson and Margolin labs, respectively, identified archaeal homologs of the central, and near universal, bacterial cell division protein FtsZ (Margolin et al. 1996; Baumann and Jackson 1996). Subsequently, the Lutkenhaus laboratory revealed that the *Haloferax volcanii* FtsZ formed a ring-like structure at mid-cell, coincident with the site of membrane constriction during cell division (Wang and Lutkenhaus 1996). As complete archaeal genome sequences became available, all of euryarchaea initially, further FtsZ homologs were identified, revealing that this protein was found in many archaea as well as bacteria. Subsequent landmark structural studies by Lowe and Amos revealed the first crystal structure of FtsZ – using the protein from the euryarchaeon *Methanocaldococcus jannaschii* (Lowe and Amos 1998).

However, the apparent ubiquity of archaeal FtsZ was challenged by an analysis from the Doolittle laboratory of the first complete crenarchaeal genome, that of *Aeropyrum pernix*, in which it was recognized that there was no identifiable homolog of FtsZ (Faguy and Doolittle 1999; Kawarabayasi et al. 1999). The absence of FtsZ in other crenarchaea was confirmed with the elucidation of the genome sequences of *Pyrobaculum aerophilum*, and *Sulfolobus tokodaii* and *Sulfolobus solfataricus* (Fitz-Gibbon et al. 2002; She et al. 2001; Kawarabayasi et al. 2001). As detailed elsewhere in this book, an actin homolog, termed crenactin, was identified in *Pyrobaculum* and other members of the Thermoproteales (Yutin et al. 2009). Immunolocalization studies of crenactin were described as showing helical-like structures within the rod-shaped *Pyrobaculum* cells (Ettema et al. 2011). These structures were interpreted as providing evidence for a cytoskeletal role for the crenactin. However, studies of bacterial MreB provide a cautionary note for this interpretation. Initial work suggested that MreB formed a coherent helical cytoskeleton in rod-shaped bacteria (Jones et al. 2001). Immunolocalization and use of fluorescently tagged proteins supported this conclusion. More recently, however, it has been proposed that the helical structures may have arisen as a consequence of appending a fluorescent tag to the MreB protein (Swulius and Jensen 2012). Further, additional studies have led to the proposal that MreB forms dynamic circumferential bands around the cell, perpendicular to the long axis, and appears to play a key role in cell wall biosynthesis (Errington 2015; Garner et al. 2011; Dominguez-Escobar et al. 2011). With respect to the archaeal actin homolog, the current limited and low-resolution data derived from analyses of fixed cells cannot exclude the possibility that crenactin might play an analogous role in S-layer synthesis in the rod-shaped Thermoproteales. Despite the completion of the *Sulfolobus solfataricus* genome in 2001, it would be a further 6 years before candidates for the *Sulfolobus* cell division machinery were identified. Figure 12.2 summarizes the distribution of potential cytoskeletal and cell division proteins across the archaeal domain of life (Makarova et al. 2010).

ESCRT Proteins

In 2007, the laboratory of Roger Williams at the LMB in Cambridge, studying the eukaryotic ESCRT (**E**ndosomal **S**orting **C**omplex **R**equired for **T**ransport) apparatus, identified a *Sulfolobus* homolog of the ESCRT system component, Vps4. Their work revealed a striking structural relationship between the so-called MIT domain of the archaeal Vps4 with its counterpart in yeast Vps4 (Obita et al. 2007). Furthermore, they identified *Sulfolobus* homologs of the eukaryotic ESCRT-III proteins. As that initial study used the eukaryotic names for the archaeal homologs, we will retain that convention in this chapter. The eukaryotic ESCRT machinery has been comprehensively dissected at the molecular level and there are a number of excellent reviews that describe that work (Hurley 2015; Henne et al. 2013; McCullough et al. 2013). Accordingly, we will give an extremely brief and simplistic summary of the roles of ESCRT in eukaryotes.

First identified by virtue of its role in endosome sorting, the ESCRT pathway directs the invagination and scission of membranes. A series of elegant *in vitro* reconstitution studies revealed that the ESCRT-III proteins form filaments and are

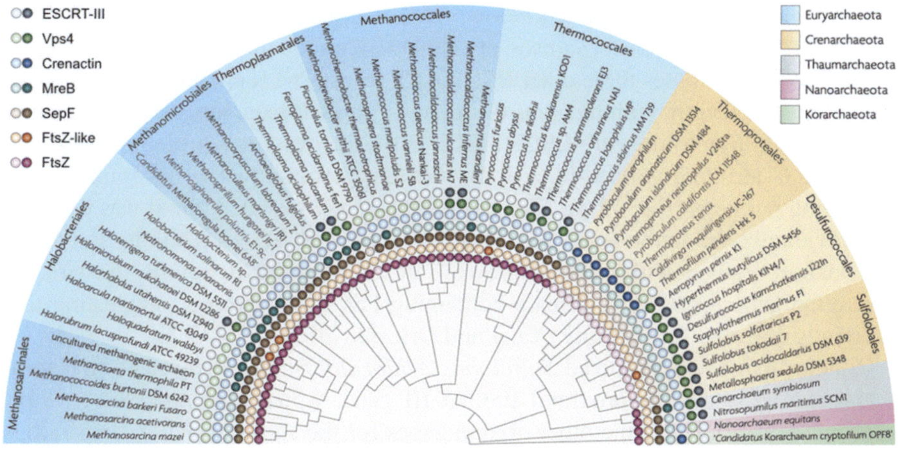

Fig. 12.2 Distribution of candidate cytoskeletal and cell division proteins across the archaeal domain of life (Figure taken from Makarova et al. 2010 with permission). Filled-in *circles* indicate the presence of one or more genes encoding homologs of the indicated proteins

necessary and sufficient to drive membrane constriction and scission in a system reconstituted with giant unilameller vesicles. Vps4 is an AAA+ ATPase that can interact with ESCRT-III, (see below for details). Dependent on its ability to hydrolyze ATP, Vps4 facilitates depolymerization and subsequent recycling of ESCRT-III to membranes. Delivery of ESCRT-III to its site of action in endosomal sorting is dependent on the sequential assembly of a protein scaffold composed of the ESCRT-0, ESCRT-I and ESCRT-II complexes. With the exception of one lineage, whose existence is inferred solely from metagenomic studies (Spang et al. 2015), these earlier ESCRT-III recruiting complexes are absent from archaea.

Following the discovery of ESCRT's role in trafficking, subsequent studies have revealed that this apparatus plays pivotal roles in a number of other processes that involve membrane scission events. These include viral egress from infected cells, plasma membrane wound healing, nuclear membrane reformation following chromosome segregation and membrane abscission – the final stage of cytokinesis in metazoan cells (Hurley 2015).

The Sulfolobus *ESCRT Machinery*

In 2008, work by our laboratory (RYS and SDB) in collaboration with the Williams lab and an independent study by Bernander and colleagues provided evidence that the *Sulfolobus* ESCRT machinery played a role in archaeal cell division (Samson et al. 2008; Lindas et al. 2008).

In *Sulfolobus*, the *vps4* gene is encoded adjacent to a gene for an ESCRT-III protein. An additional gene, called *cdvA*, is found immediately upstream of this operon (Fig. 12.3). As discussed below, *cdvA* is a distinct transcription unit from ESCRT-III and Vps4 genes (Samson et al. 2011; Wurtzel et al. 2010).

Sulfolobus possesses three other ESCRT-III genes. The four ESCRT-III paralogs are variable in length, with the gene beside *vps4* encoding the longest open-reading frame. In addition to the core ESCRT-III α-helix-rich fold, this protein possesses a signature C-terminal winged-helix-like (wH-like) domain that is not found in the remaining paralogs (Fig. 12.3). A recent genetic study revealed that it was possible to individually delete the genes for the three shorter ESCRT-III paralogs and retain cell viability, albeit with reduced growth rate. In contrast, it was not possible to delete the gene for the wH domain-containing ESCRT-III protein, suggesting this protein is essential for viability (Yang and Driessen 2014).

Studies using synchronized *Sulfolobus acidocaldarius* cells revealed that transcripts for *vps4* and its proximal ESCRT-III gene, which are expressed as a bicistronic transcript, and the other orphan genes for the shorter ESCRT-III paralogs are maximally abundant in populations enriched with cells undergoing division (Samson et al. 2008). Similarly, the *cdvA* gene transcript showed cell cycle regulation, although its levels peaked slightly before those for ESCRT-III and Vps4 (Samson et al. 2011).

As well as being cell cycle regulated, transcript levels of the genes for CdvA, ESCRT-III and Vps4 are modulated in response to a number of insults to the cell, including UV-induced DNA damage, oxidative stress caused by hydrogen peroxide, viral infection and respiratory arrest caused by acetic acid treatment (Lindas et al. 2008; Ortmann et al. 2008; Maaty et al. 2009; Gotz et al. 2007; Frols et al. 2007).

The timing of expression of the *Sulfolobus* ESCRT components is compatible with a role in cell division. Immunolocalization studies on fixed cells revealed that polyclonal antisera raised against the wH-containing ESCRT-III and Vps4 localized

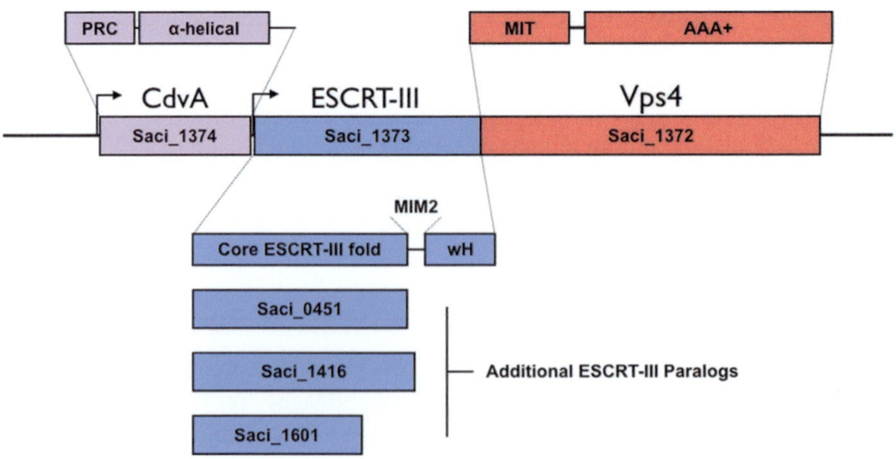

Fig. 12.3 The organization of the genes for the *Sulfolobus acidocaldarius* ESCRT-associated proteins. Principal domains of the encoded proteins are indicated. The Saci_XXXX names are from the original genome annotation (Chen et al. 2005)

to a belt between segregated nucleoids (Lindas et al. 2008; Samson et al. 2008). Furthermore, the location of the belt correlated with the site of membrane ingression in cells undergoing division. Dividing cells are very rare within asynchronous *Sulfolobus* populations, suggesting that cell division is a very rapid event. Similarly, very few cells are observed with segregated nucleoids, again suggesting that the process of chromosome segregation, the mechanism of which is entirely mysterious, is rapid and closely linked to the membrane ingression event. We note, however, that cells can be observed that have segregated nucleoids and contiguous ESCRT-III bands, and yet have no visible ingression of the membrane (Samson et al. 2008; Lindas et al. 2008). Thus, our working hypothesis is that nucleoid segregation occurs prior to ESCRT-III assembly.

A causal role for the *Sulfolobus* ESCRT system in cell division was supported by the conditional expression of a *trans*-dominant negative allele of *vps4* in *Sulfolobus solfataricus* cells (Samson et al. 2008). This allele, encoding an alanine substitution of the conserved glutamic acid residue in the so-called Walker B motif, allows ATP binding but prevents its hydrolysis. Previous work in mammalian cells revealed that expression of this and other ATPase-defective alleles impacts negatively on cell division (Carlton and Martin-Serrano 2007; Morita et al. 2007). Upon expression of this form of the protein in *Sulfolobus*, bloated cells up to four-times the usual diameter and with elevated DNA content were observed. In addition, abundant small membranous structures that lacked discernable DNA content were seen. The generation of these bloated cells and ghosts is compatible with impaired cell division processes (Samson et al. 2008).

In eukaryotes, the ESCRT-III proteins bind directly to membranes but are recruited to their sites of actions by the prior assembly of ESCRT-I and ESCRT-II complexes (Wollert and Hurley 2010). In contrast, we have been unable to detect direct membrane interaction by the *Sulfolobus* ESCRT-III proteins. Indeed, they lack the charged patch that has been shown to be important for membrane interaction by the eukaryotic proteins (Samson et al. 2011). Further, as discussed above, *Sulfolobus* lacks discernable homologs of the earlier ESCRT components.

The close linkage of the *cdvA* gene and its cell cycle regulated expression profile suggested it could be an early assembling component of the cell division machinery. In agreement with this, like ESCRT-III and Vps4, CdvA forms a ring-like structure at mid-cell that shrinks concomitant with division plane ingression. Intriguingly, mRNA levels for CdvA peak before those for ESCRT-III and Vps4, and CdvA structures can be detected prior to nucleoid segregation. Co-immunolocalization studies of late G2/early M-phase cells revealed that only a subset with CdvA structures were decorated with punctate foci of ESCRT-III. In contrast it was not possible to detect cells that were CdvA negative but ESCRT-III positive. In agreement with these observations, *in vitro* binding assays demonstrated that, while ESCRT-III was unable to bind directly to purified membranes, CdvA could. Futhermore, CdvA was capable of recruiting ESCRT-III to liposomes with consequential deformation of the membrane. Thus, taken together, these data indicate that CdvA forms a ring-like structure at mid-cell and serves as a platform for the subsequent recruitment of the ESCRT-III proteins (Samson et al. 2011).

Structures of Archaeal ESCRT Proteins

Structural studies of the ESCRT-machinery in eukaryotes and archaea have been aided by the strikingly modular nature of the proteins. As described below, this has allowed minimal interaction domains to be dissected and some highly informative structures of these domains with their partner peptides have been elucidated (Samson et al. 2008, 2011). In addition, the structures of the ATPase domain of Vps4 from *Sulfolobus solfataricus* (Sso) and *Acidianus hospitalis* (both crenarchaea of the order Sulfolobales) have been determined by the Sundquist lab using X-ray crystallography (Monroe et al. 2014) (Fig. 12.4).

Sso Vps4 is a 372 amino acid protein; the first 75 residues constitute a MIT domain, the structure of which was solved by the Williams laboratory (Obita et al. 2007). The MIT domain is separated from the AAA+ ATPase domain by a short and apparently flexible linker. As with many AAA+ proteins, Vps4 forms homo-oligomers and the Sundquist laboratory revealed Sso Vps4 forms a homohexamer in the presence of ATP or ADP following heat treatment (Monroe et al. 2014). The Sso Vps4 AAA+ domain structure does not reveal how the protein forms hexamers but modeling based on superimposition of the Sso Vps4 monomer on the AAA+ domain of P97 allowed a model to be built. The model was verified by mutation of predicted interfacial residues, resulting in a destabilization of the hexamer and consequential reduction in ATPase activity. As with many other AAA+ proteins, the ATPase active site lies at the interface between two protomers, with residues from both subunits

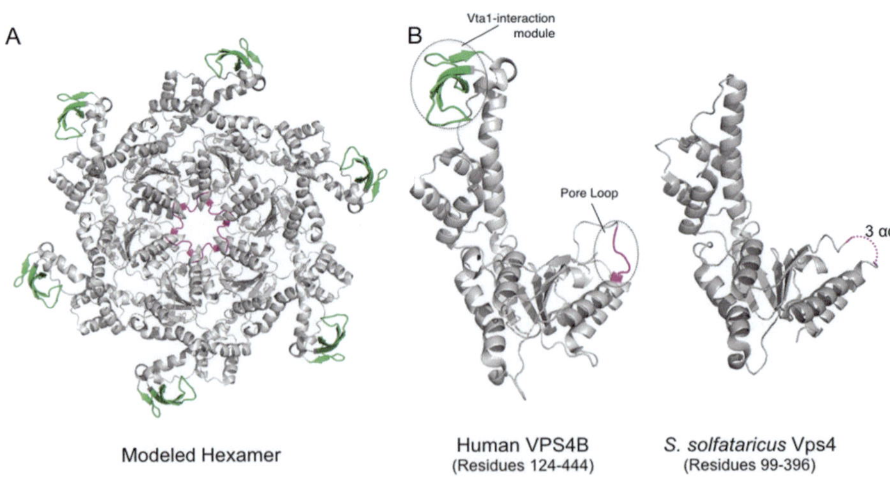

Fig. 12.4 (**a**) A modeled hexamer of human VPS4. The model was generated by superposition of the VPS4B monomer structure (PDB 1XWI) onto the symmetric hexamer structure of ClpX (PDB 4I9K). Pore loop 1 and the eukaryotic-specific Vta1 interaction module are colored magenta and green respectively (**b**) Comparison of monomers of human and *Sulfolobus* Vps4 (PDB 4LGM). 3 residues in the pore loop (*magenta*) of *S. solfataricus* Vps4 are disordered in the structure and indicated by a *dashed line*

contributing to the active site (Erzberger and Berger 2006). The predicted hexamer is ring-shaped with a central pore into which loops project (Fig. 12.4). By analogy with other AAA+ proteins, including the ClpX protein-unfoldase and DNA helicases, it is likely that these pore loops facilitate remodeling of Vps4's ESCRT-III substrate. Indeed, mutation of these loops impact upon Vps4's ability to function in human cells (Scott et al. 2005). Comparison of the structure of the AAA+ domain from Sso Vps4 with that of yeast Vps4p reveals a RMSD of 1.62 Å over 237 Cα atoms. The principal difference between the eukaryotic and archaeal structure lies in the absence or presence of a eukaryotic-specific 45-residue beta-strand-rich insertion in the smaller "lid" domain of the AAA+ module (green in Fig. 12.4). This serves as an interaction site for the eukaryotic-specific Vps4 activator protein Vta1 (Scott et al. 2005).

The N-terminal 75 residues of Vps4 form a three-helix bundle termed a MIT (**M**icrotubule **I**nteracting and **T**rafficking) domain. This domain is a versatile protein-protein interaction module that, as its name implies, has been identified in a number of proteins involved in cytoskeletal processes and trafficking, including katanin, spastin, Vps4 and Vta1 (Fig. 12.5). Studies in yeast and mammal systems revealed that the Vps4 MIT domain interacted with specific sequence motifs in

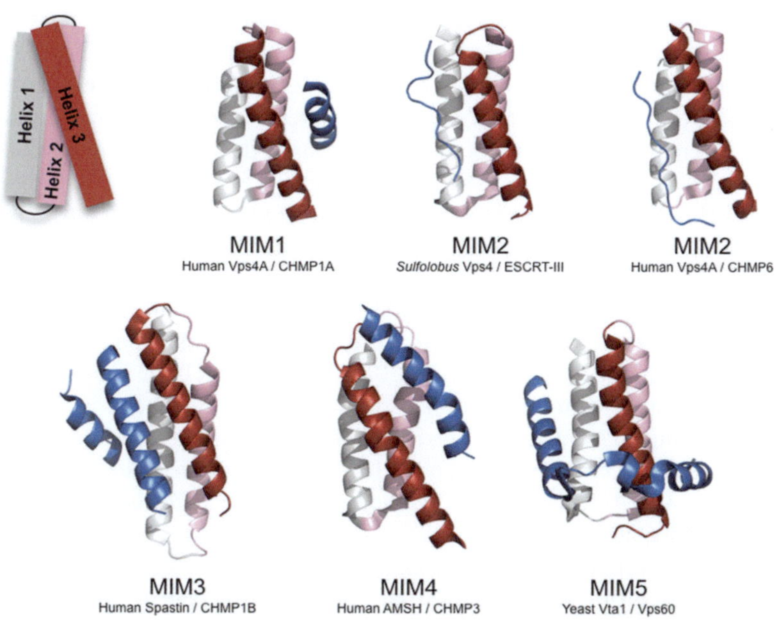

Fig. 12.5 Comparison of the structures of various MIT domains and their ESCRT-III interaction partners, demonstrating the distinct MIM1–5 binding modes. A schematic of the MIT domain is given in the *top left* and the color convention is adhered to in the subsequent images. The partner peptide is in blue. PDB codes are MIM1 – 2JQ9, *Sulfolobus* MIM2 – 2W2U, Human MIM2 – 2K3W, MIM3 – 3EAB, MIM4 – 2XZE and MIM5 – 2LUH. The CHMP and Vps60 proteins are all ESCRT-III family members

ESCRT-III proteins. Initial work identified a MIT-interacting motif (MIM) in yeast Vps2 and human CHMP1 and CHMP2B and revealed that they bound in the groove between the second and third helix of the MIT domain (Obita et al. 2007; Stuchell-Brereton et al. 2007). Subsequent work revealed that the human ESCRT-III protein CHMP6 also interacted with the MIT domain but, remarkably, utilized an alternate MIM, termed MIM2, that interacted with the groove between the first and third helix of the Vps4 MIT domain (Kieffer et al. 2008). More recently three additional modes of interaction between MIT domains and partner proteins have been described (MIM3–5; Fig. 12.5) (Yang et al. 2008, 2012; Solomons et al. 2011).

Interaction studies revealed that *Sulfolobus* Vps4 interacted specifically with the wH-containing ESCRT-III protein. The interaction interface was mapped to residues 183–193, between the core ESCRT-III fold and the C-terminal wH-like domain (Samson et al. 2008). The X-ray crystal structure of the ESCRT-III peptide•MIT domain complex revealed that the peptide from the ESCRT-III protein bound along the groove between helices 1 and 3 of the Vps4 MIT domain, utilizing the MIM2 mode of interaction (Fig. 12.5). Indeed, comparison of the archaeal and human MIM2•MIT domain interactions reveals clear conservation at the level of primary sequence (Samson et al. 2008; Kieffer et al. 2008). Thus, the ATPase Vps4 interacts with ESCRT-III subunits and ATP hydrolysis is required for Vps4 to fulfill its function. Recent work with yeast Vps4 and an artificial chimeric ESCRT-III substrate (Vps24–2) has revealed that Vps4 effects global unwinding of ESCRT-III while stripping protomers from a filament of the protein. In addition, the substrate passes through the central pore of the Vps4 during the unfolding reaction (Yang et al. 2015). While this mechanism has not yet been directly demonstrated for the archaeal proteins, given the conservation of the system and the demonstrated requirement for ATP hydrolysis by *Sufolobus* Vps4 for cell division, it seems likely that disassembly of the ESCRT-III structures seen at mid-cell will be similarly effected by Vps4.

As described above, there is currently no evidence for direct interactions of the archaeal ESCRT-III proteins with archaeal membranes. However, liposome-binding studies with Large Unilameller Vesicles (LUVs) reconstituted with purified *Sulfolobus* tetra-ether lipids revealed that the alpha-helical domain of CdvA bound directly to membranes (Samson et al. 2011). Like ESCRT-III, CdvA forms filaments, and negative stain electron microscopy provided evidence for a lattice-like structure forming on LUVs in the presence of CdvA. Strikingly, while ESCRT-III in isolation had no discernable effect on the LUVs, the addition of the wH-containing ESCRT-III to CdvA-coated LUVs resulted in dramatic deformation of the liposomes (Samson et al. 2011).

CdvA has a tri-partite structure with a ~ 70 residue beta-strand rich N-terminal domain, predicted to form a PRC barrel (Anantharaman and Aravind 2002), followed by an alpha-helix-rich region and finally a C-terminal domain that is poorly conserved in sequence apart from its final 10 amino-acids. The role of the PRC-domain remains undetermined, however, the alpha-helical region of the protein has been shown to interact with archaeal membranes. The C-terminal tail binds to the wH-like domain of the ESCRT-III paralog with a *Kd* of 6 μM. Further, disruption of this interaction *in vivo* leads to cell division defects (Samson et al. 2011).

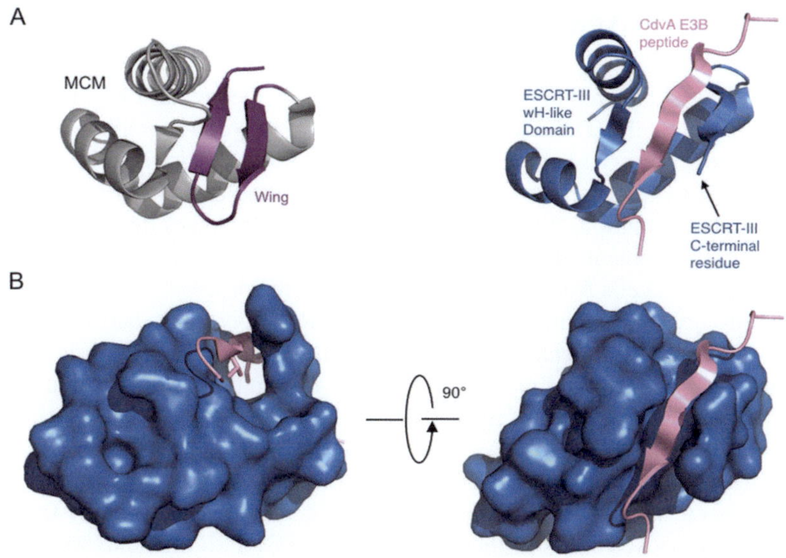

Fig. 12.6 (**a**) The *left*-hand image shows an example of the classical winged helix topology with the β-hairpin wing element highlighted in *purple*. The example shown is from *S. solfataricus* MCM (PDB 2 M45) in which this domain serves as a protein-protein interaction module (Samson et al. 2016; Wiedemann et al. 2015). The image on the *right* is the structure of the "broken wing" domain from the *S. solfataricus* ESCRT-III protein (*blue*) in complex with the E3B peptide from CdvA (*pink*), PDB 2XVC (Samson et al. 2011). (**b**) Two views, related by a 90° rotation, of a surface model of the ESCRT-III broken wing domain in complex with a cartoon view of the CdvA E3B peptide. The side-chains of two valine residues (V259 and V261) that important for the interaction are shown in stick mode in the left hand image

The crystal structure of the ESCRT-III-Binding (E3B) peptide of CdvA (residues 251–265) in complex with the wH-like domain of ESCRT-III (residues 210–259) was solved and revealed a novel mode of protein-protein interaction (Samson et al. 2011). Classical wH domains possess a three helix bundle surmounted by a pair of antiparallel β-strands that are connected by a turn of variable length (Fig. 12.6). This latter feature is the eponymous wing. Remarkably, the wH-like domain of the ESCRT-III protein is truncated, with the C-terminal residue of the protein at the end of the first of the β-strands, leaving a cleft in the surface of the protein. CdvA interacts with this cleft, in essence donating a β-strand to heal the broken wing structure. Hydrophobic residues, mutation of which impairs the interaction, dominate the inter-protein interface (Samson et al. 2011). A consequence of CdvA acting as a recruitment platform for ESCRT-III is that ESCRT-III could be spatially removed from the membrane by a distance equivalent to the depth of the CdvA lattice. As discussed below, this might have facilitated detection of the ESCRT-III membrane ingression belt by electron cryotomography (Dobro et al. 2013).

Ultrastructural Analyses of the Archaeal ESCRT Machinery

In 2013, we used electron cryotomography to explore the archaeal ESCRT machinery in *Sulfolobus* cells (Dobro et al. 2013). In electron cryotomography, samples are rapidly frozen, preserving them in a near-native, "frozen-hydrated" state. Samples are then imaged iteratively while being rotated around an axis to provide three-dimensional information. While each cryotomogram elucidates the structures present in a single moment, dynamic processes like cell division can be inferred by imaging different cells frozen at different stages of the process.

Previous images of purified eukaryotic ESCRT-III proteins had revealed that ESCRT-III proteins frequently polymerize into filaments, which then coil and spiral (Effantin et al. 2013). Adding to this body of information, we saw that purified CHMP1B, a human member of the ESCRT-III family, polymerized into helical filaments that formed 3-D bull's-eye patterns and stacked cones (Fig. 12.7). These assemblies contained layers of varying radii and angles, revealing that the lateral bonds between ESCRT filaments are non-specific, an essential property for spirals and coils. The stacked cones further revealed that ESCRT-III filaments were capable of binding each other on multiple surfaces including the top, sides, and bottom.

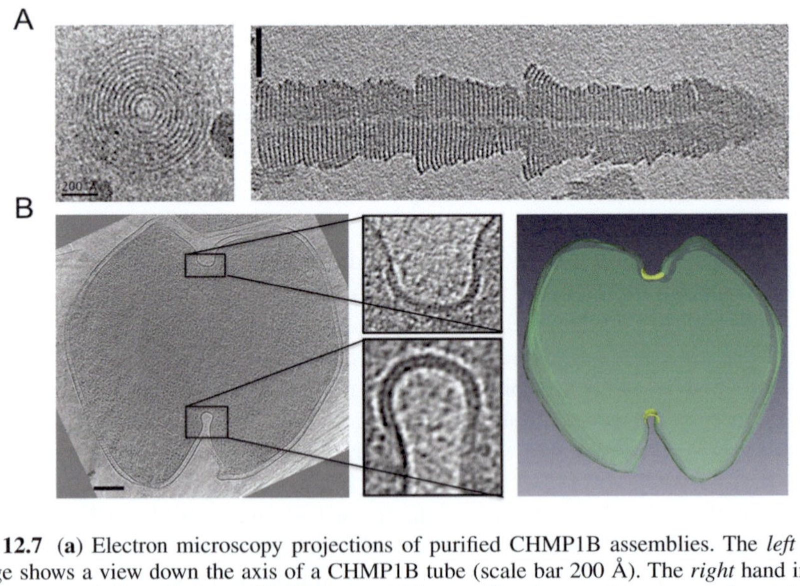

Fig. 12.7 (**a**) Electron microscopy projections of purified CHMP1B assemblies. The *left* hand image shows a view down the axis of a CHMP1B tube (scale bar 200 Å). The *right* hand image shows a central slice of a large CHMP1B assembly revealing layers of varying angles and radii. This implies that lateral bonds between filaments are non-specific and capable of binding on multiple surfaces, including the top, sides, and bottom. Scale bar is 500 Å. (**b**) A dividing *S. acidocaldarius* cell observed with electron cryotomography. The *left* hand images show a central slice of the tomogram with the division furrows magnified in the *offset* panels. The *right* hand image contains a model of that slice with the U-shaped protein coating on the division furrow colored in *yellow*. Scale bar is 200 nm (Figures adapted with permission from Dobro et al. 2013)

Moving to the archaeal proteins, we imaged mixtures of purified CdvA and liposomes made of *Sulfolobus* lipids. In this system, CdvA formed filaments of similar dimensions to ESCRT-III that wrapped around the liposomes. As described above, *Sulfolobus* ESCRT-III is recruited to membranes by CdvA. To observe these CdvA-ESCRT-III complexes in their physiological context, synchronized *Sulfolobus acidocaldarius* cells were collected at the time of division and rapidly frozen. Tomograms revealed that *every* cell displaying any membrane ingression also contained a thick, U-shaped protein belt around the constricting ring. The protein belt was also observed in cells with no visible constriction, indicating that the belt assembles before ingression begins (Dobro et al. 2013). This is unlike eukaryotic cells, which use ESCRT to divide only at the final scission stage when the diameter of the midbody site is approximately 100–200 nm (Elia et al. 2011; Agromayor and Martin-Serrano 2013).

By analogy to the properties of eukaryotic ESCRT-III's *in vitro* (Effantin et al. 2013; Dobro et al. 2013), the belts observed in dividing *Sulfolobus* cells were likely to be ESCRT-III filaments wrapped around the membrane in tight spirals. This is supported by comparing the dimensions of the belt to dimensions of known ESCRT crystal structures (Muziol et al. 2006; Bajorek et al. 2009). The belt appears dynamic in nature. During the process of constriction, the protein belt gets wider, yet the total surface area decreases. Importantly, the thickness remains the same. Together with all else that is known about the ESCRT machinery, this information supports a model we call the "hourglass" (Fig. 12.8). With mother and daughter cells stacked on top of each other, the protein belt around the middle forms an hourglass shape. This is similar to the proposed dome model for budding vesicles of eukaryotes, in which ESCRT-III forms dome-shaped spirals (Lata et al. 2008) but the symmetrical division septum forms two domes. In the hourglass model, two growing filaments (one on each side of the division plane) spiral towards each other into ever-smaller radii until the membranes are close enough for spontaneous scission (Video 12.1). While this hourglass model depends on ESCRT-III filaments being able to grow and bind each other along many surfaces as they progress from flat ribbons at the beginning to deeply invaginated U-shapes at the end, the *in vitro* assemblies demonstrate these capabilities.

The location and dimensions of the belt point to ESCRT-III as the candidate protein but, as discussed earlier, CdvA is likely linking ESCRT-III to the membrane (Samson et al. 2011). The belt was spaced about 6 nm from the membrane, which would accommodate CdvA as an adaptor (Fig. 12.8). As the membrane constrictions grew deeper in the various tomograms, the belt got wider but its total surface area decreased, suggesting Vps4 is actively re-modelling the ESCRT-III filaments during the process by removing ESCRT-III protomers. Since the diameter of the cytokinetic ring is decreasing toward zero, if the protein complexes were not being depleted, the late stage belt would have been extremely wide. Instead the belt only slightly increased in width, always at the leading edge of the division furrow.

Our images of dividing *Sulfolobus* cells do not support other models proposed in the literature These include the "purse-string" model in which Vps4 disassembles a ring filament by ESCRT-III protomer removal and resealing of the ring and the

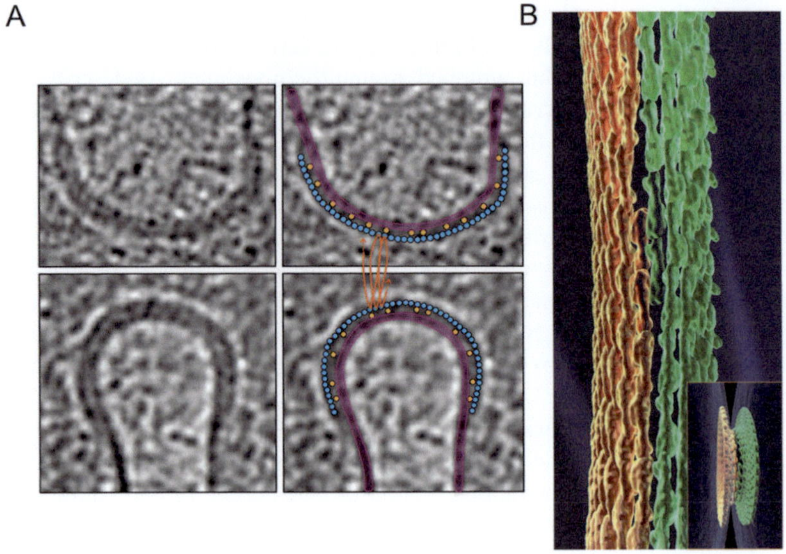

Fig. 12.8 (**a**) Interpretation of the protein belt as ESCRT helices. *Left* hand panels: central slices through opposing edges of the division furrow show the thick protein layer on the membrane. The *right* hand images show the same slices with the membrane (*purple*) and our model for how ESCRT (*blue*) filaments wrap helically (*orange lines*) around the cytokinetic ring to form the protein belt. The ESCRT filaments appear as *circles* because they are cut in cross section, but their diameter matches the width of the belt. CdvA filaments (*yellow*) link the membrane and ESCRT belt (Figures adapted with permission from Dobro et al. 2013). (**b**) Two spirals made of ESCRT monomers grow from the middle in opposite directions with a decreasing radius, pulling the membrane inward. The larger image is a close-up of the *inset* image

"whorl" model in which ESCRT-III filaments lie along the axis of a budding neck rather than spiraling around it (Saksena et al. 2009; Boura et al. 2012). Significantly, the data presented in the studies leading to those models also support the hourglass model. Filament growth in a spiraling pattern toward smaller radius may therefore drive constriction in cell division, as well as in vesicle and viral budding in eukaryotes.

Additional Roles for the Archaeal ESCRT Proteins

As described in the introduction, eukaryotic ESCRT proteins are involved in a broad range of membrane manipulation processes, both intrinsic to normal cellular metabolism but also imposed upon the cell by infectious agents (Hurley 2015). In regard to the latter, a number of enveloped viruses co-opt ESCRT proteins, including important pathogens such as HIV, Ebola, Hepatitis C, Rabies virus and Hepatitis B virus - see (Votteler and Sundquist 2013) for review. It has become apparent that there is a remarkable diversity of viruses infecting archaea (Prangishvili 2013), and studies of one, *Sulfolobus* **T**urreted **I**cosahedral **V**irus (STIV) have implicated the archaeal ESCRT machinery as an important cellular facilitator of the viral life cycle (Snyder et al. 2013). STIV is a lytic virus and utilizes a remarkable mechanism of egress from infected cells (Brumfield et al. 2009). A single 92 amino acid viral gene product, C92, assembles seven-sided pyramid structures that project through the S-layer of the host cells. Eventually the facets of the pyramids separate from one another at the vertices, like petals opening in a flower, resulting in cell lysis and virus release. Remarkably, an otherwise completely unrelated rod-shaped *Sulfolobus* virus, also encodes a C92 ortholog and directs the assembly of analogous structures (Bize et al. 2009). No further viral proteins are required for pyramid assembly and it has been demonstrated that expression of the C92 ortholog in *E. coli* or budding yeast also results in formation of pyramids, although, intriguingly, those pyramids never open in these heterologous cells (Daum et al. 2014).

STIV virus particles assemble in the *Sulfolobus* cytosol. Initially, icosahedral particles assemble and package an internal host-derived lipid envelope structure. Subsequent embellishments to the capsids occur before viral genome packaging. Ultimately the mature virus particles are released through the open pyramid structures (Fu et al. 2010; Brumfield et al. 2009). The first hints of a role for the host ESCRT proteins in the STIV life cycle came from the observation that *Sulfolobus* ESCRT-III proteins were associated with purified viral particles. In addition, analyses of the host transcriptional reprograming upon viral infection revealed up-regulation of the ESCRT genes, including CdvA, ESCRT-III paralogs and Vps4 (Ortmann et al. 2008).

Protein-protein interaction studies have revealed an interaction between one ESCRT-III paralog (SSO0619) and the major capsid protein of STIV, termed B345. While the functional significance of this interaction has not been resolved, it is tempting to speculate that it may be indicative of a role for the ESCRT machinery in the maturation of the lipid envelope within the viral particle (Snyder et al. 2013). Additionally, an interaction between the C92 pyramid protein and the wH-containing and Vps4-interacting ESCRT-III protein was detected. Immunolocalization epifluorescence microscopy and electron microscopy coupled with immunogold staining revealed Vps4 recruitment to pyramid structures both in cells infected with STIV and cells expressing only the C92 protein from a plasmid. The importance of the integrity of the host ESCRT machinery for the viral life cycle was underscored by the observation that expression of the *trans*-dominant negative, ATP hydrolysis deficient, allele of Vps4 abrogated viral replication, while not impacting the ability of STIV to enter cells and execute viral gene expression (Snyder et al. 2013). While the molecular mechanisms of host ESCRT function in the STIV life-cycle remain to be resolved, it is possible that the ESCRT apparatus could act both at the level of initial particle assembly, conceivably in maturation of the inner capsid membrane, and possibly also at the level of viral egress (Snyder et al. 2013).

In contrast to the up-regulation of the ESCRT pathway in response to STIV infection, the host response to SIRV2 revealed down-regulation of the ESCRT apparatus (Okutan et al. 2013). More specifically, mRNA levels for CdvA, ESCRT-III and Vps4 were all reduced three- to ten-fold. As described above, SIRV2 also utilizes the unusual pyramid structures to effect cell lysis. With regard to ESCRT functions in the viral life cycle, there are two potentially important differences between STIV and SIRV2. First, STIV has an internal membrane structure, SIRV2 does not. Second, the time course of infection is dramatically different for the two viruses. In the case of SIRV2 infection, pyramids appear between 6 and 9 h post-infection. This contrasts with STIV infection in which pyramids do not appear until 32 h post-infection. As SIRV2 infection leads to extensive host DNA degradation, it is possible that the down-regulation of ESCRT genes in response to SIRV2 could reflect a checkpoint-like response to inhibit cell division upon viral infection (Bize et al. 2009).

In addition to co-option by archaeal viruses, the *Sulfolobus* ESCRT apparatus may also play a role in the generation of extra-cellular vesicles. *Sulfolobus* cells secrete small vesicles during normal growth and in response to cellular stresses. Proteomic studies have demonstrated that these vesicles are associated with host ESCRT-III and Vps4 proteins (Ellen et al. 2009; Prangishvili et al. 2000). However, a causal linkage has yet to be established. As described above, a proteinaceous belt, presumed to be ESCRT-III, can be detected at mid-cell in dividing *Sulfolobus*. Intriguingly, the same tomograms reveal some cells extruding vesicles, yet examination of these structures does not provide evidence for an analogous belt-like feature at the vesicle necks (Dobro et al. 2013). Conceivably, the cell-division structure is discernable because of the role of CdvA as an adaptor between the membrane and ESCRT-III -assembly. If vesicle formation exploits a different adaptor protein, then an alternate morphology of the ESCRT-III at vesicle necks could exist. While this is complete speculation at this point, it is interesting to note that several of the *Thermococcales*, members of the euryarchaea, also produce extra-cellular vesicles and possess ESCRT-III and Vps4 homologs (Soler et al. 2008). However, they do not encode a detectable CdvA homolog and are thought to effect cell division using FtsZ (Fig. 12.2).

Future Directions

Our understanding of the biology of the archaeal ESCRT system remains very rudimentary at this time. While structural studies have illuminated the nature of the molecular handshakes that orchestrate assembly of the cytokinetic ring, how these events are appropriately coordinated in time and space remains largely unknown. The nature of the cell cycle oscillators that drive the regulated expression of the ESCRT genes is another key unresolved issue. Yet another lies in the coordinated assembly of CdvA at mid-cell. CdvA assembles prior to nucleoid segregation. Does CdvA, by polymerizing, define mid-cell and drive segregation away from its zone of assembly? Alternatively, are spatial cues established in the cell prior to CdvA assembly? What is the nature of the *Sulfolobus* DNA segregation machinery? Another utterly mysterious process lies in how S-layer assembly is coordinated with the division process. While key players in the assembly and glycosylation of S-layer proteins have been identified, essentially nothing is known about how they are regulated and localized in the cell. The tomograms of dividing cells show a dramatic restructuring of the presumptive ESCRT-III belt, from a planar to a U-shaped topography. Is this due to a belt-intrinsic process, perhaps via differential incorporation of ESCRT-III paralogs, or is it mechanically driven by S-layer growth? It is anticipated that answers to many of these questions will be forthcoming over the next few years as advances in the genetic and cell biological analyses of archaea become ever more sophisticated.

References

Agromayor M, Martin-Serrano J (2013) Knowing when to but and run: mechanisms that control cytokinetic abscission. Trends Cell Biol 23:433–441

Anantharaman V, Aravind L (2002) The PRC-barrel: a widespread, conserved domain shared by photosynthetic reaction center subunits and proteins of RNA metabolism. Genome Biol 3, RESEARCH0061

Bajorek M, Schubert HL, Mccullough J, Langelier C, Eckert DM, Stubblefield WM, Uter NT, Myszka DG, Hill CP, Sundquist WI (2009) Structural basis for ESCRT-III protein autoinhibition. Nat Struct Mol Biol 16:754–762

Bang C, Schmitz RA (2015) Archaea associated with human surfaces: not to be underestimated. FEMS Microbiol Rev 39:631–648

Baumann P, Jackson SP (1996) An archaebacterial homologue of the essential eubacterial cell division protein FtsZ. Proc Natl Acad Sci U S A 93:6726–6730

Bernander R (1998) Archaea and the cell cycle. Mol Microbiol 29:955–961

Bize A, Karlsson EA, Ekefjard K, Quax TE, Pina M, Prevost MC, Forterre P, Tenaillon O, Bernander R, Prangishvili D (2009) A unique virus release mechanism in the Archaea. Proc Natl Acad Sci U S A 106:11306–11311

Boura E, Rozycki B, Chung HS, Herrick DZ, Canagarajah B, Cafiso DS, Eaton WA, Hummer G, Hurley JH (2012) Solution structure of the ESCRT-I and -II supercomplex: implications for membrane budding and scission. Structure 20:874–886

Breuert S, Allers T, Spohn G, Soppa J (2006) Regulated polyploidy in halophilic archaea. PLoS One:1

Brumfield SK, Ortmann AC, Ruigrok V, Suci P, Douglas T, Young MJ (2009) Particle assembly and ultrastructural features associated with replication of the lytic archaeal virus sulfolobus turreted icosahedral virus. J Virol 83:5964–5970

Carlton JG, Martin-Serrano J (2007) Parallels between cytokinesis and retroviral budding: A role for the ESCRT machinery. Science 316:1908–1912

Chen L, Brugger K, Skovgaard M, Redder P, She Q, Torarinsson E, Greve B, Awayez M, Zibat A, Klenk HP, Garrett RA (2005) The genome of Sulfolobus acidocaldarius, a model organism of the Crenarchaeota. J Bacteriol 187:4992–4999

Daum B, Quax TEF, Sachse M, Mills DJ, Reimann J, Yildiz O, Hader S, Saveanu C, Forterre P, Albers SV, kuhlbrandt W, Prangishvili D (2014) Self-assembly of the general membrane-remodeling protein PVAP into sevenfold virus-associated pyramids. Proc Natl Acad Sci U S A 111:3829–3834

Dobro MJ, Samson RY, Yu Z, Mccullough J, Ding HJ, Chong PL, Bell SD, Jensen GJ (2013) Electron cryotomography of ESCRT assemblies and dividing Sulfolobus cells suggests that spiraling filaments are involved in membrane scission. Mol Biol Cell 24:2319–2327

Dominguez-Escobar J, Chastanet A, Crevenna AH, Fromion V, Wedlich-Soldner R, Carballido-Lopez R (2011) Processive movement of MreB-associated cell wall biosynthetic complexes in bacteria. Science 333:225–228

Effantin G, Dordor A, Sandrin V, Martinelli N, Sundquist WI, Schoehn G, Weissenhorn W (2013) ESCRT-III CHMP2A and CHMP3 form variable helical polymers in vitro and act synergistically during HIV-1 budding. Cell Microbiol 15:213–226

Elia N, Sougrat R, Spurlin TA, Hurley JH, Lippincott-Schwartz J (2011) Dynamics of endosomal sorting complex required for transport (ESCRT) machinery during cytokinesis and its role in abscission. Proc Natl Acad Sci U S A 108:4846–4851

Ellen AF, Albers SV, Huibers W, Pitcher A, Hobel CFV, Schwarz H, Folea M, Schouten S, Boekema EJ, Poolman B, Driessen AJM (2009) Proteomic analysis of secreted membrane vesicles of archaeal Sulfolobus species reveals the presence of endosome sorting complex components. Extremophiles 13:67–79

Errington J (2015) Bacterial morphogenesis and the enigmatic MreB helix. Nat Rev Microbiol 13:241–248

Erzberger JP, Berger JM (2006) Evolutionary relationships and structural mechanisms of AAA plus proteins. Annu Rev Biophys Biomol Struct 35:93–114

Ettema TJ, Lindas AC, Bernander R (2011) An actin-based cytoskeleton in archaea. Mol Microbiol 80:1052–1061

Faguy DM, Doolittle WF (1999) Lessons from the Aeropyrum pernix genome. Curr Biol 9:R883–R886

Fitz-Gibbon ST, Ladner H, Kim UJ, Stetter KO, Simon MI, Miller JH (2002) Genome sequence of the hyperthermophilic crenarchaeon Pyrobaculum aerophilum. Proc Natl Acad Sci U S A 99:984–989

Frols S, Gordon PM, Panlilio MA, Duggin IG, Bell SD, Sensen CW, Schleper C (2007) Response of the hyperthermophilic archaeon Sulfolobus solfataricus to UV damage. J Bacteriol 189:8708–8718

Fu CY, Wang K, Gan L, Lanman J, Khayat R, Young MJ, Jensen GJ, Doerschuk PC, Johnson JE (2010) In vivo assembly of an archaeal virus studied with whole-cell electron cryotomography. Structure 18:1579–1586

Garner EC, Bernard R, Wang W, Zhuang X, Rudner DZ, Mitchison T (2011) Coupled, circumferential motions of the cell wall synthesis machinery and MreB filaments in *B. subtilis*. Science 333:222–225

Gotz D, Paytubi S, Munro S, Lundgren M, Bernander R, White MF (2007) Responses of hyperthermophilic crenarchaea to UV irradiation. Genome Biol 8:R220

Guy L, Ettema TJ (2011) The archaeal 'TACK' superphylum and the origin of eukaryotes. Trends Microbiol 19:580–587

Henne WM, Stenmark H, Emr SD (2013) Molecular mechanisms of the membrane sculpting ESCRT pathway. Cold Spring Harb Perspect Biol 5

Hurley JH (2015) ESCRTs are everywhere. EMBO J 34:2398–2407

Jones LJ, Carballido-Lopez R, Errington J (2001) Control of cell shape in bacteria: helical, actin-like filaments in Bacillus subtilis. Cell 104:913–922

Kawarabayasi Y, Hino Y, Horikawa H, Yamazaki S, Haikawa Y, Jin-No K, Takahashi M, Sekine M, Baba S, Ankai A, Kosugi H, Hosoyama A, Fukui S, Nagai Y, Nishijima K, Nakazawa H, Takamiya M, Masuda S, Funahashi T, Tanaka T, Kudoh Y, Yamazaki J, Kushida N, Oguchi A, Kikuchi H et al (1999) Complete genome sequence of an aerobic hyper-thermophilic crenarchaeon, Aeropyrum pernix K1. DNA Res 6(83–101):145–152

Kawarabayasi Y, Hino Y, Horikawa H, Jin-No K, Takahashi M, Sekine M, Baba S, Ankai A, Kosugi H, Hosoyama A, Fukui S, Nagai Y, Nishijima K, Otsuka R, Nakazawa H, Takamiya M, Kato Y, Yoshizawa T, Tanaka T, Kudoh Y, Yamazaki J, Kushida N, Oguchi A, Aoki K, Masuda S, Yanagii M, Nishimura M, Yamagishi A, Oshima T, Kikuchi H (2001) Complete genome sequence of an aerobic thermoacidophilic crenarchaeon, Sulfolobus tokodaii strain7. DNA Res 8:123–140

Kieffer C, Skalicky JJ, Morita E, De Domenico I, Ward DM, Kaplan J, Sundquist WI (2008) Two distinct modes of ESCRT-III recognition are required for VPS4 functions in lysosomal protein targeting and HIV-1 budding. Dev Cell 15:62–73

Lata S, Schoehn G, Jain A, Pires R, Piehler J, Gottlinger HG, Weissenhorn W (2008) Helical structures of ESCRT-III are disassembled by VPS4. Science 321:1354–1357
Lindas AC, Karlsson EA, Lindgren MT, Ettema TJG, Bernander R (2008) A unique cell division machinery in the Archaea. Proc Natl Acad Sci U S A 105:18942–18946
Lowe J, Amos LA (1998) Crystal structure of the bacterial cell-division protein FtsZ. Nature 391:203–206
Lundgren M, Malandrin L, Eriksson S, Huber H, Bernander R (2008) Cell cycle characteristics of Crenarchaeota: Unity among diversity. J Bacteriol 190:5362–5367
Maaty WS, Wiedenheft B, Tarlykov P, Schaff N, heinemann J, Robison-Cox J, Valenzuela J, Dougherty A, Blum P, Lawrence CM, Douglas T, Young MJ, Bothner B (2009) Something old, something new, something borrowed; how the thermoacidophilic archaeon Sulfolobus solfataricus responds to oxidative stress. PLoS One 4:e6964
Makarova KS, Yutin N, Bell SD, Koonin EV (2010) Evolution of diverse cell division and vesicle formation systems in Archaea. Nat Rev Microbiol 8:731–741
Margolin W, Wang R, Kumar M (1996) Isolation of an ftsZ homolog from the archaebacterium Halobacterium salinarium: implications for the evolution of FtsZ and tubulin. J Bacteriol 178:1320–1327
Mccullough J, Colf LA, Sundquist WI (2013) Membrane fission reactions of the mammalian ESCRT pathway. Annu Rev Biochem 82:663–692
Monroe N, Han H, Gonciarz MD, Eckert DM, Karren MA, Whitby FG, Sundquist WI, Hill CP (2014) The oligomeric state of the active Vps4 AAA ATPase. J Mol Biol 426:510–525
Morita E, Sandrin V, Chung HY, Morham SG, Gygi SP, Rodesch CK, Sundquist WI (2007) Human ESCRT and ALIX proteins interact with proteins of the midbody and function in cytokinesis. EMBO J 26:4215–4227
Muziol T, Pineda-Molina E, Ravelli RB, Zamborlini A, Usami Y, Gottlinger H, Weissenhorn W (2006) Structural basis for budding by the ESCRT-III factor CHMP3. Dev Cell 10:821–830
Obita T, Saksena S, Ghazi-Tabatabai S, Gill DJ, Perisic O, Emr SD, Williams RL (2007) Structural basis for selective recognition of ESCRT-III by the AAA ATPase Vps4. Nature 449:735–U11
Okutan E, Deng L, Mirlashari S, Uldahl K, Halim M, Liu C, Garrett RA, She QX, Peng X (2013) Novel insights into gene regulation of the rudivirus SIRV2 infecting Sulfolobus cells. RNA Biol 10:875–885
Ortmann AC, Brumfield SK, Walther J, Mcinnerney K, Brouns SJJ, Van De Werken HJG, Bothner B, Douglas T, Van De Oost J, Young MJ (2008) Transcriptome analysis of infection of the Archaeon Sulfolobus solfataricus with Sulfolobus turreted icosahedral virus. J Virol 82:4874–4883
Prangishvili D (2013) The wonderful world of archaeal viruses. Annu Rev Microbiol 67:565–585
Prangishvili D, Holz I, Stieger E, Nickell S, Kristjansson JK, Zillig W (2000) Sulfolobicins, specific proteinaceous toxins produced by strains of the extremely thermophilic archaeal genus Sulfolobus. J Bacteriol 182:2985–2988
Preston CM, Wu KY, Molinski TF, Delong EF (1996) A psychrophilic crenarchaeon inhabits a marine sponge: Cenarchaeum symbiosum gen. nov., sp. nov. Proc Natl Acad Sci U S A 93:6241–6246
Rivera MC, Lake JA (2004) The ring of life provides evidence for a genome fusion origin of eukaryotes. Nature 431:152–155
Robinson NP, Blood KA, Mccallum SA, Edwards PAW, Bell SD (2007) Sister chromatid junctions in the hyperthermophilic archaeon Sulfolobus solfataricus. EMBO J 26:816–824
Saksena S, Wahlman J, Teis D, Johnson AE, Emr SD (2009) Functional reconstitution of ESCRT-III assembly and disassembly. Cell 136:97–109
Samson RY, Bell SD (2011) Cell cycles and cell division in the archaea. Curr Opin Microbiol 14:350–356
Samson RY, Bell SD (2014) Archaeal chromosome biology. J Mol Microbiol Biotechnol 24:420–427
Samson RY, Obita T, Freund SM, Williams RL, Bell SD (2008) A Role for the ESCRT System in Cell Division in Archaea. Science 322:1710–1713

Samson RY, Obita T, Hodgson B, Shaw MK, Chong PL, Williams RL, Bell SD (2011) Molecular and structural basis of ESCRT-III recruitment to membranes during archaeal cell division. Mol Cell 41:186–196

Samson RY, Abeyrathne PD, Bell SD (2016) Mechanism of archaeal MCM helicase recruitment to DNA replication origins. Mol Cell 61:287–296

Schink B (1997) Energetics of syntrophic cooperation in methanogenic degradation. Microbiol Mol Biol Rev 61:262–280

Scott A, Chung HY, Gonciarz-Swiatek M, Hill GC, Whitby FG, Gaspar J, Holton JM, Viswanathan R, Ghaffarian S, Hill CP, Sundquist WI (2005) Structural and mechanistic studies of VPS4 proteins. EMBO J 24:3658–3669

She Q, Singh RK, Confalonieri F, Zivanovic Y, Allard G, Awayez MJ, Chan-Weiher CCY, Clausen IG, Curtis BA, De Moors A, Erauso G, Fletcher C, Gordon PMK, Heikamp-De Jong I, Jeffries AC, Kozera CJ, Medina N, Peng X, Thi-Ngoc HP, Redder P, Schenk ME, Theriault C, Tolstrup N, Charlebois RL, Doolittle WF, Duguet M, Gaasterland T, Garrett RA, Ragan MA, Sensen CW, Van Der Oost J (2001) The complete genome of the crenarchaeon Sulfolobus solfataricus P2. Proc Natl Acad Sci U S A 98:7835–7840

Snyder JC, Samson RY, Brumfield SK, Bell SD, Young MJ (2013) Functional interplay between a virus and the ESCRT machinery in archaea. Proc Natl Acad Sci U S A 110:10783–10787

Soler N, Marguet E, Verbavatz JM, Forterre P (2008) Virus-like vesicles and extracellular DNA produced by hyperthermophilic archaea of the order Thermococcales. Res Microbiol 159:390–399

Solomons J, Sabin C, Poudevigne E, Usami Y, Hulsik DL, Macheboeuf P, Hartlieb B, Gottlinger H, Weissenhorn W (2011) Structural basis for ESCRT-III CHMP3 recruitment of AMSH. Structure 19:1149–1159

Spang A, Saw JH, Jorgensen SL, Zaremba-Niedzwiedzka K, Martijn J, Lind AE, Van Eijk R, Schleper C, Guy L, Ettema TJG (2015) Complex archaea that bridge the gap between prokaryotes and eukaryotes. Nature 521:173–179

Stahl DA, De La Torre JR (2012) Physiology and diversity of ammonia-oxidizing archaea. Annu Rev Microbiol 66:83–101

Stuchell-Brereton MD, Skalicky JJ, Kieffer C, Karren MA, Ghaffarian S, Sundquist WI (2007) ESCRT-III recognition by VPS4 ATPases. Nature 449:740–744

Swulius MT, Jensen GJ (2012) The helical MreB cytoskeleton in *Escherichia coli* MC1000/pLE7 is an artifact of the N-Terminal yellow fluorescent protein tag. J Bacteriol 194:6382–6386

Votteler J, Sundquist WI (2013) Virus budding and the ESCRT pathway. Cell Host Microbe 14:232–241

Wang X, Lutkenhaus J (1996) FtsZ ring: the eubacterial division apparatus conserved in archaebacteria. Mol Microbiol 21:313–319

Wiedemann C, Szambowska A, Hafner S, Ohlenschlager O, Guhrs KH, Gorlach M (2015) Structure and regulatory role of the C-terminal winged helix domain of the archaeal minichromosome maintenance complex. Nucleic Acids Res 43:2958–2967

Williams TA, Foster PG, Cox CJ, Embley TM (2013) An archaeal origin of eukaryotes supports only two primary domains of life. Nature 504:231–236

Woese CR, Fox GE (1977) Phylogenetic structure of the prokaryotic domain: the primary kingdoms. Proc Natl Acad Sci U S A 74:5088–5090

Wollert T, Hurley JH (2010) Molecular mechanism of multivesicular body biogenesis by ESCRT complexes. Nature 464:864–U73

Wurtzel O, Sapra R, Chen F, Zhu YW, Simmons BA, Sorek R (2010) A single-base resolution map of an archaeal transcriptome. Genome Res 20:133–141

Yang N, Driessen AJ (2014) Deletion of cdvB paralogous genes of Sulfolobus acidocaldarius impairs cell division. Extremophiles 18:331–339

Yang D, Rismanchi N, Renvoise B, Lippincott-Schwartz J, Blackstone C, Hurley JH (2008) Structural basis for midbody targeting of spastin by the ESCRT-III protein CHMP1B. Nat Struct Mol Biol 15:1278–1286

Yang Z, Vild C, Ju J, Zhang X, Liu J, Shen J, Zhao B, Lan W, Gong F, Liu M, Cao C, Xu Z (2012) Structural basis of molecular recognition between ESCRT-III-like protein Vps60 and AAA-ATPase regulator Vta1 in the multivesicular body pathway. J Biol Chem 287:43899–43908
Yang B, Stjepanovic G, Shen Q, Martin A, Hurley JH (2015) Vps4 disassembles an ESCRT-III filament by global unfolding and processive translocation. Nat Struct Mol Biol 22:492–498
Yutin N, Wolf MY, Wolf YI, Koonin EV (2009) The origins of phagocytosis and eukaryogenesis. Biol Direct 4:9
Zerulla K, Soppa J (2014) Polyploidy in haloarchaea: advantages for growth and survival. Front Microbiol 5:274

Chapter 13
Archaeal Actin-Family Filament Systems

Ann-Christin Lindås, Karin Valegård, and Thijs J.G. Ettema

Abstract Actin represents one of the most abundant and conserved eukaryotic proteins over time, and has an important role in many different cellular processes such as cell shape determination, motility, force generation, cytokinesis, amongst many others. Eukaryotic actin has been studied for decades and was for a long time considered a eukaryote-specific trait. However, in the early 2000s a bacterial actin homolog, MreB, was identified, characterized and found to have a cytoskeletal function and group within the superfamily of actin proteins. More recently, an actin cytoskeleton was also identified in archaea. The genome of the hyperthermophilic crenarchaeon *Pyrobaculum calidifontis* contains a five-gene cluster named Arcade encoding for an actin homolog, Crenactin, polymerizing into helical filaments spanning the whole length of the cell. Phylogenetic and structural studies place Crenactin closer to the eukaryotic actin than to the bacterial homologues. A significant difference, however, is that Crenactin can form single helical filaments in addition to filaments containing two intertwined proto filaments. The genome of the recently discovered Lokiarchaeota encodes several different actin homologues, termed Lokiactins, which are even more closely related to the eukaryotic actin than Crenactin. A primitive, dynamic actin-based cytoskeleton in archaea could have enabled the engulfment of the alphaproteobacterial progenitor of the mitochondria, a key-event in the evolution of eukaryotes.

Keywords Actin superfamily in archaea • Crenarchaea • *Pyrobaculum calidifontis* • Crenactin • MreB - Arcadin • Archaeal cytoskeleton • Helical filaments

---

A.-C. Lindås (✉)
Department of Molecular Biosciences, The Wenner-Gren Institute, Stockholm University, Svante Arrhenius v. 20C, SE-106 91 Stockholm, Sweden
e-mail: ann.christin.lindas@su.se

K. Valegård
Department of Cell and Molecular Biology/Molecular Biophysics, Uppsala University, Box 596, SE-751 24 Uppsala, Sweden
e-mail: karin.valegard@icm.uu.se

T.J.G. Ettema
Department of Cell and Molecular Biology/Molecular Evolution, Uppsala University, Box 596, SE-751 24 Uppsala, Sweden
e-mail: thijs.ettema@icm.uu.se

J. Löwe, L.A. Amos (eds.), *Prokaryotic Cytoskeletons*, Subcellular Biochemistry 84, DOI 10.1007/978-3-319-53047-5_13

Introduction

Actin is not only the most abundant protein in eukaryotic cells but also the most conserved protein over time. The amino acid sequence of actin from rabbit and yeast share 88% identity (Erickson 2007) and there are no changes in amino acid sequence between skeletal muscle actin from chicken compared to human (Hennessey et al. 1993). In the eukaryotic cell, actin has a plethora of functions such as cell shape determination, motility, cytokinesis, cellular trafficking of organelles and vesicles etc. (Carlier et al. 2015; Davidson and Wood 2016; Köster and Mayor 2016; Pollard and Cooper 2009). These activities are possible due to the dynamic nature of the actin filament. Monomeric actin (G-actin) polymerizes into protofilaments, which in a parallel manner, form a right-handed double helical filament known as F-actin. Polymerization is regulated by the binding and hydrolysis of ATP bound to actin (Holmes et al. 1990; Kabsch et al. 1990; Fujii et al. 2010). Furthermore, the filament dynamics are modulated by different actin binding proteins (ABPs), e.g. profilin, formins, the Arp2/3 complex, which act on the polymerization rate and branching of the actin filament (Goodson and Hawse 2002; Vorobiev et al. 2003).

The eukaryotic actin cytoskeleton has been the subject of investigations for decades (Frixione 2000) whereas prokaryotic cells until recently were though to lack a cytoskeleton (Erickson 2007). In 2001 the bacterial actin homolog MreB was seen to form helical filaments spanning the whole cell and MreB was shown to have a role in cell shape determination (Jones et al. 2001). Later the same year, the three dimensional crystal structure was solved for MreB and confirmed the affiliation of MreB to the superfamily of actin proteins (van den Ent et al. 2001). A more comprehensive review of the bacterial actin cytoskeleton, including MreB, can be found in Chap. 8.

Cytoskeletal components of cells from the third domain of life – the Archaea – have not been studied until recently. Protein homologs of bacterial MreB (Ettema et al. 2011) as well as an actin homolog, Ta0583, with unknown function, have been identified (Roeben et al. 2006) but the first archaeal actin cytoskeleton identified and investigated was the Arcade system (actin-related cytoskeleton in Archaea involved in shape determination), from the hyperthermophilic crenarchaeon *Pyrobaculum calidifontis* (Ettema et al. 2011).

An Archaeal Actin Homolog – Crenactin

Pyrobaculum calidifontis is a hyperthermophilic crenarchaeon of the order Thermoproteales. The cells are rod-shaped and grow heterotrophically at 90–95 °C and neutral pH (Amo et al. 2002). Its genome is approximately 2 Mbp and contains the Arcade cluster encompassing five genes, of which one encodes for an actin homolog named Crenactin (Ettema et al. 2011). Phylogenetic analysis places Crenactin in a group of ATPases including eukaryotic actin, sugar kinases, heat

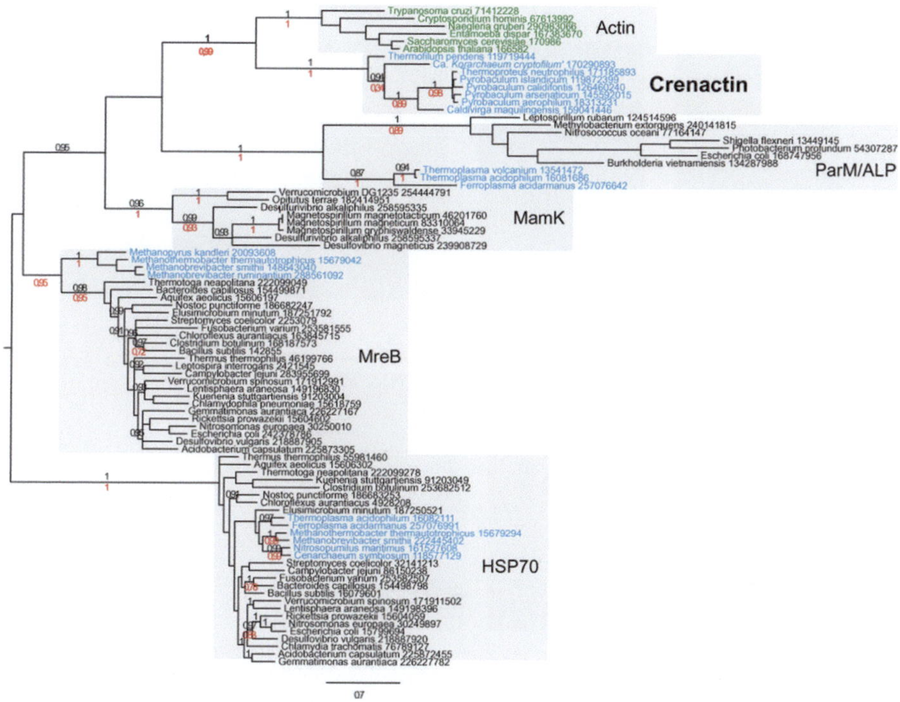

Fig. 13.1 A phylogenetic tree based on Bayesian analysis of the actin family of ATPases. The HSP70 family of proteins was used as outgroup. Bacterial, eukaryotic and archaeal sequences are shown in *black*, *green* and *blue*, respectively (The figure is taken from Ettema et al. 2011)

shock proteins and bacterial actin homologues (MreB, MamK, ParM) (Fig. 13.1). Crenactin shows the closest relationship with eukaryotic actin, whereas the bacterial actin homologues are more distantly related (Ettema et al. 2011).

Crenactin Forms Spiral Structures Spanning the Cell

The size of the rod-shaped *P. calidifontis* is approximately 0.6 μm in diameter and up to 8 μm in length (Amo et al. 2002). The presence of Crenactin correlates with rod-shaped cell morphologies within the crenarchaeal order Thermoproteales and also with that of *Ca.* Korarchaeum cryptophilum, an unrelated archaeon belonging to the Korarchaeota (Elkins et al. 2008). The cellular localization of Crenactin, determined with immunofluorescence microscopy, shows a similar structure as the bacterial MreB with spirals spanning the whole length of the cell (Fig. 13.2a) (Ettema et al. 2011).

In agreement with the position of Crenactin within the group of actin ATPases, Crenactin displays an NTP hydrolysis activity, with the highest activity against ATP

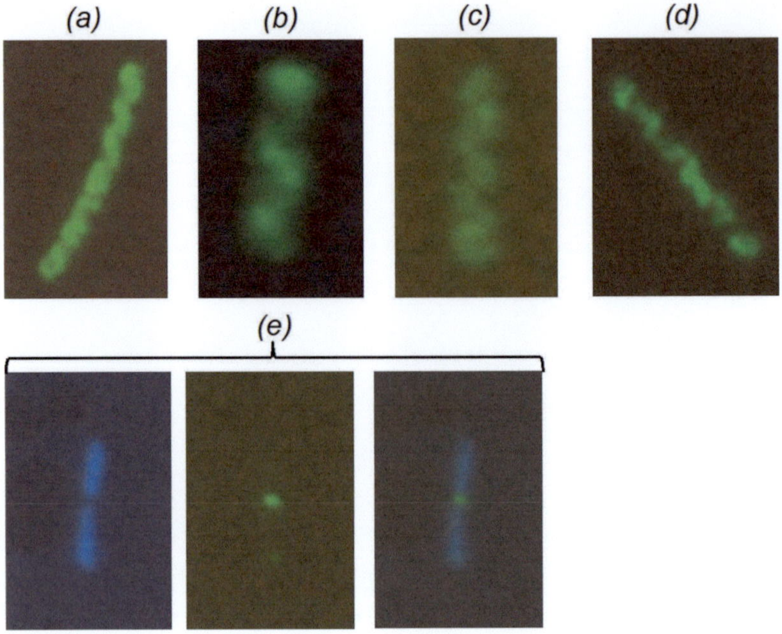

Fig. 13.2 Cellular localization of Crenactin and respective Arcadin determined by immunofluorescence microscopy. (**a**) The Crenactin protein arranged in spiral-shaped structures traversing the cell of *P. calidifontis*. (**b**) Arcadin-1, (**c**) Arcadin-3, (**d**) Arcadin-4, (**e**) Arcadin-2. *Left panel*: two segregated chromosomes stained with DAPI (*blue*); *middle panel*: Arcadin-2 stained with anti-Arcadin-2 antibodies (*green*) and *right panel*: overlay of *left* and *middle panel* (The figure is adapted from Ettema et al. 2011)

and GTP. In addition, Crenactin also displays a low activity against CTP, UTP and dNTPs, similar to bacterial MreB (van den Ent et al. 2001), whereas eukaryotic actins only hydrolyse ATP (Pollard and Cooper 1986).

Exposure to Cytochalasins B or D, which inhibit polymerization of eukaryotic actin, has no effect on growing cells in terms of cell morphology *in vivo* or on ATPase activity *in vitro* (Ettema et al. 2011). Furthermore, the bacterial MreB inhibitor A22 (*S*-(3,4-dichlorobenzyl)isothiourea) did not result in any growth defects or loss of cell shape *in vivo*. However, elevated levels of A22 (in mM range, compared to μM-range for MreB) significantly decreased the ATPase activity of Crenactin *in vitro*. These results may indicate that the mechanism of filament formation for Crenactin is different compared to eukaryotic actin and bacterial MreB (Ettema et al. 2011), as will be further discussed in section “The Crenactin filament”.

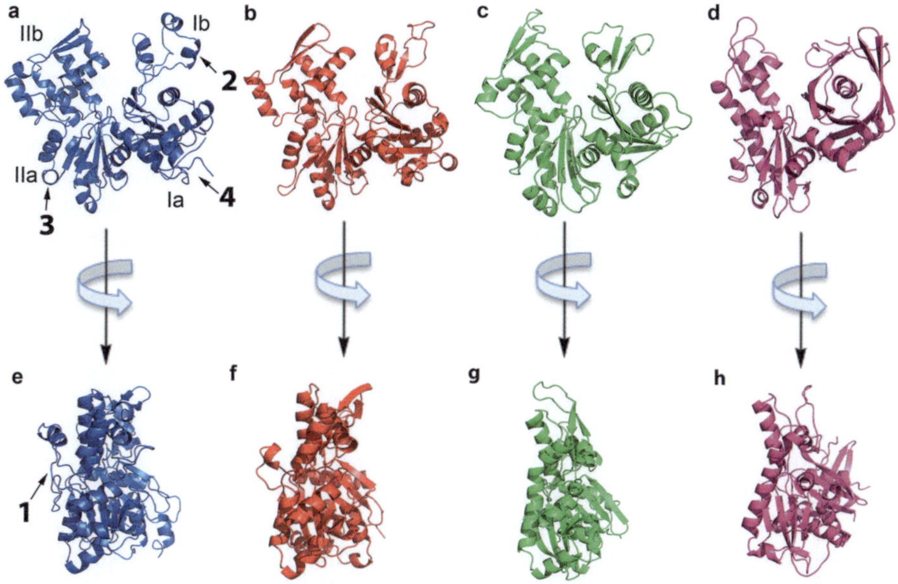

Fig. 13.3 Crystal structures of the monomer of (**a**) Crenactin from *P. calidifontis* bound to ADP (PDB 4bql), (**b**) actin from *S. cervisiae* bound to ATP (PDB 1yag), (**c**) MreB from *T. maritima* bound to AMP-PNP (PDB 1jcg) and (**d**) Ta0583 from *Thermoplasma acidophilum* (PDB 2fsj). Four major differences between Crenactin and the other actin family members are indicated by *arrows* and *numbers* in the Crenactin structure (The figure is adapted from Lindås et al. 2014)

The Structure of Crenactin

The Crystal Structure of the Crenactin Monomer

Both structures of actin and MreB have been extensively investigated (Mishra et al. 2014; Eun et al. 2015 and references therein) and recently, also the structure of Crenactin was determined (Lindås et al. 2014; Izoré et al. 2014). Two different methods were applied for resolving the Crenactin structure: (i) single-wavelength anomalous dispersion (SAD) on selenomethionine-substituted protein (Lindås et al. 2014) and (ii) molecular replacement using eukaryotic actin (PDB 1yvn) as a starting model followed by multi-crystal averaging (Izoré et al. 2014). The structure was solved to 3.35 and 3.2 Å, respectively. In the study by Lindås et al., Crenactin crystallized into an orthorhombic space group $P2_12_12_1$ with the unit-cell dimensions a = 73.9, b = 88.2 and c = 421.6 Å, respectively. There are four Crenactin monomers in the asymmetric unit. The dimensions of a monomer are 56 × 56 × 38 Å, which is similar to actin and MreB. Crenactin shares less than 20% amino acid sequence identity with eukaryotic actin but the overall structure of Crenactin confirms its belonging to the actin protein family (Fig. 13.3) (Lindås et al. 2014; Izoré et al. 2014).

The actin monomer consists of two domains, I and II, between which a cleft is formed providing a binding site for ATP and also the coordination of a divalent metal ion, Ca^{2+}or Mg^{2+}. The two domains are further divided into two subdomains (Ia, Ib, IIa and IIb, respectively) where the Ia and IIa fold into a five-stranded β-sheet flanked by α-helices, the hallmark of the actin family proteins. Subdomains Ib and IIb, on the other hand, vary more in size and structure. Subdomain Ib in Crenactin consists of three short helices while both actin and MreB contains a β-sheet (Fig. 13.3). The N- and C-termini are both located in subdomain Ia of Crenactin as well as in actin, MreB and Ta0583 (Lindås et al. 2014; Roeben et al. 2006). Compared with actin, MreB and Ta0583, there are four major differences in the Crenactin structure highlighted in Fig. 13.3 and in the alignment (Fig. 13.4). The most significant difference is a long insertion (residues 293–325) between subdomain IIa and IIb in Crenactin that is significantly shorter or absent in the other three (Fig. 13.3, Number 1) (Lindås et al. 2014; Izoré et al. 2014). The extension is called the 'hydrophobic plug' in actin and its function has been proposed in interactions between protofilaments in F-actin. The second difference is a loop at the top of subdomain Ib, present in Crenactin and actin but absent in MreB and Ta0583, responsible for the interaction between the monomers in the actin filament (Fig. 13.3, Number 2). Thirdly, the presence of an extra loop (residue 358–370) in Crenactin not present in actin, MreB or Ta0853 (Fig. 13.3, Number 3) and finally, the folding of the C-terminus of Crenactin (Fig. 13.3, Number 4) (Lindås et al. 2014).

In the bottom of the cleft, formed between domain I and II, there are three conserved phosphate-binding loops (P-loops, P1, P2 and P3) that coordinate the binding of the nucleotide (Fig. 13.5).

The residues Gly353, Gly3554, Ala355, Trp358 and Arg239, in Crenactin, form a hydrophobic pocket that binds the purine base of the nucleotide and the ribose and phosphate make several interactions with residues in the active site. A comparison of the structure of the active site of Crenactin, actin and MreB show extensive similarities with two conserved glycines, Gly353 and Gly354, and a stabilizing α-helix (Fig. 13.5) (Lindås et al. 2014).

The Crenactin Filament

Crenactin monomers polymerize into right-handed helical filaments observed in the crystal (Fig. 13.6) (Lindås et al. Lindås et al. 2014; Izoré et al. 2014) but also in cryo-electron microscopy (cryo-EM) (Braun et al. 2015). It requires eight subunits in order to complete a full turn of the helix. Consequently, the longitudinal spacing between subunits is 53 Å (Lindås et al. 2014; Izoré et al. 2014), comparable to 55 Å and 51 Å in actin and MreB, respectively. The rotation, on the other hand, of the Crenactin filament is approximately 45° per monomer (Izoré et al. 2014), as an average of several filaments, in contrast to F-actin which rotates 27° per monomer (Egelman et al. 1982).

The monomers form interactions at two sites (Fig. 13.6c). At the first site, subdomain IIa interacts with IIb of the next monomer and, at the second site, subdomain

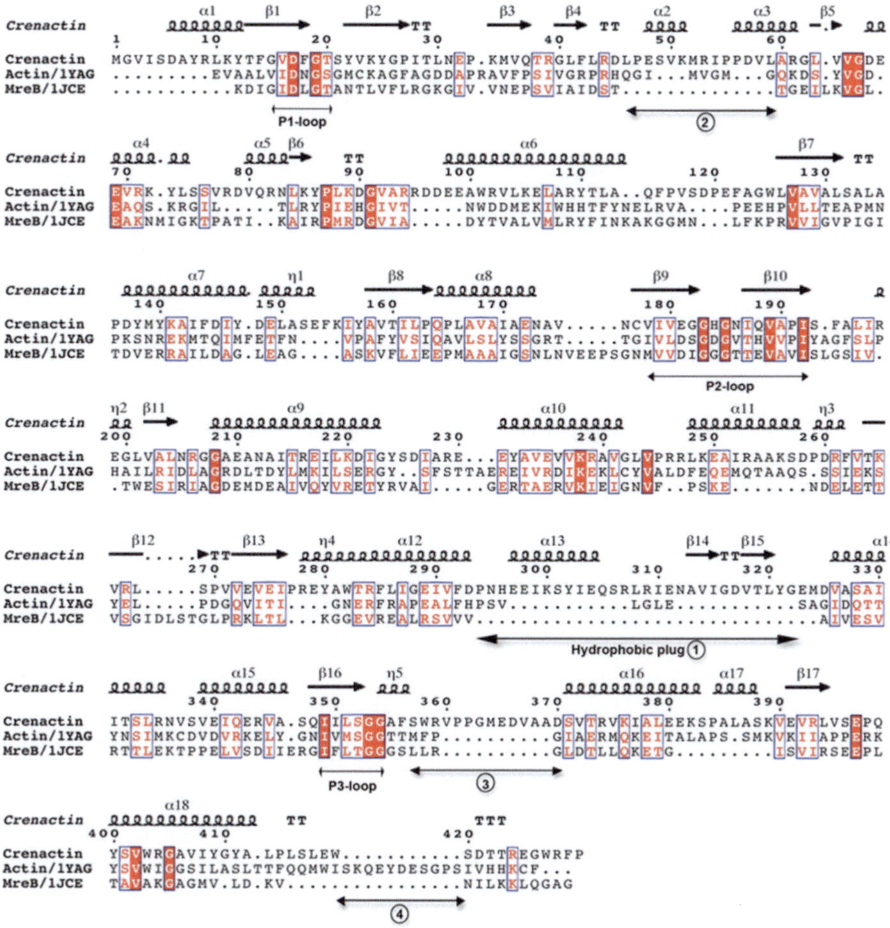

Fig. 13.4 A structure-based sequence alignment of Crenactin (*P. calidifontis*, PDB 4bql), actin (*S. cerevisiae,* PDB 1yag) and MreB (*T. maritima*, PDB 1jcg) using DALI (Holm and Rosenström 2010). The major structure differences between Crenactin, actin and MreB are indicated with numbers 1–4, also shown in Fig. 13.3. The three conserved P-loops are also marked as well as the hydrophobic plug. *Boxes* indicate homologous regions, *white* letters on *red* background indicate identical amino acid residues and *red* letters indicate functionally equivalent amino acid residues. Residues interacting with ADP are shown on cyan background and residues interacting with A22 in MreB are marked with an *asterisk* (The figure is taken from Lindås et al. 2014)

IIb interacts with both subdomains IIa and Ia of the next monomer (Lindås et al. 2014; Izoré et al. 2014). The amino acid residues involved in the interactions between monomers are highly conserved comparing Crenactin and actin, even though the rotation of the monomers varies. A mutation of Val339 to Lysine in the interface between monomers of Crenactin abrogated the filament formation (Izoré et al. 2014). The so-called "binding hot spots" (residues contributing to the free energy) in the interfaces between monomers are less conserved between Crenactin

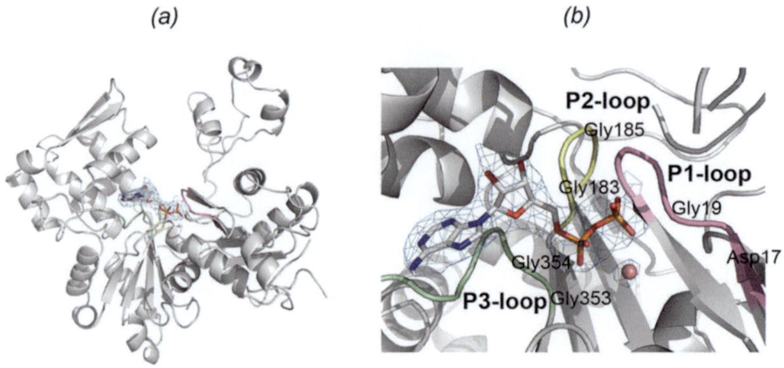

Fig. 13.5 The crystal structure of Crenactin bound to ADP. (**a**) ADP and a divalent metal ion bound in the nucleotide binding site. (**b**) A close-up of the binding pocket with the three conserved P-loops highlighted, *pink* (P1), *yellow* (P2) and *green* (P3). A selection of conserved amino acid residues interacting with the nucleotide is also indicated (The figure is taken from Lindås et al. 2014)

and actin (Braun et al. 2015). Furthermore, in these interfaces, Crenactin has 13–17 hydrogen bonds but actin has only 5 and 8. The number of salt bridges, on the other hand, is high in actin whereas Crenactin has almost none. This difference in binding energies is the most probable reason for the Crenactin filament being stable at 90 °C *in vivo* and also *in vitro*. It also explains the ability of Crenactin to form single filaments, seen both in the crystals and also with cryo-EM, and not requiring double filaments as in F-actin (Braun et al. 2015).

As mentioned earlier, Crenactin polymerization *in vivo* and ATPase hydrolysis *in vitro* were not affected by the actin polymerization inhibitors cytochalasin B and D and ATPase hydrolysis was affected by the MreB-polymerization-inhibitor A22 only at elevated concentrations (Ettema et al. 2011). This may be explained by comparing the structure of Crenactin with those of actin and MreB. In actin, cytochalasin binds into a hydrophobic cleft between subdomain Ia and IIa (Cooper 1987), which is not conserved in Crenactin in terms of amino acid residues and charge (Lindås et al. 2014). In MreB, A22 has been shown to interact with the residues Glu131, Thr158 and Val315 in the nucleotide-binding site (Bean et al. 2009; van den Ent et al. 2014). Corresponding residues in Crenactin are Gln164, Asn186 and Val402, which may explain the decreased inhibitory effect of A22 on ATPase hydrolysis in Crenactin (Lindås et al. 2014).

The Arcadins

As mentioned previously, the Crenactin gene, *cren1*, is part of an cluster encompassing four other genes, *rkd-1, -2, -3* and *-4*, encoding the proteins Arcadin-1 to -4 (Ettema et al. 2011; Bernander et al. 2011). Arcadin-4 shows similarity to archaeal

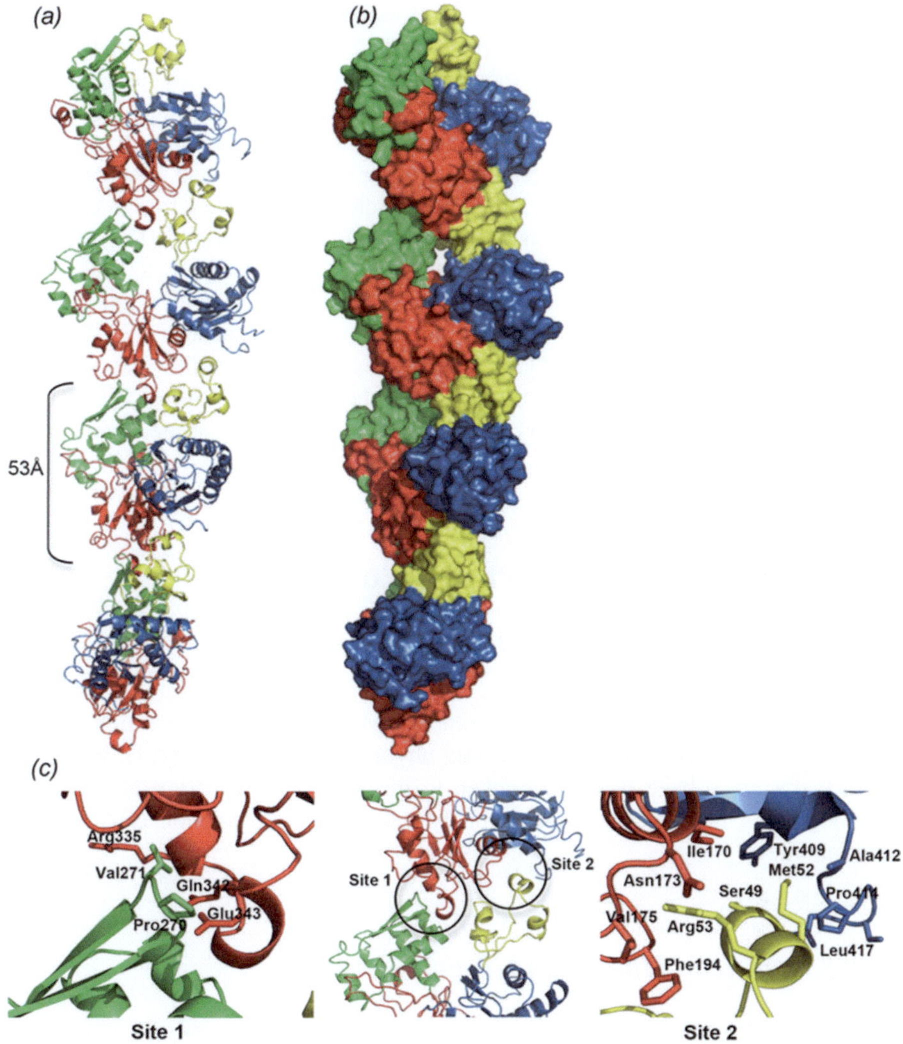

Fig. 13.6 Structure of the Crenactin filament. (**a**) A ribbon representation with subdomain Ia in *blue*, Ib in *yellow*, IIa in *red* and IIb in *green*. (**b**) A surface model with the subdomains coloured as in (**a**). (**c**) A close-up of the monomer-monomer interaction interface (The figure is taken from Lindås et al. 2014)

and bacterial SMC (structural maintenance of chromosomes) proteins, whereas Arcadin-1, -2 and -3 are found only within the crenarchaeal order Thermoproteales, and show no homology to any protein with known function. Immunostaining and microscopy of Arcadins displayed a similar distribution of these proteins, except Arcadin-2, as for Crenactin (Fig. 13.2b–e) (Ettema et al. 2011). However, Arcadin-1, -3, -4 do not display any ATPase activity, and do not seem to polymerize on their own and, therefore, the most likely scenario is that these arcadins interact with the

Crenactin filament (Ettema et al. 2011). The function of the Arcadins is not known to date, but both eukaryotic actin and bacterial actin homologs are associated with actin binding proteins involved in modifying and regulating the dynamics of the respective cytoskeleton structures. Therefore, these are also potential functions of the Arcadins in terms of regulating Crenactin polymerization dynamics.

Arcadin-2 showed a different pattern, forming distinct foci and, in a subset of cells, located between segregated nucleoids, which may indicate a role in chromosome remodelling and/or cell division (Fig. 13.2e) (Ettema et al. 2011). The cell division machinery is still unknown for organisms belonging to the order Thermoproteales since they do not harbour genes for the Cdv system, found in other crenarchaeal species, or for FtsZ, found among the euryarchaeaota (Lindås et al. 2008). It cannot be excluded that Crenactin and associated Arcadins play a role in cell division (Makarova et al. 2010). However, the presence of an FtsZ-encoding gene in the *Ca.* Korarchaeum cryptophilum genome would argue against this scenario.

Actin Homologs Within the Phylum Euryarchaeota

An Archaeal Homolog of the Bacterial MreB

Genes coding for the bacterial actin homolog MreB were first identified in rod-shaped methanogens of the euryarchaeota phylum (Ettema et al. 2011). However, no further studies have been undertaken to investigate the function of this actin homolog, even though it is most probably a cytoskeleton function in cell-shape determination.

Ta0583, an Actin Homolog in **T. acidophilum**

In 2006, Roeben et al. reported an actin homolog Ta0583 in *Thermoplasma acidophilum*, a euryarchaeon of the order Thermoplasmatales which grows optimally at a temperature of 56 °C and at pH 2. Similar to actin, MreB and Crenactin, Ta0583 displays ATPase activity but, like Crenactin, Ta0583 is also able to hydrolyse GTP, CTP and UTP, though at considerably lower activity compared to ATP hydrolysis (Roeben et al. 2006; Hara et al. 2007). The polymerization rate increases with temperature, reaching a maximum at 65 °C (Hara et al. 2007). The MreB inhibitor A22 is also a competitive inhibitor of Ta0583, but addition of A22 to *T. acidophilum* cultures does not affect the growth of the cells (Roeben et al. 2006). The crystal structure and phylogenetic analyses place Ta0583 closer, in particular, to the bacterial plasmid segregation protein ParM than to actin (Ettema et al. 2011).

The Crystal Structure of Ta0583

The crystal structure of Ta0583 was determined to 2.1 Å resolution using SAD data collected on selenomethionine-substituted protein (Roeben et al. 2006) (Fig. 13.3d). Ta0583 crystallizes in the space group $P2_12_12_1$ with unit-cell dimensions a = 51.3, b = 50.8 and c = 123.4 Å. Similar to actin, the Ta0583 monomer consists of two domains, referred to as lobe I and lobe II, with the overall monomer dimensions 60 Å × 55 Å × 30 Å. Each lobe is further divided into two subdomains, Ia, Ib, IIa and IIb, respectively. Sequence comparison of Ta0583 with actin, MreB and the bacterial plasmid segregation protein ParM reveals a poor preservation with only four amino acid residues completely conserved between the four proteins. These residues are, however, all located in the nucleotide binding site. Ta0583 displays a simpler domain structure compared to the other actin family members, which implicates a lack of regulatory mechanisms by, for example, actin binding proteins.

The crystal structure of Ta0583 bound to ADP was also determined to 2.9 Å resolution (Roeben et al. 2006). The structure showed that upon binding of ADP, lobe II undergoes an extensive rotation, facilitating interaction with the nucleotide. This major conformational change is also seen for ParM upon nucleotide binding but not for MreB, actin or Crenactin.

Two parallel protofilaments of Ta0583 interact to form filaments similar to MreB in *T. maritima*. The two parallel protofilaments only make contacts with one atom per monomer but considerably stronger contacts are made with neighbouring anti-parallel filaments, resulting in the formation of crystalline sheets (Roeben et al. 2006).

The cellular abundance of Ta0583, approximately 0.04% of total protein content in the cell compared with 8% actin in unicellular eukaryotes, is considered too low to provide a cytoskeletal function (Roeben et al. 2006). However, Ta0583 is up-regulated during certain conditions in the cell, suggesting a role in segregation of cell components during fission or in maintaining cell organization.

Evolution of the Actin Cytoskeleton

The initial identification of a Crenactin-based cytoskeleton in archaea belonging to the TACK (Thaumarchaeota, Aigarchaeota, Crenarchaeota, Korarhaeota) superphylum (Guy and Ettema 2011) suggests that the actin-based cytoskeleton present in eukaryotic cells was already established before the split of Archaea and Eukarya (Ettema et al. 2011). This idea was reinforced with the recent discovery of the Lokiarchaeota (Spang et al. 2015). Analysis of the lokiarchaeal composite genome resulted in the identification of several different actin homologues, which, based on phylogenetic analysis, turned out to be more closely related to eukaryotic actin and actin-like proteins than the previously identified Crenactins (Fig. 13.7). In addition, the composite genome was also found to encode genes for short proteins containing Gelsolin-like domains. In eukaryotes these domains are found in various Actin

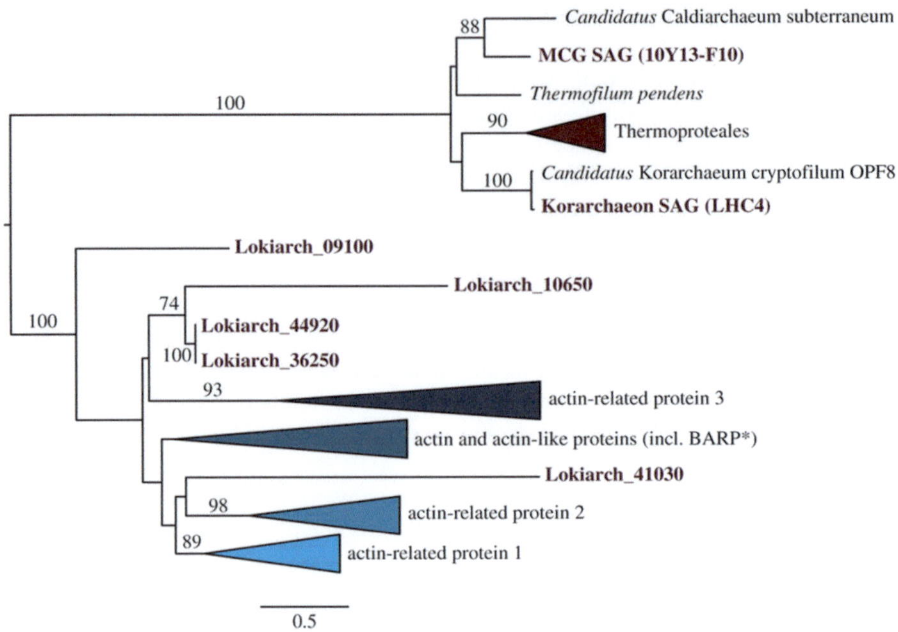

Fig. 13.7 Maximum-likelihood phylogeny of 378 aligned amino acid residues of eukaryotic actins, Arp1-3 homologues and Crenactin, including homologues identified in Lokiarchaeum as well as the MCG (Miscellaneous Crenarchaeotal Group) SAG (single-cell amplified genome) (10Y13-F10) and Korarchaeon SAG (LHC4). Only bootstrap values larger than 70 are shown. *BARP, bacterial actin related protein (The figure is taken from Saw et al. 2015)

Binding Proteins of the villin/gelsolin protein family, which are involved in regulation of actin filament assembly and disassembly. Despite that these domains have previously not been identified in prokaryotes, and that the function of the proteins in Lokiarchaeota is still unknown, this suggests that Lokiarchaeota may, like eukaryotes, have a dynamic actin cytoskeleton (Spang et al. 2015). The presence of an actin-based cytoskeleton in the archaeal ancestor of eukaryotes holds relevance for the mechanism via which the eukaryotic cell might have evolved. The evolution of phagocytosis is believed to have played a key role in eukaryogenesis. The presence of a dynamic actin-based cytoskeleton could have provided the archaeal ancestor of eukaryotes with primitive phagocytic capabilities, which could have enabled the engulfment of the alphaproteobacterial progenitor of the mitochondria, a key-event in the evolution of eukaryotes.

References

Amo T, Paje MLF, Inagaki A et al (2002) *Pyrobaculum calidifontis* sp. nov., a novel hyperthermophilic archaeon that grows in atmospheric air. Archaea 1:113–121

Bean GJ, Flickinger ST, Westler WM et al (2009) A22 disrupts the bacterial actin cytoskeleton by directly binding and inducing a low-affinity state in MreB. Biochemistry 48:4852–4857
Bernander R, Lind AE, Ettema TJG (2011) An archaeal origin for the actin cytoskeleton. Commun Integr Biol 4:664–667
Braun T, Orlova A, Valegård K et al (2015) Archaeal actin from a hyperthermophile forms a single-stranded filament. Proc Natl Acad Sci U S A 112:9340–9345
Carlier MF et al (2015) Control of polarized assembly of actin filaments in cell motility. Cell Mol Life Sci 72:3051–3067
Cooper JA (1987) Effects of cytochalasin and phalloidin on actin. J Cell Biol 105:1473–1478
Davidson AJ, Wood W (2016) Unravelling the actin cytoskeleton: a new competitive edge. Trends Cell Biol 26(8):569–576
Egelman EH, Francis N, DeRosier DJ (1982) F-actin is a helix with a random variable twist. Nature 298:131–135
Elkins JG, Podar M, Graham DE et al (2008) A korarchaeal genome reveals insights into the evolution of the Archaea. Proc Natl Acad Sci U S A 105:8102–8107
Erickson HP (2007) Evolution of the cytoskeleton. Bioessays 29:668–677
Ettema TJG, Lindås AC, Bernander R (2011) An actin-based cytoskeleton in archaea. Mol Microbiol 80:1052–1061
Eun YJ, Kapoor M, Hussain S et al (2015) Bacterial filament systems: toward understanding their emergent behaviour and cellular functions. J Biol Chem 290:17181–17189
Frixione E (2000) Recurring views on the structure and function of the cytoskeleton: a 300-year epic. Cell Motil Cytoskeleton 46:73–94
Fujii T, Iwane AH, Yanagida T et al (2010) Direct visualization of secondary structures of F-actin by electron cryomicroscopy. Nature 467:724–728
Goodson HV, Hawse WF (2002) Molecular evolution of the actin family. J Cell Sci 115:2619–2622
Guy L, Ettema TJG (2011) The archaeal 'TACK' superphylum and the origin of eukaryotes. Trends Microbiol 19:580–587
Hara F, Yamashiro K, Nemoto N et al (2007) An actin homolog of the archaeon *Thermoplasma acidophilum* that retains the ancient characteristics of eukaryotic actin. J Bacteriol 189:2039–2045
Hennessey ES, Drummond DR, Sparrow JC (1993) Molecular genetics of actin function. Biochem J 291:657–671
Holm L, Rosenström P (2010) Dali server: conservation mapping in 3D. Nucleic Acids Res 38:545–549
Holmes KC, Popp D, Gebhard W et al (1990) Atomic model of the actin filament. Nature 347:44–49
Izoré T, Duman R, Kureisaite-Ciziene D et al (2014) Crenactin from *Pyrobaculum calidifontis* is closely related to actin in structure and forms steep helical filaments. FEBS Lett 588:776–782
Jones LJF, Carballido-López R, Errington J (2001) Control of cell shape in bacteria: helical, actin-like filaments in *Bacillus subtilis*. Cell 104:913–922
Kabsch W, Mannherz HG, Suck D et al (1990) Atomic structure of the actin:DNase I complex. Nature 347:37–44
Köster DV, Mayor S (2016) Cortical actin and the plasma membrane: inextricably intertwined. Curr Opin Cell Biol 38:81–89
Lindås AC, Karlsson EA, Lindgren MT et al (2008) A unique cell division machinery in the Archaea. Proc Natl Acad Sci U S A 105:18942–18946
Lindås AC, Chruszcz M, Bernander R et al (2014) Structure of crenactin, an archaeal actin homologue active at 90 °C. Acta Crystallogr D Biol Crystallogr D70:492–500
Makarova KS, Yutin N, Bell SD et al (2010) Evolution of diverse cell division and vesicle formation systems in Archaea. Nat Rev Microbiol 8:731–741
Mishra M, Huang J, Balasubramanian MK (2014) The yeast actin cytoskeleton. FEMS Microbiol Rev 38:213–227

Pollard TD, Cooper JA (1986) Actin and actin-binding proteins. A critical evaluation of mechanisms and functions. Annu Rev Biochem 55:987–1035
Pollard TD, Cooper JA (2009) Actin, a central player in cell shape and movement. Science 326:1208–1212
Roeben A, Kofler C, Nagy I et al (2006) Crystal structure of an archaeal actin homolog. J Mol Biol 358:145–156
Saw JH, Spang A, Zaremba-Niedzwiedzka K et al (2015) Exploring microbial dark matter to resolve the deep archaeal ancestry of eukaryotes. Philos Trans R Soc Lond Ser B Biol Sci 370:20140328
Spang A, Saw JH, Jørgensen SL et al (2015) Complex archaea that bridge the gap between prokaryotes and eukaryotes. Nature 521:173–179
van den Ent F, Amos LA, Löwe J (2001) Prokaryotic origin of the actin cytoskeleton. Nature 413:39–44
van den Ent F, Izoré T, Bharat TA, Johnson CM, Löwe J (2014) Bacterial actin MreB forms antiparallel double filaments. Elife 3:e02634
Vorobiev S, Strokopytov B, Drubin DG et al (2003) The structure of nonvertebrate actin: implications for the ATP hydrolytic mechanism. Proc Natl Acad Sci U S A 100:5760–5765

Chapter 14
The Tubulin Superfamily in Archaea

Christopher H.S. Aylett and Iain G. Duggin

Abstract In comparison with bacteria and eukaryotes, the large and diverse group of microorganisms known as archaea possess a great diversity of cytoskeletal proteins, including members of the tubulin superfamily. Many species contain FtsZ, CetZ and even possible tubulins; however, some major taxonomic groups do not contain any member of the tubulin superfamily. Studies using the model archaeon, *Halferax volcanii* have recently been instrumental in defining the fundamental roles of FtsZ and CetZ in archaeal cell division and cell shape regulation. Structural studies of archaeal tubulin superfamily proteins provide a definitive contribution to the cytoskeletal field, showing which protein-types must have developed prior to the divergence of archaea and eukaryotes. Several regions of the globular core domain – the "signature" motifs – combine in the 3D structure of the common molecular fold to form the GTP-binding site. They are the most conserved sequence elements and provide the primary basis for identification of new superfamily members through homology searches. The currently well-characterised proteins also all share a common mechanism of GTP-dependent polymerisation, in which GTP molecules are sandwiched between successive subunits that are arranged in a head-to-tail manner. However, some poorly-characterised archaeal protein families retain only some of the signature motifs and are unlikely to be capable of dynamic polymerisation, since the promotion of depolymerisation by hydrolysis to GDP depends on contributions from both subunits that sandwich the nucleotide in the polymer.

Keywords CetZ • FtsZ • Tubulin superfamily in archaea • TubZ • Archaeal cell shape • Archaeal cell division • Haloarchaea • Thaumarchaeota • Korarchaeota • Euryarchaeota • Halobacteria • Thermococci

C.H.S. Aylett
Department of Biology, Institute for Molecular Biology and Biophysics, Eidgenössische Technische Hochschule (ETH) Zürich, Zürich, Switzerland
e-mail: christopher.aylett@mol.biol.ethz.ch

I.G. Duggin (✉)
The iThree Institute, University of Technology Sydney, Sydney, NSW 2007, Australia
e-mail: Iain.Duggin@uts.edu.au

J. Löwe, L.A. Amos (eds.), *Prokaryotic Cytoskeletons*, Subcellular Biochemistry 84, DOI 10.1007/978-3-319-53047-5_14

Introduction

Archaea are a large and diverse group of microorganisms known for their remarkable metabolic diversity and capacity to thrive in harsh environments. They represent one of the three domains (or superkingdoms) of life, and have distinct relationships with eukaryotes and bacteria. They share a closer phylogenetic relationship to eukaryotes than bacteria, because the primary biosynthetic machineries needed in all cells/for DNA, RNA and protein synthesis – are relatively closely-related in eukaryotes and archaea compared to the distantly related machineries found in bacteria. On the other hand, archaea and bacteria share a similar prokaryotic cellular organisation, lacking the membrane-bound organelles and nucleus of the larger and more complex eukaryotic cells. Archaea also have unique and interesting characteristics, including alternative metabolic pathways, different membrane lipid and cell envelope structures, and archaea-specific families of cytoskeletal proteins. These organisms offer great potential to reveal new insights into the functions, evolution and diversity of cytoskeletal systems. Archaea encode a great diversity of proteins homologous to the well-known eukaryotic and bacterial cytoskeletal proteins, as well as families specific for archaea.

Whereas some proteins relating to basic cellular structure and function have been conserved in most prokaryotes since the early divergence of the archaeal and bacterial lineages (*e.g.* FtsZ), or in other cases acquired from one another later, recent studies have also identified archaeal homologs of proteins fundamental to eukaryotic cell structure and function. These include, among others, the primary filament-forming proteins of the cytoskeleton, actin (Chap. 13) and tubulin, and certain proteins of the Endosomal Sorting Complexes Required for Transport (ESCRT-III), which is a membrane remodelling system involved in cell division in Crenarchaeota (see Chap. 12). Some of these protein families must have developed prior to the divergence of archaea and eukaryotes, with some lineages of archaea retaining these where function has dictated.

This chapter describes the distribution, functions and structures of cytoskeletal proteins belonging to the tubulin superfamily in archaea, including FtsZ, CetZ, tubulin and several other diverse groups of unknown function. Proteins from the tubulin superfamily share a common molecular fold related to GTP-binding. In all superfamily members, there are several regions within the globular core domain—the "signature" motifs—that combine in the 3D structure to form the GTP-binding site. These most conserved sequence elements provide the primary basis for identification of new superfamily members through homology searches. Across diverse superfamily members characterized so far, these proteins also share a common mechanism of GTP-dependent polymerization, in which GTP molecules are sandwiched between successive subunits that are arranged in a head-to-tail manner. However, some poorly characterised archaeal protein families described further below are unlikely to be capable of this mode of polymerisation, and retain only some of the signature motifs of the core domain. Depolymerisation is favoured upon hydrolysis to GDP, which normally occurs through the contributions of both subunits that sandwich the nucleotide in the polymer. Regulated polymerisation and depolymerisation provide the basis for the assembly and movement of structural filaments that are the foundation of many cytoskeletal systems across the three domains of life.

The superfamily includes the numerous tubulin families that function to establish the microtubule-based cytoskeleton in eukaryotes, the FtsZs that provide the foundation for the bacterial, archaeal and some eukaryotic organelle division machineries, and several diverse relatives involved in microbial plasmid and viral segregation processes or with currently unknown functions. In comparison with bacteria and eukaryotes, the archaea possess a great diversity of these proteins, and yet some major taxonomic groups do not contain a tubulin superfamily member. Studies using the model archaeon *Haloferax volcanii* were instrumental in defining the fundamental roles of FtsZ and CetZ in archaeal in cell division and cell shape regulation, respectively, and we explain why this species is well-suited as a model for future studies of the archaeal cytoskeleton and archaeal cell biology. We also outline the original contributions that structural studies of archaeal tubulin superfamily proteins have provided to the broader cytoskeletal field, and compare the recently-described CetZ family with other well-studied tubulin-superfamily proteins.

Distribution, Diversity and Functions of the Tubulin Superfamily in Archaea

Members of the tubulin superfamily are widespread in archaea. Figure 14.1a shows a phylogenetic tree of the tubulin superfamily in the currently known major archaeal taxa, with representative groups of diverse FtsZ proteins (from bacteria and plants) and tubulins (eukaryotes) also included for comparison. A great diversity of archaeal homologs has now been identified. Despite this, archaeal tubulin superfamily proteins generally show a more limited distribution amongst the sub-groups of Archaea, when compared to the near-ubiquity of FtsZ and tubulin within Bacteria and Eukaryota, respectively. Figure 14.1b shows particular homologs of the named families that have been identified in the indicated archaeal taxa. Some archaea do not possess a clear tubulin superfamily homolog, including the major groupings of the Crenarchaeota, which use ESCRT-III (CdvB) based cell division machineries instead (Lindas et al. 2008; Samson et al. 2008). Many other archaea encode multiple tubulin superfamily homologs. Several major taxa of archaea contain members of a third family of cellular tubulin-superfamily proteins, called CetZ, whose members are clearly distinguishable from FtsZ and tubulin, and they have so far been found to control archaeal cell shape. Some archaea also possess the highly divergent FtsZ-like (FtsZL1) proteins, and other non-canonical members, of currently unknown function (Fig. 14.1a, grey branches).

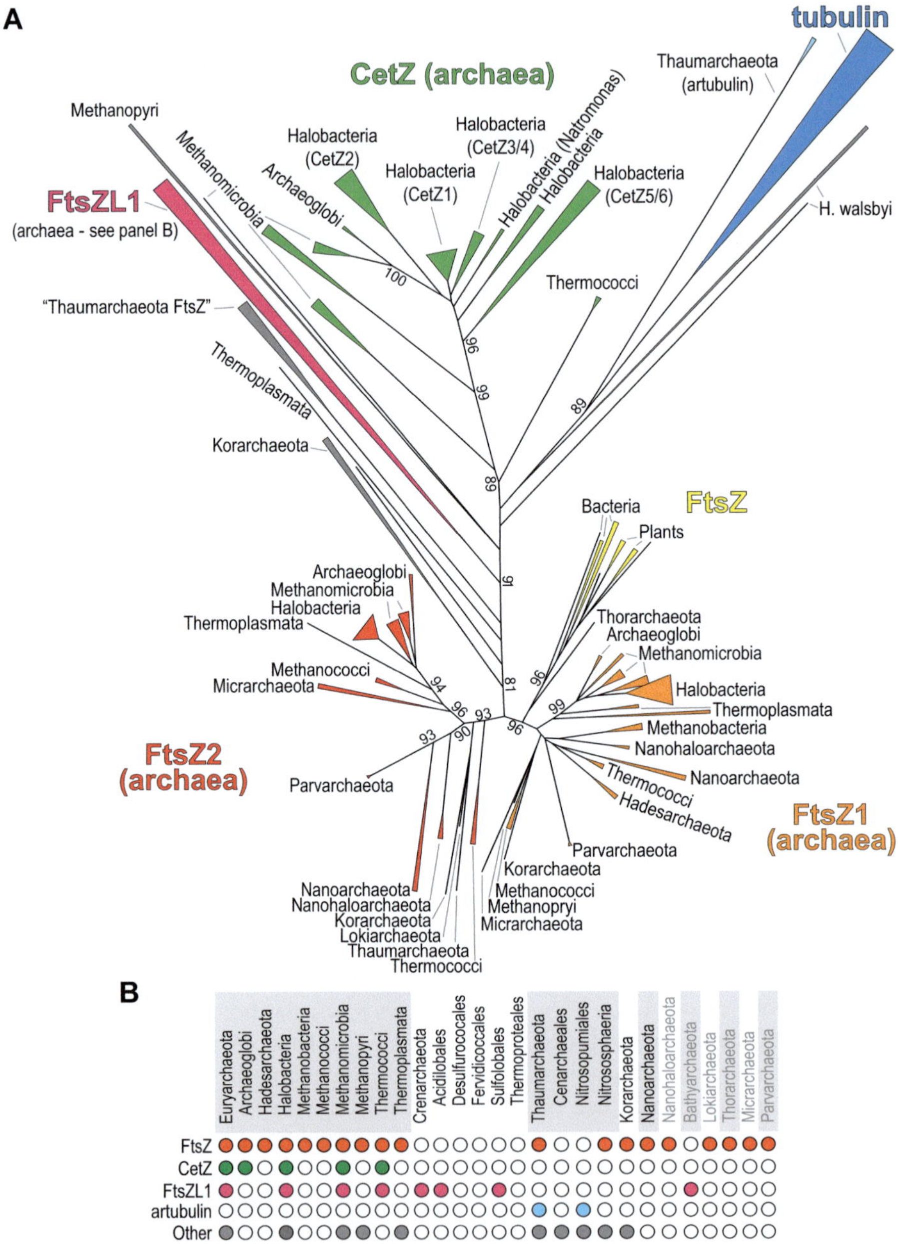

Fig. 14.1 Molecular phylogeny of the tubulin superfamily in archaea. (**a**) Radial phylogram of the tubulin superfamily, generated (by *FastTree* (Price et al. 2010)) from the conserved blocks of a multiple sequence alignment of 271 tubulin superfamily members from across the major archaeal taxa, including representatives of tubulin and FtsZ from eukaryotes and bacteria. Bootstrap support is shown for selected branches. The previously-described families are represented by colour, as shown. (**b**) Identification of the indicated protein families in currently-known major archaeal taxa. A *filled circle* indicates that a family member has been identified in at least one species of this taxon. Taxa shown in *grey* text indicate. Candidate taxa or taxa for which only incomplete or scaffold-status genome sequence data are available

Archaeal FtsZ

FtsZ proteins are tubulin-superfamily proteins whose primary function is in forming the contractile ring for division of prokaryotic cells (bacterial and archaea), and organelles (specifically plastids, and mitochondria of some protists). They are the most wide-spread of the major groups of the tubulin superfamily, and generally show the greatest sequence similarity across diverse taxa (Fig. 14.1a). We refer to FtsZ-family proteins as those that sit within the branches indicated in Fig. 14.1a, for which there is evidence that they are involved in division, *i.e.* FtsZ (bacterial type, yellow), FtsZ1 (archaea, orange) or FtsZ2 (archaea, red).

Identification of FtsZ in Archaea

The first tubulin superfamily proteins identified in archaea were from the FtsZ family. DNA hybridization approaches were used to clone *ftsZ* genes from two model species from the Halobacteria Class, *Haloferax volcanii* (Wang and Lutkenhaus 1996) and *Halobacterium salinarum* (Margolin et al. 1996). At the same time, isolation of GTP-binding proteins from cell lysates was used to identify an FtsZ from *Pyrococcus woesei* (Baumann and Jackson 1996), a hyperthermophilic archaeon from the Class Thermococci. Not long after these discoveries, the first complete archaeal genome sequence, of *Methanocaldococcus jannaschii*, revealed that there were two *ftsZ* genes present in this organism (Bult et al. 1996), in contrast to most bacteria, which contain only one *ftsZ*. More recent genome sequencing has confirmed that many archaea with FtsZ have two homologs from separate families called FtsZ1 and FtsZ2 (Vaughan et al. 2004) (Fig. 14.1a). The original clones of *ftsZ* from *H. volcanii* (Wang and Lutkenhaus 1996) and *P. woesei* (Baumann and Jackson 1996) were FtsZ1 family members, whereas the gene originally isolated from *H. salinarum* (Margolin et al. 1996) happened to be this organism's FtsZ2 family member.

Some archaea have only one FtsZ family member (Fig. 14.1a), including species of the Euryarchaeota Classes Methanobacteria and Methanopyri, as well as recently described candidate phyla Lokiarchaeota (Spang et al. 2015), Hadesarchaea (Baker et al. 2016), Thorarchaeota (Seitz et al. 2016), and certain groups of the Thaumarchaeota (*Candidatus Caldiarchaeum* and *Nitrososphaera*)(Nunoura et al. 2011; Zhalnina et al. 2014). Furthermore, FtsZ has not been specifically identified in the Crenarchaeota, Bathyarchaeota (also known as the Miscellaneous Crenarchaeota Group (MCG) organisms (Webster et al. 2015)), and the majority of sequenced species of Thaumarchaeota. Some of these species encode divergent relatives with unknown functions (discussed in section "Archaeal tubulin and FtsZ-like homologs"). In these organisms, the ESCRT-III-based system, first characterized in Crenarchaeota (Lindas et al. 2008; Samson and Bell 2009) is probably required for cell division (Ng et al. 2013; Pelve et al. 2011).

Biochemical Activities of Archaeal FtsZ

In the original study of *H. volcanii* FtsZ1, a recombinant tagged version of the protein was overproduced in *E. coli* and purified (Wang and Lutkenhaus 1996). GTPase activity of this protein was demonstrated in buffers containing at least 2 M KCl (3 M KCl showed greater activity). The GTPase specific activity was dependent on FtsZ1 concentration (>200 μg/mL), suggesting that self-association (polymerisation) is required for GTPase activity, as seen with bacterial FtsZ and eukaryotic tubulin. Work with *Haloarcula japonica* FtsZ1 revealed that polymerization in a sedimentation assay was dependent on GTP and at least 2.25 M KCl (Ozawa et al. 2005). In both studies, it was found that NaCl could not substitute for KCl. Potassium is the predominant intracellular cation utilized at high concentration by haloarchaea for survival in hypersaline environments, so haloarchaeal proteins are well adapted to these conditions (Reed et al. 2013). Potassium is also specifically required for efficient GTPase activity of other bacterial and archaeal FtsZ proteins, although generally at much lower concentrations in non-halophiles (Mendieta et al. 2009; Mukherjee and Lutkenhaus 1998).

FtsZ homologs from thermophilic archaea have also been characterized. After the original identification of *P. woesei* FtsZ1, in which GTP binding was demonstrated (Baumann and Jackson 1996), the *M. jannaschii* FtsZ1 protein became a focus for biochemical characterization after its crystal structure was solved (Lowe and Amos 1998). This allowed some of the first high-resolution biochemical analysis of FtsZ proteins generally (Andreu et al. 2002; Diaz et al. 2001; Huecas and Andreu 2004; Lowe and Amos 1999; Lowe and Amos 2000), outlined in section "Structures of tubulin superfamily proteins from archaea" of this chapter. Other thermophilic archaeal FtsZ proteins have also been investigated. One study isolated the *ftsZ1* gene from the moderate thermophile *Thermoplasma acidophilum* (Yaoi et al. 2000). Remarkably, expression of this gene in *E. coli* resulted in strong inhibition of cell division, which was attributed to interaction of the *T. acidophilum* FtsZ with the native FtsZ in *E. coli* and/or interference with its function. Another FtsZ-related protein was isolated from the hyperthermophile *Thermococcus kodakaraensis* (Nagahisa et al. 2000), however, this protein is now recognised as a CetZ-family protein (section "The euryarchaeota CetZ family of proteins").

The Role of Archaeal FtsZ in Cell Division

The first cytological evidence to suggest that archaeal FtsZ functions in cell division utilized immunofluoresence microscopy with *H. volcanii.* FtsZ1 was seen to localize at the mid-cell region, appearing as a ring around the middle of the cell that corresponded with mid-cell constrictions in cells that appeared to be dividing (Wang and Lutkenhaus 1996). Similar results were obtained with the closely related *Haloferax mediterranei*, where it was also seen that constricted FtsZ1-rings were associated with cells of a variety of different shapes, including rods and the irregular plate-shaped cells that are common in these pleomorphic haloarchaea (Poplawski et al. 2000). Recently, *H. volcanii* FtsZ1 tagged with GFP was observed in live cells (Duggin et al. 2015), and the localization pattern was consistent with the earlier immunofluorescence results. FtsZ1-GFP was produced at a low level from a regulated gene promoter in a wild-type genetic background, which resulted in cells of normal size, indicating minimal disturbance to the normal cell division process by the presence of FtsZ1-GFP. Interestingly, almost all cells in a growing culture were found to have a mid-cell ring, indicating that this structure is maintained throughout the cell cycle. As most cells in these cultures do not contain noticeable division constrictions, FtsZ ring assembly occurs well in advance of constriction, similar to bacteria (Sun and Margolin 1998).

The fluorescence microscopy approaches revealed obvious similarities between the basic subcellular localization behaviour of archaeal and bacterial FtsZ, strongly implicating FtsZ in archaeal cell division. However, the original study of FtsZ in *H. salinarum* found that supplementary expression of *ftsZ2* did not greatly affect cell division, since cell size was similar to the wild-type (Margolin et al. 1996). In contrast, overexpression of *ftsZ* in *E. coli* causes division defects that affect cell size (Dai and Lutkenhaus 1992). The *H. salinarum* cells instead appeared pleomorphic, quite similar to the wild-type strains of some other haloarchaea, whereas wild-type *H. salinarum* is typically rod-shaped in laboratory culture. The basis for this morphological effect is yet to be determined, however, haloarchaeal cell shapes are generally sensitive to experimental conditions and it is possible that FtsZ overexpression does not have an inhibitory effect on division in all species, as it does in *E. coli*.

Direct evidence for the importance of FtsZ in archaeal cell division came only recently (Duggin et al. 2015). Expression of a point mutant of *H. volcanii ftsZ1* (D250A), which targets the GTPase activation region (T7 loop) essential for the normal polymerization-depolymerisation cycle in tubulin superfamily proteins (Richards et al. 2000; Scheffers et al. 2002), caused the hallmark cell division defect, in which cells cease division but continue to grow, resulting in the generation of extremely large cells (*e.g.* Fig. 14.2b). Thus, a functional GTPase activation domain in FtsZ1 is essential for normal cell division.

The genetic amenability of haloarchaea, particularly *H. volcanii* (Allers and Mevarech 2005), and the availability of functional fluorescent fusion proteins (Duggin et al. 2015; Reuter and Maupin-Furlow 2004) combined with their relatively large, flattened cell morphologies, show that these organisms are excellent models for understanding archaeal cell division, morphology, and cell biology generally (Duggin et al. 2015). Future studies with this system are anticipated to provide significant insights into functional diversity of cell division and structure, providing a much-needed contrast with the well-studied bacterial systems. This is expected to highlight the most fundamental features critical to both systems, and reveal a very different FtsZ-based system for prokaryotic cell division.

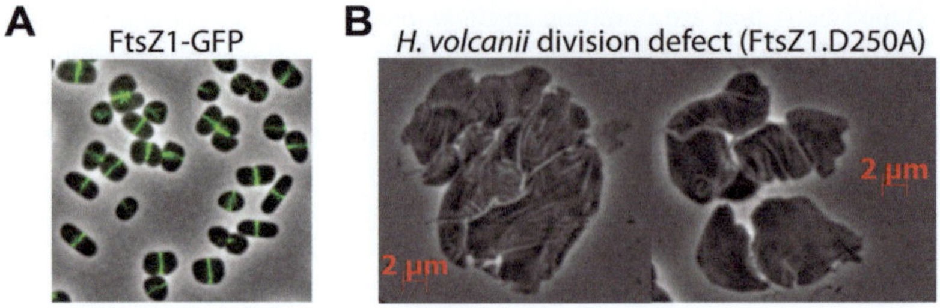

Fig. 14.2 FtsZ1 and cell division in *H. volcanii.* (**a**) 2D images of *H. volcanii* cells (phase-contrast and GFP fluorescence overlay) that are expressing FtsZ1-GFP. An approximately central band of fluorescence is clear in all cells of various shapes, and appears brighter in cells undergoing division constriction. The band is seen as a ring in 3D (not shown). (**b**) Examples of the *H. volcanii* cell-division defect. Cells were induced to express the FtsZ1.D250A protein, a GTPase-domain mutant that strongly inhibits division, during mid-log phase growth. Cells continue to grow in a 2D manner, but fail to divide, forming giant plate cells in culture (After Duggin et al. 2015)

Archaeal Tubulin and FtsZ-Like Homologs

As may be seen in Fig. 14.1a, archaea house a variety of deeply-branching tubulin superfamily members, many of which are not classifiable into the FtsZ, tubulin or CetZ families, and their functions are essentially unknown. Many of these proteins have only recently been identified owing to the recent vast increase in genome sequence availability, brought about by major advances in high-throughput environmental DNA sequencing.

The FtsZ-Like Group 1 Homologs (FtsZL1)

Members of the unusual FtsZ-like group 1 (FtsZL1) were identified in certain bacteria and archaea on the basis of homology to the GTP-binding core (N-terminal) domain, which includes the signature motifs of the tubulin superfamily (Makarova and Koonin 2010). There are only a small number of known FtsZL1 sequences in archaea, although they may be found in diverse species including the Methanomicrobia, Halobacteria and Thermococci Classes. These proteins form a phylogenetic subgroup distinct from the related bacterial FtsZL1 proteins, suggesting that recent horizontal transfer between domains is an unlikely mechanism to explain their patchy distribution.

Interestingly, a transposon-interrupted FtsZL1 gene was identified in *Sulfolobus solfataricus* (SSO1376), representing the first tubulin-superfamily relative found in the Crenarchaeota (Makarova and Koonin 2010). Another FtsZL1 member (undisrupted) in the Crenarchaeote *Acidilobus saccharovorans* was recently sequenced (Uniprot D9PZB8). Both Crenarchaeote FtsZL1 proteins have unusually-long homologous N-terminal extensions to the core GTP-binding domain, suggesting that these have a particular (unknown) function in Crenarchaeota FtsZL1.

The domain arrangement of FtsZL1 proteins more generally is highly unusual amongst the tubulin superfamily, as the central GTPase activation region and C-terminal domain, normally involved in polymerization (Fig. 14.4), are replaced with a much larger and unrelated domain of unknown structure and function. Thus, the capacity of these proteins for polymerization and GTP hydrolysis is currently unknown. A noticeable conservation of the FtsZL1 genomic neighbourhood across diverse species led to the hypothesis that these proteins are involved in membrane remodelling processes in conjunction with the products of the linked genes (Makarova and Koonin 2010).

Thaumarchaeota Tubulin-Superfamily Members and Other Deeply-Branching Archaeal Homologs

The Thaumarchaeota show quite a complex distribution of tubulin-superfamily proteins, with specific homologs of FtsZ and tubulin seen in some species, and another family called "Thaumarchaeota FtsZ" (Yutin and Koonin 2012) present in most of the major taxa of the Thaumarchaeota (Fig. 14.1). Despite the name currently used for this family, the Thaumarchaeota-FtsZ proteins are distinct from the major FtsZ, CetZ and tubulin branches, and they are at least as far removed from the FtsZ family as another small group of proteins found in the Korarchaeota (Fig. 14.1a). In addition to these, Korarchaeota possess specific FtsZ family sequences (but not ESCRT-III), suggesting that Korarchaeota division is based on FtsZ, and that the other deeply branching homologs in these organisms may have other non-cell division roles. We would suggest that newly identified tubulin-superfamily groups, such as Thaumarchaeota-FtsZ, be renamed where appropriate once a prototypical member has a clear biological role determined, so that the term "FtsZ" is limited to proteins that primarily function in division.

The Thaumarchaeota-FtsZ group shows significant sequence divergence in the GTP-binding T4-loop, with a consensus of AGKAG in this region instead of the superfamily consensus GGGTG. Most Thaumarchaeota-FtsZ proteins also appear to completely lack the catalytic residues of the T7 loop required for GTP hydrolysis. Thus, the normal mode of polymerization and depolymerisation is very unlikely to occur in these proteins. Consistent with this, the Thaumarchaeota-FtsZ protein in *Nitrosopumilus maritimus* does not appear to be involved in cell division, as it was localized diffusely in these cells, whereas the CdvB (ESCRT-III) homolog was located at mid-cell (division) sites (Pelve et al. 2011). Furthermore, an *N. maritimus* FtsZ-GFP fusion showed diffuse localization when expressed in yeast, whereas CdvB-GFP formed obvious filaments (Ng et al. 2013). Interestingly, when the unusual T4 loop sequence of *N. maritimus* FtsZ was replaced by the *E. coli* FtsZ T4 loop, this resulted in localization as foci and short filaments in yeast, much like the localization pattern of the *E. coli* FtsZ (GTPase-deficient mutant) in this system (Ng et al. 2013). The substituted *N. maritimus* FtsZ might therefore have gained nucleotide binding but not hydrolysis capacity, promoting polymerization and assembly into localized structures. The wild-type Thaumarchaeota-FtsZ proteins therefore might not bind/hydrolyse GTP or polymerise, so they remain one of the more mysterious groups in the tubulin superfamily.

Another group of tubulin-superfamily proteins in Thaumarchaeota, called artubulin, was identified in the genomes of two members of the genus *Nitrosoarchaeum*, in the order Nitrosopumiales (Fig. 14.1b) (Yutin and Koonin 2012). Thaumarchaeota-FtsZ genes are also present in these organisms. The artubulins form a sister group to the eukaryotic tubulin family (Fig. 14.1a), raising questions surrounding the origin of this family within the eukaryotic lineage. The artubulin genes are located adjacent to a component of the ESCRT-III machinery, hinting at a possible role of artubulin in ESCRT-III mediated division in these organisms (Yutin and Koonin 2012). Future functional studies of artubulins may provide insights into the early evolutionary events that led to the expansion of tubulin-superfamily roles in the Eukaryota/Archaea lineage.

There are a number of other tubulin-superfamily proteins in archaea that are currently uncharacterized (grey branches, Fig. 14.1a). These could be expected to have specialized roles, and of particular note are the numerous highly-divergent tubulin-superfamily members present in *Haloquadratum walsbyi* (Bolhuis et al. 2006; Burns et al. 2007). This organism forms remarkably thin, square cell shapes that are maintained in this shape during cell division (Stoeckenius 1981; Walsby 1980). The shapes would appear to provide a substantial selective advantage in their hypersaline lake environments (Burns et al. 2007). One might predict that the divergent tubulin-superfamily proteins form cytoskeletal structures required for generating and maintaining the square shapes, while the FtsZ and CetZ proteins that also exist in this organism may have roles in cell division and morphological changes, respectively. The future study of the cytoskeletal systems of this and related organisms is a fascinating prospect.

The Euryarchaeota CetZ Family of Proteins

The CetZ family were first recognised as a phylogenetically distinct group, originally called FtsZ paralog 3 (FtsZ3), after the cloning and sequencing of an *ftsZ* homolog from *Thermococcus kodakaraensis* (Nagahisa et al. 2000), a hyperthermophilic species of the Class Thermococci. The purified protein showed GTPase activity and polymerisation that increased continuously with temperature, up to 90°C, which is near the growth optimum of this organism. The protein sequence showed high similarity to one of three homologs in the available genome sequence of *Pyrococcus horikoshii* (Class Thermococci) (Kawarabayasi et al. 1998), and was only distantly related to the two other homologs, FtsZ1 and FtsZ2. Other members of the CetZ family were then noticed in the genomes of *Pyrococcus abyssi* (Chinen et al. 2000) and *Archaeoglobus fulgidus* (Class Archaeoglobi) (Klenk et al. 1997), raising the possibility that the multiple conserved paralogs in these archaea could have differing cytoskeletal roles (Gilson and Beech 2001). A thorough survey of tubulin-superfamily sequences (Vaughan et al. 2004) further identified two *cetZ* genes in the genome of *Halobacterium NRC-1* (Class Halobacteria) (Ng et al. 2000). It was also observed that the Thermococci CetZ (FtsZ3) proteins branch from near the base of the group (Vaughan et al. 2004). The large quantity of sequence data now available has shown that CetZ proteins are present in many other diverse archaea from the phylum Euryarchaeota, including the Class Methanomicrobia (Duggin et al. 2015). CetZ proteins form a distinct branch (family) within the tubulin-superfamily (Fig. 14.1a).

The first *in vivo* functional study of members of the CetZ family found that they control cell morphology in *H. volcanii* (Duggin et al. 2015). The family was named cell-structure-related Euryarchaeota tubulin/FtsZ homologs (CetZ) to reflect the non-cell-division role seen for *H. volcanii* CetZ1. *cetZ* genes are especially abundant in genomes of Halobacteria (*e.g.* six paralogs in *H. volcanii*), although the number of paralogs varies amongst related species. Some of the haloarchaeal CetZ proteins cluster phylogenetically into several distinct orthologous groups, suggesting that proteins in each group may have a common function in these species (Fig. 14.1a). The most conserved of these is the CetZ1 group, which was present in all species of Halobacteria analysed (Duggin et al. 2015), and members typically shared ~80% sequence identity in this Class. CetZ2 orthologues were identified in many of these. The *H. volcanii* CetZ3-6 proteins were located in weakly-clustered groups and might be expected to have more specific or redundant roles in this species. Generally, the Methanomicrobia and Halobacteria Classes possess multiple CetZ homologs, whereas the Archaeoglobi and Thermococci carry one CetZ each. CetZ has so far only been identified in species that also contain FtsZ (Fig. 14.1b).

H. volcanii is described as pleomorphic (Mullakhanbhai and Larsen 1975), displaying a variety of cell sizes and shapes in complex growth medium (*e.g.* Hv-YPC (Allers et al. 2004)). However, as in many haloarchaea, the cells remain uniformly thin (~0.4 μm) during growth and division, growing either two-dimensionally (the plate morphotype) or one-dimensionally (the rod morphotype). *H. volcanii* cells lye on their flat, larger surfaces when placed on a gel-pad, and they can be grown for many generations on a gel-pad; the relatively large, flat cells are well-suited for visualization by fluorescence microscopy (*e.g.* by using GFP or mCherry) and other approaches including electron cryotomography (Duggin et al. 2015). Time-lapse imaging of live cells showed that they can transition between the two basic modes of cell growth, and some cells appear to actively narrow during the plate-to-rod transition (Duggin et al. 2015). The CetZ1 protein was found to be essential for the development of the rod morphotype, which was needed for efficient swimming motility (Duggin et al. 2015).

When *H. volcanii* is in steady-state growth (mid-log growth for >10 generations) in a defined medium, "Hv-Cab" (*i.e.* Hv-Ca (Allers et al. 2004) with additional trace-elements (Duggin et al. 2015)), it grows consistently as the plate morphotype (Duggin et al. 2015). However, moderate overexpression of CetZ1, via the use of the finely-regulated *tnaA* gene promoter (Allers et al. 2010; Large et al. 2007), caused many of the cells to adopt an obvious rod morphology. Interestingly, over-expression of the CetZ1.E218A mutant (to prevent GTPase activity and filament depolymerisation) caused the cells to adopt a blocky, jagged morphology instead (a similar phenotype was seen with CetZ2.E212A (Duggin et al. 2015)). This appears to be caused by hyperstable filaments of CetZ1.E218A, since a GFP fusion protein to this mutant localized very stably to the cell envelope at the stalk-like projections and inward puckers seen in these cells. In contrast, CetZ1-GFP showed very dynamic/patchy or diffuse localization in the log-phase plate cells. During rod development, however, CetZ1-GFP sometimes appeared as faint, dynamic filaments aligning with the direction of rod cell development, suggesting that such cytoskeletal structures are specifically related to rod development. It was concluded that CetZ1 forms a key part of an archaeal cytoskeletal system involved in regulation of cell morphology, and represents a prototypical member of this family of cytoskeletal proteins.

The great amenability of *H. volcanii* to genetic manipulation (Allers and Mevarech 2005), combined with its ease of growth, controllable cellular differentiation processes, and excellent properties for many live-cell microscopy approaches (Duggin et al. 2015) show that it is an excellent model for further studies into the archaeal cytoskeleton and cell biology.

Structures of Tubulin Superfamily Proteins from Archaea

In this section, we describe the original role structural studies of archaeal tubulin-superfamily proteins have played in helping identify and characterize cytoskeletal proteins in archaea and other organisms, and we describe the known archaeal protein structures and aspects of the structure-function relationship that have been obtained from these studies. Finally, we compare structures of the recently-described CetZ family to the other major groups of the tubulin superfamily.

Archaeal proteins can provide technical advantages in structural studies, such as high thermostability, which in many cases have enabled new insights and clues into the structure and function of less amenable homologs in the intensively studied eukaryotic and bacterial species. An archaeal FtsZ featured prominently amongst the first high-resolution structural studies of the tubulin superfamily (Lowe and Amos 1998). We will describe the general features of the protein structures that have been obtained from studies of archaeal tubulin superfamily proteins, and compare them with other members of the superfamily. Atomic-resolution structures that are currently available for archaeal tubulin superfamily proteins are listed in Table 14.1.

Table 14.1 Available crystal structures of archaeal tubulin superfamily proteins

Protein	Organism[a]	PDB code[b]	Resolution	Notes
FtsZ1	*M. jannaschii*	1FSZ Lowe and Amos (1998)	2.8 Å	GDP-bound monomer
FtsZ1	*M. jannaschii*	2VAP Oliva et al. (2007)	1.7 Å	GDP-bound monomer (different crystal form)
FtsZ1	*M. jannaschii*	1W58 Oliva et al. (2004)	2.5 Å	Mg^{2+} and GMPCPP-bound[c] monomer
FtsZ1	*M. jannaschii*	1W59 Oliva et al. (2004)	2.7 Å	Nucleotide-free dimer
FtsZ1	*M. jannaschii*	1W5A Oliva et al. (2004)	2.4 Å	Mg^{2+} and GTP-bound dimer
FtsZ1	*M. jannaschii*	1W5B Oliva et al. (2004)	2.2 Å	GTP-bound dimer
FtsZ1	*M. jannaschii*	1W5E Oliva et al. (2004)	3.0 Å	W319Y mutant (GTPase defect), GTP-bound
CetZ1	*H. volcanii*	4B46 Duggin et al. (2015)	1.9 Å	GDP-bound monomer
CetZ2	*H. volcanii*	4B45 Duggin et al. (2015)	2.1 Å	GTPγS-bound protofilament (trimer)
CetZ	*M. thermophila*	3ZID Duggin et al. (2015)	2.0 Å	GDP-bound

[a]Full organism names: *Methanocaldococcus jannaschii*; *Haloferax volcanii*; *Methanosaeta thermophila*

[b]Protein Data Bank identification code (www.rcsb.org). The associated reference is given in parentheses

[c]GMPCPP is a GTP analog: phosphomethylphosphonic acid guanylate ester

FtsZ Structures

In 1998, an FtsZ protein from the archaeon *Methanocaldococcus jannaschii* became the first member of the tubulin superfamily to have its structure solved at high resolution by X-ray crystallography (Lowe and Amos 1998). When compared to the electron crystallographic structure of polymerized tubulin reported in the same issue of *Nature* (Nogales et al. 1998b), debate as to the homology between FtsZ and tubulin ended – their structural folds were remarkably similar (Nogales et al. 1998a). Thus, structural comparisons of these proteins from highly diverse organisms highlighted the usefulness of protein 3D structure in verifying weak homology identified through alignment of a limited number of sequences. Additional structures of the *M. jannaschii* FtsZ1 protein, including different crystal forms and ligand-bound states have helped provide further insights into protofilament formation and structural variability that underpin the basic functions of these proteins (Lowe and Amos 1999, 2000; Oliva et al. 2004, 2007).

The original structure of *M. jannaschii* FtsZ1 bound to GDP showed that it is a globular protein (39 kDa) containing two closely-associated but independently-folded domains (Lowe and Amos 1998; Oliva et al. 2004), essentially the same as tubulin (Nogales et al. 1998b). These proteins form a distinct superfamily of GTPases (Nogales et al. 1998a), which nevertheless retain some structural similarities to the Rossmann-fold di-nucleotide binding proteins and small GTPases of the Ras superfamily (Lowe and Amos 1998). This similarity is seen in the N-terminal GTP-binding domain, which contains most of the secondary structural elements that establish the nucleotide-binding site (shown in green; Fig. 14.3). However, within this domain the tubulin superfamily is distinguished by the presence of four highly conserved loops (T1, T2, T3 and T4, on the upper surface as depicted in Fig. 14.3) that primarily engage the phosphates of the nucleotide, whist allowing another FtsZ subunit to bind to this surface in a "head-to-tail" orientation (*i.e.* via the lower surface of the additional subunit), sandwiching the GTP molecule (Oliva et al. 2004). Thus, both subunits supply residues required for GTP hydrolysis, explaining why GTP hydrolysis is polymerization-dependent (Mukherjee and Lutkenhaus 1998).

The C-terminal domain (shown in purple; Fig. 14.3) is separated from the N-terminal GTP-binding domain by an important central region, consisting of the

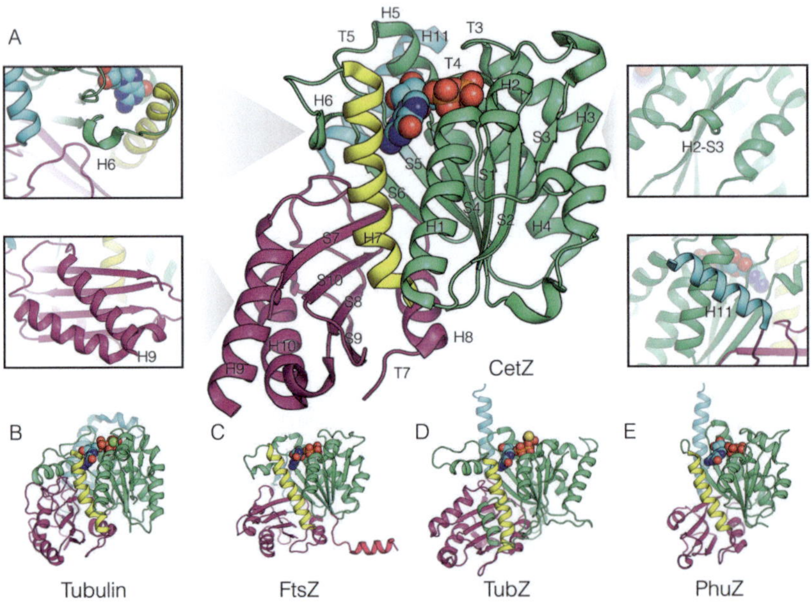

Fig. 14.3 Crystal structures of CetZ proteins. (**a**) The crystal structure of the prototypical CetZ (*Haloferax volcanii* CetZ1; PDB code 4B46) is shown with the secondary structural elements indicated according to the tubulin/FtsZ nomenclature (Lowe et al. 2001). The *inset panels* indicate; the reduction of α-helix H6, helix H11 from a viewpoint rotated 180°, helix H9 and loop H1-S2 respectively. Structural superimpositions of CetZ2 upon; (**b**) α-tubulin, (**c**) FtsZ, (**d**) *Bacillus thuringiensis* TubZ and (**e**) bacteriophage ΦKZ TubZ are shown in the side panels. All full structures are shown from the same viewpoint in cartoon (**a**) or ribbon (**b–e**) representation, with GTPase domains coloured *green*, activation domains coloured *purple*, the principal core helix, H7, linking the two elements coloured *yellow* and the remaining amino- and carboxy-terminal extensions to the canonical tubulin/FtsZ fold shown in *reds* and *blues* respectively

long H7 helix (yellow; Fig. 14.2) followed by the T7/H8 region. Although not usually considered a distinct domain, this region nonetheless has the specific function of delivering key acidic residues to the GTP-binding region of the adjacent subunit, and instigates a nucleophilic attack at the gamma-phosphate of GTP, resulting in hydrolysis to GDP. The GTPase activation function of the H7/T7/H8 region has the important role of promoting filament disassembly and turnover (Nogales et al. 1998a).

Hydrolysis and filament disassembly are very important processes in the spatial regulation and cytomotive activity of cytoskeletal structures (Lowe and Amos 2009). The early studies of *M. jannaschii* FtsZ1 raised the question of how hydrolysis leads to filament disassembly; a comparison of numerous crystal structures of FtsZ proteins, including several guanosine-nucleotide bound states of *M. jannaschii* FtsZ1, revealed no large conformational changes associated with hydrolysis (Oliva et al. 2007). Recent high-resolution electron cryo-microscopy (cryoEM) of tubulin in various states would suggest that hydrolysis causes a more subtle compaction around the region of the GTP-binding domain that surrounds the nucleotide's phosphate groups (*i.e.* the T1-T4 loops). This creates strain within the subunit and moderate conformational shifts that destabilize the longitudinal interaction between subunits (Alushin et al. 2014). It therefore appears that the flexible glycine-rich T1-T4 loops, which include the tubulin superfamily "signature" sequence (T4), sense the number of phosphates of the nucleotide and transduce this information to the whole molecule in order to promote disassembly. It remains to be determined whether this mechanism occurs in the same manner in FtsZ filaments, but the strong conservation of these loops would suggest it is very likely.

The presence and structure of the C-terminal domain (shown in purple; Fig. 14.3) is another defining characteristic of the tubulin superfamily amongst GTPases, and shows a conserved fold despite the very sparse sequence conservation amongst the superfamily's most diverse members. Crystal structures of *M. jannaschii* FtsZ1 dimers that sandwich GTP in the expected manner (Oliva et al. 2004) showed that the C-terminal domain provides the majority of the contact surface area between subunits (Fig. 14.4c). This is also observed in the other families (Fig. 14.4). The C-terminal domain can also mediate lateral association of protofilaments (Alushin et al. 2014; Duggin et al. 2015; Monahan et al. 2009), fulfilling another essential characteristic of the superfamily—the ability to form bundles, sheets and tubes that build the cytoskeleton in its myriad of forms across the domains of life. One of the C-terminal domain's main purposes therefore appears to be to mediate self-association.

In addition to the two main domains of the globular fold of FtsZ, these proteins frequently have N- and C-terminal extensions that have specific functions, generally in mediating interactions with other molecules. They protrude from the "front", and "rear", (inner and outer, relative to a microtubule) surfaces of the molecule, respectively (as shown in Figs. 14.2 and 14.3), and are thus located on the two exposed surfaces of protofilament sheets. These regions are rarely fully resolved in crystal structures owing to their flexibility and lack of secondary structure in isolation. *M. jannaschii* FtsZ1 showed an N-terminal extension (which is generally absent in the

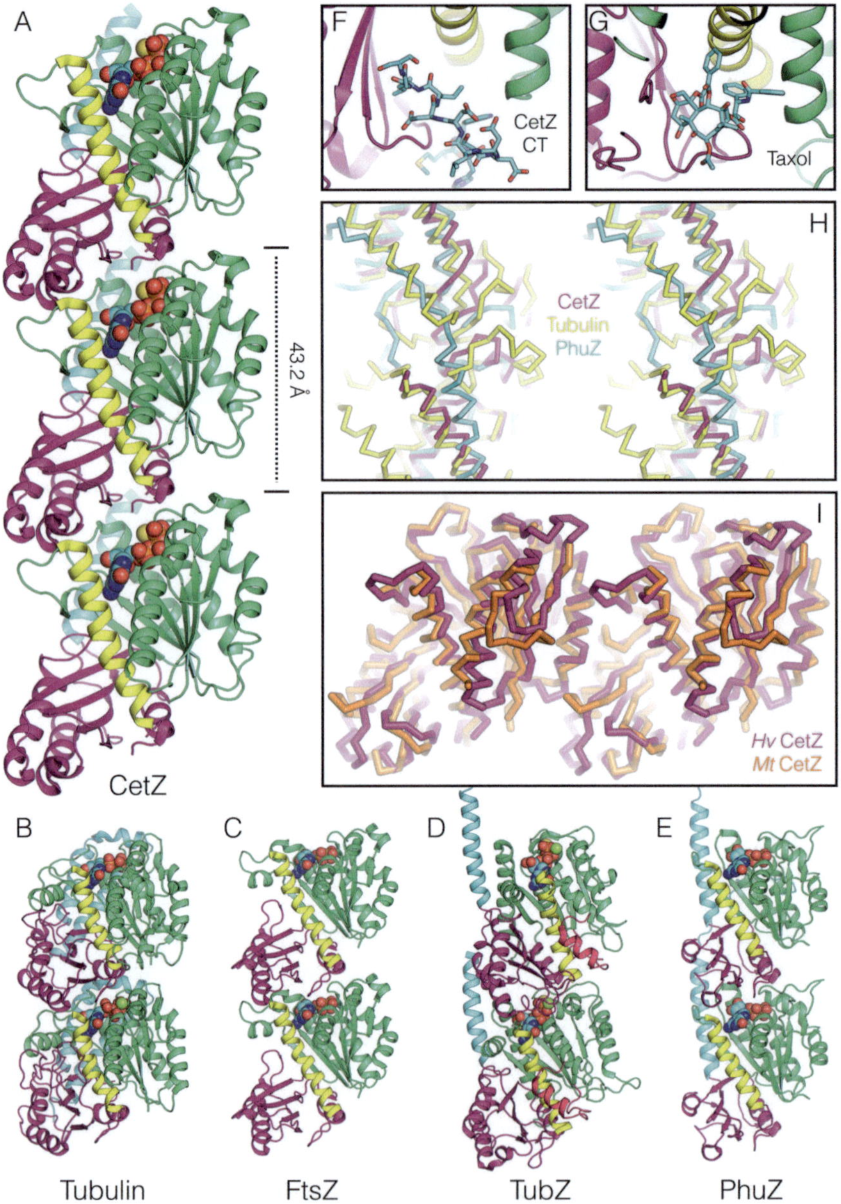

Fig. 14.4 CetZ subunit-subunit interactions and comparisons. (**a**) The structure of a *H. volcanii* CetZ2 protofilament is shown according to an identical colour scheme and representation as that in Fig. 14.2. Protofilaments of tubulin (**b**), FtsZ (**c**), TubZ (**d**), and bacteriophage ΦKZ TubZ (**e**) are shown for the purposes of comparison. The *inset panels* show: the interaction site of the carboxy-terminus of a crystallographically related molecule in the CetZ2 crystal structure (**f**), compared to the binding mode of taxol within the same site on β-tubulin (**g**), the contribution to the polymerisation interface by the central helix H11 of CetZ compared to that of similar regions in tubulin and PhuZ (**h**), and a comparison of the lateral interactions between *H. volcanii* CetZ2 protofilaments and those between adjacent monomers within the crystal structure of *Methanosaeta thermophila* CetZ (**i**)

tubulin or CetZ families) that has limited secondary structure apart from one helix at the very N-terminus (Lowe and Amos 1998). The function of this helix and remaining N-terminal extension is still unknown. The C-terminal extension in *M. jannaschii* FtsZ1 is a β-hairpin, comprising strands S11 and S12, beyond which the protein is typically extended without significant structural elements, and differs from the primarily helical C-terminal extensions seen in tubulin and CetZ proteins (discussed below). The length of the C-terminal extension is highly variable amongst different species and in bacteria this is thought to play an important role as a flexible spacer (Buske et al. 2015; Evans 2006), while the more sequence-conserved few residues immediately preceding the C-terminus bind other molecules with important roles in cell division (Buske and Levin 2012). In archaea, molecules that interact with FtsZ are unknown, however the conservation of the C-terminal few residues amongst other archaea would suggest that these residues have a common function in archaea too.

CetZ Structures

Three crystal structures of CetZ family proteins recently solved (Duggin et al. 2015) show some interesting similarities and differences to the FtsZ and tubulin family proteins that may eventually help to reconcile structure-function relationships in the superfamily.

***H. volcanii* CetZ1**

The structure of CetZ1 from *H. volcanii* represents the prototypical CetZ, having been the first structure solved, and to date it is the only CetZ protein to which a definitive function in the regulation of cellular morphology has been ascribed (Duggin et al. 2015). The structure revealed CetZ1 in a monomeric state, with a GDP molecule that had co-purified retained in the active site. CetZ exhibits the two principle domains present in other tubulin superfamily proteins, however several notable modifications to the typical structural features are apparent. CetZ1, and indeed most CetZs, lack a N-terminal extension to the core GTP-binding domain. By comparison, an N-terminal extension is generally present in FtsZ, but absent in tubulins. Within the GTP-binding domain, helix H6 is substantially shorter than usual, leaving only a single helical turn to form this section of the subunit-subunit interface (Fig. 14.4a). Helix H9, on the other hand, is considerably extended, increasing the size and prominence of the C-terminal domain. Furthermore, the nearby M-loop, which links S7 and H9 (Fig. 14.3a), is unresolved in the crystal structure but is unusually long in CetZ proteins compared to FtsZ and tubulins. The M-loop mediates lateral association of tubulin protofilaments within microtubules (Alushin et al. 2014; Nogales et al. 1998a, 1998b), but is of unknown function in CetZ proteins. Finally, the C-terminal extension of CetZ1 is a long helix (H11), which aligns with the expected axis of polymerisation of protofilaments. These three regions of notable difference (*i.e.* H6, H9 and H11) compared to FtsZ and tubulin are also the only regions showing notable variations in structure that were observed between *H. volcanii* CetZ1 and CetZ2.

H. volcanii CetZ2

CetZ2 was crystallized and its structure solved bound to GTPγS (Duggin et al. 2015), which resulted in the generation of a lattice incorporating crystallographic protofilaments. The subunit structure of *H. volcanii* CetZ2 is remarkably similar to CetZ1; the Cα r.m.s.d between the two structures is only 0.85 Å, as might be expected given their relatively high sequence similarity (51% identical). CetZ2 protofilaments in the crystal lattice had a repeat unit of 43.2 Å, very similar to tubulin (41.8 Å), and even closer to FtsZ and TubZ proteins (43.5 Å). Furthermore, the subunit-subunit interface formed within CetZ2 protofilaments is very similar to that conserved within the tubulin and FtsZ groups, although it differs more from that in the TubZ proteins, which are tilted differently across the active site. This directly affects the conformation of protofilaments, and CetZ2 recapitulated the straight protofilaments typically seen in crystal structures of FtsZ and tubulin protofilaments. This contrasts some of the known bacterial TubZ protofilaments, which show twisted, helical arrangements (Aylett et al. 2010). The presence and degree of helical structure directly affects the higher-order filament or sheet/tube-like structures that are critical to the function of these proteins, and it will be informative to see the conformation of further CetZ filaments, ideally with methods that have minimal influence on the subtle quaternary arrangements important to tubulin superfamily proteins.

The C-terminal helix (H11) in CetZ2 occupies a similar space to that of the tubulins and TubZ proteins, and aligns almost perfectly with H11 from tubulin protofilaments (1.0 Å Cα r.m.s.d, CetZ2 C-terminus versus tubulin C-terminus). H11 also extends along the surface, generally in the longitudinal direction of polymerization, and contacts the base of the adjacent subunit, similar to tubulins and TubZ proteins. Although regions of the C-terminal tail from H11 onwards are not fully resolved in the structure, it would appear possible that the H11 helix of CetZ proteins may continue along the surface of the adjacent subunit in a manner similar to the TubZ proteins, with H11 filling the space that would otherwise be occupied by H12 in the multi-helix folded C-terminal tails seen in eukaryotic tubulin protofilaments.

Interestingly, the CetZ2 protofilaments were arranged as crystallographic sheets, with lateral contacts between protofilaments generally comparable to the lateral interactions of tubulin in the microtubule wall and FtsZ lateral associations, suggesting that such interactions might be biologically relevant in CetZ proteins too. The loops of C-terminal domain, including an extended S7-H9, and those on the same face of the GTPase domain, including the altered loop H1-S2 as well H3. Further support for the significance of these contacts comes from the observation that the lateral crystal contacts within the more distantly related *Methanosaeta thermophila* CetZ crystal structure recapitulate this interaction almost perfectly (2.6 Å Cα r.m.s.d over adjacent chains; *H. volcanii* CetZ2 versus *M. thermophila* CetZ).

The aforementioned sheets within CetZ2 crystals are bridged by the C-terminal tail of the protein. An interesting observation, however, was the association of the C-terminal amino acids of one subunit within a pocket abutting helix H7 of an adjacent subunit. This pocket in tubulin is where the anticancer drug, taxol, binds to β-tubulin in protofilaments. The concomitant slight gap in the closely related CetZ1 crystal structure, in an unpolymerised state, is closed somewhat by a small rotation of the activation domain and remains unoccupied; it is not possible to say from our results whether protofilament formation is responsible for this rotation and intrusion of the C-terminal peptide, or whether it might be a constitutive difference between the folds of the two CetZ proteins, however we note that a similar rotation of the activation domain was observed in microtubules stabilised in the presence of taxol or GTP analogues. It is yet to be determined whether such docking of the C-terminal residues on the adjacent protofilament sheet of CetZ2 has any relevance to the physiological form and function of CetZ proteins.

Methanosaeta thermophila CetZ

The structure of a divergent CetZ homolog, which branches from near the root of the tubulin superfamily phylogenetic tree (Fig. 14.1), was also determined (Duggin et al. 2015). Superimposition of this *M. thermophila* CetZ on *H. volcanii* CetZ2 leads to an overall Cα r.m.s.d of only 1.8 Å, and the structural features observed within both chains are also very well conserved from those in the two *H. volcanii* CetZ structures. The only clear difference between the *M. thermophila* and *H. volcanii* proteins is the substantial extension of the carboxy-terminal helix H11 resolved in *M. thermophila* CetZ and a slight shortening of helix H9, the latter being more like the H9 of other tubulin superfamily proteins. Although there are many small surface differences between the proteins from the two organisms, the deviation between surface loops is extremely low, supporting the conclusion that the proteins should be grouped cleanly into the single (CetZ) family. Since these proteins branch from near the base of the superfamily tree, it will be useful to determine the function of *M. thermophila* CetZ, or similar "non-canonical" CetZ proteins, to determine whether these proteins have functions in cell shape regulation like the *H. volcanii* CetZ1 and CetZ2 proteins (Duggin et al. 2015) or as yet unknown functions.

Structural Comparisons of CetZs with Other Tubulin Superfamily Proteins

As noted above, the archaeal FtsZ and CetZ families exhibit many features in common with eukaryotic tubulins as well as bacterial FtsZ and TubZ proteins. Here, we will compare structures of CetZ proteins to those of the FtsZ protofilament structure from *Staphylococcus aureus (Tan et al.* 2012*)* (PDB code 4DXD), the taxol/Zinc-stabilised sheets of *Bos taurus* tubulin protofilaments (Lowe et al. 2001) (PDB code 1JFF), and *Bacillus thuringiensis* (Aylett et al. 2010) and *Pseudomonas aeruginosa* bacteriophage ΦKZ (Aylett et al. 2013) TubZ protofilament structures (PDB codes 2XKB and 3ZBQ respectively) as prototypes.

Comparison of CetZ and FtsZ

The CetZ protein structures show several notable structural features specifically in common with FtsZ. In particular, the conformation of the key γ-phosphate contact loop T3 and proximal regions (S3, H3) in both *H. volcanii* CetZ1 and CetZ2 is almost identical to that in the majority of FtsZ structures resolved (Cα r.m.s.d 1.1 Å FtsZ against CetZ1). The conformation of loop T5 is also strongly conserved between these two groups, although it typically exhibits lower overall deviation in many cases, and the numerous surface loops of the CetZ proteins are closer in conformation to those in FtsZ than tubulins. The overall backbone difference between CetZs and FtsZs (overall Cα r.m.s.d 2.9 Å for FtsZ and CetZ1) is also significantly lower than that observed on comparison with tubulins (overall Cα r.m.s.d 3.6 Å for α-tubulin and CetZ1). However, as we note below, a series of conserved catalytic residues shared between the tubulins and CetZ proteins provide a counterargument to this narrative.

Comparison of CetZ and Tubulin

The tubulins are the most well understood proteins in the superfamily. They build the microtubule network present in all known eukaryotes, which is a major component of the cytoskeleton. Microtubules control cellular organisation and have well-known role in providing the spindle for chromosome segregation. The polymerising unit for tubulin includes two different monomers, α- and β-tubulin, which are pre-assembled around a molecule of GTP which is bound essentially irreversibly within a catalytically inactive subunit-subunit interface. In addition to the canonical N-terminal GTP-binding and C-terminal domain structure, tubulins incorporate a distinct C-terminal extension comprising a helical hairpin (H11 and H12) that provides the binding platform for processive motor proteins, and several notable surface loops (H1-S2, H2-S3, M- and N-loops) that form the lateral interactions which stabilise the contacts between the thirteen protofilaments incorporated into each cylindrical microtubule (Lowe et al. 2001).

In the CetZ proteins, the active site shows great similarity to that found in α-tubulin. Guanine base recognition is achieved through an asparagine residue protruding from H7 in all three CetZ structures, a similar situation to that in tubulins and bacterial TubZ proteins (described below), whereas the equivalent residue is an aspartate in the FtsZs. Furthermore, the essential catalytic carboxylate within T7/H8, which activates water for γ-phosphate hydrolysis, is also provided by a glutamate residue in CetZ and tubulin, but is an aspartate in FtsZ. All three CetZ structures share a glutamine residue at the apex of loop T1 which forms an important protofilament contact with loop T7 in tubulin protofilaments, but which is absent in the known FtsZ structures and replaced in many TubZs. Finally, several regions of the CetZs form loops extending from the core fold, the location of which are identical to those responsible for lateral interactions between tubulin protofilaments (S7-H9 in CetZ2, and H1-S2 in *M. thermophila* CetZ). However, the loop structures are not homologous between CetZs and tubulins, suggesting that they may differ substantially in function.

Comparison of CetZ and TubZ

The TubZ group of proteins represent some of the most diverse members of the established tubulin superfamily, and they cannot clearly be classified into their own distinct subgroup. Two functional sub-classes exist, one responsible for partitioning plasmids and prophages within species of Gram-positive *Bacilli*, and the other associated with bacteriophage of the genus *Pseudomonas* and closely-related bacteria (Kraemer et al. 2012; Larsen et al. 2007). TubZ proteins typically have a long α-helical C-terminal extension that makes extensive contacts with the adjacent subunit and is also believed to be responsible for the recruitment of cognate binding partners. They also show a reduction in helix H6, particularly in the bacteriophage-borne members of the family associated with the application of a tilt across the subunit-subunit interface, and in the relative rotation of the GTPase and C-terminal domains against one another, yielding twisted-helical protofilaments and filaments (Aylett et al. 2010, 2013). All three archaeal CetZ structures resolved to date possess only a single turn of H6, substantially shorter than in FtsZ and tubulin. Some TubZ proteins lack H6 altogether, with a similar structural volume to CetZs in this region. Another commonality among all TubZ and CetZ proteins is the elongated single α-helical C-terminal tail that proceeds along the axis of the protofilament and contacts the adjacent subunit, suggesting that these might have common functions (*e.g.* stabilising the protofilament).

Acknowledgment I.G. Duggin's research is funded by the Australian Research Council.

References

Allers T, Mevarech M (2005) Archaeal genetics – the third way. Nat Rev Genet 6 (1):58–73. doi:nrg1504 [pii]1038/nrg1504 [doi]

Allers T, Ngo HP, Mevarech M, Lloyd RG (2004) Development of additional selectable markers for the halophilic archaeon *Haloferax volcanii* based on the *leuB* and *trpA* genes. Appl Environ Microbiol 70(2):943–953

Allers T, Barak S, Liddell S, Wardell K, Mevarech M (2010) Improved strains and plasmid vectors for conditional overexpression of His-tagged proteins in *Haloferax volcanii*. Appl Environ Microbiol 76(6):1759–1769. doi:10.1128/aem.02670-09

Alushin GM, Lander GC, Kellogg EH, Zhang R, Baker D, Nogales E (2014) High-resolution microtubule structures reveal the structural transitions in alphabeta-tubulin upon GTP hydrolysis. Cell 157(5):1117–1129. doi:10.1016/j.cell.2014.03.053

Andreu JM, Oliva MA, Monasterio O (2002) Reversible unfolding of FtsZ cell division proteins from archaea and bacteria. Comparison with eukaryotic tubulin folding and assembly. J Biol Chem 277(45):43262–43270. doi:10.1074/jbc.M206723200

Aylett CH, Wang Q, Michie KA, Amos LA, Lowe J (2010) Filament structure of bacterial tubulin homologue TubZ. Proc Natl Acad Sci U S A 107(46):19766–19771. doi:10.1073/pnas.1010176107

Aylett CH, Izore T, Amos LA, Lowe J (2013) Structure of the tubulin/FtsZ-like protein TubZ from *Pseudomonas* bacteriophage PhiKZ. J Mol Biol 425(12):2164–2173. doi:10.1016/j.jmb.2013.03.019

Baker BJ, Saw JH, Lind AE, Lazar CS, Hinrichs K-U, Teske AP, Ettema TJG (2016) Genomic inference of the metabolism of cosmopolitan subsurface *Archaea, Hadesarchaea*. Nat Microbiol 1:16002. doi:10.1038/nmicrobiol.2016.2. http://www.nature.com/articles/nmicrobiol20162#supplementary-information

Baumann P, Jackson SP (1996) An archaebacterial homologue of the essential eubacterial cell division protein FtsZ. Proc Natl Acad Sci U S A 93(13):6726–6730

Bolhuis H, Palm P, Wende A, Falb M, Rampp M, Rodriguez-Valera F, Pfeiffer F, Oesterhelt D (2006) The genome of the square archaeon *Haloquadratum* walsbyi: life at the limits of water activity. BMC Genomics 7:169. doi:10.1186/1471-2164-7-169

Bult CJ, White O, Olsen GJ, Zhou L, Fleischmann RD, Sutton GG, Blake JA, FitzGerald LM, Clayton RA, Gocayne JD, Kerlavage AR, Dougherty BA, Tomb JF, Adams MD, Reich CI, Overbeek R, Kirkness EF, Weinstock KG, Merrick JM, Glodek A, Scott JL, Geoghagen NS, Venter JC (1996) Complete genome sequence of the methanogenic archaeon, *Methanococcus jannaschii*. Science 273(5278):1058–1073

Burns DG, Janssen PH, Itoh T, Kamekura M, Li Z, Jensen G, Rodriguez-Valera F, Bolhuis H, Dyall-Smith ML (2007) *Haloquadratum walsbyi gen. nov., sp. nov.*, the square haloarchaeon of Walsby, isolated from saltern crystallizers in Australia and Spain. Int J Syst Evol Microbiol 57(Pt 2):387–392. doi:10.1099/ijs.0.64690-0

Buske PJ, Levin PA (2012) Extreme C terminus of bacterial cytoskeletal protein FtsZ plays fundamental role in assembly independent of modulatory proteins. J Biol Chem 287(14):10945–10957. doi:10.1074/jbc.M111.330324

Buske PJ, Mittal A, Pappu RV, Levin PA (2015) An intrinsically disordered linker plays a critical role in bacterial cell division. Semin Cell Dev Biol 37:3–10. doi:10.1016/j.semcdb.2014.09.017

Chinen A, Uchiyama I, Kobayashi I (2000) Comparison between Pyrococcus horikoshii and Pyrococcus abyssi genome sequences reveals linkage of restriction-modification genes with large genome polymorphisms. Gene 259(1–2):109–121

Dai K, Lutkenhaus J (1992) The proper ratio of FtsZ to FtsA is required for cell division to occur in *Escherichia coli*. J Bacteriol 174(19):6145–6151

Diaz JF, Kralicek A, Mingorance J, Palacios JM, Vicente M, Andreu JM (2001) Activation of cell division protein FtsZ. Control of switch loop T3 conformation by the nucleotide gamma-phosphate. J Biol Chem 276(20):17307–17315. doi:10.1074/jbc.M010920200

Duggin IG, Aylett CH, Walsh JC, Michie KA, Wang Q, Turnbull L, Dawson EM, Harry EJ, Whitchurch CB, Amos LA, Lowe J (2015) CetZ tubulin-like proteins control archaeal cell shape. Nature 519(7543):362–365. doi:10.1038/nature13983

Evans P (2006) Scaling and assessment of data quality. Acta Crystallogr D Biol Crystallogr 62 (Pt 1):72-82. doi:S0907444905036693 [pii]1107/S0907444905036693

Gilson PR, Beech PL (2001) Cell division protein FtsZ: running rings around bacteria, chloroplasts and mitochondria. Res Microbiol 152(1):3–10

Huecas S, Andreu JM (2004) Polymerization of nucleotide-free, GDP- and GTP-bound cell division protein FtsZ: GDP makes the difference. FEBS Lett 569(1–3):43–48. doi:10.1016/j.febslet.2004.05.048

Kawarabayasi Y, Sawada M, Horikawa H, Haikawa Y, Hino Y, Yamamoto S, Sekine M, Baba S, Kosugi H, Hosoyama A, Nagai Y, Sakai M, Ogura K, Otsuka R, Nakazawa H, Takamiya M, Ohfuku Y, Funahashi T, Tanaka T, Kudoh Y, Yamazaki J, Kushida N, Oguchi A, Aoki K, Kikuchi H (1998) Complete sequence and gene organization of the genome of a hyper-thermophilic archaebacterium, *Pyrococcus horikoshii* OT3. DNA Res 5(2):55–76

Klenk HP, Clayton RA, Tomb JF, White O, Nelson KE, Ketchum KA, Dodson RJ, Gwinn M, Hickey EK, Peterson JD, Richardson DL, Kerlavage AR, Graham DE, Kyrpides NC, Fleischmann RD, Quackenbush J, Lee NH, Sutton GG, Gill S, Kirkness EF, Dougherty BA, McKenney K, Adams MD, Loftus B, Peterson S, Reich CI, McNeil LK, Badger JH, Glodek A, Zhou L, Overbeek R, Gocayne JD, Weidman JF, McDonald L, Utterback T, Cotton MD, Spriggs T, Artiach P, Kaine BP, Sykes SM, Sadow PW, D'Andrea KP, Bowman C, Fujii C, Garland SA, Mason TM, Olsen GJ, Fraser CM, Smith HO, Woese CR, Venter JC (1997) The complete genome sequence of the hyperthermophilic, sulphate-reducing archaeon *Archaeoglobus fulgidus*. Nature 390(6658):364–370. doi:10.1038/37052

Kraemer JA, Erb ML, Waddling CA, Montabana EA, Zehr EA, Wang H, Nguyen K, Pham DS, Agard DA, Pogliano J (2012) A phage tubulin assembles dynamic filaments by an atypical mechanism to center viral DNA within the host cell. Cell 149(7):1488–1499. doi:10.1016/j.cell.2012.04.034

Large A, Stamme C, Lange C, Duan Z, Allers T, Soppa J, Lund PA (2007) Characterization of a tightly controlled promoter of the halophilic archaeon *Haloferax volcanii* and its use in the analysis of the essential *cct1* gene. Mol Microbiol 66 (5):1092-1106. doi:MMI5980 [pii]1111/j.1365-2958.2007.05980.x [doi]

Larsen RA, Cusumano C, Fujioka A, Lim-Fong G, Patterson P, Pogliano J (2007) Treadmilling of a prokaryotic tubulin-like protein, TubZ, required for plasmid stability in *Bacillus thuringiensis*. Genes Dev 21(11):1340–1352. doi:10.1101/gad.1546107

Lindas AC, Karlsson EA, Lindgren MT, Ettema TJ, Bernander R (2008) A unique cell division machinery in the *Archaea*. Proc Natl Acad Sci U S A 105(48):18942–18946

Lowe J, Amos LA (1998) Crystal structure of the bacterial cell-division protein FtsZ. Nature 391(6663):203–206. doi:10.1038/34472

Lowe J, Amos LA (1999) Tubulin-like protofilaments in Ca_2+-induced FtsZ sheets. EMBO J 18(9):2364–2371. doi:10.1093/emboj/18.9.2364

Lowe J, Amos LA (2000) Helical tubes of FtsZ from *Methanococcus jannaschii*. Biol Chem 381(9–10):993–999. doi:10.1515/BC.2000.122

Lowe J, Amos LA (2009) Evolution of cytomotive filaments: the cytoskeleton from prokaryotes to eukaryotes. Int J Biochem Cell Biol 41(2):323–329

Lowe J, Li H, Downing KH, Nogales E (2001) Refined structure of alpha/beta-tubulin at 3.5 Å resolution. J Mol Biol 313(5):1045–1057. doi:10.1006/jmbi.2001.5077

Makarova KS, Koonin EV (2010) Two new families of the FtsZ-tubulin protein superfamily implicated in membrane remodeling in diverse bacteria and archaea. Biol Direct 5:33. doi:10.1186/1745-6150-5-33

Margolin W, Wang R, Kumar M (1996) Isolation of an *ftsZ* homolog from the archaebacterium *Halobacterium salinarium*: implications for the evolution of FtsZ and tubulin. J Bacteriol 178(5):1320–1327

Mendieta J, Rico AI, Lopez-Vinas E, Vicente M, Mingorance J, Gomez-Puertas P (2009) Structural and functional model for ionic (K(+)/Na(+)) and pH dependence of GTPase activity and polymerization of FtsZ, the prokaryotic ortholog of tubulin. J Mol Biol 390(1):17–25. doi:10.1016/j.jmb.2009.05.018

Monahan LG, Robinson A, Harry EJ (2009) Lateral FtsZ association and the assembly of the cytokinetic Z ring in bacteria. Mol Microbiol 74 (4):1004-1017. doi:MMI6914 [pii]1111/j.1365-2958.2009.06914.x [doi]

Mukherjee A, Lutkenhaus J (1998) Dynamic assembly of FtsZ regulated by GTP hydrolysis. EMBO J 17(2):462–469. doi:10.1093/emboj/17.2.462

Mullakhanbhai MF, Larsen H (1975) Halobacterium volcanii spec. nov., a Dead Sea halobacterium with a moderate salt requirement. Arch Microbiol 104(3):207–214

Nagahisa K, Nakamura T, Fujiwara S, Imanaka T, Takagi M (2000) Characterization of FtsZ homolog from hyperthermophilic archaeon Pyrococcus kodakaraensis KOD1. J Biosci Bioeng 89(2):181–187

Ng WV, Kennedy SP, Mahairas GG, Berquist B, Pan M, Shukla HD, Lasky SR, Baliga NS, Thorsson V, Sbrogna J, Swartzell S, Weir D, Hall J, Dahl TA, Welti R, Goo YA, Leithauser B, Keller K, Cruz R, Danson MJ, Hough DW, Maddocks DG, Jablonski PE, Krebs MP, Angevine CM, Dale H, Isenbarger TA, Peck RF, Pohlschroder M, Spudich JL, Jung KW, Alam M, Freitas T, Hou S, Daniels CJ, Dennis PP, Omer AD, Ebhardt H, Lowe TM, Liang P, Riley M, Hood L, DasSarma S (2000) Genome sequence of Halobacterium species NRC-1. Proc Natl Acad Sci U S A 97(22):12176–12181. doi:10.1073/pnas.190337797

Ng KH, Srinivas V, Srinivasan R, Balasubramanian M (2013) The *Nitrosopumilus maritimus* CdvB, but not FtsZ, assembles into polymers. 2013:104147. doi:10.1155/2013/104147

Nogales E, Downing KH, Amos LA, Lowe J (1998a) Tubulin and FtsZ form a distinct family of GTPases. Nat Struct Biol 5(6):451–458

Nogales E, Wolf SG, Downing KH (1998b) Structure of the alpha/beta-tubulin dimer by electron crystallography. Nature 391(6663):199–203. doi:10.1038/34465

Nunoura T, Takaki Y, Kakuta J, Nishi S, Sugahara J, Kazama H, Chee GJ, Hattori M, Kanai A, Atomi H, Takai K, Takami H (2011) Insights into the evolution of Archaea and eukaryotic protein modifier systems revealed by the genome of a novel archaeal group. Nucleic Acids Res 39(8):3204–3223. doi:10.1093/nar/gkq1228

Oliva MA, Cordell SC, Lowe J (2004) Structural insights into FtsZ protofilament formation. Nat Struct Mol Biol 11(12):1243–1250. doi:10.1038/nsmb855

Oliva MA, Trambaiolo D, Lowe J (2007) Structural insights into the conformational variability of FtsZ. J Mol Biol 373(5):1229–1242. doi:10.1016/j.jmb.2007.08.056

Ozawa K, Harashina T, Yatsunami R, Nakamura S (2005) Gene cloning, expression and partial characterization of cell division protein FtsZ1 from extremely halophilic archaeon *Haloarcula japonica* strain TR-1. Extremophiles 9(4):281–288. doi:10.1007/s00792-005-0443-6

Pelve EA, Lindas AC, Martens-Habbena W, de la Torre JR, Stahl DA, Bernander R (2011) Cdv-based cell division and cell cycle organization in the thaumarchaeon *Nitrosopumilus maritimus*. Mol Microbiol 82(3):555–566. doi:10.1111/j.1365-2958.2011.07834.x

Poplawski A, Gullbrand B, Bernander R (2000) The *ftsZ* gene of *Haloferax mediterranei*: sequence, conserved gene order, and visualization of the FtsZ ring. Gene 242 (1–2):357-367. doi:S0378-1119(99)00517-X [pii]

Price MN, Dehal PS, Arkin AP (2010) FastTree 2 – approximately maximum-likelihood trees for large alignments. PLoS One 5(3):e9490. doi:10.1371/journal.pone.0009490

Reed CJ, Lewis H, Trejo E, Winston V, Evilia C (2013) Protein adaptations in archaeal extremophiles. 2013:373275. doi:10.1155/2013/373275

Reuter CJ, Maupin-Furlow JA (2004) Analysis of proteasome-dependent proteolysis in *Haloferax volcanii* cells, using short-lived green fluorescent proteins. Appl Environ Microbiol 70(12):7530–7538. doi:10.1128/AEM.70.12.7530-7538.2004

Richards KL, Anders KR, Nogales E, Schwartz K, Downing KH, Botstein D (2000) Structure-function relationships in yeast tubulins. Mol Biol Cell 11(5):1887–1903

Samson RY, Bell SD (2009) Ancient ESCRTs and the evolution of binary fission. Trends Microbiol 17(11):507–513. doi:10.1016/j.tim.2009.08.003

Samson RY, Obita T, Freund SM, Williams RL, Bell SD (2008) A role for the ESCRT system in cell division in archaea. Science 322(5908):1710–1713

Scheffers DJ, de Wit JG, den Blaauwen T, Driessen AJ (2002) GTP hydrolysis of cell division protein FtsZ: evidence that the active site is formed by the association of monomers. Biochemistry 41(2):521–529

Seitz KW, Lazar CS, Hinrichs KU, Teske AP, Baker BJ (2016) Genomic reconstruction of a novel, deeply branched sediment archaeal phylum with pathways for acetogenesis and sulfur reduction. Isme J. doi:10.1038/ismej.2015.233

Spang A, Saw JH, Jorgensen SL, Zaremba-Niedzwiedzka K, Martijn J, Lind AE, van Eijk R, Schleper C, Guy L, Ettema TJ (2015) Complex archaea that bridge the gap between prokaryotes and eukaryotes. Nature 521(7551):173–179. doi:10.1038/nature14447

Stoeckenius W (1981) Walsby's square bacterium: fine structure of an orthogonal procaryote. J Bacteriol 148(1):352–360

Sun Q, Margolin W (1998) FtsZ dynamics during the division cycle of live *Escherichia coli* cells. J Bacteriol 180(8):2050–2056

Tan CM, Therien AG, Lu J, Lee SH, Caron A, Gill CJ, Lebeau-Jacob C, Benton-Perdomo L, Monteiro JM, Pereira PM, Elsen NL, Wu J, Deschamps K, Petcu M, Wong S, Daigneault E, Kramer S, Liang L, Maxwell E, Claveau D, Vaillancourt J, Skorey K, Tam J, Wang H, Meredith TC, Sillaots S, Wang-Jarantow L, Ramtohul Y, Langlois E, Landry F, Reid JC, Parthasarathy G, Sharma S, Baryshnikova A, Lumb KJ, Pinho MG, Soisson SM, Roemer T (2012) Restoring methicillin-resistant *Staphylococcus aureus* susceptibility to beta-lactam antibiotics. Sci Transl Med 4(126):126ra135. doi:10.1126/scitranslmed.3003592

Vaughan S, Wickstead B, Gull K, Addinall SG (2004) Molecular evolution of FtsZ protein sequences encoded within the genomes of archaea, bacteria, and eukaryota. J Mol Evol 58(1):19–29. doi:10.1007/s00239-003-2523-5 [doi]

Walsby AE (1980) Square bacterium. Nature 283(5742):69–71. doi:10.1038/283069a0

Wang X, Lutkenhaus J (1996) FtsZ ring: the eubacterial division apparatus conserved in archaebacteria. Mol Microbiol 21(2):313–319

Webster G, O'Sullivan LA, Meng Y, Williams AS, Sass AM, Watkins AJ, Parkes RJ, Weightman AJ (2015) Archaeal community diversity and abundance changes along a natural salinity gradient in estuarine sediments. FEMS Microbiol Ecol 91(2):1–18. doi:10.1093/femsec/fiu025

Yaoi T, Laksanalamai P, Jiemjit A, Kagawa HK, Alton T, Trent JD (2000) Cloning and characterization of ftsZ and pyrF from the archaeon *Thermoplasma acidophilum*. Biochem Biophys Res Commun 275 (3):936–945. doi:10.1006/bbrc.2000.3401 [doi] S0006-291X(00)93401-6 [pii]

Yutin N, Koonin EV (2012) Archaeal origin of tubulin. Biol Direct 7:10. doi:10.1186/1745-6150-7-106150-7-10 [pii]

Zhalnina KV, Dias R, Leonard MT, Dorr de Quadros P, Camargo FA, Drew JC, Farmerie WG, Daroub SH, Triplett EW (2014) Genome sequence of Candidatus Nitrososphaera evergladensis from group I.1b enriched from Everglades soil reveals novel genomic features of the ammonia-oxidizing archaea. PLoS One 9(7):e101648. doi:10.1371/journal.pone.0101648

Chapter 15
Reconstitution of Protein Dynamics Involved in Bacterial Cell Division

Martin Loose, Katja Zieske, and Petra Schwille

Abstract Even simple cells like bacteria have precisely regulated cellular anatomies, which allow them to grow, divide and to respond to internal or external cues with high fidelity. How spatial and temporal intracellular organization in prokaryotic cells is achieved and maintained on the basis of locally interacting proteins still remains largely a mystery. Bulk biochemical assays with purified components and *in vivo* experiments help us to approach key cellular processes from two opposite ends, in terms of minimal and maximal complexity. However, to understand how cellular phenomena emerge, that are more than the sum of their parts, we have to assemble cellular subsystems step by step from the bottom up. Here, we review recent *in vitro* reconstitution experiments with proteins of the bacterial cell division machinery and illustrate how they help to shed light on fundamental cellular mechanisms that constitute spatiotemporal order and regulate cell division.

Keywords MinC • MinD • MinE • FtsZ • FtsA • ZipA • Self-organised spatial patterns • Turing • Oscillation • Reconstitution of Min protein patterns in vitro • MinD dimerization • Rotating Z-rings • FtsZ-YFP-MTS

M. Loose (✉)
Institute for Science and Technology Austria (IST Austria), Klosterneuburg, Austria
e-mail: martin.loose@ist.ac.at

K. Zieske
Stanford University, Stanford, CA, USA
e-mail: kzieske@stanford.edu

P. Schwille
Max Planck Institute of Biochemistry, Munich, Germany
e-mail: schwille@biochem.mpg.de

J. Löwe, L.A. Amos (eds.), *Prokaryotic Cytoskeletons*, Subcellular Biochemistry 84, DOI 10.1007/978-3-319-53047-5_15

Introduction

It has long been assumed that prokaryotic cells do not have much internal structural organization. Instead, bacteria were considered to be just simple containers for biochemical reactions. The initial avenue of understanding the intracellular organization of *Escherichia coli*, as with many other complex biological processes, has been through genetics. For example, the discovery of the Minicell phenotype allowed identification of genes responsible for spatial regulation of cell division ("Min") (Adler et al. 1967). However, knowing the genes and their corresponding proteins alone cannot provide a mechanistic model of their function.

The use of fluorescent proteins and their visualization with fluorescence microscopy revolutionized our understanding of how bacterial cells are organized. It turned out that even a relatively simple cell like the bacterium *Escherichia coli* has a precisely regulated cellular anatomy, emerging from the dynamic behavior of proteins. For example, the tubulin-homolog FtsZ and the actin-homolog FtsA form a ring-like structure in the middle of the cell to initiate cell division (Adams and Errington 2009). The actin-related protein ParM forms a spindle-like apparatus that segregates plasmid DNA (Moller-Jensen et al. 2002; Salje et al. 2010), the membrane-bound, actin-like protein MreB regulates cell wall synthesis, playing an important role for the determination of cell morphology (Garner et al. 2011; Domínguez-Escobar et al. 2011; van Teeffelen et al. 2011; Errington 2015), while MamK organizes membranous bacterial organelles called magnetosomes (Komeili et al. 2006). In the absence of obvious counterparts of the molecular motors involved in cytoplasmic motility throughout the eukaryotic kingdom, some filaments were recognized to facilitate motility themselves; these structures are now referred to as cytomotive filaments (Löwe and Amos 2009) and include, a set of ATPases with structural similarity to actin or tubulin that assemble into dynamic filaments in order to segregate DNA (Gerdes et al. 2010). These proteins were found to form dynamic large-scale structures that organize the intracellular space of bacteria. But understanding the underlying mechanisms from *in vivo* behavior alone has been hampered by a number of complications:

- First, a quantitative and reliable characterization of the spatiotemporal dynamics of proteins *in vivo* has been frustrated by the small size of most bacteria, at the limit of optical resolution.
- Second, although super-resolution studies have recently been introduced that can provide deeper insight in the intracellular organization of bacteria, functional fluorescence labeling is a major challenge. In fact, it has been found that labeling proteins with genetically encoded fluorescent proteins sometimes dramatically perturbs their functionality or affect their intracellular organization (Landgraf et al. 2012; Margolin 2012).
- Third, the functional characterization of some proteins has remained challenging, due to apparent functional redundancies: Their deletion can often be compensated by a number of backup systems, which limits the phenotypic range

shown by these mutants. Consequently, observed *in vivo* dynamics often are not explained by the activity of one protein alone.

- And fourth, our knowledge of large-scale protein structures in bacteria often relies on static snapshots gained through electron microscopy. While this technique provides detailed structural information, cellular functions often arise through the dynamics of complex biochemical networks (Typas and Sourjik 2015).

Accordingly, many of the mechanisms leading to biological activities may not be conclusively testable in a living system. Thus the question arises how a more complete and insightful picture of the collective protein interactions in dynamic cellular networks might be achieved.

In recent years, a number of studies have demonstrated the power of a reductionist *in vitro* approach, where a minimal set of proteins is used to rebuild cellular functions *de novo*, i.e., from the bottom-up. In this approach, purified components such as nucleic acids, lipids, and proteins are studied outside of the complexity of the living cell, but brought together in a biomimetic environment, which reproduces similar boundary conditions as found *in vivo*. The idea is to reduce the system under investigation to the smallest possible combination of molecules that is still able to reproduce a specific process or cellular function, and subject this system to highly quantitative analytics that are impossible to apply with the same rigor to living systems.

In contrast to many live-cell studies, reconstitution approaches offer full control over the molecular composition, as well as the ability to systematically perturb the experimental system parameter by parameter. It is therefore possible to yield a more detailed understanding of the molecular mechanisms underlying cellular functions otherwise hidden in the complexity of the cell. Of course, the *in vitro* reconstitution approach can be conveniently complemented by further studies both in cells and also at the chemical and molecular biology levels in order to connect to the *in vivo* situation and also to our understanding of the molecules at the atomic level.

In this book chapter, we want to highlight two specific processes that organize bacterial cells during cell division: In Part A, we will address how the Min oscillations from *E. coli* spatially organize cell division. In Part B, we discuss the membrane-based polymerization dynamics of the tubulin-homolog FtsZ. Finally, in Part C, we illustrate the recent advances that have been made to understand how the Min system direct FtsZ polymerization using a bottom-up approach. Together, we aim to illustrate how *in vitro* reconstitution experiments make it possible to obtain a much more detailed biochemical understanding of the molecular processes of cellular self-organization, previously not accessible using either classical bulk experiments and cellular assays alone.

Part A: The Min Protein Machinery as an Archetype for Protein Pattern Formation

Towards a Molecular Understanding of Pattern Formation in Biology

Already in 1952, Alan Turing (1952) lined out the conditions for a system of two interacting species to self-organize into spatial patterns. As an example, he suggested a combination of an autocatalytic reaction that generates a self-replicating "activator" species and an "inhibitor" species. The inhibitor is produced by the activator but suppresses it by a negative feedback by catalyzing its back-reaction into a non-activated state. Both species need to be subject to random motion on a molecular scale, i.e. diffusion, to result in spatial and temporal concentration fluctuations. These concentration fluctuations are amplified by nonlinear reaction schemes into large-scale patterns. Turing envisaged these reactions to provide the mechanical basis for embryogenesis, i.e. how an animal can develop starting from a single cell.

In spite of these simple formal requirements for pattern formation, and although Turing's concepts were welcomed and further extended by theoretical biophysicists in the following decades (Gierer and Meinhardt 1972), it was challenging to identify a pattern-forming system that operates in the exact same way that Turing had predicted. One reason for this is that most of the patterns found in higher organisms root in cascades of genetic regulation processes, which are much more complex than the system assumed by Turing.

With the advent of fluorescent proteins and video fluorescence microscopy it became possible to study protein distributions inside of individual cells. Now complex cellular processes could be evidenced and quantified directly in space and time, paving the way to use the concepts of self-organization originally developed to explain morphogenesis to describe the formation of intracellular order.

One fascinating example of the self-organization of proteins was found in the rod-shaped *Escherichia coli* cell: the proteins MinC, MinD and MinE repetitively oscillated between its two poles (Fig. 15.2b), presumably to position the polymerization of FtsZ and the assembly of the cell division machinery. In this first part of this chapter, we will highlight the Min protein system and the fundamental insights gained so far from its *in vitro* reconstitution.

Min Protein Oscillations in E. coli *and Their Physiological Relevance*

Cell division in *E. coli* is surprisingly precise, giving rise to two daughter cells of almost identical size (Yu and Margolin 1999). Investigations into the spatial regulation of *E. coli* cell division follow an exemplary path of cell biological research (Fig.

15.1). The first mutants pointing out the importance of a spatial regulator of cell division were identified already in the 1960s (Adler et al. 1967). In contrast to wild-type cells, these mutant cells divide asymmetrically, which results in one long daughter cell obtaining two copies of the duplicated chromosome and one much smaller "mini-cell", containing only cytoplasm. While these early studies demonstrated that the spatial organization of cell division must be genetically encoded, it took another 30 years until the cloning of the Min locus allowed studying the molecular mechanism underlying this complex task (de Boer et al. 1988, 1989).

Early phenotypic analyses already suggested that MinD and MinC are located at the cell poles (de Boer et al. 1989), where they block FtsZ assembly and thereby confine cell division to mid-cell. However, initially it was assumed that the Min proteins have a static distribution in the cell: MinE was first described as a static ring close to the cell center and suggested to keep the mid-cell free of the cell division inhibitor MinCD (Raskin and de Boer 1997). In contrast, time-resolved studies demonstrated that none of the Min proteins is localized statically, but that MinD, MinC as well as MinE dynamically oscillate between the two ends of the rod-shaped cell (Fig. 15.2b) (Raskin and de Boer 1999; Hu and Lutkenhaus 1999; Hale et al. 2001).

Bulk biochemical assays have helped to understand the requirements of Min oscillations. Already in 1991 it has been shown that MinD is an ATPase (de Boer et al. 1991), but how energy consumption and the spatial organization of cell division are functionally connected was not yet clear. In 2005, it was found that MinD has a short, C-terminal nascent helix that interacts with phospholipid membranes (Szeto et al. 2005). Importantly, this so-called membrane targeting sequence (MTS) can only promote membrane binding if MinD is in its ATP-bound form. In turn, MinE activates ATPase hydrolysis by membrane-bound MinD, increasing its rate of ATP hydrolysis about tenfold (Lackner et al. 2003) and triggering membrane detachment. MinC, the third Min protein, is not required for the oscillations. These data suggested that MinD and MinE give rise to ATP-driven, reversible binding to the membrane and that MinC binds to MinD as a passenger of the Min oscillations. While MinC does not directly contribute to the formation of the spatial Min protein patterns, it inhibits the formation of polar cell division sites by inhibiting filaments of the cell division protein FtsZ (de Boer et al. 1992a). Accordingly, by the turn of the millennium, the elementary biochemical interactions and intracellular behavior of MinC, MinD and MinE in *E. coli* were known. However, while these reactions explained the ATP-dependent, switch-like binding and detachment of MinD to the membrane, a detailed explanation for how reactions on a molecular level could lead to large-scale spatial patterns much larger than their components was still missing.

Theoretical models were able to bridge this knowledge gap and made specific predictions about the mechanisms involved. Most importantly, they were able to demonstrate conceptually that only two proteins, MinD and MinE, ATP as an energy source and the cell membrane could be sufficient to establish cellular protein oscillations, provided that the system encodes for a sufficient degree of non-linearity in the reaction steps, such as cooperativity, or positive and negative feedbacks. The simplicity of the Min protein system excited many theoretical biophysicists and

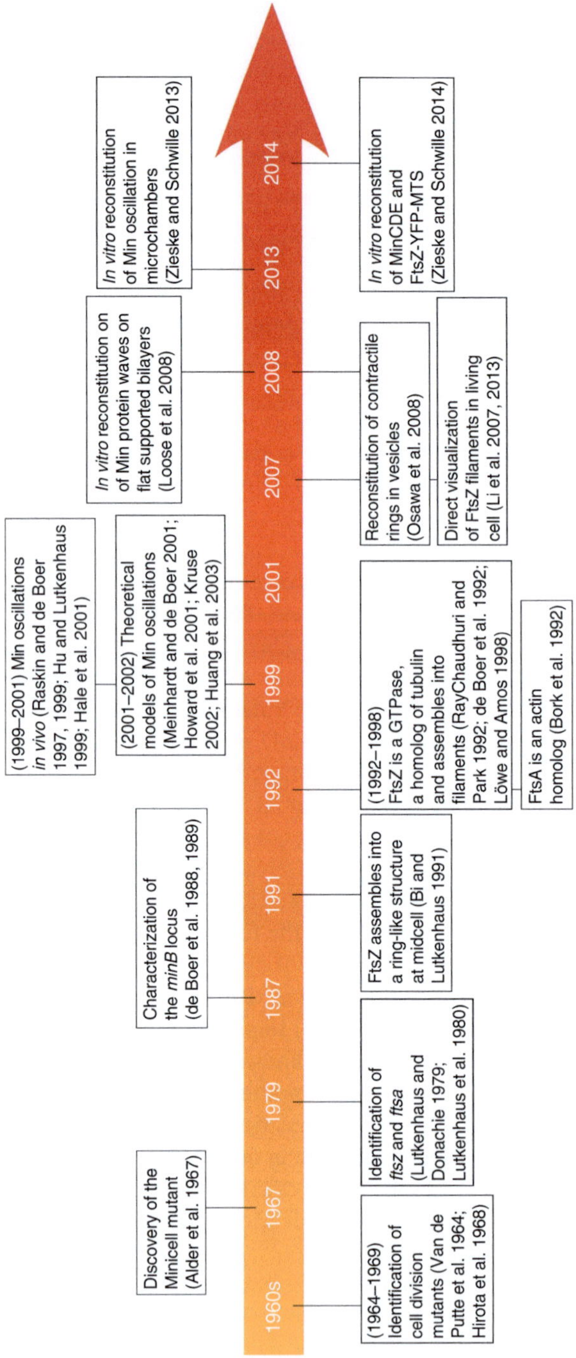

Fig. 15.1 Timeline: Important milestones for our understanding of the self-organization of the Min and FtsZ-FtsA system of *Escherichia coli*

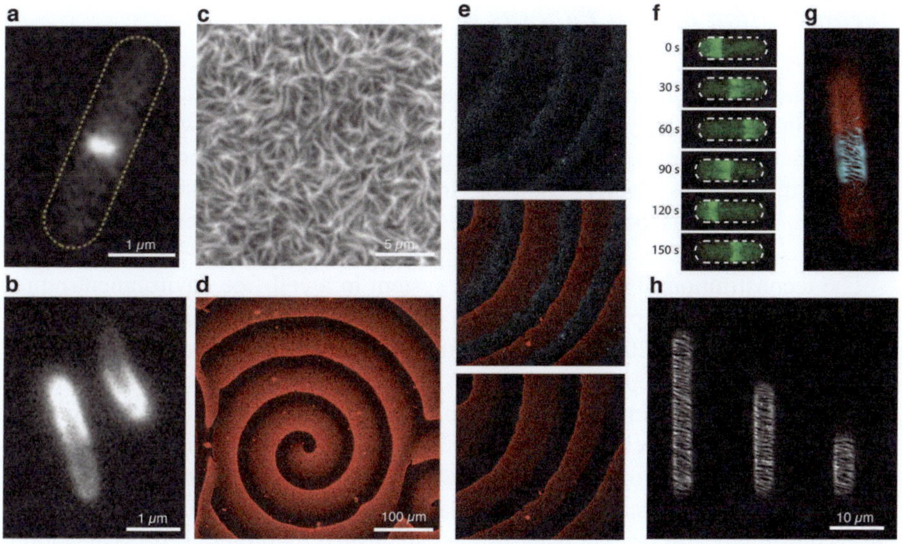

Fig. 15.2 *In vitro* reconstitution of minimal biochemical systems is a powerful approach to understand self-organized processes in the living cell. Fluorescent fusion proteins of FtsZ (GFP-FtsZ) (**a**) and MinD (GFP-MinD) (**b**) allow to image the intracellular dynamics of proteins in the living cell. (**c**) and (**d**): Using flat, supported lipid bilayer, which mimic the intracellular leaflet of the inner cell membrane and purified proteins, it is possible to reconstitute their emergent properties outside of the living cell. The micrograph in (**c**) shows the polymerization of membrane-bound FtsZ-YFP-MTS, (**d**) shows traveling waves of the two proteins MinD and MinE labeled with Cy5. (**e**) When both protein systems are combined on a flat supported membrane, anti-correlated waves of MinCDE and membrane-bound FtsZ-YFP are formed. (**f, g**) Geometric confinement can have a strong impact on the self-organization of protein systems. Using micrometer-sized chambers as membrane support, the Min systems shows a pole-pole oscillation as observed in vivo instead of traveling waves (**f**). When combined with FtsZ-YFP-MTS, FtsZ polymerization is restricted to the center of the reaction compartment (**g**). Without the Min system, FtsZ-YFP-polymers cover the entire membrane in the microchamber (**h**)

triggered a wave of publications suggesting appropriate models to describe this phenomenon (Meinhardt and de Boer 2001; Howard et al. 2001; Kruse 2002; Huang et al. 2003).

According to these models, the Min proteins create intracellular patterns in the absence of pre-established spatial cues. For example, nonlinear effects allow the system to collectively switch between different states, when a specific parameter assumes a critical value, leading to the breaking of symmetry. In this case, spontaneous local fluctuations, e.g., variations in the concentrations of the players involved, have the tendency to get amplified, and can lead to local accumulation of proteins. Diffusion of the molecules involved causes them to spread over time. Thus, spatio-temporal patterns can result from the nonlinear properties of the reaction network. Due to the absence of molecular motors in bacterial cells, such reaction systems could represent an important general mechanism to create spatial cellular order in the bacterial world without directed transport.

Reconstitution of Min Patterns in vitro

While many computational models were able to reproduce the oscillating behavior of the Min proteins, they differed in many aspects, particularly with regard to the actual reaction schemes. Quantitative experimental data to support or falsify the different models was clearly required. However, rigorously testing the models in live *E. coli* cells has been challenging, due to the complexities of the *in vivo* system.

The reconstitution of the Min protein system in a cell-free environment represented an important step forward for the quantitative understanding of oscillatory behavior (Loose et al. 2008). Two proteins, MinD and MinE, and ATP in a buffer solution on top of a supported membrane comprising *E. coli* lipids were sufficient to form dynamic protein patterns *in vitro*. After an initial phase of local binding-unbinding dynamics in circular "hot spots", planar protein wave fronts formed, and waves of high and low concentrations of MinD and MinE propagated over the membrane surface at a constant speed, length scale and direction (Fig. 15.2d). Importantly, the reconstituted Min protein patterns are established by protein self-organization, which means that energy in the form of ATP is constantly dissipated. This first reconstitution of self-organizing Min protein patterns not only confirmed that proteins can create large-scale order using a reaction diffusion mechanism, but also paved the way to studying the behavior of the Min proteins in much greater detail than in the living cell, and under fully controlled experimental conditions.

In spite of their largely reduced molecular complexity, the Min protein waves *in vitro* share a number of properties characteristic for the *in vivo* oscillations. First, just as in the cell, the Min protein waves represent a biochemical oscillation where proteins repetitively bind and detach from the membrane surface. Second, the *in vitro* patterns show a protein distribution similar to that of *in vivo* patterns, with the maximal of MinE concentration located behind the maximum of MinD. Third, MinC is the passenger of the pattern formed by MinD and MinE (Loose et al. 2011a).

However, while these initial experiments reproduced the self-organized formation of protein pattern as a key aspect of the *in vivo* oscillation, they also showed important differences and several key processes still remain obscure to date. First, while on a flat membrane the protein waves consistently travel in one direction, the proteins in the cell change direction to oscillate between the two cell poles. How spatial confinement can influence the spatial patterns formed by the Min proteins is explained below. In addition, the wavelength in the cell-free system was about an order of magnitude larger than the characteristic cell length of *E. coli*, and thus, of the wavelength *in vivo* (50 μm instead of ca. 5 μm). While one possible explanation for the larger dimensions *in vitro* could be the lack of an unidentified component that limits the mobility of the proteins *in vivo* was missing in the *in vitro* system, other reasons are also conceivable. Note that the lipid concentration in cellular membranes is only about 20%, while the rest are embedded proteins and that this difference might result in a different length scale *in vivo*. Moreover, the cell poles could be enriched in the highly negatively charged lipid cardiolipin (Renner and Weibel 2011), which can have an additional effect on membrane binding and diffusion of the membrane-bound proteins. While it was found that molecular crowding in the reaction buffer has only very limited effect on the size scale of the protein pattern, increasing the salt concentration enlarges the width of the traveling protein band up to sixfold (Vecchiarelli et al. 2014). In addition, increasing the ratio of the negatively charged lipid phosphatidylglycerol leads to up to eight times narrower proteins bands (Zieske and Schwille 2014).

Origin of Non-linearity and the Relevance of the Membrane

In order to create large-scale order, the reaction kinetics of a protein system needs to contain sufficient non-linearity (Novák and Tyson 2008). What are the regulatory mechanisms of the Min system, giving rise to its biochemical oscillation?

MinD has a C-terminal 10 amino acid long amphipathic helix (membrane targeting sequence, MTS), which confers its membrane affinity. However, a single MTS is insufficient to promote membrane binding of proteins, as GFP-MTS alone is found soluble when expressed *in vivo* (Szeto 2003). Instead, MinD dimerization is thought to provide the switch initiating membrane binding. Importantly, this dimerization can not only explain the cooperativity during membrane binding found in *in vitro* experiments (Mileykovskaya et al. 2003), but might contribute to the nonlinearity required for pattern formation. The concentration of membrane-bound MinD-ATP initially increases exponentially with time within a traveling wave (Loose et al. 2011a), suggesting an additional self-enforcing mechanism during membrane binding. The exact molecular origin of this behavior has yet to be elucidated, but conformational rearrangements of the membrane-bound protein might favor higher order MinD multimerization. Recent data indicate that MinC could play a role during the formation of large-scale structures (Ghosal et al. 2014), but this mechanism is still under debate (Park et al. 2015). Alternatively, the alteration of the membrane head-group area and lipid packing by insertion of an amphipathic helix may also lead to a higher affinity of further MinD helices towards the membrane (Hatzakis et al. 2009).

Detachment of MinD from the membrane is triggered by MinE, which catalyzes the inherently slow ATP hydrolysis of the membrane-bound MinD dimer. Detachment of both proteins is much faster than membrane binding, with MinE showing an almost instantaneous drop of protein density. Interestingly, stimulation of ATP hydrolysis by MinE was found to be cooperative, with a high Hill coefficient of 2–4 (Ghasriani et al. 2010). What could be the reason for this cooperative behavior? MinD forms a symmetric dimer with two binding sites for MinE, however it was found that MinE binding to one side of MinD is sufficient to stimulate ATP hydrolysis and membrane detachment (Park et al. 2012). In addition, structural analyses revealed that MinE performs a dramatic molecular rearrangement during the interaction and activation of MinD (Park et al. 2011): While MinE exists in a closed, autoinhibited conformation when soluble in the cytoplasm, it releases a sequestered anti-MinCD domain upon interaction with MinD. In the membrane-bound MinD-MinE complex, a N-terminal helix of MinE is released and oriented toward the membrane, positioning a cryptic membrane targeting sequence (MTS) in proximity to the membrane surface. This nascent MinE-MTS would allow for a transient membrane interaction, even after MinD has detached from the membrane and before MinE has found a new binding partner. Indeed, single-molecule fluorescence studies on Min protein waves confirmed that MinE remains about 40% longer in the traveling wave than MinD (Loose et al. 2011a).

The transient membrane binding of MinE has important consequences for the dynamic, spatial behavior of the Min system. Because MinE is able to transiently interact with the membrane after activation and detachment of MinD, the density of membrane-bound MinE increases while MinD is leaving the membrane. This gives rise to a higher [MinE/MinD] ratio, which further accelerates MinD detachment. As a result, this leads to a positive reinforcement that ensures the release of all proteins from the membrane after one oscillation cycle. Together these two studies illustrate how structural approaches and *in vitro* reconstitution of protein dynamics provide valuable information about the molecular choreography underlying the emergent behavior of the Min system.

In summary, several aspects of the activation of MinD by MinE could provide non-linearity to the biochemical network required for its oscillatory behavior (Loose et al. 2011b): the autoinhibited conformation of MinE in solution, asymmetric MinD ATPase stimulation, and persistent MinD activation due to transient membrane interaction. For these reactions, the role of the membrane in the emergence of MinDE self-organization and pattern formation cannot be overestimated. The reversible attachment of MinD and MinE to the membrane, operated and "powered" though nucleotide hydrolysis, is a constitutive step in breaking the diffusion-range symmetry between activator and inhibitor, and thus plays a vital role in the initiation of concentration gradients. Indeed, many polarity-inducing and gradient-forming systems rely on such a reversible membrane switch in at least one of their key biological elements (Altschuler et al. 2008; Hoege and Hyman 2013).

The Role of Compartment Geometry in Min Pattern Formation

A significant advantage of *in vitro* experiments is the possibility to systematically change not only the biochemical conditions, but also the geometric boundaries of the biomimetic system. Of particular importance is the role of the membrane as a two-dimensional platform for protein-protein interactions. To elucidate how the generated Min patterns react to the geometry of the environment, cell-free Min protein assays were developed to change the spatial dimensions and shape of lipid membranes.

In 2012, Schweizer et al. showed that the two-dimensional geometry of simple membrane patches of defined shape can orient and "guide" MinDE waves, if this membrane patch has similar dimensions as the wavelength *in vitro* (Schweizer et al. 2012). Using photolithographically patterned substrates it was found that the protein waves faithfully propagated along the long axes of a membrane patch, even if they had to turn by 90 degrees or were following a serpentine track, and oriented themselves along the diagonal of small rectangles with an aspect ratio of two to three. On squares or structures where all dimensions exceeded the wavelength several times, no preferential propagation direction could be observed.

Min protein waves can also respond to gaps between membrane patches, i.e. narrow lateral interruptions of the membrane substrate, depending on protein mobility and gap size. In crowded solutions with lower diffusional mobility, the gaps that could be bridged by propagating waves were significantly smaller than in regular buffer solution. Overall, it was shown that the two-dimensional shape of the membrane template could establish a distinct orientation of the resulting protein patterns.

In addition to the two-dimensional shape of the membrane patch, the three-dimensional topology of the membrane could also act as a cue to orient wave propagation directions (Zieske et al. 2014). On microfabricated membrane supports with micron-scale grooves that were either arranged as concentric rings or as alternating zones of parallel and perpendicular lines, Min protein wave propagation was always established perpendicularly to the grooves.

Finally, to determine how volume confinement reflecting the three-dimensional shape of a bacterium might influence pattern formation of the Min protein system, an assay was developed to reconstitute proteins in micro-compartments of well-defined geometry (Zieske and Schwille 2013). Cell-shaped micro-compartments were produced using micro-fabrication technology, clad with lipid membranes and filled with buffer and proteins. Air above the compartments ensured the confinement of the system to the microchamber. In contrast to the traveling protein waves on flat supported lipid membranes, the Min proteins in these compartments now oscillated between the two poles of the rod-shaped cell resembling their behavior in the living cell (Fig. 15.2f). In addition, the well-defined, custom-made geometry of the reaction compartments allowed to systematically investigate the influence of spatial boundaries. For example, the behavior of Min proteins found in round and filamentous cell-shape mutants, rotating patterns or oscillation with multiple maxima, could be reproduced in the cell-free system. This assay was also applied to test how cell geometry might help to distribute the Min proteins to both daughter cells during cell division. By reconstituting Min protein patterns in compartments mimicking distinct degrees of septum closure, it was found that the changing geometry of a dividing cell influences the dynamic patterns of Min proteins, ensuring the equal distribution of Min protein to both daughter cells similar to the observations made *in vivo* (Zieske and Schwille 2014; Di Ventura and Sourjik 2011).

All these experiments provided the experimental evidence that no additional factors located at the cell poles are required for the Min proteins to oscillate, but that it is the interplay between the biochemical reactions and their three-dimensional confinement that leads to the characteristic protein pattern of the Min proteins.

In summary, the *in vitro* reconstitution of Min protein dynamics not only demonstrated that proteins do form large-scale patterns by use of a reaction-diffusion mechanism, it also allowed systematic perturbations of the system, biochemically and geometrically, which is very difficult to achieve in a living cell.

Part B: Rotating Rings: How FtsZ Filaments Assemble into a Large-Scale Machine

The Discovery of the Bacterial Cytoskeleton

During cell division, the cell diameter at mid-cell continuously decreases until the division septum is fully closed, giving rise to two daughter cells. This is a difficult task, which is performed by a complex, yet highly dynamic protein machinery that controls the concerted invagination of a layered cell envelope while simultaneously remodeling the surrounding peptidoglycan layer.

Although many individual players involved in this process have been identified, how they act together to accomplish the mechanical transformation from one into two cellular compartments is still not understood. In particular, the actual forces governing the constriction of the division zone and their molecular origins remain obscure. In this part, we will give a brief historical sketch of how the protein machinery required for cell division has been discovered, and some new insights gained on functional reconstitution that govern our present understanding. We describe emergent phenomena that can still be observed after dramatically simplifying the molecular system, and hope to convince the reader that the *in vitro* reconstitution of a simplified cytokinetic machinery is a highly valuable complementary approach to the essential live cell studies.

The Bacterial Divisome

The genes responsible for bacterial cell division were primarily discovered through identification of a group of thermosensitive mutants, which failed to divide and instead produced very long, filamentous cells at elevated temperatures (filamentous thermosensitive, fts) (Fig. 15.1). Using this approach, several mutants that were unable to divide could be identified (Van de Putte et al. 1964; Hirota et al. 1968), providing a preliminary list of genes involved in cell division. In 1972, Wijsman (1972) mapped several of the originally identified mutants to the *ftsA* locus. However, the exact identity of the genes affected was not yet clear, in fact it was not yet known whether these mutations affected one or multiple genes.

The gene product of *ftsA* was identified in 1979, and a newly designated gene encoding for *ftsZ* was identified 1 year later (Lutkenhaus and Donachie 1979;

Lutkenhaus et al. 1980). Although it became clear that both proteins have to act very early during cell division, a close characterization of the elongated cells revealed that these filamentous mutants were not identical. Mutations in *ftsZ* usually give a smooth morphology with no signs of constriction along the cells, whereas mutations in *ftsA* result in cells with an indented morphology indicative of attempts to form septa (Begg and Donachie 1985; Taschner et al. 1988). While these studies provided strong evidence that the gene products of both *ftsA* and *ftsZ* play crucial but distinct roles for septum formation and cell division, they could not yet provide information about the intracellular localization of the corresponding proteins nor the biochemical basis explaining how they are able to fulfill their cellular roles.

The functional characterization of the divisome proteins followed a common workflow. First, immunolabeling of chemically fixed cells, combined with bright-field fluorescence and conventional electron microscopy was used to visualize the location of FtsZ in a dividing cell (Bi and Lutkenhaus 1991). With this approach it was found that FtsZ is distributed in a ring-like structure at the site of cell invagination and that it remained localized at the leading edge of the division septum during constriction. According to this early concept of FtsZ function, the first step of cell division was the self-assembly of FtsZ into a ring-like structure on the inner side of the cytoplasmic membrane. The formation of this so-called Z-ring was suggested to be the key event from which temporal and spatial control of cell division by FtsZ was exerted. This seminal study not only provided data for a functional role of FtsZ, but also questioned the up to then common belief that in contrast to eukaryotic cells, bacteria do not exhibit a cytoskeleton. However, the relation of FtsZ to the cytoskeleton of eukaryotes, as well as the biochemical basis of its cellular function, was still unknown.

The possibility that FtsZ could be related to eukaryotic tubulin was originally raised upon recognition of a glycine-rich motif GGGTGTG in its sequence, which resembles the tubulin signature motif (GGGTGS/TG). Three independent studies discovered that FtsZ, just like tubulin, hydrolyses GTP (RayChaudhuri and Park 1992; de Boer et al. 1992b; Mukherjee and Lutkenhaus 1998) and accumulating data supported the idea that FtsZ is able to form cytoskeletal filaments. FtsZ was found to polymerize in the presence of GTP into a variety of different structures (Bramhill and Thompson 1994; Mukherjee and Lutkenhaus 1994). Apart from short, single-stranded filaments, which could either be straight or curved, depending on the nucleotide bound (Erickson et al. 1996), FtsZ was also found to form tubules, sheets and minirings *in vitro* (Mukherjee and Lutkenhaus 1994; Erickson et al. 1996; González et al. 2003). Although the crystal structure of FtsZ confirmed its relation to tubulin (Löwe and Amos 1998), what kind of filament structures are physiologically relevant for cell division was still unclear.

In the case of FtsA, a bioinformatic approach using available sequence information helped to identify it as a member of an ATPase family that shows structural similarities to actin (Bork et al. 1992). This hypothesis was confirmed in 2001, when the crystal structure of FtsA was solved (van den Ent et al. 2001).

Using GFP-fusion proteins, the dynamics of these two cytoskeletal proteins were imaged in *E. coli* cells, and FtsZ as well as FtsA were both found to rapidly assemble into a ring-like structure at midcell (Fig. 15.2a), very early during the cell cycle (Addinall et al. 1996). In *E. coli* and closely related species, the formation of the Z-ring depends on the presence of either FtsA or ZipA. Both proteins bind to a short conserved sequence at the extreme C-terminus of FtsZ, which allows them to recruit FtsZ to the inner side of the cytoplasmic membrane. While ZipA has a transmembrane domain (Hale and de Boer 1997), FtsA binds to membrane with the help of a C-terminal amphipathic helix (Pichoff and Lutkenhaus 2005) similar to MinD. The localization of ZipA and FtsA to the division septum depends on the activity of FtsZ. In the absence of either one of these two proteins, Z-rings could still form but these rings are not functional and cell division cannot proceed.

As a relative of actin, it has been a longstanding question whether FtsA was able to polymerize as well (Löwe et al. 2004) and a large number of studies had set out to confirm this hypothesis. First, FtsA was found to self-associate based on two-hybrid experiments (Yan et al. 2000) and the overexpression of GFP-tagged C-terminal truncations leads to filamentous structures in cells, as seen by fluorescence microscopy (Pichoff and Lutkenhaus 2007). The important role of FtsA-FtsA interaction during the assembly of the cell division machinery became clear when self-interaction mutants of FtsA were discovered that were able to bypass proteins normally required for cytokinesis (Beuria et al. 2009; Pichoff et al. 2011). These findings gave rise to a model where FtsA monomers are able to recruit proteins downstream of its assembly at midcell (Liu et al. 2014; Tsang and Bernhardt 2015), in contrast to its oligomers. Eventually, FtsA from *Streptococcus pneumoniae* has been demonstrated to form helical filaments in the presence of ATP i*n vitro* (Szwedziak et al. 2012; Lara et al. 2005). However, a detailed quantitative characterization of FtsA polymerization as well as a direct demonstration of how it regulates the protein-protein interaction network underlying cytokinesis is still missing.

Recent Approaches to Understand the Mechanism of Z-Ring Assembly and Function in vivo

Conventional electron or fluorescence microscopy did not permit direct visualization of individual FtsZ protofilaments in the crowded and biochemically complex environment of the living cell. Accordingly, little was known about the organization of individual FtsZ filaments and how they coordinate cell division in space and time. In fact, it also remained unclear whether the cytoskeletal element provided by FtsZ generates the contractile force itself or whether it mainly acts as a dynamic scaffold that recruits and coordinates the assembly of other proteins, or both.

In recent years, two complementary approaches have been followed to further elucidate the role of FtsA and FtsZ: First, the improvements in electron cryotomography and super-resolution fluorescence microscopy made it possible to visualize individual FtsZ filaments in dividing bacterial cells. Second, as genetic screens identified ever more genes and gene-gene interactions (Typas et al. 2012; Brochado and Typas 2013), we now have a probably complete list of the proteins required for cell division. Importantly, this comprehensive information represents the first step for the reconstitution of these complex biochemical processes *in vitro*, using a minimal set of necessary molecular components.

The first direct visualization of individual FtsZ filaments in living cells was achieved by Li et al. in 2007. Using cryotomography, the authors discovered short, arc-like filaments at the constriction sites of dividing *C. crescentus* cells, just underneath the inner cell membrane. Based on the observation that FtsZ filaments are straight in the GTP-bound state and more curved in the GDP-bound state (Erickson et al. 2010), an 'iterative pinching' mechanism was proposed to explain the constriction of the Z-ring. According to this model, FtsZ itself generates the force that constricts the membrane in a GTP-hydrolysis-driven cycle of polymerization, membrane attachment, conformational change, depolymerization, and nucleotide exchange.

Although electron cryotomography provides superior resolution, it lacks labeling technology, limiting the potential of imaging protein assemblies *in vivo*. When novel super-resolution fluorescence microscopy techniques were developed, a more specific visualization of fluorescently labeled proteins in living cells along with greater spatial resolution than conventional fluorescence microscopy was achieved. For example, Fu et al. visualized filament structures at midcell in *E. coli*. Their results suggested that the Z-ring is composed of a loose bundle of FtsZ protofilaments of variable lengths that randomly overlap with each other (Fu et al. 2010). This is in contrast to the short FtsZ arcs seen at the division sites of *Caulobacter crescentus* cells imaged with electron tomography (Li et al. 2007). Following this initial study, high-resolution images of FtsZ in four different organisms were obtained, all depicting the Z-ring as a heterogeneous assembly of short protofilaments (Fu et al. 2010; Strauss et al. 2012; Biteen 2012). In all these studies, the Z-ring appears to be a multi-layered structure composed of short (50–500 nm) filaments, rather than a continuous, helical ring. Some studies also identified a substantial portion of FtsZ filaments aligned along the cell axis, supporting neither the sliding nor the iterative-pinching model. Furthermore, it became possible to visualize the spatial distribution of FtsZ-associated proteins, such as those supposed to be involved in filament bundling. This shed light not only on the orientation of FtsZ filaments, but also on the molecular architecture of the Z-ring associated proteins (Buss et al. 2013).

In vitro *Reconstitution Experiments Help to Elucidate Z-Ring Architecture and FtsZ Filament Dynamics*

Despite the progress of fluorescence imaging and electron microscopy, there is still no agreement on the mechanism of Z-ring assembly and constriction, showing that *in vivo* data alone cannot provide all the information required to provide mechanistic understanding of bacterial cell division. One reason for this lack of knowledge could be that electron microscopy and super-resolution fluorescence microscopy data rely on static pictures that only indirectly provide information about the fast dynamics of most intracellular processes.

The first direct evidence for a potential force-generating role of FtsZ filaments came from *in vitro* reconstitution experiments using a modified, autonomously membrane-binding version of FtsZ (Osawa et al. 2008). In this engineered construct, called FtsZ-YFP-MTS, FtsZ's C-terminal FtsA-and ZipA-binding peptide was replaced with YFP and a membrane targeting sequence from MinD (MTS). Therefore, it did no longer depend on the native membrane anchors to bind to the membrane. Importantly, as for MinD, the affinity of a single MTS to the membrane was too low for attaching single FtsZ monomers to the membrane, only in the polymeric state could a quantitative membrane binding be observed.

When FtsZ-YFP-MTS was seemingly encapsulated into multilamellar lipid vesicles, it was able to polymerize into a ring-like structure encircling the vesicle and forming visible constrictions (Osawa et al. 2008), indicating that membrane-bound FtsZ filaments might be sufficient to generate force. Supporting this conclusion, FtsZ-YFP-MTS on the outside of liposomes were found to induce concave depressions and membrane tubules (Milam et al. 2012). However, an alternative explanation for the observation of vesicle constriction could also be that the local accumulation of a protein with an amphipathic helix can deform the phospholipid membrane (Stachowiak et al. 2012). Harold Erickson's lab corroborated their hypothesis of FtsZ filament curvature as driver of membrane deformation by switching the MTS to the N-terminus of FtsZ, which is on the opposite face of the molecule from the C-terminus: MTS-FtsZ-YFP induces convex bulges on the surface of spherical liposomes. These observations may be explained by FtsZ filaments or filament bundles having an preferred diameter, which either gives rise to membrane curvature sensing or deformation. This notion was further supported by Arumugam et al. (Arumugam et al. 2012), who found that proto-ring like filament bundles aligned perpendicularly on membrane-coated micrometer-sized troughs or rods (for FtsZ-YFP-MTS or MTS-FtsZ-YFP, respectively) if the surface curvature was close to the curvature of non-constricted cells. At smaller curvatures, the FtsZ proto-rings were no longer perpendicularly aligned, but rather tilted themselves to conserve their larger intrinsic curvatures. From analyzing the preferred angles at which the filaments tilted with respect to the long axis of the trough or rod, they could be best modeled assuming that FtsZ assembles into intrinsically helically twisted but flexible filaments.

Osawa et al.'s initial model for the force generation of FtsZ favors a mechanism where the force is generated by a concerted conformational change of bundled FtsZ filaments driven by GTP hydrolysis (Erickson et al. 1996; Li et al. 2007; Lu et al. 2000) and a recent structural study supported this idea (Li et al. 2013). However, this mechanism would only allow constriction down to a 24-nm inside diameter, too large to allow for complete fission of the *E. coli* cell (Erickson et al. 2010). Furthermore, Lan et al. pointed out that FtsZ filaments are too soft to generate sufficient force for membrane deformation (Lan et al. 2009) and it has been shown that GTP hydrolysis is not the rate limiting step for cell division (Coltharp et al. 2016).

More recently, Szwedziak et al. were able to reconstitute the co-assembly of FtsA and FtsZ from *Thermotoga maritima* inside of phospholipid vesicles made of *E. coli* lipids and similar to what they have observed in living *E. coli* and *C. crescentus* cells, the authors found a continuous helix, often made from several overlapping shorter filaments at the site of constriction, supporting the idea of FtsZ and FtsA being sufficient to generate mechanical force (Szwedziak et al. 2014). Instead of an "iterative pinching" model, however, the authors prefer a filament sliding mechanism to explain the constriction, where a continuous helix or connected filaments could slide along each other, thereby continuously decreasing the diameter of the Z-ring. In this case, the thermodynamic driving force for the constriction would be the increase in lateral contact area, which would strive to reach an energetic minimum until a critical density of FtsZ filaments is reached and the ring would not constrict further (Lan et al. 2009; Hörger et al. 2008). Alternatively, a model was suggested where force is generated in a cycle of gap formation and ring closure (Osawa and Erickson 2011), but a direct confirmation of this hypothesis is still missing.

The Intriguing Self-Organization of Reconstituted FtsZ with FtsA

The experiments described above were able to demonstrate that membrane-bound FtsZ filaments can generally deform a phospholipid membrane. However, this radically simplified assay may not mimic all the essential properties of the initial steps of cell division. For example, the role of the membrane anchors during cell constriction remained unclear and an autonomously binding version of FtsZ might neglect their important influence. In fact, most previous models for Z-ring formation mainly assumed that both FtsA and ZipA are merely passive membrane anchors for FtsZ (Erickson et al. 2010; Kirkpatrick and Viollier 2011; Lutkenhaus et al. 2012), e.g. that they do not influence the polymerization dynamics of FtsZ. In contrast, earlier studies had already pointed out that FtsA might not be just a simple membrane anchor for FtsZ, but that it could actively take part in the formation of the Z-ring and function of the cell division machinery. For example, in 1992 Dewar et al. found that overexpression of FtsA blocks cell division at a very early stage, resulting in a

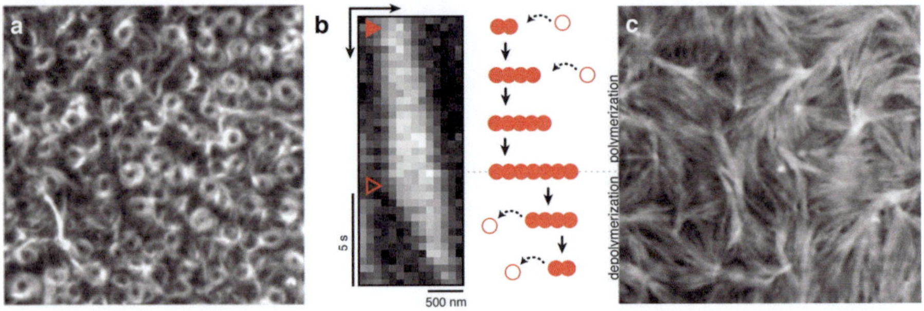

Fig. 15.3 The two different membrane anchors of FtsZ give rise to different emergent behavior in vitro. With FtsA, FtsZ organizes in rapidly reorganizing cytoskeletal structures at high protein concentrations (**a**). At lower concentrations FtsZ forms treadmilling filaments (**b**). With ZipA, FtsZ assembles to thick, stationary bundles (**c**)

mutant that was similar to an FtsZ mutant (Dewar et al. 1992). Two papers by the Margolin group corroborated this finding by showing that a hypermorphic mutant FtsA* (Beuria et al. 2009; Geissler et al. 2007) leads to more stable FtsZ rings and faster ring assembly. Using yeast two hybrid assays, it was possible to show that this mutant binds stronger to FtsZ. In fact, more recent data by Osawa et al. (Osawa and Erickson 2013) suggest that FtsA can support FtsZ-driven constriction of vesicles and even help to complete membrane fission.

To study the mechanism underlying Z-ring assembly and to find out how the membrane anchors might influence the organization of FtsZ filaments, Loose and Mitchison developed an assay to reconstitute FtsZ polymerization mediated by native FtsA or ZipA on supported lipid bilayers *in vitro*. With FtsA, FtsZ assembled into highly dynamic cytoskeletal patterns resembling fast moving streams and rapidly rotating rings of FtsZ filament bundles (Fig. 15.3a). Interestingly, the dimensions of the rings formed in this experimental system were similar to the dimensions of the Z-ring found *in vivo* (Loose and Mitchison 2014). Further experiments indicated that the large-scale dynamics of FtsZ bundles likely originates from treadmilling FtsZ filaments, where one end of a filament grows while the opposite end shrinks (Fig. 15.3b). In contrast, the presence of ZipA in the membrane gave rise to thick bundles of FtsZ that did not show any large-scale reorganization (Fig. 15.3c).

What can explain these differences in the emergent behavior of FtsZ? In agreement with previous observations that FtsA can break FtsZ filaments (Beuria et al. 2009), the dynamics of the membrane-bound FtsZ filament network might arise from a dual role of FtsA: On the one hand, it is required to recruit FtsZ filaments to the membrane, on the other it can destabilize these polymers, providing a negative feedback regulation that can explain the rapid reorganization observed. Unlike FtsA, ZipA stabilizes FtsZ filaments (Hale et al. 2000) and therefore promotes the formation of thick filament bundles on the membrane. Interestingly, FtsZ-YFP-MTS, which binds autonomously to the membrane, formed a network of filament bundles on the membrane (Fig. 15.2c), which did not show large-scale reorganization or dynamic vortices.

Together, these data showed that in contrast to previous assumptions, the membrane anchor does influence the properties of FtsZ filaments. Since FtsA and ZipA give rise to very different behavior of FtsZ filaments, this study also illustrates that the emergent behavior of a protein system critically depends on the properties of its individual components and that *in vivo* their activities need to precisely balanced (Loose and Mitchison 2014). This data also favor the idea that a rapidly reorganizing Z-ring would predominantly play a signaling role to guide peptidoglycan synthesis. In this case, the FtsZ would act as a spatial scaffold coordinating the activity of other divisome proteins, while the dynamics of the ring allows it to follow the inward growth of the division septum.

To summarize, although we made great progress in understanding bacterial cell division in the last two decades, some of the most fundamental questions still remain: what triggers and what drives FtsZ ring constriction? Surprisingly, the best answer today is very much the same as 20 years ago: "The function of the FtsZ ring could be twofold. It could function as a cytoskeletal element that provides the energy for constriction of the cytoplasmic membrane, act as a mooring to recruit other division proteins to the division site, or both." (Lutkenhaus and Addinall 1997; Meier and Goley 2014).

Part C: Min Oscillations Regulate FtsZ Ring Localization *in vitro*

After the Min protein oscillations were discovered as a mechanism to regulate cell division, the question arose how exactly the Min system limits FtsZ polymerization to the center of the cell. In a test tube, FtsZ assembles into filament networks, which can be visualized by electron microscopy (Mukherjee and Lutkenhaus 1994). Addition of MinC rapidly disassembles these filaments, showing that MinC is the component of the Min system that directly inhibits FtsZ polymerization (Hu and Lutkenhaus 1999; Dajkovic et al. 2008). However, the concentration ratio used in these experiments was tenfold higher than in the *E. coli* cell (de Boer et al. 1991; Szeto et al. 2001). Thus, while these experiments provided evidence for the role of MinC, it revealed discrepancies to the *in vivo* situation. One possible explanation raised was that although the mean cellular concentration of MinC is lower than the critical concentration for FtsZ disassembly, MinC is enriched at the cell periphery due to its interaction with MinD, which could provide for sufficiently high subcellular MinC concentrations at the cell poles (Hu and Lutkenhaus 1999; Hu et al. 1999). To experimentally test how membrane recruitment might affect FtsZ filaments, the autonomously membrane binding protein FtsZ-YFP-MTS was combined with Min protein patterns on a flat, supported lipid membrane. In these experiments, Min protein waves and a polymer network of FtsZ-YFP-MTS formed an anti-correlated protein pattern on the membrane, where FtsZ could polymerize in the valleys between the travelling Min proteins bands (Arumugam et al. 2014) (Fig. 15.2e).

From this co-reconstitution study, it could be shown that MinC acts on the FtsZ-YFP-MTS filaments primarily by interfering with the detachment and re-attachment of FtsZ monomers from and to the surface-assembled filament bundles. Similar results were obtained when ZipA and native FtsZ was used (Martos et al. 2015). Importantly, in both studies, much lower concentrations of MinC than in the bulk assay were required for this localized FtsZ inhibition, verifying that enrichment of MinC close to the membrane indeed accounts for FtsZ disassembly by low cellular MinC concentrations (Raskin and de Boer 1999; Hale et al. 2001).

While the co-reconstitution of Min protein waves and filament networks of FtsZ-YFP-MTS on flat supported membranes revealed the minimal biochemical conditions accounting for FtsZ network disassembly, it did not yet reproduce the spatial constraints found *in vivo*. In the *E. coli* cell, the Min oscillations act on a faster time-scale than the assembly of the Z-ring. Accordingly, the Min proteins can restrict FtsZ assembly to midcell because their time-averaged concentration profile represents a protein gradient with the highest concentration of MinC located at the cell poles (Loose et al. 2011b). To test if protein oscillations *in vitro* can direct the polymerization of FtsZ filaments, Min oscillations and FtsZ-YPF-MTS were co-reconstituted in membrane-covered microcompartments (Zieske and Schwille 2014). Indeed, in this experimental system consisting only of the Min proteins (MinD, MinE, MinC), a membrane-binding version of FtsZ and membrane-coated reaction compartments, it was found that the Min protein oscillations alone can achieve spatial regulation and targeting of FtsZ assembly, where FtsZ-YFP-MTS filaments were only found in the center of the rod-shaped microstructure (Fig. 15.2h). Intriguingly, the reconstitution of the regulated localization of FtsZ in a cell-free environment does not only allow to verify the minimal biochemical components and required physical boundary conditions for division site placement, but in our opinion also provides a starting point to reconstitute the complete cell division machinery from purified components (Zieske and Schwille 2014).

Summary and Outlook

It has become increasingly clear that complex and diverse mechanisms organize the bacterial cell, but we are only beginning to identify the underlying biochemical networks that achieve this task. In this book chapter, we have illustrated the power of *in vitro* reconstitution experiments, which may not only help to identify the minimal requirements for cellular processes, but also to elucidate the molecular mechanism of intracellular organization.

Although *E. coli* and other bacteria display relatively simple intracellular architecture, they rely on complex biochemical networks to grow and divide in various environmental conditions, nutrient sources and temperatures. Since this complexity is the result of many millions of years of evolution, bottom-up approaches can shed light on the minimal motifs and modules able to fulfill a cellular function and this minimal machine might closely resemble the cellular machines present in distant ancestors of *E. coli*. It may be too complex and tedious to eventually re-assemble the full *E. coli* divisome using purified components, but it is certainly conceivable to eventually arrive at a basic set of functional elements that allow for a controlled division of primitive vesicles, and thus come closer to a understanding of how cells emerged as the amazing reproduction machines that we know today.

References

Adams DW, Errington J (2009) Bacterial cell division: assembly, maintenance and disassembly of the Z ring. Nat Rev Microbiol 7:642–653. doi:10.1038/nrmicro2198

Adler H, Fisher W, Cohen A, Hardigree A (1967) Miniature *Escherichia coli* cell deficient in DNA. PNAS 57:321–326

Altschuler SJ, Angenent SB, Wang Y, Wu LF (2008) On the spontaneous emergence of cell polarity. Nature 454:886–889. doi:10.1038/nature07119

Arumugam S, Chwastek G, Fischer-Friedrich E, Ehrig C, Mönch I, Schwille P (2012) Surface topology engineering of membranes for the mechanical investigation of the tubulin homologue FtsZ. Angew Chem Int Ed Eng. doi:10.1002/anie.201204332

Arumugam S, Petrasek Z, Schwille P (2014) MinCDE exploits the dynamic nature of FtsZ filaments for its spatial regulation. PNAS 111:E1192–E1200. doi:10.1073/pnas.1317764111

Begg KJ, Donachie WD (1985) Cell shape and division in *Escherichia coli*: experiments with shape and division mutants. J Bacteriol 163:615–622

Beuria TK, Mullapudi S, Mileykovskaya E, Sadasivam M, Dowhan W, Margolin W (2009) Adenine nucleotide-dependent regulation of assembly of bacterial tubulin-like FtsZ by a hypermorph of bacterial actin-like FtsA. J Biol Chem 284:14079–14086. doi:10.1074/jbc.M808872200

Bi EF, Lutkenhaus J (1991) FtsZ ring structure associated with division in *Escherichia coli*. Nature 354:161–164. doi:10.1038/354161a0

Biteen JS (2012) Extending the tools of single-molecule fluorescence imaging to problems in microbiology. Mol Microbiol. doi:10.1111/j.1365-2958.2012.08089.x

Bork P, Sander C, Valencia A (1992) An ATPase domain common to prokaryotic cell cycle proteins, sugar kinases, actin, and hsp70 heat shock proteins. PNAS 89:7290–7294

Bramhill D, Thompson CM (1994) GTP-dependent polymerization of *Escherichia coli* FtsZ protein to form tubules. Proc Natl Acad Sci U S A 91:5813–5817

Brochado AR, Typas A (2013) High-throughput approaches to understanding gene function and mapping network architecture in bacteria. Curr Opin Microbiol 16:199–206. doi:10.1016/j.mib.2013.01.008

Buss J, Coltharp C, Huang T, Pohlmeyer C, Wang S-C, Hatem C, Xiao J (2013) In vivo organization of the FtsZ-ring by ZapA and ZapB revealed by quantitative super-resolution microscopy. Mol Microbiol 89:1099–1120. doi:10.1111/mmi.12331

Dajkovic A, Lan G, Sun SX, Wirtz D, Lutkenhaus J (2008) MinC spatially controls bacterial cytokinesis by antagonizing the scaffolding function of FtsZ. Curr Biol 18:235–244. doi:10.1016/j.cub.2008.01.042

de Boer PA, Crossley RE, Rothfield LI (1988) Isolation and properties of minB, a complex genetic locus involved in correct placement of the division site in *Escherichia coli*. Journal of Bacteriology 170:2106–2112

de Boer PA, Crossley RE, Rothfield LI (1989) A division inhibitor and a topological specificity factor coded for by the minicell locus determine proper placement of the division septum in *E. coli*. Cell 56:641–649 de Boer PA, Crossley RE, Hand AR, Rothfield LI (1991) The MinD protein is a membrane ATPase required for the correct placement of the *Escherichia coli* division site. EMBO J 10:4371–4380

de Boer PA, Crossley RE, Rothfield LI (1992a) Roles of MinC and MinD in the site-specific septation block mediated by the MinCDE system of *Escherichia coli*. J Bacteriol 174:63–70

de Boer P, Crossley R, Rothfield L (1992b) The essential bacterial cell-division protein FtsZ is a GTPase. Nature 359:254–256. doi:10.1038/359254a0

Dewar SJ, Begg KJ, Donachie WD (1992) Inhibition of cell division initiation by an imbalance in the ratio of FtsA to FtsZ. J Bacteriol 174:6314–6316

Di Ventura B, Sourjik V (2011) Self-organized partitioning of dynamically localized proteins in bacterial cell division. Mol Syst Biol 7:1–13. doi:10.1038/msb.2010.111

Domínguez-Escobar J, Chastanet A, Crevenna AH, Fromion V, Wedlich-Soldner R, Carballido-López R (2011) Processive movement of MreB-associated cell wall biosynthetic complexes in bacteria. Science 333:225–228. doi:10.1126/science.1203466

Erickson HP, Taylor DW, Taylor KA, Bramhill D (1996) Bacterial cell division protein FtsZ assembles into protofilament sheets and minirings, structural homologs of tubulin polymers. PNAS 93:519–523

Erickson HP, Anderson DE, Osawa M (2010) FtsZ in bacterial cytokinesis: cytoskeleton and force generator all in one. Microbiol Mol Biol Rev 74:504–528. doi:10.1128/MMBR.00021-10

Errington J (2015) Bacterial morphogenesis and the enigmatic MreB helix. Nat Rev Microbiol. doi:10.1038/nrmicro3398

Fu G, Huang T, Buss J, Coltharp C, Hensel Z, Xiao J (2010) In vivo structure of the *E. coli* FtsZ-ring revealed by photoactivated localization microscopy (PALM). PLoS One 5:e12682. doi:10.1371/journal.pone.0012680

Garner EC, Bernard R, Wang W, Zhuang X, Rudner DZ, Mitchison T (2011) Coupled, circumferential motions of the cell wall synthesis machinery and MreB filaments in *B. subtilis*. Science 333:222–225. doi:10.1126/science.1203285

Geissler B, Shiomi D, Margolin W (2007) The ftsA* gain-of-function allele of *Escherichia coli* and its effects on the stability and dynamics of the Z ring. Microbiology (Reading, Engl) 153:814–825. doi:10.1099/mic.0.2006/001834–0

Gerdes K, Howard M, Szardenings F (2010) Pushing and pulling in prokaryotic DNA segregation. Cell 141:927–942. doi:10.1016/j.cell.2010.05.033

Ghasriani H, Ducat T, Hart CT, Hafizi F, Chang N, Al-Baldawi A, Ayed SH, Lundström P, Dillon J-AR, Goto NK (2010) Appropriation of the MinD protein-interaction motif by the dimeric interface of the bacterial cell division regulator MinE. PNAS 107:18416–18421. doi:10.1073/pnas.1007141107

Ghosal D, Trambaiolo D, Amos LA, Löwe J (2014) MinCD cell division proteins form alternating copolymeric cytomotive filaments. Nat Commun 5:5341. doi:10.1038/ncomms6341

Gierer A, Meinhardt H (1972) A theory of biological pattern formation. Kybernetik 12:30–39. doi:10.1007/BF00289234

González JM, Jiménez M, Vélez M, Mingorance J, Andreu JM, Vicente M, Rivas G (2003) Essential cell division protein FtsZ assembles into one monomer-thick ribbons under conditions resembling the crowded intracellular environment. J Biol Chem 278:37664–37671. doi:10.1074/jbc.M305230200

Hale CA, de Boer PAJ (1997) Direct binding of FtsZ to ZipA, an essential component of the septal ring structure that mediates cell division in *E. coli*. Cell 88:175–185. doi:10.1016/S0092-8674(00)81838-3

Hale CA, Rhee AC, de Boer PA (2000) ZipA-induced bundling of FtsZ polymers mediated by an interaction between C-terminal domains. J Bacteriol 182:5153–5166

Hale CA, Meinhardt H, de Boer PA (2001) Dynamic localization cycle of the cell division regulator MinE in *Escherichia coli*. EMBO J 20:1563–1572. doi:10.1093/emboj/20.7.1563

Hatzakis NS, Bhatia VK, Larsen J, Madsen KL, Bolinger P-Y, Kunding AH, Castillo J, Gether U, Hedegård P, Stamou D (2009) How curved membranes recruit amphipathic helices and protein anchoring motifs. Nat Methods 5:835–841. doi:10.1038/nchembio.213

Hirota Y, Ryter A, Jacob F (1968) Thermosensitive mutants of *E. coli* affected in the processes of DNA synthesis and cellular division. Cold Spring Harb Symp Quant Biol 33:677–693

Hoege C, Hyman AA (2013) Principles of PAR polarity in *Caenorhabditis elegans* embryos. Nat Rev Mol Cell Biol 14:315–322. doi:10.1038/nrm3558

Hörger I, Velasco E, Mingorance J, Rivas G, Tarazona P, Vélez M (2008) Langevin computer simulations of bacterial protein filaments and the force-generating mechanism during cell division. Phys Rev E Stat Nonlinear Soft Matter Phys 77:011902

Howard M, Rutenberg A, de Vet S (2001) Dynamic compartmentalization of bacteria: accurate division in *E. coli*. Phys Rev Lett. doi:10.1103/PhysRevLett.87.278102

Hu Z, Lutkenhaus J (1999) Topological regulation of cell division in *Escherichia coli* involves rapid pole to pole oscillation of the division inhibitor MinC under the control of MinD and MinE. Mol Microbiol 34:82–90

Hu Z, Mukherjee A, Pichoff S, Lutkenhaus J (1999) The MinC component of the division site selection system in *Escherichia coli* interacts with FtsZ to prevent polymerization. PNAS 96:14819–14824

Huang KC, Meir Y, Wingreen NS (2003) Dynamic structures in *Escherichia coli*: spontaneous formation of MinE rings and MinD polar zones. PNAS 100:12724–12728. doi:10.1073/pnas.2135445100

Kirkpatrick CL, Viollier PH (2011) New(s) to the (Z-)ring. Curr Opin Microbiol 14:691–697. doi:10.1016/j.mib.2011.09.011

Komeili A, Li Z, Newman DK, Jensen GJ (2006) Magnetosomes are cell membrane invaginations organized by the actin-like protein MamK. Science 311:242–245. doi:10.1126/science.1123231

Kruse K (2002) A dynamic model for determining the middle of *Escherichia coli*. Biophys J 82:618–627. doi:10.1016/S0006-3495(02)75426-X

Lackner LL, Raskin DM, de Boer PAJ (2003) ATP-dependent interactions between *Escherichia coli* Min proteins and the phospholipid membrane in vitro. J Bacteriol 185:735–749

Lan G, Daniels BR, Dobrowsky TM, Wirtz D, Sun SX (2009) Condensation of FtsZ filaments can drive bacterial cell division. PNAS 106:121–126. doi:10.1073/pnas.0807963106

Landgraf D, Okumus B, Chien P, Baker TA, Paulsson J (2012) Segregation of molecules at cell division reveals native protein localization. Nat Methods. doi:10.1038/nmeth.1955

Lara B, Rico AI, Petruzzelli S, Santona A, Dumas J, Biton J, Vicente M, Mingorance J, Massidda O (2005) Cell division in cocci: localization and properties of the Streptococcus pneumoniae FtsA protein. Mol Microbiol 55:699–711. doi:10.1111/j.1365-2958.2004.04432.x

Li Z, Trimble MJ, Brun YV, Jensen GJ (2007) The structure of FtsZ filaments in vivo suggests a force-generating role in cell division. EMBO J 26:4694–4708. doi:10.1038/sj.emboj.7601895

Li Y, Hsin J, Zhao L, Cheng Y, Shang W, Huang KC, Wang H-W, Ye S (2013) FtsZ protofilaments use a hinge-opening mechanism for constrictive force generation. Science 341:392–395. doi:10.1126/science.1239248

Liu B, Persons L, Lee L, de Boer PAJ (2014) Roles for both FtsA and the FtsBLQ subcomplex in FtsN-stimulated cell constriction in *Escherichia coli*. Mol Microbio n/a–n/a. doi:10.1111/mmi.12906

Loose M, Mitchison TJ (2014) The bacterial cell division proteins FtsA and FtsZ self-organize into dynamic cytoskeletal patterns. Nat Cell Biol 16:38–46. doi:10.1038/ncb2885

Loose M, Fischer-Friedrich E, Ries J, Kruse K, Schwille P (2008) Spatial regulators for bacterial cell division self-organize into surface waves in vitro. Science 320:789–792. doi:10.1126/science.1154413

Loose M, Fischer-Friedrich E, Herold C, Kruse K, Schwille P (2011a) Min protein patterns emerge from rapid rebinding and membrane interaction of MinE. Nat Struct Mol Biol 18:577–583. doi:10.1038/nsmb.2037

Loose M, Kruse K, Schwille P (2011b) Protein self-organization: lessons from the Min system. Annu Rev Biophys 40:315–336. doi:10.1146/annurev-biophys-042910-155332
Löwe J, Amos LA (1998) Crystal structure of the bacterial cell-division protein FtsZ. Nature 391:203–206. doi:10.1038/34472
Löwe J, Amos LA (2009) Evolution of cytomotive filaments: the cytoskeleton from prokaryotes to eukaryotes. Int J Biochem Cell Biol 41:323–329. doi:10.1016/j.biocel.2008.08.010
Löwe J, van den Ent F, Amos LA (2004) Molecules of the bacterial cytoskeleton. Annu Rev Biophys 33:177–198. doi:10.1146/annurev.biophys.33.110502.132647
Lu C, Reedy M, Erickson H (2000) Straight and curved conformations of FtsZ are regulated by GTP hydrolysis. J Bacteriol 182:164–170
Lutkenhaus J, Addinall SG (1997) Bacterial cell division and the Z ring. Annu Rev Biochem 66:93–116. doi:10.1146/annurev.biochem.66.1.93
Lutkenhaus JF, Donachie WD (1979) Identification of the ftsA gene product. J Bacteriol 137:1088–1094
Lutkenhaus JF, Wolf-Watz H, Donachie WD (1980) Organization of genes in the ftsA-envA region of the *Escherichia coli* genetic map and identification of a new fts locus (ftsZ). J Bacteriol 142:615–620
Lutkenhaus J, Pichoff S, Du S (2012) Bacterial cytokinesis: from Z ring to divisome. Cytoskeleton (Hoboken). doi:10.1002/cm.21054
Margolin W (2012) The price of tags in protein localization studies. J Bacteriol 194:6369–6371. doi:10.1128/JB.01640-12
Martos A, Raso A, Jiménez M, Petrasek Z, Rivas G, Schwille P (2015) FtsZ polymers tethered to the membrane by ZipA are susceptible to spatial regulation by Min waves. Biophys J 108:2371–2383. doi:10.1016/j.bpj.2015.03.031
Meier EL, Goley ED (2014) Form and function of the bacterial cytokinetic ring. Curr Opin Cell Biol 26:19–27
Meinhardt H, de Boer PAJ (2001) Pattern formation in *Escherichia coli*: a model for the pole-to-pole oscillations of Min proteins and the localization of the division site. PNAS 98:14202–14207. doi:10.1073/pnas.251216598
Milam SL, Osawa M, Erickson HP (2012) Negative-stain electron microscopy of inside-out FtsZ rings reconstituted on artificial membrane tubules show ribbons of protofilaments. Biophys J 103:59–68. doi:10.1016/j.bpj.2012.05.035
Mileykovskaya E, Fishov I, Fu X, Corbin BD, Margolin W, Dowhan W (2003) Effects of phospholipid composition on MinD-membrane interactions in vitro and in vivo. J Biol Chem 278:22193–22198. doi:10.1074/jbc.M302603200
Moller-Jensen J, Jensen RB, Lowe J, Gerdes K (2002) Prokaryotic DNA segregation by an actin-like filament. EMBO J 21:3119–3127. doi:10.1093/emboj/cdf320
Mukherjee A, Lutkenhaus J (1994) Guanine nucleotide-dependent assembly of FtsZ into filaments. J Bacteriol 176:2754–2758
Mukherjee A, Lutkenhaus J (1998) Dynamic assembly of FtsZ regulated by GTP hydrolysis. EMBO J 17:462–469. doi:10.1093/emboj/17.2.462
Mukherjee A, Saez C, Lutkenhaus J (2001) Assembly of an FtsZ mutant deficient in GTPase activity has implications for FtsZ assembly and the role of the Z ring in cell division. Journal of Bacteriology 183:7190–7197. doi:10.1128/JB.183.24.7190-7197.2001
Novák B, Tyson JJ (2008) Design principles of biochemical oscillators. Nat Rev Mol Cell Biol 9:981–991. doi:10.1038/nrm2530
Osawa M, Erickson HP (2011) Inside-out Z rings – constriction with and without GTP hydrolysis. Mol Microbiol 81:571–579. doi:10.1111/j.1365-2958.2011.07716.x
Osawa M, Erickson HP (2013) Liposome division by a simple bacterial division machinery. PNAS. doi:10.1073/pnas.1222254110
Osawa M, Anderson DE, Erickson HP (2008) Reconstitution of contractile FtsZ rings in liposomes. Science 320:792–794. doi:10.1126/science.1154520

Park K-T, Wu W, Battaile KP, Lovell S, Holyoak T, Lutkenhaus J (2011) The Min oscillator uses MinD-dependent conformational changes in MinE to spatially regulate cytokinesis. Cell 146:396–407. doi:10.1016/j.cell.2011.06.042
Park K-T, Wu W, Lovell S, Lutkenhaus J (2012) Mechanism of the asymmetric activation of the MinD ATPase by MinE. Mol Microbiol. doi:10.1111/j.1365-2958.2012.08110.x
Park K-T, Du S, Lutkenhaus J (2015) MinC/MinD copolymers are not required for Min function. Mol Microbiol 98:895–909. doi:10.1111/mmi.13164
Pichoff S, Lutkenhaus J (2005) Tethering the Z ring to the membrane through a conserved membrane targeting sequence in FtsA. Mol Microbiol 55:1722–1734. doi:10.1111/j.1365-2958.2005.04522.x
Pichoff S, Lutkenhaus J (2007) Identification of a region of FtsA required for interaction with FtsZ. Mol Microbiol 64:1129–1138. doi:10.1111/j.1365-2958.2007.05735.x
Pichoff S, Shen B, Sullivan B, Lutkenhaus J (2011) FtsA mutants impaired for self-interaction bypass ZipA suggesting a model in which FtsA's self-interaction competes with its ability to recruit downstream division proteins. Mol Microbiol. doi:10.1111/j.1365-2958.2011.07923.x
Raskin DM, de Boer PA (1997) The MinE ring: an FtsZ-independent cell structure required for selection of the correct division site in *E. coli*. Cell 91:685–694
Raskin DM, de Boer PA (1999) Rapid pole-to-pole oscillation of a protein required for directing division to the middle of *Escherichia coli*. PNAS 96:4971–4976
RayChaudhuri D, Park JT (1992) *Escherichia coli* cell-division gene ftsZ encodes a novel GTP-binding protein. Nature 359:251–254. doi:10.1038/359251a0
Renner LD, Weibel DB (2011) Cardiolipin microdomains localize to negatively curved regions of *Escherichia coli* membranes. PNAS 108:6264–6269. doi:10.1073/pnas.1015757108
Salje J, Gayathri P, Löwe J (2010) The ParMRC system: molecular mechanisms of plasmid segregation by actin-like filaments. Nat Rev Microbiol 8:683–692. doi:10.1038/nrmicro2425
Schweizer J, Loose M, Bonny M, Kruse K, Mönch I, Schwille P (2012) Geometry sensing by self-organized protein patterns. PNAS 109:15283–15288. doi:10.1073/pnas.1206953109
Stachowiak JC, Schmid EM, Ryan CJ, Ann HS, Sasaki DY, Sherman MB, Geissler PL, Fletcher DA, Hayden CC (2012) Membrane bending by protein-protein crowding. Nat Cell Biol 14:944–949. doi:10.1038/ncb2561
Strauss MP, Liew ATF, Turnbull L, Whitchurch CB, Monahan LG, Harry EJ (2012) 3D-SIM super resolution microscopy reveals a bead-like arrangement for FtsZ and the division machinery: implications for triggering cytokinesis. PLoS Biol 10:e1001389. doi:10.1371/journal.pbio.1001389
Szeto TH (2003) The MinD membrane targeting sequence is a transplantable lipid-binding helix. J Biol Chem 278:40050–40056. doi:10.1074/jbc.M306876200
Szeto TH, Rowland SL, King GF (2001) The dimerization function of MinC resides in a structurally autonomous C-terminal domain. J Bacteriol 183:6684–6687. doi:10.1128/JB.183.22.6684-6687.2001
Szeto J, Eng N, Acharya S, Rigden M, Dillon J (2005) A conserved polar region in the cell division site determinant MinD is required for responding to MinE-induced oscillation but not for localization within coiled arrays. Res Microbiol 156:17–29. doi:10.1016/j.resmic.2004.07.009
Szwedziak P et al (2012) FtsA forms actin-like protofilaments. EMBO J 31:2249–2260
Szwedziak P, Wang Q, Bharat TAM, Tsim M, Löwe J (2014) Architecture of the ring formed by the tubulin homologue FtsZ in bacterial cell division. eLife. doi:10.7554/eLife.04601
Taschner PE, Huls PG, Pas E, Woldringh CL (1988) Division behavior and shape changes in isogenic ftsZ, ftsQ, ftsA, pbpB, and ftsE cell division mutants of *Escherichia coli* during temperature shift experiments. J Bacteriol 170:1533–1540
Trusca D, Scott S, Thompson C, Bramhill D (1998) Bacterial SOS checkpoint protein SulA inhibits its polymerization of purified FtsZ cell division protein. Journal of Bacteriology 180:3946–3953

Tsang M-J, Bernhardt TG (2015) A role for the FtsQLB complex in cytokinetic ring activation revealed by an ftsL allele that accelerates division. Mol Microbiol 95:925–944. doi:10.1111/mmi.12905

Turing A (1952) The chemical basis for morphogenesis. Philos Trans Royal Soc B Biol Sci 237:37

Typas A, Sourjik V (2015) Bacterial protein networks: properties and functions. Nat Rev Microbiol 13:559–572. doi:10.1038/nrmicro3508

Typas A, Banzhaf M, Gross CA, Vollmer W (2012) From the regulation of peptidoglycan synthesis to bacterial growth and morphology. Nat Rev Microbiol 10:123–136. doi:10.1038/nrmicro2677

Van de Putte P, Van Dillewijn J, Rörsch A (1964) The selection of mutants of *Escherichia coli* with impaired cell division at elevated temperature. Mutat Res Fundam Mol Mech Mutag 1:121–128. doi:10.1016/0027-5107(64)90014-4

van den Ent F, Amos LA, Löwe J (2001) Prokaryotic origin of the actin cytoskeleton. Nature 413:39–44. doi:10.1038/35092500

van Teeffelen S, Wang S, Furchtgott L, Huang KC, Wingreen NS, Shaevitz JW, Gitai Z (2011) The bacterial actin MreB rotates, and rotation depends on cell-wall assembly. PNAS 108:15822–15827. doi:10.1073/pnas.1108999108

Vecchiarelli AG, Li M, Mizuuchi M, Mizuuchi K (2014) Differential affinities of MinD and MinE to anionic phospholipid influence Min patterning dynamics in vitro. Mol Microbiol 93:453–463. doi:10.1111/mmi.12669

Wijsman HJ (1972) A genetic map of several mutations affecting the mucopeptide layer of *Escherichia coli*. Genet Res 20:65–74

Yan K, Pearce KH, Payne DJ (2000) A conserved residue at the extreme C-terminus of FtsZ is critical for the FtsA-FtsZ interaction in *Staphylococcus aureus*. Biochem Biophys Res Commun 270:387–392. doi:10.1006/bbrc.2000.2439

Yu XC, Margolin W (1999) FtsZ ring clusters in min and partition mutants: role of both the Min system and the nucleoid in regulating FtsZ ring localization. Mol Microbiol 32:315–326

Zieske K, Schwille P (2013) Reconstitution of pole-to-pole oscillations of min proteins in micro-engineered polydimethylsiloxane compartments. Angew Chem Int Ed Eng 52:459–462. doi:10.1002/anie.201207078

Zieske K, Schwille P (2014) Reconstitution of self-organizing protein gradients as spatial cues in cell-free systems. eLife. doi:10.7554/eLife.03949

Zieske K, Schweizer J, Schwille P (2014) Surface topology assisted alignment of Min protein waves. FEBS Lett 588:2545–2549. doi:10.1016/j.febslet.2014.06.026

Erratum to

Prokaryotic Cytoskeletons

Filamentous Protein Polymers Active in the Cytoplasm of Bacterial and Archaeal Cells

Jan Löwe and Linda A. Amos

J. Löwe, L.A. Amos (eds.), *Prokaryotic Cytoskeletons*, Subcellular Biochemistry 84, DOI 10.1007/978-3-319-53047-5

DOI 10.1007/978-3-319-53047-5_16

Chapter 8
The full name of the author in chapter 8 is missing. The author's full name is "**Thierry Izoré**".

The online version of the original chapter can be found under
DOI 10.1007/978-3-319-53047-5_8

Chapter 9
An incorrect figure legend has been used for this figure 9.1b on page 269. The corrected figure legend is as follows:

(b) The TEM of FtsZ is from *Caulobacter* (Current Opinion 2014 (https://www.ncbi.nlm.nih.gov/pubmed/24529242)).

A new reference has been included in the reference list of chapter 9. The reference is as follows:

Meier EL, Goley ED (2014) Form and function of the bacterial cytokinetic ring. Curr Opin Cell Biol 26:19–27. doi:10.1016/j.ceb.2013.08.006

The online version of the original chapter can be found under
DOI 10.1007/978-3-319-53047-5_9

The online version of the original book can be found under
DOI 10.1007/978-3-319-53047-5

J. Löwe, L.A. Amos (eds.), *Prokaryotic Cytoskeletons*, Subcellular Biochemistry 84, DOI 10.1007/978-3-319-53047-5_16

Index

J. Löwe, L.A. Amos (eds.), *Prokaryotic Cytoskeletons*, Subcellular Biochemistry 84, DOI 10.1007/978-3-319-53047-5

D

E

MIX
Papier aus verantwortungsvollen Quellen
Paper from responsible sources
FSC® C105338

If you have any concerns about our products,
you can contact us on
ProductSafety@springernature.com

In case Publisher is established outside the EU,
the EU authorized representative is:
Springer Nature Customer Service Center GmbH
Europaplatz 3, 69115 Heidelberg, Germany

Printed by Libri Plureos GmbH
in Hamburg, Germany